Ecological Studies

Analysis and Synthesis

Volume 231

Ecological Studies is Springer's premier book series treating all aspects of ecology. These volumes, either authored or edited collections, appear several times each year. They are intended to analyze and synthesize our understanding of natural and managed ecosystems and their constituent organisms and resources at different scales from the biosphere to communities, populations, individual organisms and molecular interactions. Many volumes constitute case studies illustrating and synthesizing ecological principles for an intended audience of scientists, students, environmental managers and policy experts. Recent volumes address biodiversity, global change, landscape ecology, air pollution, ecosystem analysis, microbial ecology, ecophysiology and molecular ecology.

More information about this series at http://www.springer.com/series/86

Mariano M. Amoroso • Lori D. Daniels
Patrick J. Baker • J. Julio Camarero
Editors

Dendroecology

Tree-Ring Analyses Applied to Ecological Studies

Springer

Editors
Mariano M. Amoroso
Instituto de Investigaciones en Recursos
Naturales, Agroecología y Desarrollo Rural
Universidad Nacional de Río Negro
El Bolsón, Río Negro, Argentina

CCT CONICET Patagonia Norte
Consejo Nacional de Investigaciones
Científicas y Técnicas
San Carlos de Bariloche
Rio Nego, Argentina

Patrick J. Baker
School of Ecosystem and Forest Sciences
University of Melbourne
Melbourne, VIC, Australia

Lori D. Daniels
Department of Forest and Conservation
Sciences
University of British Columbia
Vancouver, BC, Canada

J. Julio Camarero
Pyrenean Institute of Ecology (IPE-CSIC)
Zaragoza, Spain

ISSN 0070-8356 ISSN 2196-971X (electronic)
Ecological Studies
ISBN 978-3-319-61668-1 ISBN 978-3-319-61669-8 (eBook)
DOI 10.1007/978-3-319-61669-8

Library of Congress Control Number: 2017950485

Printed on acid-free paper

This Springer imprint is published by Springer Nature
The registered company is Springer International Publishing AG
The registered company address is: Gewerbestrasse 11, 6330 Cham, Switzerland

Foreword

The connection between dendrochronology and ecology runs deep. Ecology as a field of study was in great ferment in the American Southwest during the 1910s to 1930s, especially in Tucson, Arizona, where the Carnegie Institute of Washington supported the establishment of the first laboratory in North America dedicated to the study of ecosystems, *and* the first laboratory dedicated to the use of tree rings in the study of climate and archaeology. The ecologists working at the Desert Botanical Laboratory included luminaries such as Daniel T. MacDougal and Frederick E. Clements, and the Laboratory of Tree-Ring Research was established by Andrew E. Douglass. Scientists at the Desert Lab were centrally involved in founding the Ecological Society of America in 1915. The first issue of their flagship journal *Ecology* (Volume 1, Issue 1, January 1920) included an article by Douglass titled "Evidence of Climatic Effects in the Annual Rings of Trees." Although the paper focused mainly on using tree rings as a proxy for rainfall, it also included illustrations of one of Douglass' seminal—and fundamentally ecological—insights regarding the effects of site conditions on climatic "sensitivity" of tree rings. This is a foundational principle of dendrochronology.

The European roots of dendrochronology also grew from tree biology and forest ecology. Bruno Huber, working as a professor of forest botany at the Technical University of Dresden and the University of Munich in Germany in the 1930s, came to tree-ring studies as a tree biologist. Like Douglass, his most famous work involved tree-ring chronology development and applications in archaeological dating, but along the way he laid foundations for subsequent wood anatomy and eco-physiological investigations in dendrochronology.

Throughout the last century dendrochronologists have expanded the applications of tree rings in environmental studies. Dendroclimatology, the application of tree rings in the study of atmospheric processes, has undoubtedly been a driving and overarching endeavor, stimulating and contributing to the myriad other applications, including dendroecology. Indeed, the ready availability of dendroclimatic reconstructions of rainfall, temperature, and drought indices has revolutionized disturbance studies in dendroecology. Fire history and insect outbreak studies, for example, have commonly combined their analyses of disturbance and climate chronologies derived from tree rings, resulting in new insights into climate-disturbance relationships.

Two of the all-time most influential dendrochronologists, Harold C. Fritts and Fritz Schweingruber, published seminal reviews of dendroecology in 1971, 1989, and 1996. Over the past several decades there has been a surge of publications in ecological and forest science journals with dendroecological themes. A Google Scholar search (in March 2017) using the terms "dendroecology" and "dendroclimatology" in the titles returned 5150 and 7080 papers, respectively. Since 1980, the 500 most-cited papers with the term "dendroecology" in the searchable text were cited a total of 14,707 times, while the 500 most-cited papers with the term "dendroclimatology" were cited 13,434 times. Today, many ecological and forest science journals include dendrochronologists on their editorial boards.

Along with the expansion of dendrochronology in ecological studies has been an extension of geographical coverage. In addition to many studies in North and South America, Europe, Asia, Australia, New Zealand and elsewhere, dendroecological studies have followed the example of dendroclimatology in exploiting regional to continental scale networks. Patterns of synchrony and asynchrony of fires, insect outbreaks, forest natality, and mortality, in concert with regional climatic variability, have demonstrated dendroecology as a key tool in macroecology.

One the key reasons that dendroecology has flourished in recent decades is the direct relevance of the insights provided for environmental management. Forest managers especially have embraced the concept that knowledge of past ecological dynamics and structures is valuable for understanding current conditions, and for deciding how to manage for future desired conditions. Recent extreme responses of ecosystems to climate change, including wildfires, insect and pathogen outbreaks, and drought-induced mortality, have been major stimuli for dendroecological investigations.

Given the great fluorescence of interest, publications, and relevance of dendroecology in recent decades, the compilation of a volume of chapters covering broad topical and geographic examples worldwide is most timely and welcome. The editors and authors have here assembled a diverse set of conceptual and methodological reviews, and examples ranging from disturbance ecology, ecophysiology, and forest dynamics to human-environment interactions. The geographical coverage highlights several regions that have benefited most from recent expansion of dendroecology, including South America, Australia, and Asia. Studies in temperate, subtropical, and tropical forests are also included.

A most engaging and potentially far reaching development in dendroecology in recent years is a focus on human-environment interactions. This advance is nicely included in chapter examples from Asia, Europe, and North America. These landscape and regional-scale studies demonstrate the broad, multidisciplinary nature of dendrochronology, weaving together and exploiting multiple lines of tree-ring based and other forms of evidence from archaeology, history, paleoecology, and paleoclimatology. In this sense, dendroecology is now embracing the full scope of ecology in its broadest definition: the study of organisms—including humans—and their interactions with the environment. Perhaps dendroecology will continue to develop and evolve as the most synthetic of dendrochronology applications, and if so, this volume will serve as foundational.

Laboratory of Tree-Ring Research
Tucson, AZ, USA
5 March 2017

Thomas W. Swetnam

Preface

Dendroecological studies have fundamentally shaped contemporary views of forest ecology and forest dynamics, particularly in temperate forests. While ecological studies using tree rings have been published for well over a century, the past several decades have seen a rapid acceleration in the rate of publication of dendroecological studies. Most syntheses and reviews of dendrochronology over the past half century have focused on climatology, archaeology, or wood formation (e.g., Fritts 1971, 1976, Eckstein 1984, Cook and Kairiukstis 1989, Schweingruber 1996, Hughes et al. 2001, Vaganov et al. 2006, and to a lesser extent Speer 2010). While dendrochronology has made important contributions in each of these areas, the ecological lessons that dendrochronological research has provided are fundamental to understanding how forested ecosystems will respond to the many threats posed by global environmental change. These threats are not limited to changing climatic conditions, but also include the impacts of invasive species, biodiversity loss, changing disturbance regimes, and the multitude of potential interactions among them. Dendroecology is a significant component of applied research, providing important historical context for adapting existing forest management strategies to mitigate and respond to current and future global environmental change.

The application of tree-ring analyses has emerged as a powerful approach to address complex ecological questions and to quantify environmental change through time. This book presents state-of-the-science reviews and application of tree-ring analyses to ecological problems using examples from tropical, temperate, and boreal forests around the globe. It is organized around four broad topical themes: tree growth and forest dynamics, disturbance regimes, forest decline, and human-environment interactions. Each theme is composed of chapters highlighting recent advances in dendroecology, often illustrated by case studies to demonstrate novel methods and approaches.

Introducing tree growth, Deslauriers et al. (Chap. 2) review the processes of tree-ring formation highlighting novel research on the impacts of resource availability and environmental variation on xylem and phloem growth and ring attributes. Their global perspective contrasts tree growth in boreal through tropical forests, demonstrating how the timing, duration, and rates of cell formation affect wood anatomy and ecophysiology. Deep understanding of these mechanisms and processes provides a strong foundation for the dendroecological interpretations presented in subsequent chapters.

Collectively, the next three chapters highlight advances in dendroecological research of forest dynamics in tropical and temperate forests. Research on more than 200 neotropical tree species has identified a suite of environmental factors that trigger annual ring formation (Schöngart et al., Chap. 3). New knowledge on tree ages and growth trajectories is applied to evaluate and adapt timber harvesting to ensure sustainability of tropical ecosystem functions and services. Similarly, dendroecological advances in seasonal dry tropical forests include the first reconstructions of the historical dynamics of mixed deciduous and seasonal evergreen forests in western Thailand (Baker and Bunyavejchewin, Chap. 4). Tree recruitment and growth releases indicate complex gap dynamics, as well as extensive, synchronous disturbances driven by regional drought. Pederson et al. (Chap. 5) emphasize the importance of multiscale disturbances in temperate mesic forests. Their critical analysis identifies limitations and potential biases inherent to contemporary research, which tends to focus on spatial and temporal scales that are unable to detect large, infrequent disturbances. They present a conceptual model and dendroecological solutions to test alternative hypotheses and address urgent questions on macroecological dynamics of temperate forests.

In their review on altitudinal and latitudinal treelines, the ecotones between forests and tundra, Lloyd et al. (Chap. 6) shed light on the surprisingly complex responses of treeline ecosystems to climate warming. They demonstrate the necessity of integrating dendroecology with spatially explicit measurements of environmental variability, allowing researchers to understand the underlying environmental gradients to which trees are responding. Their integrated research approaches allow us to understand trees not simply as the sum of their rings, but as complex organisms whose overall growth reflects the impact of multiple limiting factors filtered through a number of simultaneously occurring physiological processes.

Although most research on forest dynamics focuses on live trees, the field of dendroecology is well suited to research on dead trees, including snags and fallen logs, collectively known as woody debris. Dendroecology provides novel methods to estimate dates of tree mortality or fall at an annual level, which are used to calculate decay rates using chronosequences or decay-class transition models. Fraver et al. (Chap. 7) present a case study on three old-growth *Picea rubens* stands in the northeastern USA to demonstrate continuous recruitment of woody debris over the twentieth century, with pulses corresponding to reconstructed gap- and meso-scale disturbances. Given the critical role coarse woody debris plays in forest ecosystems—carbon storage, nutrient cycling, soil development, and maintenance of biodiversity—it is imperative to understand how its abundance changes through time.

The second theme of this book illustrates how dendroecological research has advanced our understanding of disturbance as a vital attribute of forests. Dendroecology also has been instrumental for understanding the drivers of forest decline. Five chapters illustrate how dendroecology allows a unique temporal perspective on disturbance by fire, insects, pathogens, and geomorphic processes. They are followed by two reviews of the intricacies of forest decline in drought-prone climates.

Daniels et al. (Chap. 8) review dendroecological contributions toward understanding of low-, mixed-, and high-severity fire regimes. Four case studies on reconstructions of historical fire frequency, severity, spatial variability, and extent demonstrate the complexity of mixed-severity fire regimes in montane forests in western North America. Similarly, Mundo et al.'s (Chap. 9) dendroecological reconstructions in five different Patagonian forests show diverse and heterogeneous patterns of fire frequency related to climatic variability and human impacts. Collectively these two chapters show how combining dendroecological evidence with independent, corroborating evidence and models of fire extent, behavior, and effects provides a powerful framework for comparing historical and contemporary fire regimes.

Research on insects and pathogens and their effects on forest resilience is an active frontier in dendroecology. Reconstructions of insect outbreaks include millennial length chronologies and research networks facilitating broad-scale spatio-temporal analyses (Speer and Kulakowski, Chap. 10). Moving beyond the reductionist approach of examining only parts of an ecosystem, dendroecologists embrace complexity by studying disturbance interactions and their signals recorded in tree rings. Building on this theme, Lewis et al. (Chap. 11) show how forest ecosystems have evolved to be resilient to native pathogens and insects, but human-induced climate change has contributed to severe epidemics, for example *Dothistroma* needle blight and mountain pine beetle. Concurrently, increasing global trade increases chances of exotic pathogens interacting with native hosts. These human influences interact with other agents often yielding synergistic negative effects.

Stoffel et al. (Chap. 12) review dendroecological analyses of trees affected by earth-surface processes, in which rings preserve valuable archives of past events on timescales of decades to centuries. Their case studies illustrate the breadth and diverse applications of contemporary dendrogeomorphology to understand processes such as flood, soil erosion, debris flows, rockfall, and landslides. Understanding the distribution, timing, and controls of geomorphic processes provides valuable information to assist prediction, mitigation, and defence against these natural hazards and their effects on society.

The instrumental role of dendroecology for understanding forest decline in drought-prone climates around the globe is the third theme of this book. Research on two continents illustrate that forest decline is caused by complex interactions among abiotic and biotic stress factors acting at different spatial and temporal scales. In Spain, drought-triggered forest dieback is predisposed by past forest use and management and increased tree-to-tree competition, which increase vulnerability of individual trees or species to death (Camarero et al., Chap. 13). In these forests, mistletoe and other biotic stressors contribute to drought-induced dieback. Similarly, the decline of native forests in northern Patagonia is complex and driven by multiple factors (Amoroso et al., Chap. 14). Dendroecological analyses of radial-growth patterns show extreme droughts incite immediate growth reductions, followed by persistent growth decline over time. Amoroso et al. (Chap. 14) present a framework for testing alternative hypotheses to a single causal agent, which emphasizes the importance of climatic variability as a key driver of forest decline.

Given ubiquitous human impacts and cumulative effects of global environmental change, understanding human-environment relationships is the fourth and final theme. Dendrochronology is at the nexus of archaeology, climatology, and ecology, making it uniquely positioned to study past human-environment interactions. Tree-ring records are ideal for exploring these relationships because they integrate climatic information at temporal and spatial scales relevant to human livelihoods. Hessl et al. (Chap. 15) use a growing network of long tree-ring records from Asia to contextualize human-environment relationships over the last 3000 years. A wide diversity of economic, political, and cultural features in historical Asia filtered, dampened, and amplified the effects of climate on society, suggesting that relationships may not be stable or consistent over time and space. Trouet et al. (Chap. 16) present a long-term perspective, using dendrochronology to quantify Holocene-era land-use changes and forest dynamics in Europe and North America. Their critical analysis highlights limitations when relying on living forests because the majority of lowland forests have been cleared throughout human history so that contemporary tree-ring chronologies might not fully represent past environments. They introduce the concept of "dendro-archeo-ecology" in which ecological information is derived from existing dendroarcheological collections to quantify past land-use changes and their impact on the carbon cycle and Earth's climate.

In the final chapter of this volume we highlight several overarching themes that have emerged from these state-of-the-science reviews, identify areas of substantial progress in dendroecology that have been made in recent decades, and emphasize several pressing questions that dendroecology is well positioned to address.

We envision this book as a waypoint in the science of dendroecology that brings together much of the excellent dendroecological research that has come before us, provides a contemporary overview of the breadth and depth of the existing research, and generates new ideas for the many possible directions that dendroecology might take going forward. We hope that this volume will provide a useful resource for dendrochronologists, ecologists, foresters, and others interested in the conservation and sustainable management of the world's forests.

El Bolsón, Río Negro, Argentina — Mariano M. Amoroso
Vancouver, BC, Canada — Lori D. Daniels
Melbourne, VIC, Australia — Patrick J. Baker
Zaragoza, Spain — J. Julio Camarero

Acknowledgements

We greatly appreciate the contributions of each author in this book. Their willingness to represent a wide array of subdisciplines of dendroecology by detailing key theoretical concepts, illustrating innovative research methods, and demonstrating practical applications through their diverse and insightful case studies from around the globe has made this book possible.

Each lead author provided a peer review of another chapter of this book and we are thankful for their thoughtful comments and helpful suggestions.

Contents

Contributors

Tuomas Aakala Department of Forest Sciences, University of Helsinki, Helsinki, FI, Finland

Mariano M. Amoroso Instituto de Investigaciones en Recursos Naturales, Agroecología y Desarrollo Rural, Universidad Nacional de Río Negro, El Bolsón, Río Negro, Argentina

CCT CONICET Patagonia Norte, Consejo Nacional de Investigaciones Científicas y Técnicas, San Carlos de Bariloche, Rio Nego, Argentina

Uyanga Ariya Department of Environment and Forest Engineering, School of Engineering and Applied Sciences, National University of Mongolia, Ulaanbaatar, Mongolia

Patrick J. Baker School of Ecosystem and Forest Sciences, University of Melbourne, Melbourne, VIC, Australia

Juan A. Ballesteros-Cánovas Climatic Change and Climate Impacts, Institute for Environmental Sciences, University of Geneva, Geneva, Switzerland

Dendrolab.ch, Department of Earth Sciences, University of Geneva, Geneva, Switzerland

Ana Carolina Maioli Campos Barbosa Departamento de Ciências Florestais, Universidade Federal de Lavras, Campus Universitário, Lavras, Brazil

Achim Bräuning Institut für Geographie, Friedrich-Alexander Universität Erlangen-Nürnberg, Erlangen, Germany

Andrew G. Bunn Environmental Sciences, Western Washington University, Bellingham, WA, USA

Sarayudh Bunyavejchewin Thai Long-term Forest Ecological Research, CTFS-Forest GEO, Bangkhen, Bangkok, Thailand

Oyunsanaa Byambasuren National University of Mongolia, Ulaanbaatar, Mongolia

J. Julio Camarero Pyrenean Institute of Ecology (IPE-CSIC), Zaragoza, Spain

José Antonio Carreira Centro de Estudios Avanzados de la Tierra (CEACTierra), Universidad de Jaén, Jaén, Spain

Christophe Corona Geolab, UMR 6042 CNRS, Université Blaise Pascal, Clermont-Ferrand, France

Anthony W. D'Amato Rubenstein School of Environment and Natural Resources, University of Vermont, Burlington, VT, USA

Lori D. Daniels Department of Forest and Conservation Sciences, University of British Columbia, Vancouver, BC, Canada

Annie Deslauriers Département des Sciences Fondamentales, Université du Québec à Chicoutimi, Chicoutimi, QC, Canada

Marta Domínguez-Delmás Department of Botany, University of Santiago de Compostela, Lugo, Spain

Patrick Fonti Swiss Federal Research Institute WSL, Birmensdorf, Switzerland

Shawn Fraver School of Forest Resources, University of Maine, Orono, ME, USA

Antonio Gazol Pyrenean Institute of Ecology (IPE-CSIC), Zaragoza, Spain

Mauro E. González Laboratorio Ecología de Bosques, Instituto de Conservación, Biodiversidad y Territorio, Facultad de Ciencias Forestales y Recursos Naturales, Universidad Austral de Chile, Campus Isla Teja, Valdivia, Chile

Center for Climate and Resilience Research (CR2), Santiago, Chile

Jozica Gričar Department of Forest Yield and Silviculture, Slovenian Forestry Institute, Ljubljana, Slovenia

Amy E. Hessl Department of Geology and Geography, West Virginia University, Morgantown, WV, USA

Emily K. Heyerdahl Rocky Mountain Research Station, Fire Sciences Laboratory, US Forest Service, Missoula, MT, USA

Andrés Holz Department of Geography, Portland State University, Portland, OR, USA

Dominik Kulakowski Graduate School of Geography, Clark University, Worcester, MA, USA

Caroline Leland Lamont-Doherty Earth Observatory, Columbia University, Palisades, NY, USA

Kathy J. Lewis Ecosystem Science and Management Program, University of Northern British Columbia, Prince George, BC, Canada

Juan Carlos Linares Department of Physical, Chemical and Natural Systems, Universidad Pablo de Olavide, Seville, Spain

Claudio Sergio Lisi Centro de Ciências Biológicas e da Saúde, Universidade Federal de Sergipe, São Cristóvão, Brazil

Andrea H. Lloyd Department of Biology, Middlebury College, Middlebury, VT, USA

Dario Martin-Benito Institute of Terrestrial Ecosystems Science ETH Zürich, Zürich, Switzerland

Department of Silviculture and Forest Management, Forest Research Centre, INIA, Madrid, Spain

Ignacio A. Mundo Laboratorio de Dendrocronología e Historia Ambiental, IANIGLA-CONICET, CCT CONICET Mendoza, CC330-M5502IRA, Mendoza, Argentina

Facultad de Ciencias Exactas y Naturales, Universidad Nacional de Cuyo, Mendoza, Argentina

Rafael M. Navarro-Cerrillo Department of Forest Engineering, ERSAF, University of Córdoba, Campus de Rabanales, Córdoba, Spain

Juliano Morales de Oliveira Universidade do Vale do Rio dos Sinos, Cristo Rei, São Leopoldo, Brazil

Juan Paritsis Laboratorio Ecotono, INIBIOMA, CONICET-Universidad Nacional del Comahue, Bariloche, Río Negro, Argentina

Charlotte Pearson Laboratory of Tree-Ring Research, University of Arizona, Tucson, AZ, USA

Neil Pederson Harvard Forest, Harvard University, Petersham, MA, USA

Cyrille B.K. Rathgeber Laboratoire d'Étude des Ressources Forêt-Bois (LERFoB), UMR INRA-AgroParisTech 1092, Centre INRA de Nancy, Champenoux, France

Milagros Rodríguez-Catón Instituto Argentino de Nivología, Glaciología y Ciencias Ambientales (IANIGLA), CONICET, CCT-Mendoza, Mendoza, Argentina

Sergio Rossi Département des Sciences Fondamentales, Université du Québec à Chicoutimi, 555 Boulevard de l'Université, Chicoutimi, QC, Canada

Darrin Rubino Department of Biology, Hanover College, Hanover, IN, USA

Karel Šilhán Department of Physical Geography and Geoecology, University of Ostrava, Ostrava, Czech Republic

Thomas Saladyga Department of Geography, Concord University, Athens, WV, USA

Raúl Sánchez-Salguero Pyrenean Institute of Ecology (IPE-CSIC), Zaragoza, Spain

Department of Physical, Chemical and Natural Systems, Universidad Pablo de Olavide, Seville, Spain

Gabriel Sangüesa-Barreda Pyrenean Institute of Ecology (IPE-CSIC), Zaragoza, Spain

Jochen Schöngart Instituto Nacional de Pesquisas da Amazônia, Coordenação de Dinâmica Ambiental, Petrópolis, Manaus, Brazil

Rosemary L. Sherriff Department of Geography, Humboldt State University, Arcata, CA, USA

James H. Speer Department of Environmental Systems, Indiana State University, Terre Haute, IN, USA

Amanda B. Stan Department of Geography, Planning and Recreation, Northern Arizona University, Flagstaff, AZ, USA

Markus Stoffel Climatic Change and Climate Impacts, Institute for Environmental Sciences, Geneva, Switzerland

Dendrolab.ch, Department of Earth Sciences, University of Geneva, Geneva, Switzerland

Patrick F. Sullivan Environment and Natural Resources Institute, University of Alaska Anchorage, Anchorage, AK, USA

Valerie Trouet Laboratory of Tree-Ring Research, University of Arizona, Tucson, AZ, USA

Ricardo Villalba Instituto Argentino de Nivología, Glaciología y Ciencias Ambientales (IANIGLA), CONICET, CCT-Mendoza, Mendoza, Argentina

Cedar Welsh Department of Geography, University of Victoria, Victoria, BC, Canada

Carmen M. Wong Yukon Field Unit, Parks Canada, Whitehorse, YT, Canada

Larissa L. Yocom Kent School of Forestry, Northern Arizona University, Flagstaff, AZ, USA

Amanda B. Young Department of Geography, Pennsylvania State University, University Park, PA, USA

Chapter 1
Introduction

Mariano M. Amoroso, Patrick J. Baker, Lori D. Daniels, and J. Julio Camarero

1.1 Introduction to Dendroecology

Forests blanket large areas of the Earth's land masses. They host a large fraction of global biodiversity and terrestrial carbon reserves, and the economic development of most countries has relied directly or indirectly on forests. Forests vary widely in structure and composition—from the low diversity expanses of boreal forest that stretch across the high latitudes of the northern hemisphere to the hyperdiverse tropical rain forests that grow on or near the equator. These differences in structure and composition reflect latitudinal and elevational bioclimatic variation, biogeographic history, and the influence of various historical disturbance regimes. Despite the diversity in physiognomy and its causes, a common feature of all

M.M. Amoroso (✉)
Instituto de Investigaciones en Recursos Naturales, Agroecología y Desarrollo Rural, Universidad Nacional de Río Negro, El Bolsón, Río Negro, Argentina

CCT CONICET Patagonia Norte, Consejo Nacional de Investigaciones Científicas y Técnicas, San Carlos de Bariloche, Rio Negro, Argentina
e-mail: mariano.amoroso@gmail.com

P.J. Baker
School of Ecosystem and Forest Sciences, University of Melbourne, Melbourne, VIC 3121, Australia
e-mail: patrick.baker@unimelb.edu.au

L.D. Daniels
Department of Forest and Conservation Sciences, University of British Columbia, Vancouver, BC, Canada
e-mail: lori.daniels@ubc.ca

J.J. Camarero
Pyrenean Institute of Ecology (IPE-CSIC), Avenida Montañana 1005, Zaragoza, 50059 Spain
e-mail: jjcamarero@ipe.csic.es

M.M. Amoroso et al. (eds.), *Dendroecology*, Ecological Studies 231,
DOI 10.1007/978-3-319-61669-8_1

forests is the arboreal growth form. The ability of trees to establish a tall canopy through competition for light with their neighbors yields majestic and often crowded woody landscapes. To reach the canopy, trees must grow in height and support large amounts of biomass. Simultaneously, trees draw water and nutrients from the soil up considerable distances into their crowns, which are coupled with the atmosphere through photosynthesis and transpiration. The need to form wood, a tissue with multiple functions (e.g., support, hydraulic conductivity, storage, defense against biotic agents), is a key characteristic of trees and forests. The secondary, or radial, growth processes that drive wood formation are fundamentally the same for all trees. This uniformity of process means that trees and forests should respond in relatively predictable ways to environmental conditions and disturbances. Over the past century, dendrochronology (the study of tree rings) applied to ecological questions has been fundamental in shaping our understanding of how forests function across a broad range of spatial and temporal scales.

Ecology is the study of how organisms interact with one another and their physical environment. At the heart of ecology is the question of how these interactions change over time to influence population and community structure and composition. Dendroecology, which uses precisely dated annual rings of trees and other woody plants to study past changes in ecological systems, has shaped contemporary understanding of forest development and dynamics, forest responses to extreme environmental conditions and disturbances, and, in some cases, long-term interactions between humans and their environments.

1.2 An Idiosyncratic History of Dendroecology

The term "dendroecology" is a relatively new term that was originally used in a much narrower sense. Serre-Bachet and Tessier (1989) cite Vins (1963) as the first to use the term in an attempt to expand dendrochronology to include a more direct focus on forest ecology and the many tree species and forest types overlooked in dendroclimatology studies. Within the dendrochronology research community, it took many years to move away from an explicitly climate-focused perspective and internalize the potential and importance of tree rings for ecological studies. For example, in Fritts' (1971) paper entitled "Dendroclimatology and dendroecology," dendroecology received a single paragraph of text in 28 pages and was introduced with the following statement (p. 446): "... some ecological studies dealing with past history of the environment may find useful information in what the dendroclimatologist may regard as 'noise'." For a long period, dendroecology was seen as the reverse side of the dendroclimatology "coin." While dendroclimatology used past tree growth patterns to reconstruct climate, dendroecology focused on how climate and other factors directly and indirectly influenced past tree growth patterns.

Today, "noise" persists when dendroecologists and dendroclimatologists communicate about their respective sides of the dendrochronology coin. For instance, dendroecologists have demonstrated that many characteristics (e.g., size, age, local density) shape the individualistic response of trees to climate variability. To address the idiosyncrasy of individuals, dendroclimatologists either focus on a subset of trees such as large, presumably old, trees growing on resource-limited sites or they compile large datasets that include trees with diverse characteristics but regard the individualistic growth patterns as statistical "noise" in standard tree-ring chronologies, which represent a site- and species-specific climatic "signal." For dendroecologists this "noise" is often the "signal," reflecting one or more of the many potential non-climatic influences on tree growth.

Dendroecology began to enter the mainstream of dendrochronological research in the 1970s and early 1980s due to growing public concerns about the impacts of pollution on forests in North America and Europe. Acid rain and forest decline, in particular, were the subjects of extensive research by teams that included forest ecologists, soil scientists, and dendrochronologists (Le Blanc 1990, Skelly and Innes 1994). In many ways, Fritts and Swetnam's (1989) paper was the catalyst for recognizing dendroecology as an independent subdiscipline of dendrochronology. In their paper, they reviewed the foundational principles of dendrochronology and provided several case studies in which dendrochronology was being applied to pressing environmental issues, such as pollution impacts, regional forest decline, and the impacts of insect outbreaks. Their focus was on the use of tree-ring time series to understand ecological and environmental variability and its impacts on forests. In so doing, they validated the core concepts that define the modern scope of dendroecology. Methodologically, however, their review was limited to ecological and environmental information available from ring-width or ring-density data—that is, they mainly focused on variation in growth patterns and their relationship to environmental patterns. Schweingruber et al. (1990, p. 31) echoed this perspective, stating "the aim of dendroecology is the determination of the year-by-year interplay of relationships among climate, site conditions, and tree growth to assess exogenous and endogenous factors that influence the growth of a plant community." However, many of the early ecological studies using tree rings to understand forest dynamics were more focused on tree establishment dates and ages than on growth patterns.

Many proponents of dendroecology in the 1960–1980s came from a dendroclimatology background and viewed ecology in the context of the ecophysiology of tree growth. They were interested in understanding the drivers of variation in tree growth and were building on centuries of research in forest science that had linked tree growth to environmental conditions. Leonardo da Vinci's fifteenth century observation that tree rings vary between wet and dry years is the earliest documented example of this (Schweingruber 1996). Significant early contributions that are in many ways the foundation of modern dendrochronological thinking were proposed by German and French forest scientists in the eighteenth and nineteenth centuries (e.g., Henri-Louis Duhamel de Monceau; George-Louis Leclerc, Comte de Buffon; Robert Hartig). During the twentieth century, Russian research on tree-ring structure and function (Vaganov et al. 2006) helped bridge the gap between dendroclimatology and dendroecology by explaining the anatomical and physiological bases of tree-ring formation. This bridge facilitated the shift in focus from climate to a forest dynamics perspective. In recent decades, many iconic publications on forest ecology and dynamics have included tree-ring data to advance their arguments for the importance of past disturbances in shaping contemporary forest structure and composition (e.g., Lorimer 1980; West et al. 1981; Pickett and White 1985; Brubaker 1986; Harmon et al. 1986; Fritts and Swetnam 1989; Glenn-Lewin et al. 1992) and provided the basis for modern views of forest stand dynamics (Oliver and Larson 1996; Franklin et al. 2002).

The use of tree rings to understand forest dynamics also has its origins in the eighteenth and nineteenth centuries, but primarily from the perspective of forest growth and yield. European scientists and foresters had long recognized the utility of annual growth rings for estimating growth rates. For example, Varennes de Fenille (1791, p. 48) described the influence of annual "accidents" (e.g., hot and dry summer, insect infestation, spring frost) on individual growth rings, growth releases due to competition, and releases following the death of neighboring trees. He highlighted the differences between high- and low-frequency variability in tree-ring series *versus* the age-related decline in ring width as trees grew. However, the use of tree rings to understand the ecology and dynamics of unmanaged forests and forested landscapes did not emerge until the late nineteenth and early twentieth centuries. In the 1850s, Sir Dietrich Brandis, then Superintendent of the teak (*Tectona grandis* L.) forests of Pegu, Burma (now Myanmar), and future Inspector-General of Indian Forests for the British Empire, described regional variation in teak growth patterns across parts of Southeast Asia. Using data on growth rates and tree ages derived from tree rings, Brandis (1956) established a quantitative basis for the management of teak forests in southern Burma. The principles that he established for stand- and landscape-scale management of these teak forests served as the basis for forest management practices across the tropics for the next century.

As Gifford Pinchot and John Muir established the foundations of modern forest management and conservation practices in western North America in the early twentieth century, ecologists were beginning to use tree rings to address questions regarding the underlying nature of forests. For example, Cooper (1913) used tree rings to characterize the age structure of the "climax" forest of Isle Royale in Lake Superior, Michigan. He found most trees were relatively young, despite the perception that these forests were untouched by humans. Similarly, Haasis (1923) used tree rings sampled in 1915–1916 to demonstrate that a large proportion of trees in the "many-aged virgin hardwood stands" of southeastern Kentucky were 250–255 years old. He attributed the origin of this distinct age class to an extreme drought in 1662 that led to fires burning through these forests and creating a discrete pulse of post-fire regeneration. Some surviving larger oaks showed reduced growth in the two decades preceding 1662. Marshall (1928) used tree rings to identify the years in which fires occurred in western white pine forests in northeastern Washington and the Idaho panhandle. The oldest fires burned in 1610, the year that "Oliver Cromwell was celebrating the close of the Thirty Years War" and "the infant Louis XIV was starting his long and infamous reign in France" (Marshall 1928, p. 48).

From a modern dendrochronological perspective, early studies of forest growth and ecological dynamics using tree rings were limited because they were largely based on ring counts. Crossdating was still in its infancy and was being used primarily by dendroclimatologists and dendroarchaeologists (Douglass 1909; 1929;

1937). However, it ultimately paved the way for more quantitative analyses of tree-ring data (Glock 1937). Perhaps the first "true" dendroecological study belongs to Bailey (1925), one of the great wood anatomists of the twentieth century. In his 1925 paper on the "spruce budworm biocoenose," he recognized that spruce budworm defoliation had the potential to create missing rings, which would compromise the ability of ring counts to accurately date outbreaks. To address this problem, he identified distinct frost rings in the tree-ring series and used those as marker years to correctly date the rings and determine the years of spruce budworm outbreaks over the preceding two centuries.

1.3 Dendroecology and Shifting Paradigms on Forest Dynamics

Dendroecological studies have been central to the shifting ideas about forest development and responses to disturbances over the past century. In the first half of the twentieth century, Clements' (1916, 1936) ideas on succession and climax dominated ecological thought on plant community composition, structure, and dynamics. However, a growing body of dendroecological studies was revealing the role of contingency in forest dynamics and the potential for multiple developmental pathways in the wake of unpredictable disturbances. These studies provided much of the empirical basis for challenging, and eventually overturning, the Clementsian notion of succession and climax. Cooper's (1923) research used tree rings to describe the stochastic patterns of disturbance and post-disturbance forest development that followed glaciation in Glacier Bay, Alaska. This built on his earlier work studying forest responses to fire on Isle Royale, Michigan (Cooper 1913) and avalanches and rock slides at Robson Pass, British Columbia, Canada (Cooper 1916). In a "virgin" forest in the northeastern US, Hough and Forbes (1940) demonstrated how individual stands in the landscape had been subjected to a complex series of interacting disturbances over the past 300 years and how these idiosyncratic, site-specific disturbance histories interacted with edaphic conditions to generate a variety of forest stand structures.

Over the ensuing decades dendroecological studies of forests enhanced understanding of the importance of unpredictable historical events and disturbances in shaping present-day forest structure and composition. The ubiquitous effects of natural disturbances on forests emerged from tree-ring studies conducted in the 1940–1950s in the USA with signal contributions from Buell (Buell and Cain 1943; Buell and Catlon 1951; Buell and Catlon 1950; Isaak et al. 1959), Spurr (1954), and Stephens (1955). In the 1970–1980s, the number of dendroecological publications increased dramatically, describing research on forest dynamics driven by wind (e.g., Estes 1970; Henry and Swan 1974; Oliver and Stephens 1977; Lorimer 1980), insects (Brubaker and Greene 1979), and fire (e.g., Heinselmann

1973; Dietrich and Swetnam 1984). Simultaneously, studies conducted in other parts of the world reinforced the global importance of disturbance and stochastic events as drivers of forest dynamics (e.g., Kassas 1951 in the United Kingdom; Blais 1962 and Payette and Gagnon 1979 in Canada; Veblen et al. 1977 in south-central Chile; Veblen and Steward 1980 in New Zealand; Morrow and LaMarche 1978 in Australia; and Engelmark 1984 in Sweden). These foundational dendroecological studies contributed to a paradigm shift in forest ecology that arose in the 1970–1980s emphasizing nonequilibrium dynamics, multiscale dynamics, and the importance of humans as agents of disturbance and drivers of forest dynamics.

1.4 The Geography of Dendroecology

The expansion of dendroecological research to represent all forests around the globe has been uneven (Worbes 2002, 2004). As the examples cited above suggest, dendroecological studies have heavily emphasized temperate and high-latitude forests, particularly in the Northern Hemisphere. Southern Hemisphere forests have received less attention, although important contributions from Argentina, Chile, and New Zealand have highlighted common patterns in forest dynamics across hemispheric boundaries. In contrast, dendroecological studies have only recently penetrated tropical forests. Encyclopedic assessments of wood quality and the properties of regional tree floras in the late nineteenth and early twentieth centuries indicated the presence of annual growth rings in many tropical tree species, particularly in tropical areas subjected to certain seasonality (e.g., flooding, drought) from Africa and southern America (Mariaux 1967) and monsoon-influenced regions such as South and Southeast Asia (e.g., Gamble 1904, Chowdhury 1939). However, the lack of distinct temperature seasonality, poorly understood phenological patterns, complex wood anatomies, and often challenging logistical issues have greatly limited the potential for dendrochronological studies of tropical tree species (Bormann and Berlyn 1980). Nevertheless, significant advances have been made over the past two decades to identify hundreds of tropical tree species in South America, Africa, and South and Southeast Asia (Worbes 2002; 2004) that possess annual growth rings. This has provided the necessary foundation for dendroecological methods to address long-standing questions regarding the dynamics of species-rich tropical forests and expands the geographical reach of dendroecology into all of the major forest biomes of the world.

1.5 Contemporary Applications of Dendroecology

Dendroecological studies have fundamentally shaped contemporary views of forest ecology and forest dynamics. While ecological studies using tree rings have been published for well over a century, the past several decades have seen a rapid acceleration in the rate of publication of dendroecological studies. Most syntheses and reviews of dendrochronology over the past half century have focused on climatology, archaeology, or wood formation (e.g., Fritts 1971, 1976, Eckstein 1984, Cook and Kairiukstis 1989, Schweingruber 1996, Hughes et al. 2001, Vaganov et al. 2006, and to a lesser extent Speer 2010). While dendrochronology has made important contributions in each of these areas, the ecological lessons that dendrochronological research has provided are fundamental to understanding how forested ecosystems will respond to the many threats posed by global environmental change. These threats are not limited to changing climatic conditions; they include the impacts of invasive species, biodiversity loss, changing disturbance regimes, and their myriad potential interactions. Dendroecology is a significant component of applied ecological research, providing important historical context for adapting existing forest management strategies to mitigate and respond to current and future global environmental change.

We envision this book as a waypoint in the science of dendroecology that brings together much of the excellent dendroecological research that has come before us, provides a contemporary overview of the breadth and depth of the existing research, and generates new ideas for the many possible directions that dendroecology might take going forward. We hope that this volume will provide a useful resource for dendrochronologists, ecologists, foresters, and others interested in the conservation and sustainable management of the world's forests.

References

Bailey IW (1925) The "spruce budworm" biocoenose. I. Forest rings as indicators of the specific biological events. Bot Gaz 80:93–101

Blais JR (1962) Collection and analysis of radial-growth data from trees for evidence of past spruce budworm outbreaks. For Chron 38(4):474–484

Bormann FH, Berlyn G (1980) Age and growth rate of tropical trees: new directions for research. In: Bormann FH, Berlyn G (eds) Proceedings of the workshop on age and growth rate determination for tropical trees. Petersham

Brandis D (1859) Rules for the administration of forests in the Province of Pegu. Superintendent of Government Printing and Stationery, Rangoon

Brubaker LB (1986) Responses of tree populations to climatic change. Vegetatio 67:119–137

Brubaker LB, Greene SK (1979) Differential effects of Douglas-fir tussock moth and western spruce budworm defoliation on radial growth of grand fir and Douglas-fir. Can J For Res 9:95–105

Buell MF, Cain RL (1943) The successional role of southern white cedar, *Chamaecyparis thyoides*, in southeastern North Carolina. Ecology 24:85–93

Buell MF, Cantlon JE (1950) A study of two communities of the New Jersey Pine Barrens and a comparison of methods. Ecology 31:576–586

Buell MF, Cantlon JE (1951) A study of two forest stands in Minnesota, with an interpretation of the prairie-forest margin. Ecol 32:294–316

Chowdhury KA (1939) The formation of growth rings in Indian trees, part I. Indian Fore Rec Utilizat 1:1–34

Clements FE (1916) Plant succession: an analysis of the development of vegetation. Carnegie Institution of Washington Publication, Washington

Clements FE (1936) Nature and structure of climax. J Ecol 24:252–284
Cook ER, Kairiukstis LA (1989) Methods of dendrochronology: applications in the environmental sciences. Kluwer Academic, Dordrecht
Cooper WS (1913) The climax forest of Isle Royale, Lake Superior, and its development. I. Bot Gaz 55:1–44
Cooper WS (1916) Plant successions in the Mount Robson region, British Columbia. Plant World 19:221–238
Cooper WS (1923) The recent ecological history of Glacier Bay. Alaska: The interglacial forests of Glacier Bay. Ecology 4:93–128
Dieterich JH, Swetnam TW (1984) Dendrochronology of a fire-scarred ponderosa pine. For Sci 30(1):238–247
Douglass AE (1909) Weather cycles in the growth of big trees. Mon Wea Rev 37:225–237
Douglass AE (1929) The secret of the southwest solved by talkative tree rings. Nat Geog Mag 56(6):736–770
Douglass AE (1941) Crossdating in dendrochronology. J For 39(10):825–831
Eckstein D (1984) Handbooks for archaeologists, 2: dendrochronological dating. European Science Foundation, Strasbourg
Engelmark O (1984) Forest fires in the Muddus National Park (northern Sweden) during the past 600 years. Can J Bot 62:893–898
Estes ET (1970) Dendrochronology of black oak (*Quercus velutina* Lam.), white oak (*Quercus alba* L.), and shortleaf pine (*Pinus echinata* Mill.) in the Central Mississippi Valley. Ecol Monogr 40:295–316
Franklin JF, Spies TA, Van Pelt R et al (2002) Disturbances and structural development of natural forest ecosystems with silvicultural implications, using Douglas-fir forests as an example. For Ecol Manage 155:399–423
Fritts HC (1971) Dendroclimatology and dendroecology. Quaternary Res 1:419–449
Fritts HC (1976) Tree-rings and climate. Academic, New York, 567 pp
Fritts HC, Swetnam TW (1989) Dendroecology: a tool for evaluating variations in past and present forest environments. Adv Ecol Res 19:111–188
Gamble JS (1920) A manual of Indian timbers, 2nd edn. Sampson, Low, Marston & Company, London
Glenn-Lewin DC, Peet RK, Veblen TT (1992) Plant succession: theory and prediction. Chapman and Hall, New York
Glock WS (1937) Principles and methods of tree-ring analysis. Carnegie Institution of Washington, Publication No. 486, Washington
Haasis FW (1923) Frost heaving of western yellow pine seedlings. Ecology 4:378–390
Harmon ME, Franklin JF, Swanson FJ et al (1986) Ecology of coarse woody debris in temperate ecosystems. Adv Ecol Res 15:133–302
Heinselman M (1973) Fire in the virgin forests of the Boundary Waters Canoe Area, Minnesota. Quaternary Res 3:329–382
Henry JD, Swan MA (1974) Reconstructing forest history from live and dead plant material—an approach to the study of forest succession in southwest New Hampshire. Ecology 55:772–783
Hough AF, Forbes RD (1943) The ecology and silvics of forests in the high plateaus of Pennsylvania. Ecol Monogr 13:299–320
Hughes MK, Kuniholm PI, Eischeid JK et al (2001) Aegean tree-ring signature years explained. Tree-Ring Res 57:67–73
Isaak D, Marshall WH, Buell MF (1959) A record of reverse plant succession in a tamarack bog. Ecology 40:317–320
Kassas M (1951) Studies in the ecology of Chippenham Fen: II. Recent history of the Fen, from evidence of historical records, vegetational analysis and tree-ring analysis. J Ecol 39:19–32
LeBlanc DC (1990) Relationships between breast-height and whole-stem growth indices for red spruce on Whiteface Mountain, New York. Can J For Res 20:1399–1407.
Lorimer CG (1980) Age structure and disturbance history of a southern Appalachian virgin forest. Ecology 61:1169–1184

Mariaux A (1967) Les cernes dan les bois tropicaux Africains, nature et périodicité. Boi et Forêts des Tropiques 113:3–14

Marshall R (1928) The life history of some western white pine stands on the Kaniksu National Forest. Northwest Sci 2(2):48–53

Morrow PA, LaMarche VC (1978) Tree ring evidence for chronic insect suppression of productivity in subalpine Eucalyptus. Science 201:1244–1246

Oliver CD and Stephens EP (1977) Reconstruction of a mixed species forest in central New England. Ecology 58:562–572

Oliver CD, Larson BC (1996) Forest stand dynamics. McGraw-Hill, New York

Payette S, Gagnon R (1979) Tree-line dynamics in Ungava peninsula, northern Quebec. Holarctic Ecol 2:239–248

Pickett STA, White PS (1985) The ecology of natural disturbance and patch dynamics. Academic, Florida

Schweingruber FH, Kairiukstis L, Shiyatov S (1990) Sample selection. In Cook ER, Kairiukstis L (eds). Methods of dendrochronology: applications in the environmental sciences. Kluwer Academic, Dordrecht

Schweingruber FH (1996) Tree rings and environment dendroecology. Birmensdorf, Swiss Federal Institute for Forest, Snow and Landscape Research. Berne, Stuttgart, Vienna, Haupt, 609 pp

Serre-Bachet F, Tessier L (1989) Response functions for ecological study. In: Cook E, Kairiukstis L (eds) Methods of tree ring analysis: applications in the environmental sciences. Kluwer Academic, Dordrecht

Skelly JM, Innes JL (1994) Waldsterben in the forests of Central Europe and Eastern North America: fantasy or reality? Plant Dis 78(11):1021–1032

Speer JH (2010) Fundamentals of tree ring research. University of Arizona Press

Spurr SH (1954) The forests of Itasca in the nineteenth century as related to fire. Ecology 35:21–25

Stephens EP (1955) The historical-developmental method of determining forest trends. Dissertation, Harvard University Library

West DC, Shugart HH, Botkin DB (1981) Forest succession: concepts and application. Springer, New York

Vaganov EA, Hughes MK, Shashkin AV (2006) Growth dynamics of tree rings: images of past and future environments. Springer, New York

Varenne-de-Fenille M (1791) Observations sur l'Amenagement des Bois, et particulierement des Forest nationales. Imprimerie de la Feuille du Cultivateur, Paris

Veblen TT, Ashton DH, Schlegel FM et al (1977) Plant succession in a timberline depressed by vulcanism in south-central Chile. J Biogeogr 4:275–294

Veblen TT, Stewart G (1980) Comparison of forest structure and regeneration on Bench and Stewart Islands, New Zealand. New Zealand J Ecol 3:50–68

Vins B (1963) Tree-ring studies in Czechoslovakia. Communicationes Instituti forestralis Cechosloveniae 3:192–196

Worbes M (2002) One hundred years of tree ring research in the tropics: a brief history and an outlook to future challenges. Dendrochronologia 20:217–231

Worbes M (2004) Mensuration. Tree ring analysis. In: Evands J, Youngquist HJ (eds) Encyclopedia of forest sciences Volume 1. Elsevier Academic Press, Oxford

Part I
Tree Growth and Forest Dynamics

Credit: M. M. Amoroso

Chapter 2
Ecophysiology and Plasticity of Wood and Phloem Formation

Annie Deslauriers, Patrick Fonti, Sergio Rossi, Cyrille B.K. Rathgeber, and Jozica Gričar

Abstract Long-lived plants need to continuously adjust their structure to fit the changing environmental constraints. Xylem and phloem growth represent both the means used to achieve these structural adjustments and a good indicator of plant success and performance. This chapter reviews the growth dynamics of the xylem and phloem and explores how resource availability and environmental variation provides important context for understanding the impacts of global change. We first illustrate the intra-annual patterns of xylem and phloem growth in different areas of the world to highlight the variations caused by the environment. Then, the components of the growth process, from timing (phenology) to the rates and the characteristics of xylem anatomy are presented to explain how these components are marking valuable indications of the way plants growth respond to changing environmental conditions. Finally we discuss how wood formation can contribute to advancements in ecophysiology and dendroecology.

Keywords Cambium activity • Xylem • Phloem • Xylogenesis • Anatomy • Growth pattern • Climate

---

A. Deslauriers (✉) • S. Rossi
Département des Sciences Fondamentales, Université du Québec à Chicoutimi 555 Boulevard de l'Université, Chicoutimi, QC, Canada G7H 2B1
e-mail: adeslaur@uqac.ca; sergio.rossi@uqac.ca

P. Fonti
Swiss Federal Research Institute WSL, Zürcherstrasse 111, 8903, Birmensdorf, Switzerland
e-mail: patrick.fonti@wsl.ch

C.B.K. Rathgeber
Laboratoire d'Étude des Ressources Forêt-Bois (LERFoB), UMR INRA-AgroParisTech 1092, Centre INRA de Nancy 54280, Champenoux, France
e-mail: cyrille.rathgeber@inra.fr

J. Gričar
Department of Forest Yield and Silviculture, Slovenian Forestry Institute, Vecna Pot 2, 1000, Ljubljana, Slovenia
e-mail: jozica.gricar@gozdis.si

M.M. Amoroso et al. (eds.), *Dendroecology*, Ecological Studies 231,
DOI 10.1007/978-3-319-61669-8_2

2.1 Introduction

Tree growth—defined here as biomass production—is principally related to the activity of the secondary meristem, the cambium. This meristem is responsible for the radial increase of stem, branches and roots, with cell division producing both phloem and xylem cells. While the phloem is responsible for the transport of assimilate, the xylem is a fundamental element of the hydraulic and mechanical architecture of plants (Tyree and Zimmermann 2002), and represents an important pathway through which the water and carbon cycles are connected (Fonti and Jansen 2012; Steppe et al. 2015). The anatomical properties of the xylem influence the process of transpiration, which transfers water to the atmosphere, and carbon fixation, which is responsible for the tree's overall long-term carbon sequestration (Cuny et al. 2015).

Terrestrial plants accumulate approximately 5×10^{10} t of atmospheric carbon in biomass each year (Zhao and Running 2010) and release 4×10^4 km^3 of water into the atmosphere (Holbrook and Zwieniecki 2003), thus contributing significantly to the global water cycle and heat flux dynamics (Seneviratne et al. 2006). Variations in tree growth due to environmental changes can have important implications for the regional and global climates (Bonan 2008). On the other hand, variations in climate require plants, as sedentary organisms, to be sufficiently flexible to acclimate to and endure the surrounding environment and its variability. Wood formation is one key process that plants need to adjust its conductive system to synchronize with a changing environment.

Cambium activity and post cambial growth, the division and differentiation of cells that will form a distinctive layer of wood, has been studied and known for more than a century (Hartig 1892). Early studies mainly focused on description of the process of wood formation (Priestley 1930; Wilson 1964), while later studies explored how cambial activity shaped wood anatomy (Skene 1969) and, more recently, how wood formation responds to the environment (Wodzicki 1971; Denne 1976; Ford et al. 1978).

The formation of a new tree-ring is a complex process of cell division, growth and maturation. As a result, the annual resolution of tree ring widths is insufficient to infer intra-annual growth metrics such as the timing, duration, and the rates of growth within the growing season (Deslauriers et al. 2003). Such metrics supply relevant information and form a solid basis for the understanding of how the wood production responds to environmental changes (Cuny et al. 2015). With the growing awareness of the ongoing impact of climate change on forest ecosystems (Breshears et al. 2009; Allen et al. 2010; McDowell 2011) knowledge of the phenology and physiology of water conduit formation is crucial to understand tree reaction to stress (Balducci et al. 2015, 2016; Vieira et al. 2015; Deslauriers et al. 2009) and how trees adjust water transport to the new conditions. In particular, a more process-based understanding of plant responses to stress and changing environmental stimuli is required in order to understand the impact of xylem plasticity on the biogeochemical interactions of the terrestrial biosphere with the atmosphere (Cuny et al. 2015; Schiestl-Aalto et al. 2015).

Wood formation only reveals one part of the story of cambial production as cambium is a bifacial lateral meristem that produces secondary xylem in the inside and secondary phloem towards the outside (Larson 1994). Compared with xylem formation, fewer studies on phloem formation and its structure can be partly explained by its relatively low economic value compared with the use of timber (Turley et al. 2006). In addition, older phloem is exposed to relatively fast secondary changes (e.g., collapse, sclerification and inflation of parenchyma). Therefore, only the youngest phloem increments are available for histological observations. As older non-conducting bark tissues eventually collapse in a radial direction, deform and later often fall off the tree, the older phloem conduits are not suitable yet for dendrochronological studies (Gričar et al. 2015a). Although their main functions differ, xylem and phloem are closely associated both spatially and functionally (Evert 2006; Steppe et al. 2015). Due to their tight coupling, any changes in water-carbon relations (Zweifel et al. 2006) have impact on xylem and phloem formation processes (Gričar et al. 2015b; Jyske and Hölttä 2015; Hinckley and Lassoie 1981). Therefore, carbon gain and whole-tree survival depend on the functioning and interplay between these two vascular subsystems (Sevanto et al. 2014). Wood and phloem-anatomical features thus offer a direct link for understanding physiological processes in a tree (Savage et al. 2016).

This chapter reviews the growth dynamics of the xylem and phloem and explores how resource availability and environmental variation provide important context for understanding the impacts of global change. The specific aims of this chapter are (1) to describe the general patterns of xylem and phloem growth, (2) to explain the *timing* and *growth rates* of the cambium and the resulting wood *anatomy*, and, (3) to present how wood formation can contribute to advancements in ecophysiology and dendroecology.

2.2 Patterns of Growth in Xylem and Phloem Over a Growing Season

The growth dynamics of the secondary meristems are dependent on the environment, as all metabolic and physiological processes involved in growth are limited by external factors (Muller et al. 2011; Deslauriers et al. 2016). Under optimal conditions, the tree produces tracheids—also named xylem conduits—with structural characteristics that best fit the functional needs of transport, storage, defense, and mechanical support. However, to produce new cells, the cambium requires water to exert the necessary turgor pressure for cell expansion (Pantin et al. 2012; Steppe et al. 2015), assimilates to build and thicken secondary cell walls, heat to allow the metabolic reactions to take place, and time enough to complete the maturation of the new conduits (Schiestl-Aalto et al. 2015). Cambium is a sink for carbon with low ranking priority for carbon allocation within a tree (Minchin and Lacointe 2005), and the amount and structure of the final xylem and phloem is the result of a trade-off between the available resources and structural and functional needs (Spicer and Groover 2010).

The ability to alter secondary vascular growth in response to external stimuli is highly adaptive but requires the coordination of the sequential developmental events during cell formation (Lupi et al. 2010; Rossi et al. 2012) to produce appropriate tissues and physiological outcomes (Hölttä et al. 2010; von Arx et al. 2012). Thus, as growth occurs and is modulated by the availability of resources (e.g., water and carbon) and environmental cues (e.g., temperature and photoperiod), the variability in the timing and dynamics of secondary growth represents a very sensitive bioindicator of plant response to the environment. This variability is expressed in intra-annual patterns of xylem and phloem production, which also vary across vegetation types (Fig. 2.1).

Secondary growth patterns are sensitive to temporal variation in the environment during the growing season and differ among forests growing in different climates (Fig. 2.1). In temperate and boreal climates with abundant soil moisture during the growing season and a dormant season in fall and winter, trees form a distinct annual ring. Growth occurs over one uninterrupted period of time, the growth rate forms a unimodal curve and cumulative growth over the growing season forms a sigmoidal curve (Pattern 1 in the top panel of Fig. 2.1) (Rossi et al. 2007, 2009a). In contrast, the growing season is longer in Mediterranean regions where summer drought limits moisture availability. When growth is uninterrupted, growth rates form a unimodal curve over a longer period but with a lower maximum rate (Pattern 1 in the bottom panel of Fig. 2.1) than in temperate and boreal climates. When drought limits water availability during the growing season, growth is temporarily suspended, but growth can resume following precipitation. Interrupted growth results in an intra-annual fluctuation (IADF) in wood density (Novak et al. 2016a), commonly called a “false ring” (Pattern 2 in the bottom panel of Fig. 2.1). In this scenario, the growth rate forms a bimodal curve (Camarero et al. 2010; de Luis et al. 2011; Vieira et al. 2015) and cumulative growth over the growing season forms an irregular sigmoidal curve. During severe drought, growth may be limited to a short period early in the growing season which is insufficient to form an annual ring, commonly called a “missing ring” (Pattern 3 in the bottom panel of Fig. 2.1). Similar to Mediterranean regions, different patterns of growth occur in subtropical and tropical climates. Unimodal wood production occurs in mesic and moderately wet forests in tropical and subtropical regions (Oliveira et al. 2009; Bosio et al. 2016); however, without a dormant season distinguishing the rings formed in one calendar year from the next is difficult based on wood morphology alone. In some subtropical climates, the alternation of dry and rainy seasons or regular floods (Callado et al. 2001) can lead to multimodal growth patterns (Schmitz et al. 2008; Krepkowski et al. 2011).

All these informations are essential for dendrochronology as the formation of annual rings and their inter-annual variation are crucial for crossdating, which is the core principle of the discipline. Moreover, knowing when and how the tree-ring is formed will also help for properly understanding and inferring the causal agents of intra and inter-ring variations.

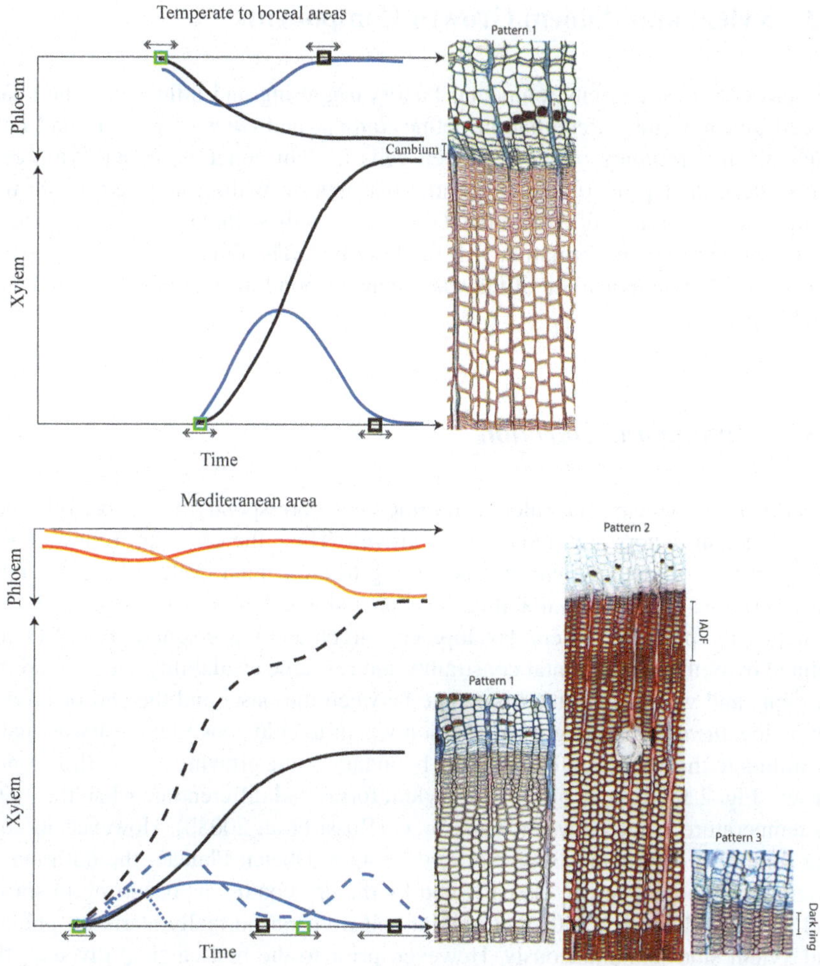

Fig. 2.1 Plasticity of growth patterns in two contrasting biomes. Xylem and phloem growth over one growing season resulting from cambium activity in boreal or temperate (*top*) and Mediterranean (*bottom*) forests. In each case, the period(s) of growth is defined by the *onset* (*green squares*) and *cessation* (*black squares*) while the arrow represents the time variability in the *onset* and *cessation*. For both xylem and phloem, the progress of *growth over time* (*black lines*), quantified either as the number of cells or amount of produced ring width along a year. A general *rate of growth* (*blue lines*), which indicates how fast the process occurs in terms of the measured quantity per time unit, can be represented as the first derivative of the progress of growth (*black lines*). Only one pattern of xylem and phloem growth over time was represented for temperate-boreal areas illustrated as the S-shaped (*pattern 1*) growth pattern (*black line*). The general *rate of growth* of this pattern is typically bell shaped (*blue line*, in *pattern 1*). For Mediterranean areas, three patterns of growth over time were represented: (*pattern 1*) S-shaped growth, such as for temperate-boreal conifers with its bell-shaped rate of growth (*solid black and blue lines*); (*pattern 2*) Intra-annual density fluctuation (*IADF, dotted black line*) with its bimodal rate of growth (*dotted blue line*); (*pattern 3*) missing rings or dark rings (illustrated as the *rate of growth* only, *dotted blue line*) where there is a local formation of few xylem cells (less than five cells), which are slowly differentiating and can look like latewood cells. Phloem growth rate (in *red*) is hypothetical and has been shown to be bimodal (in *red*) like the xylem of pattern 2

2.3 Xylem and Phloem Growth Components

The physiological and environmental factors triggering and influencing the chain of cell division and differentiation define ***timings*** and ***rates*** of growth, and consequently the ***anatomy*** of the produced cells (xylem conduits, parenchyma cell, fibers, etc.), their proportions and spatial distribution within the tree-ring. In this section, the importance of environmental impacts on these three growth components for ***xylem*** and ***phloem*** formation will be described. The three growth components provide a valuable indication of the way plants respond to changing environmental conditions.

2.3.1 Timing and Duration

The ***timing*** of xylem and phloem development corresponds to the period when environmental factors are favourable to growth. Thus, the cambium follows alternating periods of activity and rest, according to the annual cycles of temperature, photoperiod, and water availability. The phenological phases of onset and end, bounding the period of xylem development (green and black square, Fig. 2.1), are defined by both environmental constraints and resource availability such as carbon, nutrients and water. The time difference between the onset and the end of growth define the ***duration*** of the growing season which in cold ecosystems vary annually according to the temperature during the boundary of the growing season (horizontal arrows, Fig. 2.1, (Rossi et al. 2016)). Xylem forms and differentiate when the mean air temperature range between 6 and 8 °C (Rossi et al. 2008b). However, in cold and semi-arid environments of the north-eastern Tibetan Plateau, the duration of xylem development can be also affected by the dry conditions occurring in spring and summer (Ren et al. 2015). Cambial cell division normally starts on phloem and xylem sides simultaneously. However, prior to the first cambial division, the differentiation of one to two layers of phloem cells can occur without previous divisions (Alfieri and Evert 1968, 1973; Kutscha et al. 1975). This is the reason why the onset of phloem formation begins earlier than xylem (see onset of phloem and xylem, Fig. 2.1, temperate to boreal areas) (Gričar and Čufar 2008; Swidrak et al. 2014; Jyske et al. 2015). Indeed, the onset of phloem formation precede xylem formation on average by 3–5 weeks in *Picea abies* (Gričar and Čufar 2008; Swidrak et al. 2014), by 5 weeks in *Larix decidua* and by 3 weeks in *Pinus sylvestris* (Swidrak et al. 2014).

In ecosystems at high latitudes or altitude, where the meristems are active during a short period of time, period of xylem development reflects a balance between avoidance of frost damage and maximization of growth and carbon assimilation (Chuine 2010; Rossi and Isabel 2016). Under the coldest conditions, xylem requires 83 days to complete the maturation of all tracheids produced by cambium, which can be considered the minimum growing season of xylem in boreal and timberline regions (Rossi et al. 2016). Coniferous stands located at the lower distribution limit of larch, Norway spruce and silver fir exhibit a period of cambial activity, i.e. cell production, up to 134 days, and a period of xylem differentiation, i.e. cell enlargement and wall thickening, reaching 178 days (Rossi et al. 2016). The duration of the growing season is a key factor in xylem cell production in some species like black spruce where the growing season is short (Lupi et al. 2010; Rossi et al. 2014), although such result is not confirmed in species of other environments, where the ***rate*** of growth exhibited a more relevant importance (Rathgeber et al. 2011; Cuny et al. 2012).

Accordingly, in Mediterranean climates, xylem formation occurs from winter/spring until the severe summer drought and resuming after fall precipitation (Giovannelli et al. 2007; Vieira et al. 2014). Such an autumnal reactivation of xylem growth, which is observed in some trees of a stand, results in the formation of false rings or latewood intra-annual density fluctuations (pattern 2, Fig. 2.1). By contrast, the lack of a distinct phenology in the phloem production in conifers in Mediterranean regions (unknown onset and end, Fig. 2.1) suggests that variation in the timing of phloem formation is mainly defined by species and site in such area (Barnett 1971; Gričar et al. 2016). Therefore, the observed differences in the phloem phenology among different areas suggest that phloem development is at least partly affected by local environmental conditions (Gričar et al. 2014b).

In tropical forests such as the Brazilian Cerrado, xylem formation lasts from October to June, when temperature is higher and precipitation events are frequent (Bosio et al. 2016). In some ecosystems, precipitation and soil water content can play an important role in defining the timings of xylem formation during spring (Ren et al. 2015; Ziaco et al. 2016). Tropical environments can show species with anomalous xylem growth. *Avicennia marina* is a mangrove species characterized by multiple-cambial layers, aclimatic growth rhythms of xylem growth, and a potentially continuous cambial activity throughout the year (Schmitz et al. 2008).

2.3.2 Rates of Growth

Recent work on the ***rates*** of xylem cell differentiation have provided novel quantification and mechanistic models of wood formation kinetics that explain tree ring formation (Schiestl-Aalto et al. 2015). Xylem cell differentiation kinetics can be characterized by the duration the cells spend in the different phases (i.e. cell enlargement and cell wall thickening) and the rate at which the different sub-processes (primary wall expansion, secondary cell wall deposition) occur (Cuny et al. 2013; Balducci et al. 2016). The duration and rate of the phase of cell enlargement determine its size, while the duration and rate of the thickening phase determine the quantity of deposited wall material (e.g., cell wall area). The relative proportion of cell wall to cell exterior dimensions are the basis for commonly used measures of tree-ring structure (e.g., cell wall thickness, tree-ring micro-density profile). At the beginning of the growing period, good growth conditions generally lead to high rates of enlargement and thickening. These rates then decrease towards the end of the season. This general pattern of variation in duration and rate of xylogenesis sub-processes is at the origin the conifers' tree-ring structure of light earlywood composed of wide cells followed by denser latewood composed of narrower cells (Cuny et al. 2014). Importantly, Cuny et al. (2015) refute the long-standing assumption that the increase in wall thickness and wood density along the ring is driven by the fixation of more biomass during the wall-thickening process. Instead, they demonstrate that the amount of wall material per cell is almost constant along a ring except for the last few cells of the ring.

Consequently, changes in wall thickness and wood density are principally driven by changes in cell size. The reduction of the cell size along the ring is mainly due to decreasing duration of cell enlargement, whereas decreasing rate of cell enlargement only contributes marginally. Recent works (Balducci et al. 2016; Cuny and Rathgeber 2016) show a very interesting mechanism of coupling between the rates and duration of cell wall deposition: as the season progresses and the rate of cell wall deposition is declining, the duration increase in such proportion that it perfectly counter balance the decrease of the rate, stabilising the production of the amount of cell wall material per cell. However, at the very end of the growing season when temperature drops (Cuny and Rathgeber 2016), or when water conditions deteriorates too much (Balducci et al. 2016), durations increase level off, breaking the coupling between rate and duration, and climatic conditions are directly written into tree-ring structure.

The highest ***rate*** of phloem cell production of *P. abies* was detected 2–6 weeks earlier than that of xylem (Gričar et al. 2014b) because the formation of the annual phloem increment is crucial for tree survival. The translocation pathways for photosynthates and biomolecules must be maintained because sieve cells function for only one to two growing seasons (Evert 2006; Steppe et al. 2015). While missing or discontinuous rings in xylem have been regularly reported, particularly in trees from extreme environments (Wilmking et al. 2012; Novak et al. 2016b), or in trees with reduced vitality (Torelli et al. 1999), phloem rings are formed every year, even if very narrow (Gričar et al. 2014a). The narrower phloem rings are caused by an asynchrony between the production of xylem and phloem derivative: cambium divides more frequently on the xylem side than on the phloem side and, consequently, xylem accumulates at a faster rate than phloem (e.g., Panshin and de Zeeuw 1980; Jyske and Hölttä 2015). However, in the case of reduced radial growth, a tree's preference is for phloem formation, leading to a progressively smaller ratio between xylem and phloem widths (Bauch 1986; Gričar et al. 2014a). The pattern of phloem formation has a S-shaped growth pattern (pattern 1 in black, Fig. 2.1) in coniferous and deciduous tree species in temperate to boreal regions resulting in the typical bell shaped rates of growth (in blue, Fig. 2.1).

2.3.3 Anatomy

The environmental influence on the process of wood formation results in xylem ***anatomy*** changes that can also have functional implications (Fonti and Jansen 2012). The xylem structure determines the functional balance between efficient water transport to achieve optimal growth, minimum investment of construction costs to secure the xylem plumbing system, mechanical support for the assimilating leaves, and storage of water and nonstructural carbohydrates for defense and resistance to stress (von Arx et al. 2012). So for example, the capacity to transport water, which is an important functional trait connected with water use and net primary productivity, can be modulated by modifying the arrangement and structure of the wooden cells (Fig. 2.2). Unmistakable examples of altered structure and functions are frost rings (Glerum and Farrar 1966), light rings (Wang et al. 2000), blue rings (Piermattei et al. 2015) or even the intra-annual density fluctuations (Zalloni et al. 2016). However, any environmental factors that alter the normal course of formation of the wooden cells can leave a characteristic signature within the anatomical characteristic of the annual ring (Fig. 2.2). Indeed, each wooden cell first needs to enlarge before thickening the wall (Rathgeber et al. 2016). Thus, both the cell size and the wall thickness of each cell in the ring can provide important information related to the environmental drivers influencing both these processes (Castagneri et al. 2017). Several studies based on tree-ring anatomical methods (Fonti et al. 2010) have already demonstrated the possibility to use time-series of cell anatomical features for a deeper insight on the factors affecting xylogenesis (Bryukhanova and Fonti 2013; Fonti et al. 2013; Martin-Benito et al. 2013; Carrer et al. 2016) and how these changes might affect future xylem functioning (Pacheco et al. 2016; Pérez-de-Lis et al. 2016).

Variations in the phloem conduits anatomy exhibit much less plasticity than that of xylem. Variations in the anatomy of phloem rings in temperate tree species are mainly observed at the site level rather than on a temporal scale (Prislan et al. 2013; Gričar et al. 2014b). Intra- and inter-species structural variation in secondary tissues exists because of local adaptations and environmental conditions (Rowe and Speck 2005). Jyske and Hölttä (2015) found that phloem conduits are slightly narrower than xylem conduits in *P. abies*. Gričar et al. (2015b) further confirmed this trend but only for the early part of the annual rings, whereas in the increments formed in the second part of the growing season, it is just the opposite. Thus, tracheids in earlywood are on average 12–30% wider and in latewood 30–50% narrower than sieve cells in early and late phloem, respectively. The variation in the structure of annual xylem and phloem increments in *P. abies* clearly demonstrates that plasticity in the seasonal dynamics of cambial cell production and cell differentiation exists on both xylem and phloem sides. At the tissue and cell levels, wood anatomical traits have largely been studied in this respect (e.g., Fonti et al. 2010 and previous section on xylem anatomy), whereas the proxy potential of phloem-anatomical traits still needs to be exploited (Gričar et al. 2015b; Jyske and Hölttä 2015).

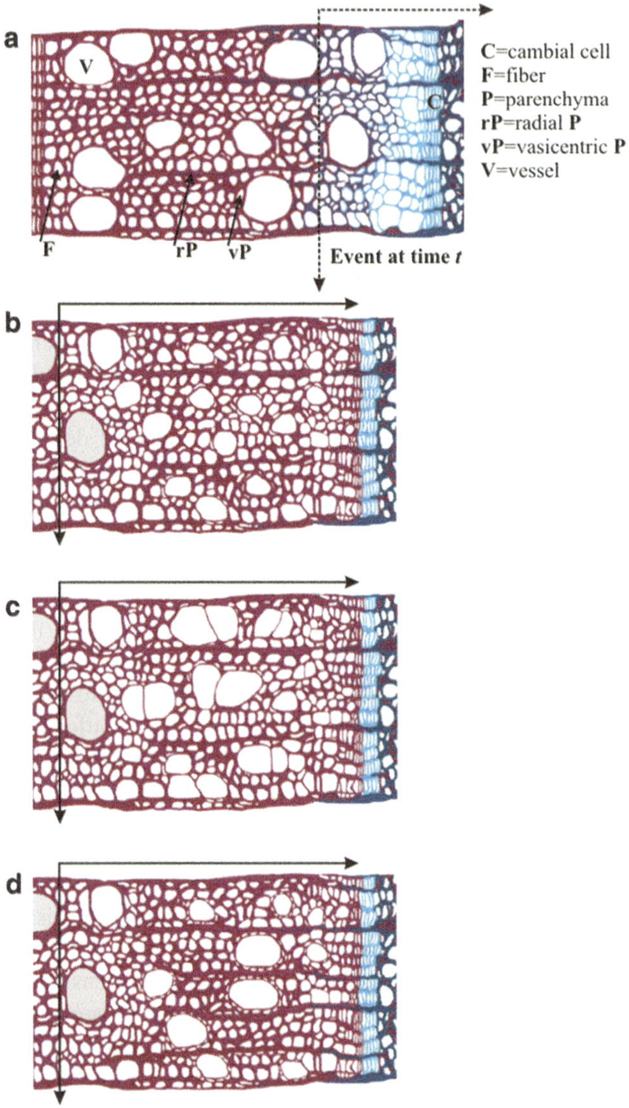

Fig. 2.2 Plasticity of the final product. An example of how the structure of the xylem hydraulic architecture of a forming *Populus* × *canadensis* annual ring from temperate environments can be plastically adjusted to respond to a new environmental constraint, e.g., a long drought period. The example is based on micro-sections observations as they might appear when performed under a light microscope (magnification 100×, staining performed with safranin and astrablue). To guarantee the fulfilment of future needs (e.g., safe and efficient water transport and mechanical stability) trees have to adjust their growth by optimizing the available resources (e.g., carbon, warmth, water, and the time to perform the process). (**a**) Shows the situation of the forming ring at the time the event is occurring. (**b–d**) Show examples that simulate possible adjustments in vessel size and arrangement as they appear visible along a ring cross-section. Efficient sap flow occurs when large and numerous vessels contribute to the transport, although large vessels are susceptible to cavitation (in *gray*) under strong negative pressure.

Intra-annual xylem and phloem formation analyses are therefore crucial for a better understanding of the mechanisms underlying the environmental response of xylem and phloem anatomy of different tree species.

2.4 Contributions to Ecophysiology and Dendroecology

Monitoring xylogenesis is contributing important mechanistic insights to disciplines that have strong ties with plant performance and survival, paleo-environmental reconstruction, and biogeochemical feed-back. In this chapter, the contribution of the studies on xylogenesis to the questions on global change and other ecophysiological fields are explained and discussed.

2.4.1 Plant Performances, Distribution and Community Composition

The ecological amplitude, and consequently the distribution of the individuals, are partly regulated by the plasticity of the species in response to spatio-temporal variability of the environmental conditions (Valladares et al. 2007). For example, the hydraulic structure of the xylem network and its plasticity is determinant, not only for the water balance of plants, but also for attaining maximal photosynthetic rates and optimal adaptations to specific microclimates. A limited capacity for phenotypic adjustment to changing conditions might have detrimental consequences for plant carbon budget, tree productivity and mortality (Martinez-Meier et al. 2008). Repeated and lengthy droughts, for example, can affect the delicate safety-efficiency trade-off in the water conduction system and trigger detrimental consequences in plant carbon allocation, which finally induces the death of the plant (McDowell 2011; Hereş et al. 2012; Pellizzari et al. 2016).

The effect of climatic factors on xylem anatomy has been extensively described (Schweingruber 2007), but particular focus has been placed to xylem anatomical responses in relation to water stress and/or plant mortality (Arend and Fromm 2007; Fonti and García-González 2008; Eilmann et al. 2009, 2011; Rossi et al.

Fig. 2.2 (continued) Drought and freezing can both cause embolisms in plants, even leading to complete system failure if extended embolisms occur. Strategies of modifications can be oriented towards increasing safety (at the cost of efficiency) (**b**), redundancy (**c**) or reparability (**d**) of the system. Other anatomical characteristics which are not shown in the figure, as for example the length of the vessels, the number, sizes and structure of pits of connecting vessels, and the division of whole plants into independent hydraulic units, can also influence these attributes. For simplicity, the responses described in the figure are shown to be rapid and to have immediate effect on the anatomy of the new forming ring. However, depending on frequency and magnitude of the events, induced modifications of the wood structure can last over the formation of more annual rings

2009b; Battipaglia et al. 2010; DeSoto et al. 2011; Fonti and Babushkina 2016). Many morphological characteristics of the xylem hydraulic architecture, such as the arrangement, frequency, size, wall thickness, pit characteristics of conduits and tissue proportion such as ray parenchyma, can be modified to regulate both the efficiency of water transport and the margins of safety against hydraulic system failures (Pittermann et al. 2006; Sperry et al. 2006; Choat et al. 2008; Olano et al. 2013). An example of such adjustments is shown in Fig. 2.2. Nevertheless, the strategies of adjustment are diversified and depend on species and conditions. The observations of wood formation can thus be used to identify the drivers (which factor), timings (when) and environmental threshold (what intensity) behind the formation of specific anatomical characteristics, as it has been performed for the intra-annual density fluctuation observed in the dry Mediterranean regions (de Luis et al. 2011; De Micco et al. 2016). In this case, xylogenesis represents an indispensable basis of understanding the link environment-structure necessary for evaluating the range of plasticity and resilience of trees growing under defined environmental conditions and for predicting plant performances, distribution and community composition under future climates (Baas and Wheeler 2011; Beeckman 2016; Sass-Klaassen et al. 2016).

2.4.2 *Paleo-environmental Reconstructions by Mean of Tree-Ring Proxy*

Dendrochronology usually dates the formation of tree rings to the exact calendar year to extract relevant paleo-environmental information integrating the whole growing season (Fritts 1976; von Arx and Dietz 2006) by means of transfer functions. Similarly, the cell characteristics fixed in the datable tree rings represent an archive of past conditions but at a much higher time resolution (Fonti et al. 2010). Thus, the assignment of a time of formation of each cell in the ring will not only allow to increase the resolution of the paleo-environmental proxies, but also put the basis for a better process-based understanding of the growth-climate interactions. The increased resolution will allow to better relate intra-ring features, as the appreciated environmental-sensitive maximum latewood density or other more novel data-types such as stable isotopes ratios (usually C, H and O) and cell anatomical features, to the physiological state of the plant at the time of cell formation. Assigning a time of formation to intra-annual wood anatomical or isotopic measurements performed in cells within tree rings will specifically allow causes and effects to be better linked and understood (Skomarkova et al. 2006; De Micco et al. 2007; Battipaglia et al. 2010; Eilmann et al. 2010; de Luis et al. 2011; Carrer et al. 2016; Castagneri et al. 2017). Questions related to the mobilisation and storage of carbon assimilates in relation to the properties of secondary growth might be finally better understood (Deslauriers et al. 2009, 2014, 2016; Simard et al. 2013) thus significantly improving the interpretation of the paleo-environmental reconstructions.

2.4.3 *Quantifying Biogeochemical Cycles*

Quantifying the intensity of carbon and water exchanges between terrestrial ecosystems and surrounding environments (atmosphere, hydrosphere, pedosphere-lithosphere) is a central challenge for understanding the biogeochemical cycle and a prerequisite for reliable future scenarios of climate change. In this complex climate system, the growth activity of plants plays a determinant role. The timings of wood formation define the period when carbon fixation into wood is activated and for how long. So, growth onset and cessation, which are usually controlled by temperature and photoperiod (Körner and Basler 2010), defines the period when growth can occur, while the environmental conditions determine the speed of the process. These conditions do not only define the properties of the ring (e.g. the thickness and the density of the wood) but also determine the future performance of the functional xylem via the anatomical structure. These properties are important in determining how much carbon is sequestered by plants in new and future growth, how much nitrogen and other nutrients are absorbed from the soil and how much water is returned to the atmosphere via transpiration. Current models estimating growth periods and intensity are usually calibrated using phenological observations from satellite remote-sensing (e.g., Pettorelli et al. 2005), or from field observations (e.g., Menzel et al. 2006). This monitoring is based on activity of the primary meristems and often assumes a direct link between primary and secondary meristems functioning (Huang et al. 2014; Rossi et al. 2009a). To increase reliability and resolution of the prediction of the impact of climate change on water and carbon cycles, more observational description on how cambial activity affects the physiology of the plant on a short and long-term perspective are necessary to improve the new generation of process-based models relying on a description of the cambial activity to simulate radial growth and related physiological processes at an intra-annual scale (Schiestl-Aalto et al. 2015).

Monitoring xylogenesis shows that environmental factors substantially modulate the duration and rate of growth processes of xylem and phloem cells. Several studies have revealed a high plasticity of wood formation to various environmental conditions in many different environments (Deslauriers et al. 2008; Camarero et al. 2010; Moser et al. 2010; Rossi et al. 2011a; Gryc et al. 2012; Vichrová et al. 2013; Treml et al. 2015; Mendivelso et al. 2016), but also between species (Rossi et al. 2007; Čufar et al. 2008; Pérez-de-Lis et al. 2016; Ziaco and Biondi 2016), age and size classes (Rossi et al. 2008a; Li et al. 2013, 2016) and social status (Rathgeber et al. 2011). Among the environmental factors studied, temperature seems to be the factor driving the onset of the growing season in temperate and boreal environments. An inter-annual variation of up to 1 month has been observed in the timing of wood formation across conifer species in Europe and Canada (Deslauriers et al. 2003, 2008; Rossi et al. 2008b; Gruber et al. 2010) and along altitudinal and latitudinal gradients (Moser et al. 2010; Boulouf-Lugo et al. 2012; Cuny et al. 2015; Rossi et al. 2016). Stem heating during the quiescent stage also demonstrated that cambium reactivation in temperate or boreal environments is extremely sensitive to temperature and xylem differentiation was induced after 2 weeks of continuous exposure independently of bud activity (Gričar et al. 2006). In the boreal forest, a model predicts longer duration of xylogenesis at higher temperatures, with an increase of 8–11 days/°C, because of an earlier onset and later ending of growth (Rossi et al. 2011b). Insufficient water availability, like in a Mediterranean climate, has also been related with the duration of the growing season. Effects of water deficit include earlier culminations of the maximum growth rate, shorter periods of growth and reductions in the number of tracheids produced (Eilmann et al. 2009; Gruber et al. 2010; Balducci et al. 2013; Vieira et al. 2015). Shifts from the S-shaped pattern (pattern 1, Fig. 2.1) to bimodal radial growth (pattern 2, Fig. 2.1) can even be observed in non-Mediterranean species, depending on the timings of the drought events during the growing season and on plant strategy (Giovannelli et al. 2007; Balducci et al. 2015).

There is now a consensus that the determination of the intra-annual growth pattern under different changing conditions yields important insights into the possible ecological response of trees by modulating how the trees can perform under current and future environment. Accounting for this more detailed information of the growth patterns and their variability is fundamental to significantly improve the process-based model of growth, with implications for up-scaled modelling of global flux changes within climate change scenarios.

2.5 Concluding Remarks

Plant growth is inextricably linked to the environment. On the shortest and smallest scales, the environmental conditions directly influence physiological processes. On longer periods and wider spatial scales, environmental changes influence species performance, distribution and community composition, which eventually influence the regional and global climate. Understanding how the growth of plants responds to these changes is a critical requirement for simulating ecosystem dynamics under future climate. The study of xylem and phloem growth, through a highly resolved and detailed description of how the environment affects the *timing*, *rates* and the *anatomy* of the cell, is providing important contributions to the major scientific challenges related to plant performance under climate change.

In the last decade, investigations on intra-annual dynamics of growth have gained in popularity and are supplying a greater and deeper understanding of the secondary growth processes with beneficial implications in many disciplines. Expanding the studies to more environments and species will provide a more complete basis for an up-scaled and up-dated process-oriented understanding of how cambium activity and secondary growth responds to and influences global and local environmental changes.

Acknowledgments This chapter was partly funded by the CRSNG Discovery Grant of A. Deslauriers. P. Fonti received support from the Swiss National Science Foundation (projects INTEGRAL-121859 and LOTFOR-150205). The authors thank the European Science Foundation supporting the LESC Strategic Workshop and the COST Action STReESS—Studying Tree Responses to extreme Events: a SynthesiS (FP1106) for creating the favorable environment for the realization of this work.

References

Alfieri FJ, Evert RF (1968) Seasonal development of the secondary phloem in *Pinus*. Am J Bot 55:518–528

Alfieri FJ, Evert RF (1973) Structure and seasonal development of the secondary phloem in the Pinaceae. Bot Gaz 134:17–25

Allen CD, Macalady AK, Chenchouni H, Bachelet D, McDowell N, Vennetier M, Kitzberger T, Rigling A, Breshears DD, Hogg EHT, Gonzalez P, Fensham R, Zhang Z, Castro J, Demidova N, Lim J-H, Allard G, Running SW, Semerci A, Cobb N (2010) A global overview of drought and heat-induced tree mortality reveals emerging climate change risks for forests. For Ecol Manag 259(4):660–684

Arend M, Fromm J (2007) Seasonal change in the drought response of wood cell development in poplar. Tree Physiol 27:985–992

von Arx G, Dietz H (2006) Growth rings in the roots of temperate forbs are robust annual markers. Plant Biol 8:224–233

von Arx G, Archer R, Hughes MK (2012) Long-term functional plasticity in plant hydraulic architecture in response to supplemental moisture. Ann Bot 109:1091–1100

Baas P, Wheeler E (2011) Wood anatomy and climate change. In: Hodkinson TR, Jones MB, Waldren S, JAN P (eds) Climate change, ecology and systematics. Cambridge University Press, Cambridge, pp 141–155

Balducci L, Deslauriers A, Giovannelli A, Rossi S, Rathgeber CBK (2013) Effects of temperature and water deficit on cambial activity and woody ring features in *Picea mariana* saplings. Tree Physiol 33:1006–1017. doi:10.1093/treephys/tpt073

Balducci L, Deslauriers A, Giovannelli A, Beaulieu M, Delzon S, Rossi S, Rathgeber CBK (2015) How do drought and warming influence plant survival and wood traits of *Picea mariana* saplings? J Exp Bot 66(1):377–389

Balducci L, Cuny H, Rathgeber C, Deslauriers A, Giovannelli A, Rossi S (2016) Compensatory mechanisms mitigate the effect of warming and drought on wood formation. Plant Cell Environ 39(6):1338–1352. doi:10.1111/pce.12689

Barnett J (1971) Winter activity in the cambium of *Pinus radiata*. N Z J For Sci 1:208–222

Battipaglia G, De Micco V, Brand WA, Linke P, Aronne G, Saurer M, Cherubini P (2010) Variations of vessel diameter and delta 13C in false rings of *Arbutus unedo* L. reflect different environmental conditions. New Phytol 188(4):1099–1112. doi:10.1111/j.1469-8137.2010.03443.x

Bauch J (1986) Characteristics and response of wood in declining trees from forests affected by pollution. IAWA Bull 7(4):269–276

Beeckman H (2016) Wood anatomy and trait-based ecology. IAWA J 37(2):127–151. doi:10.1163/22941932-20160127

Bonan GB (2008) Forests and climate change: Forcings, feedbacks, and the climate benefits of forests. Science 320:1444–1449

Bosio F, Rossi S, Marcati CR (2016) Periodicity and environmental drivers of apical and lateral growth in a Cerrado woody species. Trees 30(5):1495–1505. doi:10.1007/s00468-016-1383-8

Boulouf-Lugo J, Deslauriers A, Rossi S (2012) Duration of xylogenesis in black spruce lengthened between 1950 and 2010. Ann Bot 110:1099–1108

Breshears DD, Myers OB, Meyer CW, Barnes FJ, Zou CB, Allen CD, McDowell NG, Pockman WT (2009) Tree die-off in response to global change-type drought: mortality insights from a decade of plant water potential measurements. Front Ecol Environ 7:185–189

Bryukhanova M, Fonti P (2013) Xylem plasticity allows rapid hydraulic adjustment to annual climatic variability. Trees 27(3):485–496. doi:10.1007/s00468-012-0802-8

Callado CH, Neto SJS, Scarano FR, Costa CG (2001) Periodicity of growth rings in some flood-prone trees of Atlantic Rain Forest in Rio de Janeiro, Brazil. Trees 15:492–497

Camarero JJ, Olano JM, Parras A (2010) Plastic bimodal xylogenesis in conifers from continental Mediterranean climates. New Phytol 185:471–480

Carrer M, Brunetti M, Castagneri D (2016) The imprint of extreme climate events in century-long time series of wood anatomical traits in high-elevation conifers. Front Plant Sci 7:1–12. doi:10.3389/fpls.2016.00683

Castagneri D, Fonti P, von Arx G, Carrer M (2017) How does climate influence xylem morphogenesis over the growing season? Insights from long-term intra-ring anatomy in Norway spruce. Ann Bot. doi:10.1093/aob/mcw274

Choat B, Cobb AR, Jansen S (2008) Structure and function of bordered pits: new discoveries and impacts on whole-plant hydraulic function. New Phytol 177(3):608–625. doi:10.1111/j.1469-8137.2007.02317.x

Chuine I (2010) Why does phenology drive species distribution? Philos Trans R Soc B 365:3149–3160

Čufar K, Prislan P, de Luis M, Gričar J (2008) Tree-ring variation, wood formation and phenology of beech (*Fagus sylvatica*) from a representative site in Slovenia, SE Central Europe. Trees 22:749–758

Cuny H, Rathgeber C (2016) Xylogenesis: Coniferous trees of temperate forests are listening to the climate tale during the growing season but only remember the last words! Plant Physiol 171:306–317

Cuny HE, Rathgeber CBK, Lebourgeois F, Fortin M, Fournier M (2012) Life strategies in intra-annual dynamics of wood formation: example of three conifer species in a temperate forest in north-east France. Tree Physiol 32:612–625

Cuny HE, Rathgeber CBK, Kiesse TS, Hartmann FP, Barbeito I, Fournier M (2013) Generalized additive models reveal the intrinsic complexity of wood formation dynamics. J Exp Bot 64(7):1983–1994

Cuny HE, Rathgeber CBK, Frank D, Fonti P, Fournier M (2014) Kinetics of tracheid development explain conifer tree-ring structure. New Phytol 203:1231–1241

Cuny HE, Rathgeber CBK, Frank D, Fonti P, Mäkinen H, Prislan P, Rossi S, del Castillo EM, Campelo P, Vavrčík H, Camarero JJ, Bryukhanova MV, Jyske T, Gričar J, Gryc V, De Luis M, Vieira J, Čufar K, Kirdyanov AV, Oberhuber W, Treml V, Huang JG, Li X, Swidrak I, Deslauriers A, Liang E, Nöjd P, Gruber A, Nabais C, Morin H, Krause C, King G, Fournier M (2015) Intra-annual dynamics of woody biomass production in coniferous forests. Nat Plants 1:15160. doi:10.1038/nplants.2015.160

De Micco V, Saurer M, Aronne G, Tognetti R, Cherubini P (2007) Variations of wood anatomy and delta C-13 within-tree rings of coastal *Pinus pinaster* showing intra-annual density fluctuations. IAWA J 28(1):61–74

De Micco V, Balzano A, Čufar K, Aronne G, Gričar J, Merela M, Battipaglia G (2016) Timing of false ring formation in *Pinus halepensis* and arbutus Unedo in southern Italy: Outlook from an analysis of xylogenesis and tree-ring chronologies. Front Plant Sci 7:1–14. doi:10.3389/fpls.2016.00705

Denne MP (1976) Effects of environmental change on wood production and wood structure in *Picea sitchensis* seedlings. Ann Bot 40:1017–1028

Deslauriers A, Morin H, Bégin Y (2003) Cellular phenology of annual ring formation of *Abies balsamea* in the Québec boreal forest (Canada). Can J For Res 33:190–200

Deslauriers A, Rossi S, Anfodillo T, Saracino A (2008) Cambium phenology, wood formation and temperature thresholds in two contrasting years at high altitude in Southern Italy. Tree Physiol 28:863–871

Deslauriers A, Giovannelli A, Rossi S, Castro G, Fragnelli G, Traversi L (2009) Intra-annual cambial activity and carbon availability in stem of poplar. Tree Physiol 29:1223–1235

Deslauriers A, Beaulieu M, Balducci L, Giovannelli A, Gagnon MJ, Rossi S (2014) Impact of warming and drought on carbon balance related to wood formation in black spruce. Ann Bot 114(2):335–345

Deslauriers A, Huang JG, Balducci L, Beaulieu M, Rossi S (2016) The contribution of carbon and water in modulating wood formation in black spruce saplings. Plant Physiol 170:2072–2084

DeSoto L, De la Cruz M, Fonti P (2011) Intra-annual patterns of tracheid size in the Mediterranean tree *Juniperus thurifera* as an indicator of seasonal water stress. Can J For Res 41:1280–1294

Eilmann B, Zweifel R, Buchmann N, Fonti P, Rigling A (2009) Drought-induced adaptation of the xylem in Scots pine and pubescent oak. Tree Physiol 29(8):1011–1020. doi:10.1093/treephys/tpp035

Eilmann B, Buchmann N, Siegwolf R, Saurer M, Cherubini P, Rigling A (2010) Fast response of Scots pine to improved water availability reflected in tree-ring width and delta 13C. Plant Cell Environ 33(8):1351–1360. doi:10.1111/j.1365-3040.2010.02153.x

Eilmann B, Zweifel R, Buchmann N, Graf Pannatier E, Rigling A (2011) Drought alters timing, quantity, and quality of wood formation in Scots pine. J Exp Bot 62:2763–2771

Evert RF (2006) Esau's plant anatomy: meristems, cells, and tissues of the plant body: their structure, function, and development, 3rd edn. John Wiley & Sons, Inc., Hoboken

Fonti P, Babushkina EA (2016) Tracheid anatomical responses to climate in a forest-steppe in Southern Siberia. Dendrochronologia 39:32–41. doi:10.1016/j.dendro.2015.09.002

Fonti P, García-González I (2008) Earlywood vessel size of oak as a potential proxy for spring precipitation in mesic sites. J Biogeogr 35(12):2249–2257. doi:10.1111/j.1365-2699.2008.01961.x

Fonti P, Jansen S (2012) Xylem plasticity in response to climate. New Phytol 195:734–736

Fonti P, von Arx G, García-González I, Eilmann B, Sass-Klaassen U, Gartner H, Eckstein D (2010) Studying global change through investigation of the plastic responses of xylem anatomy in tree rings. New Phytol 185(1):42–53. doi:10.1111/j.1469-8137.2009.03030.x

Fonti P, Bryukhanova MV, Myglan VS, Kirdyanov AV, Naumova OV, Vaganov EA (2013) Temperature-induced responses of xylem structure of *Larix sibirica* (Pinaceae) from the Russian Altay. Am J Bot 100(7):1332–1343. doi:10.3732/ajb.1200484

Ford ED, Robards AW, Piney MD (1978) Influence of environmental factors on cell production and differentiation in the earlywood of *Picea sitchensis*. Ann Bot 42:683–692

Fritts HC (1976) Tree rings and climate. Academic, London

Giovannelli A, Deslauriers A, Fragnelli G, Scaletti L, Castro G, Rossi S, Crivellaro A (2007) Evaluation of drought response of two poplar clones (*Populus x canadensis* Mönch 'I-214' and *P. deltoides* Marsh. 'Dvina') through high resolution analysis of stem growth. J Exp Bot 58:2673–2683

Glerum C, Farrar JL (1966) Frost ring formation in the stems of some coniferous species. Can J Bot 44(7):879–886. doi:10.1139/b66-103

Gričar J, Čufar K (2008) Seasonal dynamics of phloem and xylem formation in silver fir and Norway spruce as affected by drought. Russ J Plant Physiol 55(4):538–543. doi:10.1134/s102144370804016x

Gričar J, Zupancic M, Čufar K, Koch G, Schmitt U, Oven P (2006) Effect of local heating and cooling on cambial activity and cell differentiation in the stem of Norway spruce (*Picea abies*). Ann Bot 97:943–951

Gričar J, Jagodic Š, Šefc B, Trajković J, Eler K (2014a) Can the structure of dormant cambium and the widths of phloem and xylem increments be used as indicators for tree vitality? Eur J For Res 133(3):551–562. doi:10.1007/s10342-014-0784-8

Gričar J, Prislan P, Gryc V, Vavrčík H, de Luis M, Čufar K (2014b) Plastic and locally adapted phenology in cambial seasonality and production of xylem and phloem cells in *Picea abies* from temperate environments. Tree Physiol 34(8):869–881. doi:10.1093/treephys/tpu026

Gričar J, Jagodic Š, Prislan P (2015a) Structure and subsequent seasonal changes in the bark of sessile oak (*Quercus petraea*). Trees 29(3):747–757. doi:10.1007/s00468-015-1153-z

Gričar J, Prislan P, De Luis M, Gryc V, Hacurova J, Vavrčík H, Čufar K (2015b) Plasticity in variation of xylem and phloem cell characteristics of Norway spruce under different local conditions. Front Plant Sci 6. doi:10.3389/fpls.2015.00730

Gričar J, Prislan P, De Luis M, Novak K, Longares LA, Martinez del Castillo E, Čufar K (2016) Lack of annual periodicity in cambial production of phloem in trees from Mediterranean areas. IAWA J 37(2):332–348

Gruber A, Strobl S, Veit B, Oberhuber W (2010) Impact of drought on the temporal dynamics of wood formation in *Pinus sylvestris*. Tree Physiol 30:490–501

Gryc V, Hacura J, Vavrčík H, Urban J, Gebauer R (2012) Monitoring of xylem formation in *Picea abies* under drought stress influence. Dendrobiology 67:15–24

Hartig R (1892) Ueber Dickenwachstum und Jahrringsbildung. Bot Zeit 50(176–180):193–196

Hereş AM, Martínez-Vilalta J, López BC (2012) Growth patterns in relation to drought-induced mortality at two Scots pine (*Pinus sylvestris* L.) sites in NE Iberian Peninsula. Trees 26(2):621–630. doi:10.1007/s00468-011-0628-9

Hinckley TM, Lassoie JP (1981) Radial growth in conifers and deciduous trees: a comparison. Mitt Forstl Bundesversuchsanst Wien 142:17–56

Holbrook NM, Zwieniecki MA (2003) Plant biology—Water gate. Nature 425(6956):361–361. doi:10.1038/425361a

Hölttä T, Mäkinen H, Nöjd P, Mäkelä A, Nikinmaa E (2010) A physiological model of softwood cambial growth. Tree Physiol 30:1235–1252

Huang J, Deslauriers A, Rossi S (2014) Xylem formation can be modeled statistically as a function of primary growth and cambium activity. New Phytol 203:831–841

Jyske T, Hölttä T (2015) Comparison of phloem and xylem hydraulic architecture in *Picea abies* stems. New Phytol 205(1):102–115. doi:10.1111/nph.12973

Jyske TM, Suuronen J-P, Pranovich AV, Laakso T, Watanabe U, Kuroda K, Abe H (2015) Seasonal variation in formation, structure, and chemical properties of phloem in *Picea abies* as studied by novel microtechniques. Planta 242(3):613–629. doi:10.1007/s00425-015-2347-8

Körner C, Basler D (2010) Phenology under global warming. Science 327:1461–1462

Krepkowski J, Brauning A, Gebrekirstos A, Strobl S (2011) Cambial growth dynamics and climatic control of different tree life forms in tropical mountain forest in Ethiopia. Trees 25:59–70

Kutscha NP, Hyland F, Schwarzmann JM (1975) Certain seasonal changes in balsam fir cambium and its derivatives. Wood Sci Technol 9:175–188

Larson PR (1994) The vascular cambium: development and structure. Springer, Berlin

Li X, Liang E, Gričar J, Prislan P, Rossi S, Čufar K (2013) Age dependence of xylogenesis and its climatic sensitivity in Smith fir on the south-eastern Tibetan Plateau. Tree Physiol 33(1):48–56. doi:10.1093/treephys/tps113

Li X, Camarero JJ, Case B, Liang E, Rossi S (2016) The onset of xylogenesis is not related to distance from the crown in Smith fir trees from the southeastern Tibetan Plateau. Can J For Res 46(6):885–889. doi:10.1139/cjfr-2016-0092

de Luis M, Novak K, Raventós J, Gričar J, Prislan P, Čufar K (2011) Climate factors promoting intra-annual density fluctuations in Aleppo pine (*Pinus halepensis*) from semiarid sites. Dendrochronologia 29:163–169

Lupi C, Morin H, Deslauriers A, Rossi S (2010) Xylem phenology and wood production: resolving the chicken-or-egg dilemma. Plant Cell Environ 33:1721–1730

Martin-Benito D, Beeckman H, Cañellas I (2013) Influence of drought on tree rings and tracheid features of *Pinus nigra* and *Pinus sylvestris* in a mesic Mediterranean forest. Eur J For Res 132(1):33–45. doi:10.1007/s10342-012-0652-3

Martinez-Meier A, Sanchez L, Pastorino M, Gallo L, Rozenberg P (2008) What is hot in tree rings? The wood density of surviving Douglas-firs to the 2003 drought and heat wave. For Ecol Manag 256(4):837–843. doi:10.1016/j.foreco.2008.05.041

McDowell NG (2011) Mechanisms linking drought, hydraulics, carbon metabolism, and vegetation mortality. Plant Physiol 155(3):1051–1059. doi:10.1104/pp.110.170704

Mendivelso HA, Camarero JJ, Gutiérrez E, Castaño-Naranjo A (2016) Climatic influences on leaf phenology, xylogenesis and radial stem changes at hourly to monthly scales in two tropical dry forests. Agric For Meteorol 216:20–36. doi:10.1016/j.agrformet.2015.09.014

Menzel A, Sparks TH, Estrella N, Koch E, Aasa A, Ahas R, Alm-Kubler K, Bissolli P, Braslavska O, Briede A, Chmielewski FM, Crepinsek Z, Curnel Y, Dahl A, Defila C, Donnelly A, Filella Y, Jatcza K, Mage F, Mestre A, Nordli O, Penuelas J, Pirinen P, Remisova V, Scheifinger H, Striz M, Susnik A, Van Vliet AJH, Wielgolaski FE, Zach S, Zust A (2006) European phenological response to climate change matches the warming pattern. Glob Chang Biol 12(10):1969–1976. doi:10.1111/j.1365-2486.2006.01193.x

Minchin PEH, Lacointe A (2005) New understanding on phloem physiology and possible consequences for modelling long-distance carbon transport. New Phytol 166:771–779

Moser L, Fonti P, Büntgen U, Esper J, Luterbacher J, Franzen J, Frank D (2010) Timing and duration of European larch growing season along altitudinal gradients in the Swiss Alps. Tree Physiol 30(2):225–233

Muller B, Pantin F, Genard M, Turc O, Freixes S, Piques M, Gibon Y (2011) Water deficits uncouple growth from photosynthesis, increase C content, and modify the relationships between C and growth in sink organs. J Exp Bot 62(6):1715–1729. doi:10.1093/jxb/erq438

Novak K, De Luis M, Gričar J, Prislan P, Merela M, Smith KT, Čufar K (2016a) Missing and dark rings associated with drougth in *Pinus halepensis*. IAWA J 37:260–274

Novak K, De Luis M, Saz MA, Longares LA, Serrano Notivoli R, Raventós J, Čufar K, Gričar J, Di Filippo A, Piovesan G, Rathgeber CBK, Papadopoulos A, Smith KT (2016b) Missing rings in *Pinus halepensis*—the missing link to relate the tree-ring record to extreme climatic events. Front Plant Sci 7. doi:10.3389/fpls.2016.00727

Olano JM, Arzac A, Garcia-Cervigon AI, Von Arx G, Rozas V (2013) New star on the stage: amount of ray parenchyma in tree rings shows a link to climate. New Phytol 198:486–495

Oliveira JM, Santarosa E, DePatta Pillar V, Roig FA (2009) Seasonal cambium activity in the subtropical rain forest tree *Araucaria angustifolia*. Trees 23:107–115

Pacheco A, Camarero JJ, Carrer M (2016) Linking wood anatomy and xylogenesis allows pinpointing of climate and drought influences on growth of coexisting conifers in continental Mediterranean climate. Tree Physiol 36(4):502–512. doi:10.1093/treephys/tpv125

Panshin AJ, de Zeeuw C (1980) Textbook of wood technology, 4th edn. McGraw-Hill, New York

Pantin F, Simonneau T, Muller B (2012) Coming of leaf age: control of growth by hydraulics and metabolics during leaf ontogeny. New Phytol 196(2):349–366. doi:10.1111/j.1469-8137.2012.04273.x

Pellizzari E, Camarero JJ, Gazol A, Sangüesa-Barreda G, Carrer M (2016) Wood anatomy and carbon-isotope discrimination support long-term hydraulic deterioration as a major cause of drought-induced dieback. Glob Chang Biol 22(6):2125–2137. doi:10.1111/gcb.13227

Pérez-de-Lis G, Rossi S, Vázquez-Ruiz RA, Rozas V, García-González I (2016) Do changes in spring phenology affect earlywood vessels? Perspective from the xylogenesis monitoring of two sympatric ring-porous oaks. New Phytol 209(2):521–530. doi:10.1111/nph.13610

Pettorelli N, Vik JO, Mysterud A, Gaillard JM, Tucker CJ, Stenseth NC (2005) Using the satellite-derived NDVI to assess ecological responses to environmental change. Trends Ecol Evol 20(9):503–510. doi:10.1016/j.tree.2005.05.011

Piermattei A, Crivellaro A, Carrer M, Urbinati C (2015) The "blue ring": anatomy and formation hypothesis of a new tree-ring anomaly in conifers. Trees 29(2):613–620. doi:10.1007/s00468-014-1107-x

Pittermann J, Sperry JS, Wheeler JK, Hacke UG, Sikkema EH (2006) Mechanical reinforcement of tracheids compromises the hydraulic efficiency of conifer xylem. Plant Cell Environ 29(8):1618–1628. doi:10.1111/1365-3040.2006.01539.x

Priestley JH (1930) Studies in the physiology of cambial activity. New Phytol 29:1618–1628

Prislan P, Gričar J, de Luis M, Smith KT, Čufar K (2013) Phenological variation in xylem and phloem formation in *Fagus sylvatica* from two contrasting sites. Agric For Meteorol 180:142–151. doi:10.1016/j.agrformet.2013.06.001

Rathgeber CBK, Rossi S, Bontemps JD (2011) Cambial activity related to tree size in a mature silver-fir plantation. Ann Bot 108:429–438

Rathgeber CBK, Cuny HE, Fonti P (2016) Biological basis of tree-ring formation: a crash course. Front Plant Sci 7. doi:10.3389/fpls.2016.00734

Ren P, Rossi S, Gričar J, Liang E, Čufar K (2015) Is precipitation a trigger of cambial reactivation in *Juniperus przewalskii* on the northeastern Tibetan Plateau? Ann Bot 115:629–639

Rossi S, Isabel N (2016) Bud break responds more strongly to daytime than night-time temperature under asymmetric experimental warming. Glob Chang Biol 23(1):446–454

Rossi S, Deslauriers A, Anfodillo T, Morin H, Saracino A, Motta R, Borghetti M (2007) Conifers in cold environments synchronize maximum growth rate of tree-ring formation with day length. New Phytol 170:301–310

Rossi S, Deslauriers A, Anfodillo T, Carrer M (2008a) Age-dependent xylogenesis in timberline conifers. New Phytol 177:199–208

Rossi S, Deslauriers A, Gričar J, Seo J-W, Rathgeber CBK, Anfodillo T, Morin H, Levanic T, Oven P, Jalkanen R (2008b) Critical temperatures for xylogenesis in conifers of cold climates. Glob Ecol Biogeogr 17:696–707

Rossi S, Rathgeber CBK, Deslauriers A (2009a) Comparing needle and shoot phenology with xylem development on three conifer species in Italy. Ann For Sci 66:206

Rossi S, Simard S, Rathgeber CBK, Deslauriers A, De Zan C (2009b) Effects of a 20-day-long dry period on cambial and apical meristem growth in *Abies balsamea* seedlings. Trees 23(1):85–93. doi:10.1007/s00468-008-0257-0

Rossi S, Morin H, Deslauriers A (2011a) Multi-scale influence of snowmelt on xylogenesis of black spruce. Arct Antarct Alp Res 43:457–464

Rossi S, Morin H, Deslauriers A, Plourde PY (2011b) Predicting timings of xylogenesis in black spruce under climate warming. Glob Chang Biol 17:614–625

Rossi S, Morin H, Deslauriers A (2012) Causes and correlations in cambium phenology: towards an integrated framework of xylogenesis. J Exp Bot 63:2117–2126

Rossi S, Girard MJ, Morin H (2014) Lengthening of the duration of xylogenesis engenders disproportionate increases in xylem production. Glob Chang Biol 20:2261–2271

Rossi S, Anfodillo T, Čufar K, Cuny HE, Deslauriers A, Fonti P, Frank D, Gričar J, Gruber A, Huang JG, Jyske T, Kašpar J, King G, Krause C, Liang E, Makinen H, Morin H, Nöjd P, Oberhuber W, Prislan P, Rathgeber C, Saracino A, Swidrak I, Treml V (2016) Pattern of xylem phenology in conifers of cold ecosystems at the northern hemisphere. Glob Chang Biol 22:3804–3813

Rowe N, Speck T (2005) Plant growth forms: an ecological and evolutionary perspective. New Phytol 166(1):61–72. doi:10.1111/j.1469-8137.2004.01309.x

Sass-Klaassen U, Fonti P, Cherubini P, Gričar J, Robert EMR, Steppe K, Brauning A (2016) A tree-centered approach to assess impacts of extreme climatic events on forests. Front Plant Sci 7(1069):1–6

Savage JA, Clearwater MJ, Haines DF, Klein T, Mencuccini M, Sevanto S, Turgeon R, Zhang C (2016) Allocation, stress tolerance and carbon transport in plants: how does phloem physiology affect plant ecology? Plant Cell Environ 39(4):709–725. doi:10.1111/pce.12602

Schiestl-Aalto P, Kulmala L, Mäkinen H, Nikinmaa E, Mäkela A (2015) CASSIA—a dynamic model for predicting intra-annual sink demand and interannual growth variation in Scots pine. New Phytol 206:647–659

Schmitz N, Robert EMR, Verheyden A, Gitundu Kairo J, Beeckman H, Koedam N (2008) A patchy growth via successive and simultaneous cambia: key to success of the most widespread mangrove species *Avicennia marina*? Ann Bot 101:49–58

Schweingruber FH (2007) Wood structure and environment. Springer, Berlin

Seneviratne SI, Lüthi D, Litschi M, Schär C (2006) Land-atmosphere coupling and climate change in Europe. Nature 443:205–209

Sevanto S, McDowell NG, Dickman LT, Pangle R, Pockman WT (2014) How do trees die? A test of the hydraulic failure and carbon starvation hypotheses. Plant Cell Environ 37(1):153–161. doi:10.1111/pce.12141

Simard S, Giovannelli A, Treydte K, Traversi ML, King GM, Frank D, Fonti P (2013) Intra-annual dynamics of non-structural carbohydrates in the cambium of mature conifer trees reflects radial growth demands. Tree Physiol 33:913–923. doi:10.1093/treephys/tpt075

Skene DS (1969) The period of time taken by cambial derivatives to grow and differentiate into tracheids in *Pinus radiata*. Ann Bot 33:253–262

Skomarkova MV, Vaganov EA, Mund M, Knohl A, Linke P, Boerner A, Schulze ED (2006) Inter-annual and seasonal variability of radial growth, wood density and carbon isotope ratios in tree rings of beech (*Fagus sylvatica*) growing in Germany and Italy. Trees 20(5):571–586. doi:10.1007/s00468-006-0072-4

Sperry JS, Hacke UG, Pittermann J (2006) Size and function in conifer tracheids and angiosperm vessels. Am J Bot 93(10):1490–1500

Spicer R, Groover A (2010) Evolution of development of vascular cambia and secondary growth. New Phytol 186:577–592

Steppe K, Sterck F, Deslauriers A (2015) Diel growth dynamics in tree stems: linking anatomy and ecophysiology. Trends Plant Sci 20:335–343

Swidrak I, Gruber A, Oberhuber W (2014) Xylem and phloem phenology in co-occurring conifers exposed to drought. Trees 28(4):1161–1171. doi:10.1007/s00468-014-1026-x

Torelli N, Shortle WC, Čufar K, Ferlin F, Smith KT (1999) Detecting changes in tree health and productivity of silver fir in Slovenia. Eur J For Pathol 29(3):187–197

Treml V, Kašpar J, Kuželová H, Gryc V (2015) Differences in intra-annual wood formation in *Picea abies* across the treeline ecotone, Giant Mountains, Czech Republic. Trees 29(2):515–526. doi:10.1007/s00468-014-1129-4

Turley DB, Chaudhry Q, Watkins RW, Clark JH, Deswarte FEI (2006) Chemical products from temperate forest tree species—Developing strategies for exploitation. Ind Crop Prod 24(3):238–243. doi:10.1016/j.indcrop.2006.06.016

Tyree MT, Zimmermann MH (2002) Xylem structure and the ascent of sap. Springer, New York

Valladares F, Gianoli E, Gomez JM (2007) Ecological limits to plant phenotypic plasticity. New Phytol 176(4):749–763. doi:10.1111/j.1469-8137.2007.02275.x

Vichrová G, Gryc V, Vavrčík H (2013) Xylem formation in young norway spruce trees in drahany highland, Czech Republic. IAWA J 34(3):231–244. doi:10.1163/22941932-00000020

Vieira J, Rossi S, Campelo F, Nabais C (2014) Are neighboring trees in tune? Wood formation in *Pinus pinaster*. Eur J For Res 133:41–50

Vieira J, Campelo F, Rossi S, Carvalho A, Freitas H, Nabais C (2015) Adjustment capacity of Maritime pine cambial activity in drought-prone environments. PLoS One 10(5):e0126223

Wang L, Payette S, Begin Y (2000) A quantitative definition of light rings in black spruce (*Picea mariana*) at the arctic treeline in Northern Québec, Canada. Arct Antarct Alp Res 32(3):324–330

Wilmking M, Hallinger M, Van Bogaert R, Kyncl T, Babst F, Hahne W, Juday GP, de Luis M, Novak K, Völlm C (2012) Continuously missing outer rings in woody plants at their distributional margins. Dendrochronologia 30(3):213–222. doi:10.1016/j.dendro.2011.10.001

Wilson BF (1964) A model for cell production by the cambium of conifers. In: Zimmerman M (ed) The formation of wood in forest trees. Academic, New York, pp 19–36

Wodzicki TJ (1971) Mechanism of xylem differentiation in *Pinus silvestris* L. J Exp Bot 22(72):670–687

Zalloni E, de Luis M, Campelo F, Novak K, De Micco V, Di Filippo A, Vieira J, Nabais C, Rozas V, Battipaglia G (2016) Climatic signals from intra-annual density fluctuation frequency in mediterranean pines at a regional scale. Front Plant Sci 7. doi:10.3389/fpls.2016.00579

Zhao M, Running SW (2010) Drought-induced reduction in global terrestrial net primary production from 2000 through 2009. Science 329(5994):940–943

Ziaco E, Biondi F (2016) Tree growth, cambial phenology, and wood anatomy of limber pine at a Great Basin (USA) mountain observatory. Trees:1–15. doi:10.1007/s00468-016-1384-7

Ziaco E, Biondi F, Rossi S, Deslauriers A (2016) Environmental drivers of cambial phenology in Great Basin bristlecone pine. Tree Physiol 36(7):818–831. doi:10.1093/treephys/tpw006

Zweifel R, Zimmermann L, Zeugin F, Newbery DM (2006) Intra-annual radial growth and water relations of trees: implications towards a growth mechanism. J Exp Bot 57(6):1445–1459. doi:10.1093/jxb/erj125

Chapter 3
Dendroecological Studies in the Neotropics: History, Status and Future Challenges

Jochen Schöngart, Achim Bräuning, Ana Carolina Maioli Campos Barbosa, Claudio Sergio Lisi, and Juliano Morales de Oliveira

Abstract We review a vast literature of Neotropical forest science and wood anatomical research that identifies 220 tree species from 46 botanical families with confirmed annual tree-ring formation. External factors that trigger annual growth rhythms include rainfall seasonality, annual long-term flooding (flood-pulse), soil water salinity (mangroves), and, with increasing latitude and altitude, photoperiod and temperature. Maximum ages for tropical angiosperms derived from tree-ring analyses generally do not exceed 400–600 years; however, at marginal sites characterized by extremely limited growth conditions individual trees might get older. Dendroecological applications provide insights into tree species' ecology and forest dynamics. Analyses of growth trajectories and age-size relationships of trees highlight considerable variability among individual trees, species, and environments. In recent decades tree-ring studies in neotropical forests have contributed new methods to project timber harvests and to evaluate and

J. Schöngart (✉)
Instituto Nacional de Pesquisas da Amazônia, Coordenação de Dinâmica Ambiental, Av. André Araújo 2936, Petrópolis 69.067-375, Manaus, Brazil
e-mail: jochen.schongart@inpa.gov.br

A. Bräuning
Institut für Geographie, Friedrich-Alexander Universität Erlangen-Nürnberg, Wetterkreuz 15, 91058, Erlangen, Germany
e-mail: achim.braeuning@fau.de

A.C.M.C. Barbosa
Departamento de Ciências Florestais, Universidade Federal de Lavras, Campus Universitário 37.200-000, Lavras, Brazil
e-mail: anabarbosa@dcf.ufla.br

C.S. Lisi
Centro de Ciências Biológicas e da Saúde, Universidade Federal de Sergipe, Av. Marechal Rondon s/n, Rosa Elze 49.100-000, São Cristóvão, Brazil
e-mail: clcslisi@gmail.com

J.M. de Oliveira
Universidade do Vale do Rio dos Sinos, Av. Unisinos 950, Cristo Rei 93.022-000, São Leopoldo, Brazil
e-mail: julianooliveira@unisinos.br

M.M. Amoroso et al. (eds.), *Dendroecology*, Ecological Studies 231,
DOI 10.1007/978-3-319-61669-8_3

adjust management practices to increase the sustainability of forest management. The better understanding of individual- and species-level growth patterns in the Neotropics provides necessary empirical information to conserve and manage tropical forests and the many ecosystem functions and services that they maintain.

Keywords Tropics • Dendrochronology, tree rings • Wood anatomy • Radiocarbon dating • Longevity • Growth trajectories • Age-size relationships • Forest management, conservation

3.1 Introduction

Tropical regions support some of the most complex and species-rich terrestrial ecosystems in the world. Tropical forests play an essential role in regional and global biogeochemical and hydrological cycles. The Neotropics—the tropical region of continental and insular America between 23°26′ N and S—include the largest area of rainforest and the largest watershed on earth, the Amazon basin. The diverse hydroclimatic conditions, topography, and soils of the Neotropics host huge complexes of savannahs, grasslands, woodlands, continental wetlands, mangroves, semi-arid, dry, moist, and humid forests that support thousands of tree species (Slik et al. 2015). Land-use changes associated with deforestation and fire due to the expansion and intensification of agriculture, livestock production and logging, as well as urban development, mining, hydroelectric dams, and other infrastructure have caused unprecedented impacts to these environments (Davidson et al. 2012; Venter et al. 2016). Ongoing anthropogenic climate change may cause further impacts to these ecosystems, although they are difficult to project due to the complexity of biosphere-atmosphere interactions with pervasive regional land-use change. In this context it is of fundamental importance to understand how tropical forests respond to global change. This requires information on their long-term dynamics.

Retrospective analyses of tree rings give information on tree ages and growth rates that can be applied in a large multi- and interdisciplinary field studying the ecology of tree species (Brienen and Zuidema 2006a, b; Schöngart et al. 2015), population dynamics and long-term dynamics of tropical forests related to disturbance regimes (Baker et al. 2005; Rozendaal et al. 2010; Baker and Bunyavejchewin 2017), and biomass productivity and mean residence time of carbon (Schöngart et al. 2010, 2011; Cintra et al. 2013). Tree-ring analyses also allow reliable estimates of felling cycles, minimum logging diameters, and yield projections that are the necessary empirical foundation of sustainable and adaptive forest management practices in the tropics (Brienen and Zuidema 2007; Schöngart 2008; De Ridder et al. 2013; López et al. 2013). These applications contribute to the conservation of tropical forest ecosystems and their multiple environmental services (Worbes 2002; Bräuning 2011; Rozendaal and Zuidema 2011).

For many decades the tropics have been a frontier for dendrochronology and its applications due to the prevailing perception of aseasonal or continuous tree growth in tropical trees. In high latitude temperate climates, the seasonal variation in temperature and photoperiod are the main trigger for cambial dormancy and the formation of anatomically distinct tree rings (e.g., Fritts 1976; Schweingruber 1983). Towards the equator climate seasonality declines and becomes almost absent. As a consequence, for many decades it was widely believed that tropical tree species experienced relatively constant growth throughout the year and therefore did not form annual growth rings (Studhalter et al. 1963; Whitmore 1975; Hallé et al. 1978; Lieberman and Lieberman 1985).

Over the past two decades a monumental effort across a wide range of research initiatives has changed this picture dramatically (Worbes 2002; Rozendaal and Zuidema 2011; Zuidema et al. 2012; Brienen et al. 2016). In this chapter we provide a short summary of the history of dendrochronology in the subtropics and tropics of the Americas. We then review the existing literature and provide an overview of Neotropical tree species that have proven annual growth rings. We describe the different mechanisms of tree-ring formation in tropical trees, considering also variability in wood anatomical patterns and the distinctiveness of ring boundaries. We discuss the longevity of tropical trees, analyses of growth patterns of tree species, age-size relationships among individuals, tree species and forest ecosystems, and their applications and relevance for forest management and conservation. We end the chapter with a summary of our main conclusions and an outlook for future dendroecological research and applications in the Neotropics.

3.2 Tree Ring Studies in the Neotropics

3.2.1 History and Development of Tree-Ring Studies in the Neotropics

Pioneering studies in the subtropics of the northern hemisphere were conducted in the 1940s and 1950s by Schulman (1944, 1956) who developed the first tree-ring chronologies on conifers in the region of Durango (northwestern Mexico) (*Abies durangensis*, *Pinus* sp., *Pseudotsuga menziesii*) and south of Mexico City (*Abies religiosa*) (see also Stahle et al. 2000). Since then a large number of multi-centennial chronologies in the subtropics of North America have been made resulting in an outstanding network of tree-ring chronologies from the southern USA and Mexico and the reconstruction of regional climate variability over the past 600 years (Stahle et al. 2016). The network includes several chronologies from tropical sites in Mexico and Guatemala mainly developed with *Taxodium mucronatum* in riparian forests and *P. menziesii* in montane regions (Stahle et al. 2000, 2011, 2012; Therrell et al. 2002).

In tropical Latin America the first dendrochronological studies only started in the 1960s (Hastenrath 1963; Tschinkel 1966), nearly a century after a series of remarkable tree-ring studies in South and Southeast Asia, mainly with *Tectona grandis* provided the first evidence of annual tree-ring formation in the tropics (Brandis 1879; Gamble 1881; Ursprung 1904; Geiger 1915; Coster 1927, 1928;

Berlage 1931). The workshop on "Age and Growth Rate of Tropical Trees" in New Haven (USA) in 1980 (Bormann and Berlyn 1981) and the follow-up workshops (Baas and Vetter 1995; Eckstein et al. 1995) gave important impetus for tree-ring analyses in the tropics by reviewing the existing knowledge on tree-ring formation and emphasizing and promoting the application of tree-ring analyses for future research. Since then dendrochronological studies have expanded rapidly to subtropical and tropical regions (Hayden 2008). In the subtropics of the southern hemisphere, dendrochronological studies started in the 1980s in northern Argentina using the angiosperms *Cedrela lilloi* and *Juglans australis* (e.g., Villalba et al. 1985) and the conifer *Araucaria angustifolia* in southern Brazil (Seitz and Kanninen 1989; Steinbrenner 1989). Dendrochronological studies performed in the central Amazonian floodplains showed evidence of annual tree ring formation as a consequence of the regular and predictable annual flood pulse (Worbes 1984, 1985, 1986, 1988, 1989), while in the non-flooded humid upland forests (terra firme) of central Amazonia and French Guyana, annual tree-ring formation was triggered by rainfall seasonality (Détienne et al. 1988; Vetter and Botosso 1989). Radiocarbon (^{14}C) dating was applied to various tree species from different tropical ecosystems across the central Amazonian floodplains, the Guyana highlands in Venezuela (Gran Sabana), and tropical mountain forests in Costa Rica (Worbes and Junk 1989). During the last 20 years an almost exponential increase in dendrochronological studies in the Neotropics has occurred (Rozendaal and Zuidema 2011). Based on the accumulating evidence for annual ring formation in Neotropical tree species, a range of tree-ring studies in ecology, climatology, archaeology, hydrology, biogeochemistry, and forest management have been performed over this period (Brienen et al. 2016).

3.2.2 Methods to Evidence Annual Growth Ring Formation

Several independent methods have been proposed and applied to evaluate the annual nature of tree rings in the tropics (Worbes 1995). The monitoring of stem diameter variations by dendrometer bands or high-resolution dendrometers during consecutive years for several tree species of different neotropical ecosystems provided empirical evidence of annual growth rhythms (Détienne et al. 1988; Breitsprecher and Bethel 1990; Vetter 1995; Bullock 1997; Worbes 1999; Schöngart et al. 2002; Dünisch et al. 2003; Lisi et al. 2008; Bräuning et al. 2009; Mehlig et al. 2010; Grogan and Schulze 2011; Lobão 2011; Volland-Voigt et al. 2011; Blagitz et al. 2016; Spannl et al. 2016). Monitoring of cambial activity either by measuring the electrical resistance of the cambial zone by a shigometer (Worbes 1995) or by wood anatomical analysis of monthly or seasonally sampled cambium tissues during different periods of the year (dry and wet seasons) has also provided information on the growth rhythms of Neotropical tree species (Marcati et al. 2006, 2008, 2016;

Santarosa et al. 2007; Oliveira et al. 2009; Worbes and Raschke 2012; Callado et al. 2013; Morel et al. 2015). Such studies are often combined with regular observations of leaf phenology, which offer insights into the physiological status of individual trees. Cambial wounding studies, based on either artificial or mechanical injury (Mariaux 1967), result in exactly datable scars in the wood and the formation of callus tissues in consecutive years (Détienne et al. 1988; Détienne 1995; Vetter 1995; Worbes 1996, 1999; Callado et al. 2001; Menezes et al. 2003; Lisi et al. 2008; Locosselli et al. 2013). Fire scars due to the occurrence of large-scale fires in known years can also create a distinct time marker (Worbes 1994; López et al. 2012). Another anthropogenic time marker was produced in the atmosphere by hundreds of aboveground thermonuclear weapon tests during the Cold War Era. This resulted in a distinct increase in the mid-1960s and subsequent exponential decrease (after the partial Nuclear Ban Treaty in 1963) in the environmental concentration of the radioisotope ^{14}C due to the uptake by the ocean's surface, terrestrial ecosystems, increasing fossil-fuel combustion, and atmospheric mixing (Stuiver et al. 1981; Nydal and Lövseth 1983; Hua et al. 2013). As trees take up ^{14}C through photosynthesis and incorporate the isotope in the cellulose, tree rings formed during the post-bomb period can be dated by comparing ^{14}C levels in the cellulose of a single tree ring to ^{14}C calibration curves describing the temporal changes in atmospheric ^{14}C concentration for different zones in the northern and southern hemisphere (Hua et al. 1999, 2013). This method has been applied for many tree species from different ecosystems of the tropical and subtropical regions of the Americas to assess the annual nature of growth rings (Worbes and Junk 1989; Lisi et al. 2001; Fichtler et al. 2003; Dezzeo et al. 2003; Menezes et al. 2003; Brienen and Zuidema 2005; Westbrook et al. 2006; Jenkins 2009; Ballantyne et al. 2011; Jiménez and del Valle 2011, Ramírez and del Valle 2011; 2012; Giraldo and del Valle 2012; Soliz-Gamboa et al. 2011; Andreu-Hayles et al. 2015; Baker et al. 2015; Santos et al. 2015; Assahira et al. 2017; Inga and del Valle 2017). The formation of annual tree rings has also been demonstrated by counting the number of rings in plantation trees of known age (Tschinkel 1966; Brienen and Zuidema 2005; López et al. 2005; Abrams and Hock 2006; Hayden 2008; Ramírez Correa et al. 2010; Costa et al. 2015; Schöngart et al. 2015; Brandes et al. 2016) and from naturally grown trees in secondary forests where local people have reliable age information (Brienen et al. 2009). The majority of Neotropical studies that have identified annual growth rings have used crossdated ring-width series from several individuals of a species to produce a mean chronology and then demonstrated a significant correlation to interannual climate variability (Villalba et al. 1985; Huante et al. 1991; Devall et al. 1995; Worbes 1999; Enquist and Leffler 2001; Dünisch et al. 2003; Morales et al. 2004; Schöngart et al. 2004, 2005; Speer et al. 2004; Brienen and Zuidema 2005, Rodriguez et al. 2005; López et al. 2006; Bräuning et al. 2009; Brienen et al. 2010a, 2011, 2013; Oliveira et al. 2010; Fichtler et al. 2010; Lobão 2011; López and Villalba 2011, 2016a; Ramírez and del Valle 2011, 2012; Soliz-Gamboa et al. 2011; Locosselli et al. 2013, 2016a, b; Espinoza et al. 2014; Pagotto et al. 2015; Paredes-Villanueva et al. 2013, 2015, 2016; Pucha-Cofrep et al. 2015).

3.2.3 Occurrence of Annual Tree Rings in the Neotropics

Zuidema et al. (2012), Schöngart (2013) and Brienen et al. (2016) have provided recent reviews of tropical tree species with confirmed annual tree rings from the voluminous forest science and wood anatomy literature. Table 3.1 presents an updated list that includes more than 100 studies identifying 220 tree species with confirmed annual tree rings from 46 botanical families from the Neotropics (Fig. 3.1). Our review does not consider wood anatomical descriptions of tree rings that do not provide independent evidence of their annual nature (e.g., Worbes 2002; Roig et al. 2005; Gutiérrez and Ramos 2013; Silva et al. 2016). The reviewed studies include diverse ecosystems and biomes, such as semi-arid, dry, semi-deciduous, moist and humid forests, savannahs, floodplains and mangroves, and premontane and montane regions, with mean annual precipitation ranging from <50 to >4200 mm. The number of tree species with confirmed annual tree ring formation in the tropics of America is higher than in the paleotropical regions of Africa, Southeast Asia and Oceania (Hayden 2008; Brienen et al. 2016). This still rather conservative estimate of species with annual growth rings in the Neotropics indicates that about one-third of the species with confirmed annual tree-ring formation belong to the Fabaceae (72 spp.), followed by the families Bignoniaceae (15 spp.), Meliaceae (11 spp.) and Malvaceae (10 spp.). At the genus level *Handroanthus* (Bignoniaceae, 8 spp.), *Cedrela* (Meliaceae) and *Tabebuia* (Bignoniaceae) (5 spp. each), *Cariniana* (Lecythidaceae), *Dipteryx* (Fabaceae), *Pinus* (Pinaceae), *Polylepis* (Rosaceae), and *Terminalia* (Combretaceae) (4 spp. each) have the highest number of tree species with annual ring formation. The largest number of records exist for *Cedrela odorata*, which has been studied in forests from Brazil, Bolivia, Peru, Venezuela, and Mexico.

3.2.4 Triggering Factors for Annual Tree-Ring Formation

Tropical trees form annual growth rings when unfavorable environmental conditions that cause cambial dormancy regularly occur in one period of the year (Worbes 2002; Rozendaal and Zuidema 2011). Vast tropical regions have a distinct rainfall seasonality, which induces an annual growth rhythm and the formation of annual tree rings (e.g., Worbes 1995; 1999; Devall et al. 1995; Vetter 1995; Enquist and Leffler 2001; Dünisch et al. 2003; Brienen and Zuidema 2005; Volland-Voigt et al. 2011; Callado et al. 2013; Marcati et al. 2016; Spannl et al. 2016). The rainfall seasonality in catchment areas of large tropical rivers results in a monomodal flood-pulse (Junk et al. 1989) that triggers annual growth rhythms of tree species in the vast floodplains along the Amazon River and its major tributaries (Worbes 1986; Schöngart et al. 2004, 2005), but also in other floodplains (Dezzeo et al. 2003; Herrera and del Valle 2011; Montanher 2012). The anoxic soil conditions during flooding hinder the uptake of water by the root system and the affected trees either shed their leaves and remain deciduous until the flood waters recede or gradually

Table 3.1 Tree species of the tropical Americas with confirmed annual tree rings (between 23°26′ North and South)

Family	Tree species	References
Anacardiaceae	*Astronium graveolens*, *Schinopsis brasiliensis*, *Tapirira* cf. *guianesis*	Worbes and Junk (1989), Maria (2002), Lisi et al. (2008), Cardoso (2014), and López and Villalba (2016a)
Annonaceae	*Annona spraguei*, *Annona* sp., *Guatteria aeruginosa*, *Rollinia jimenezii*,	Devall et al. (1995), Fichtler et al. (2003), and Brienen et al. (2009)
Apocynaceae	*Aspidosperma polyneuron*, *A. pyrifolium*, *A. tomentosum*, *Thevetia ahouai*	Lisi et al. (2008), Brienen et al. (2009), Mendivelso et al. (2013), and Pagotto (2015)
Araucariaceae	*Araucaria angustifólia*	Seitz and Kanninen (1989), Lisi et al. (2001), Santarosa et al. (2007), Oliveira et al. (2009, 2010), and Santos et al. (2015)
Betulaceae	*Alnus acuminata*	Morales et al. (2004)
Bignoniaceae	*Handroanthus barbatus*, *H. chrysanthus*, *H. heterophyllus*, *H. impetiginosus*, *H. ochraceus*, *H. pentaphyllus*, *H. serratifolius*, *H. umbellatus*, *Jacaranda copaia*, *Tabebuia aurea*, *T. billbergii*, *T. donnell-smithii*, *T. rosea*, *T. cassinoides*, *Zeyheria tuberculosa*	Worbes (1988), Mattos (1998), Callado et al. (2001), Schöngart et al. (2002), Hayden (2008), Lisi et al. (2008), Brienen et al. (2009), Chagas (2009), Jenkins (2009), Volland-Voigt et al. (2011), Leite (2012), Worbes and Raschke (2012), Mendivelso et al. (2013), and García-Cervigón et al. (2017)
Boraginaceae	*Cordia alliodora*, *C. thaisiana*	Tschinkel (1966), Devall et al. (1995), Worbes (1999), Brienen et al. (2009), and Briceño et al. (2016)
Burseraceae	*Bursera graveolens*, *B. instabilis*, *B. simaruba*, *Protium pittieri*, *P. hebetatum*	Fichtler et al. (2003), Rodriguez et al. (2005), Hayden (2008), Worbes and Raschke (2012), Pucha-Cofrep et al. (2015), Ohashi et al. (2016), and García-Cervigón et al. (2017)
Calophyllaceae	*Calophyllum angulare*, *C. brasiliense*	Vetter (1995) and Botosso et al. (2000)
Capparaceae	*Capparis indica*, *Capparis odoratissima*, *Colicodendron scabridum*	Enquist and Leffler (2001), Ramírez and del Valle (2011), and García-Cervigón et al. (2017)
Caryocaraceae	*Caryocar glabrum*	Détienne et al. (1988) and Détienne (1995)
Celastraceae	*Goupia glabra*	Détienne et al. (1988), Détienne (1995), and Vetter (1995)
Clusiaceae	*Moronobea coccinea*, *Platonia insignis*, *Symphonia globulifera*	Détienne et al. (1988), Détienne (1995), and Callado et al. (2001)
Cochlospermaceae	*Cochlospermum vitifolium*	Bullock (1997)
Combretaceae	*Laguncularia racemosa*, *Terminalia amazonica*, *T. catappa*, *T. guyanensis*, *T. quintalata*	Worbes (1999), Estrada et al. (2008), Chagas (2009), Fichtler et al. (2010), and Mendivelso et al. (2013)
Connaraceae	*Rourea* sp.	Worbes and Junk (1989)
Cupressaceae	*Taxodium mucronatum*	Therrell et al. (2002) and Stahle et al. (2011)

(continued)

Table 3.1 (continued)

Family	Tree species	References
Ebenaceae	*Diospyros salicifolia*	Worbes and Raschke (2012)
Euphorbiaceae	*Alchornea glandulosa, A. sidifolia, A. triplinervia, Cnidoscolus spinosus, Croton floribundus, C. sonderianus, Euphorbia schlechtendalii, Hura crepitans*	Bullock (1997), Callado et al. (2001), Silva (2006), Lisi et al. (2008), Brienen et al. (2009), López et al. (2012), Nascimento (2013), and van der Sleen et al. (2014)
Fabaceae	*Acacia cochliacantha, A. uzumacintla, Acosium cardenasii, A. nitens, Albizia hassleri, A. multiflora, A. niopoides, Amburana cearensis, Anadenanthera macrocarpa, Balizia elegans, Caesalpinia ferrea, C. pluviosa, Campsiandra laurifolia, Cedrelinga catenaeformis, Centrolobium microchaete, C. tomentosum, C. robustum, Clitoria cearensis, Copaifera langsdorffii, C. multijuga, Dalbergia brownei, D. nigra, Dialium guianense, Dicorynia guianensis, Dipteryx alata, D. micrantha, D. odorata, D. panamensis, Enterolobium schomburgkii, Eperua bijiga, Geoffroea spinosa, Hymenaea courbaril, H. stigonocarpa, Hymenolobium mesoamericanum, Inga edulis, I. vera, Lonchocarpus cruentus, L. emarginatus, L. torresiorum, Machaerium incorruptibile, M. scleroxylon, M. villosum, Macrolobium acaciifolium, Mimosa acantholoba, M. tenuiflora, Myroxylon balsamum, Parkia nitida, P. pendula, Parkia* sp., *Parkinsonia praecox, Peltogyne* cf. *heterophylla, Peltophorum dubium, Pentaclethra macroloba, Piptadenia gonoacantha, Piscidia carthagenensis, Platycyamus regnellii, Poincianella pyramidalis, Prioria copaifera, Prosopis ferox, P. pallida, Pterocarpus rohrii, P. vernalis, Pterodon pubescens, Recordoxylon speciosum, Schizolobium parahyba, Senna multijuga, Swartzia laevicarpa, S. polyphylla, Sweetia fruticosa, Tachigali vasquezii, Vouacapoua americana*	Détienne et al. (1988), Worbes and Junk (1989), Tsuchiya (1994), Détienne (1995), Vetter (1995), Worbes et al. (1995), Ishii (1998), Worbes (1999), Botosso et al. (2000), Maria (2002), Schöngart et al. (2002, 2005), Dezzeo et al. (2003), Fichtler et al. (2003), Morales et al. (2004), Brienen and Zuidema (2005), López et al. (2005, 2006), Lisi et al. (2008), Marcati et al. (2008), Brienen et al. (2009, 2010a, 2011, 2013), Jenkins (2009), Zumaeta (2009), Grogan and Schulze (2011), Herrera and del Valle (2011), Jiménez and del Valle (2011), Lobão (2011), López and Villalba (2011), Ramírez and del Valle (2012), Soliz-Gamboa et al. (2011), Giraldo and del Valle (2012), López et al. (2012), Montanher (2012), Worbes and Raschke (2012), Locosselli et al. (2013, 2016b), Mendivelso et al. (2013), Nascimento (2013), Paredes-Villanueva et al. (2013, 2015) van der Sleen et al. (2014), Vogado (2014), Costa (2014), Costa et al. (2015), Latorraca et al. (2015), Mattos et al. (2015), Morel et al. (2015), Pagotto et al. (2015), Brandes et al. (2016), Carmo (2016), Medeiros (2016), Vasconcellos et al. (2016), Assahira et al. (2017), and García-Cervigón et al. (2017)

(continued)

Table 3.1 (continued)

Fagaceae	*Quercus copeyensis, Q. costaricensis*	Worbes and Junk (1989)
Hypericaceae	*Vismia camparaguey*	Brienen et al. (2009)
Juglandaceae	*Juglans australis, J. neotropica*	Villalba et al. (1985), Morales et al. (2004), and Inga and del Valle (2017)
Lamiaceae	*Aegiphila sellowiana, Gmelina arborea, Vitex cymosa*	Maria (2002), Schöngart et al. (2002), Hayden (2008), and Lisi et al. (2008)
Lauraceae	*Aniba rosaeodora, Nectandra amazonum, Ocotea porosa, O. tenera, Persea americana*	Botosso and Vetter (1991), Maria (2002), Schöngart et al. (2002), Anchukaitis et al. (2008), and Chagas (2009)
Lecythidaceae	*Bertholletia excelsa, Cariniana estrellensis, C. ianeirensis, C. legalis, C. micrantha, Gustavia augusta, Lecythis corrugate*	Détienne (1995), Maria (2002), Schöngart et al. (2002, 2015), Funck (2004), Rigozzo et al. (2012), Brienen and Zuidema (2005), Lisi et al. (2008), López et al. (2012), and van der Sleen et al. (2014)
Malvaceae	*Bombacopsis quinata, Ceiba parvifolia, C. pentandra, Guazuma ulmifolia, Luehea candida, L. cymulosa, Pseudobombax marginatum, P. munguba, P. spetenatum, Scleronema micranthum*	Worbes and Junk (1989), Devall et al. (1995), Vetter (1995), Worbes (1999), Schöngart et al. (2002), Hayden (2008), Brienen et al. (2009), Jenkins (2009), Nogueira (2011), and Worbes and Raschke (2012)
Melastomataceae	*Miconia glaberrima, M. trinervia*	Brienen et al. (2009)
Meliaceae	*Cabralea canjerana, Carapa guinanesis, C. procera, Cedrela angustifolia, C. fissilis, C. lilloi, C. montana, C. odorata, Swietenia macrophylla, Trichilia casretti, T. clausserii*	Détienne et al. (1988), Villalba et al. (1998), Détienne (1995), Vetter (1995), Worbes (1999), Botosso et al. (2000), Maria (2002), Dünisch et al. (2003), Brienen and Zuidema (2005), Marcati et al. (2006), Westbrook et al. (2006), Hayden (2008), Bräuning et al. (2009), Jenkins (2009), Rosero Alvorado (2009), Fichtler et al. (2010), Ballantyne et al. (2011), Grogan and Schulze (2011), Lobão (2011), Brienen et al. (2012), Ferreira (2012), Worbes and Raschke (2012), Anholetto (2013), Barbosa et al. (2013), Nascimento (2013), Espinoza et al. (2014), Baker et al. (2015), Paredes-Villanueva et al. (2016), Pereira et al. (2016), Volland et al. (2016), and Inga and del Valle (2017)
Moraceae	*Castilla elastica, Clarisia racemosa, Ficus boliviana, F. maxima, Maclura tinctoria, Pseudolmedia laevis, Sorocea duckei*	Worbes and Junk (1989), Brienen et al. (2009), Soliz-Gamboa et al. (2011), López et al. (2012), Andreu-Hayles et al. (2015), and Pucha-Cofrep et al. (2015)
Myristicaceae	*Virola mellionii*	Détienne et al. (1988) and Détienne (1995)

(continued)

Table 3.1 (continued)

Family	Tree species	References
Myrtaceae	*Myrcia* sp., *Psidium acutangulum*, *P. ovatifolium*	Worbes (1994) and Dezzeo et al. (2003)
Picrodendraceae	*Piranhea trifoliata*	Schöngart et al. (2002, 2004)
Pinaceae	*Abies religiosa*, *Pinus caribaea*, *P. hartwegii*, *P. lagunae*, *P. occidentalis*, *Pseudotsuga menziesii*	Huante et al. (1991), Worbes (1999), Biondi (2001), Biondi et al. (2003), Therrell et al. (2002), Speer et al. (2004), and Stahle et al. (2016)
Phyllanthaceae	*Savia dictyocarpa*	Lisi et al. (2008)
Podocarpaceae	*Podocarpus lambertii*	Locosselli et al. (2016a)
Rhamnaceae	*Hovenia dulcis*	Chagas (2009)
Rhizophoraceae	*Rhizophora mangle*	Menezes et al. (2003), Mehlig et al. (2010), Ramírez Correa et al. (2010), and Souza et al. (2016)
Rosaceae	*Polylepis pepei*, *P. rugulosa*, *P. subsericans*, *P. tarapacana*, *Prunus myrtifolia*	Argollo et al. (2004), Morales et al. (2004), Christie et al. (2009), Solíz et al. (2009), Ballantyne et al. (2011), Moya and Lara (2011), Jomelli et al. (2012), Nascimento (2013), and Baker et al. (2015)
Rubiaceae	*Genipa americana*	Enquist and Leffler (2001)
Rutaceae	*Balfourodendron riedelianum*, *Citrus sinensis*, *Esenbeckia leiocarpa*, *Euxylophora paraensis*, *Zanthoxylum caribaeum*, *Z. kellermanii*	Ferreira-Fedele et al. 2004, Abrams and Hock (2006), Lisi et al. (2008), Brienen et al. (2009), and Andrade et al. (2017)
Sapotaceae	*Chrysophyllum gonocarpum*, *Manilkara bidentata*, *M. huberi*, *Pouteria orinocoensis*	Détienne (1995), Vetter (1995), Dezzeo et al. (2003), Pinto (2012), and Nascimento (2013)
Simaroubaceae	*Simarouba amara*	Détienne et al. (1988) and Détienne (1995)
Ulmaceae	*Ampelocera ruizii*	López et al. (2012) and van der Sleen et al. (2014)
Verbenaceae	*Tectona grandis*	Worbes (1999)
Vochysiaceae	*Vochysia divergens*	Ishii (1998), Fortes (2006), and Sallo et al. (2017)

Some of the genera and families changed due to taxonomic revisions since the year of publication of the indicated references

(evergreen) or rapidly (brevi-deciduous) replace them during the flooding period, resulting finally in a cambial dormancy (Schöngart et al. 2002). For mangroves there is still a controversial discussion about the formation of tree rings due to the complex dynamics of ecological drivers that may influence tree growth (Robert et al. 2011). Studies in the mangroves of Colombia (Ramírez Correa et al. 2010), southeast (Estrada et al. 2008; Souza et al. 2016) and southeast Brazil (Menezes et al. 2003) have reported the formation of annual tree rings in the wood of

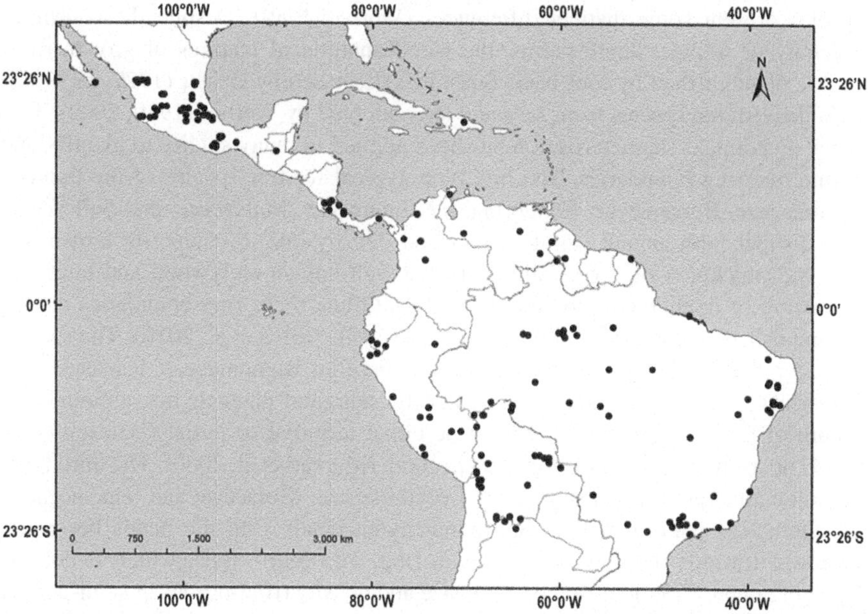

Fig. 3.1 Occurrence of tree species (see also Table 3.1) with annual tree rings in the Neotropics (between 23°26′ North and South)

Rhizophora mangle and *Laguncularia racemosa* using various dendrochronological methods. In these specific cases the annual variation in soil water salinity seems to trigger the formation of annual rings, with lower salinity concentrations during the rainy season due to fresh water input favoring tree growth (Robert et al. 2011). The mangrove species *Avicennia marina* has an uncommon secondary growth with more than one simultaneously active successive cambia, leading to a network of non-annual ring boundaries probably triggered by endogenous factors, but also related to tree size and site conditions (Schmitz et al. 2007; Robert et al. 2011). The onset and cessation of cambial activity of branches from 10 tree and shrub species growing in the savannahs (Cerrado) of southeast Brazil seemed to be best explained by variation in day length which increases with increasing latitude, although precipitation and temperature also indicated significant correlations with cambial activity (Blagitz et al. 2016; Marcati et al. 2016). These environmental factors are also considered as main triggers of cambial dormancy and tree-ring formation in subtropical mountain ecosystems of Brazil (Oliveira et al. 2009).

3.2.5 *Wood Anatomical Features and Distinctiveness of Tree Rings*

While the wood anatomical patterns of tree rings from tropical and subtropical conifers is similar to those from boreal and temperate regions, tropical broadleaved species present some distinct differences (Worbes 1989). Despite the enormous diversity of tropical angiosperms, the wood anatomical features of growth rings can be distinguished by four basic types first described by Coster (1927, 1928) for broadleaved tree species from Java and later adopted by Worbes (1989; 2002). This wood anatomical characterization has been applied in many studies to identify and define tree ring boundaries. The first type, typical for tree species of the families Annonaceae, Burseraceae, Euphorbiaceae, Lauraceae, Malvaceae, and Salicaceae, is a distinct intra-annual variation in wood density due to change in lumen and cell wall thickness of fibers, similar to the well-known earlywood and latewood formation of tracheids in conifers (Fig. 3.2). Often, these ring boundaries can be distinguished as flattened and thickened fiber walls (Silva et al. 2016). The second type, typified by many tree species from the families Bignoniaceae, Fabaceae, and Meliaceae, has ring boundaries delimited by marginal parenchyma, although the nature of this parenchyma tissue can be either terminal or initial (Wilczek et al. 2014) or, as in the case of *Cedrela fissilis*, both (Marcati et al. 2006). The third type, common amongst Chrysobalanaceae, Lecythidaceae, Moraceae, and Sapotaceae is an alternating pattern of fiber and parenchyma bands with the bands becoming narrower towards the end of the growth ring. The fourth type, which is defined by intra-annual variations in the vessel size and density (ring-porous or semi-porous rings), is less common in the tropical lowlands (Worbes 2002; Lisi et al. 2008; Silva et al. 2016) and often occurs in combination with the previous described anatomic patterns (Brienen et al. 2016). Vessel arrangements, especially ring-porosity, seem to be more frequent in environments with temperature seasonality (Alves and Angyalossy-Alfonso 2000).

The distinctiveness of growth rings depends on the interaction between genetically defined wood anatomy and abiotic factors such as rainfall, temperature and flooding influencing the tree physiology and leaf phenology (Borchert 1999; Brienen et al. 2016). In a review of 491 species from 133 genera and 22 families from different Brazilian biomes, Alves and Angyalossy-Alfonso (2000) demonstrated that 48% of the analyzed species possessed distinct tree rings. Analyzing the suitability of 139 species for tree-ring analysis across South America, Worbes (2002) showed that the distinctiveness of annual rings often varies between life stages with tree rings clear in the adult phases and absent or vague in the juvenile phases, or vice versa. From 117 tree species examined in the Brazilian Coastal Atlantic forest, 53–59% formed distinct tree rings (Silva et al. 2016). Tomazello Filho et al. (2004) observed distinct tree rings for 46% of 41 species analyzed in the rainforests and savannahs of southeast Brazil. For the Peruvian rainforest Gutiérrez and Ramos (2013) report a lower percentage (30% of 80 tree species) with distinct annual growth rings. Similar findings were highlighted by Roig et al. (2005) for the dry forests of Mexico with 35% of 52 studied tree species indicating potential for dendrochronological applications. Other studies demonstrate that the formation of distinct tree rings depends also on the seasonality of the environmental conditions (Geiger 1915; Coster 1927). Trees of a species growing at sites with a distinct seasonality form clearly visible tree rings, which may be vague or even

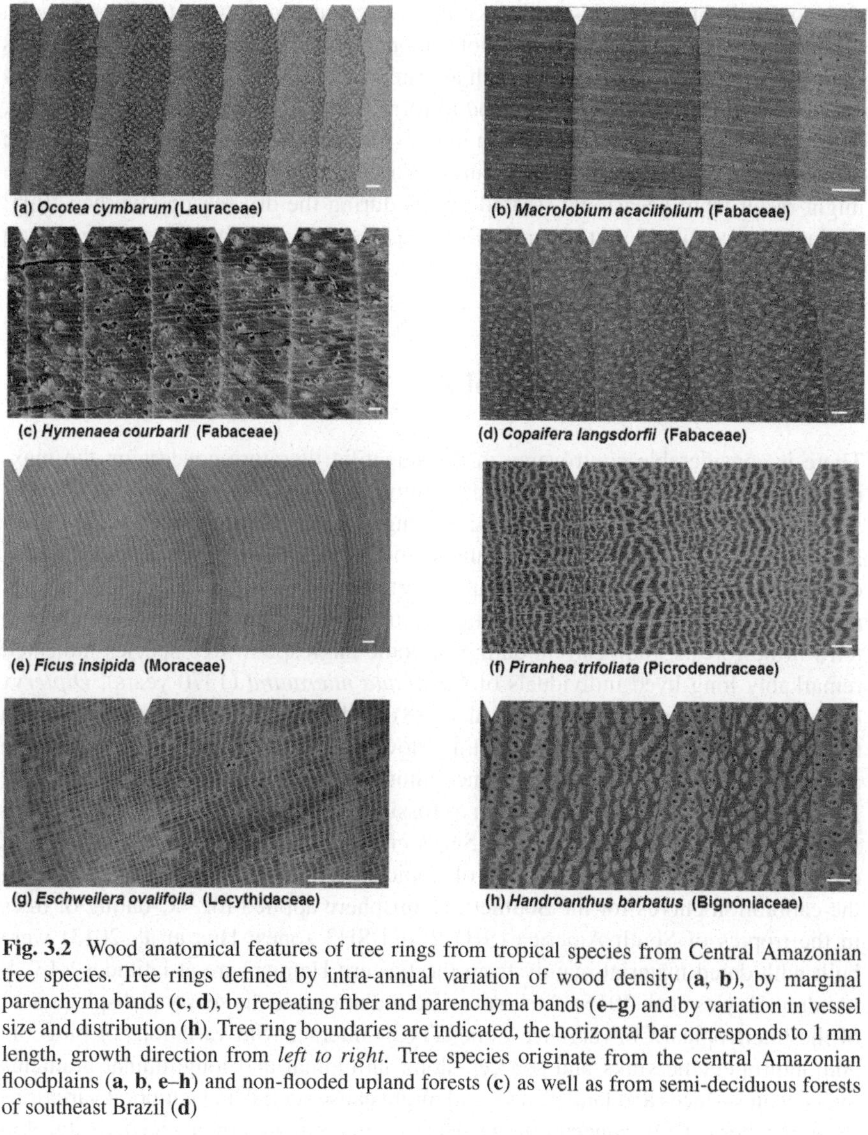

Fig. 3.2 Wood anatomical features of tree rings from tropical species from Central Amazonian tree species. Tree rings defined by intra-annual variation of wood density (**a**, **b**), by marginal parenchyma bands (**c**, **d**), by repeating fiber and parenchyma bands (**e**–**g**) and by variation in vessel size and distribution (**h**). Tree ring boundaries are indicated, the horizontal bar corresponds to 1 mm length, growth direction from *left to right*. Tree species originate from the central Amazonian floodplains (**a**, **b**, **e**–**h**) and non-flooded upland forests (**c**) as well as from semi-deciduous forests of southeast Brazil (**d**)

absent in the same species growing at sites with a weak or irregular seasonality (Geiger 1915; Brienen et al. 2016). For tree-ring studies it is therefore important to carefully examine the wood anatomy of species explored for the first time as well as previously studied species at new sites with different environmental conditions.

Tropical species frequently present tree-ring anomalies (e.g., Worbes 2002; Roig et al. 2005; Brienen et al. 2016) such as wedging rings or partially missing rings, which trees form during one or consecutive years in just one section of the circumference. These anomalies are commonly found in trees growing under poor light conditions with increased interspecific competition resulting in differences of the local supply of phytohormones, water, minerals and carbohydrates (Dünisch et al. 1999). Some taxa such as Annonaceae (Worbes 2002) or *Araucaria angustifolia* (Oliveira et al. 2010) tend to form wedging tree rings quite frequently. Another commonly reported problem in tree-ring analysis of tropical species is the occurrence of false rings and double rings (Worbes 2002; Brienen et al. 2016), which might result from occasional rainfall events during the dry season (Worbes 1995) or due to exceptional, but short-lived, drought events during the growing season (Schöngart et al. 2002).

3.3 Longevity of Neotropical Tree Species

There is considerable controversy in the scientific literature concerning the maximum age of tropical trees (Martínez-Ramos and Alvarez-Buylla 1998; Worbes and Junk 1999; Laurance et al. 2004; Schöngart et al. 2015; Brienen et al. 2016). Maximum age is an ecological parameter of broad relevance for forest ecology, dynamics and productivity, disturbance regimes, biogeochemical cycles, natural resource management, and conservation of tropical forests. In central Amazonian terra firme forests, radiocarbon dating of some angiosperm tree species indicated remarkably long-lived individuals of *Cariniana micrantha* (1370 years), *Dipteryx odorata* (1180 years) (Chambers et al. 1998), and *Bertholletia excelsa* (996 years) (Vieira et al. 2005). Radiocarbon dating, however, often has the problem of wide age ranges associated with dating uncertainties (Petrillo et al. 2016). During the Industrial Revolution, the combustion of fossil fuel resulted in highly varying atmospheric radiocarbon concentrations (Suess effect), which limits the application of direct ^{14}C-dating to examine the age of tropical trees (Worbes 2002). Furthermore, the calibration curves for the Southern Hemisphere applied for ^{14}C-dating of trees in the tropics of South America (SH1-2 and SH3 zones; Hua et al. 2013) were built with dated material mainly from the Eastern Hemisphere at latitudes >18° S (Andreu-Hayles et al. 2015). Atmospheric circulation and the seasonal displacement of the Intertropical Convergence Zone (ITCZ) interact with variations of biogenic and anthropogenic sinks and sources along latitudinal and longitudinal gradients over ocean surfaces and land masses and might cause spatial and temporal variations in atmospheric CO_2 concentrations. Hence, there is an urgent need to develop ^{14}C calibration curves from exactly dated tree rings of long-lived species from different tropical regions of South America to improve radiocarbon dating accuracy especially for the pre-bomb period (Santos et al. 2015).

Tree-ring analyses in subtropical and tropical regions of the Americas indicate longevities of more than 1000 years only for the gymnosperms *Taxodium mucronatum* and *T. distichum* growing in the swamp forests of central Mexico, Guatemala, and the southern USA (Stahle et al. 2012). However, most angiosperms in the tropical lowlands are substantially younger with reported maximum ages of not more than 400–600 years (Worbes and Junk 1999; Fichtler et al. 2003; Schöngart et al. 2005, 2015; Brienen and Zuidema 2006a, b; Worbes and Fichtler 2010; Brienen et al. 2016), similar to maximum life-spans for temperate angiosperm tree species (Loehle 1988; Schweingruber and Wirth 2009; Di Filippo et al. 2015). A recent review of maximum ages for 71 tropical tree species based on tree-ring analyses (Brienen et al. 2016) resulted in a median observed maximum age of 200 years. Only a few species in the Neotropics have confirmed maximum tree ages exceeding 400 years. Among these are *Bertholletia excelsa*, *Cariniana micrantha*, and *Dipteryx micrantha* from non-flooded Amazonian terra firme forests (Funck 2004; Brienen and Zuidema 2006a, b; Jenkins 2009; Schöngart et al. 2015), *Macrolobium acaciifolium* growing in oligotrophic black-water floodplains (igapó) of central Amazonia (Schöngart et al. 2005), and *Hymenolobium mesoamericanum* growing in humid forests of Costa Rica (Fichtler et al. 2003). Hence, there is a discrepancy between evidences of longevities for tropical angiosperms based on tree-ring studies and proposed maximum tree ages of up to 1000 years and more based on ^{14}C-dating (Chambers et al. 1998; Vieira et al. 2005).

In boreal, temperate, and semi-arid climate zones tree ages are based on ring counts and dendrochronological crossdating techniques. Ancient trees with multi-millenial ages are all gymnosperms such as *Juniperus occidentalis*, *Pinus aristata*, *P. balfouriana*, *P. longaeva*, *Sequoiadendron giganteum*, *Sequoia sempervirens*, *Taxodium distichum* and *Thuja occidentalis* in the USA or *Fitzroya cupressoides* in Chile (Brown 1996; Larson et al. 2000; Schweingruber and Wirth 2009; Stahle et al. 2012). Maximum tree ages reported for angiosperms, however, result in much lower ages; for example, *Weinmannia trichosperma* and *Liriodendron philippiana* in temperate forests of southern Chile, which reach maximum ages of more than 600 years (Lara 1991; Lusk 1999) and *Quercus douglasii* in California with ages over 500 years (Stahle et al. 2013). *Polylepis tarapacana* (Rosaceae) growing at the upper tree line in the subtropical and tropical Andean Altiplano at altitudes of more than 4500 m above sea level shows ages above 700 years (Solíz et al. 2009). These ancient trees are generally found at marginal sites where environmental factors such as temperature, inundation, fire, drought and edaphic factors strongly limit tree growth (Schweingruber and Wirth 2009; Di Filippo et al. 2015). Often, only one or a few tree species possess the morphological, anatomical, biochemical, and physiological adaptations to establish and survive under these unfavorable growing conditions at the edge of forest occurrence, leading to the formation of monodominant or even monospecific stands characterized by low intra- and interspecific competition. The constellation of unfavorable environmental growing conditions and low competition at extreme sites allows the trees to reach great ages. Furthermore, these ancient trees have a typical morphology characterized by often longitudinally twisted stems, associated with crown dieback, hollow voids or heart rot, and bark-covered knobs on the stem (Stahle 1996). Trees with such morphological characteristics have also been found in the Jaú National Park in central Amazonia, where *Eschweilera tenuifolia* forms open monospecific stands.

In these nutrient-poor black-water floodplain forests, trees grow under stable geomorphological conditions with low competition and are limited by annual flooding of more than 8 m amplitude and an average duration of more than 10 months. Preliminary data on the average annual increment of 118 trees of different size estimate long-term annual diameter increments of 1.38 ± 0.30 mm. Mean passage times through 10-cm diameter classes varied between 47 and 78 years, suggesting that trees with diameters of more than 140 cm might be more than 1000 years old (Junk et al. 2015). This leads to the hypothesis that tropical angiosperms reach maximum longevities at marginal sites characterized by extremely limited growth conditions requiring sophisticated adaptations of tree species to local site conditions. The resulting stands are monodominant formations with open structure resulting in low intra-specific competition that contrasts sharply with forest occurrence in areas more optimal for growth such as the Amazonian terra firme that are characterized by high tree species diversity and inter-tree competition.

3.4 Growth Patterns of Tropical Trees

Age and growth rates of tree species provide fundamental insights on forest ecology and dynamics. Retrospective analysis of tree-ring series allows a detailed reconstruction of individual growth over the course of an entire lifespan. Furthermore, it provides insights into long-term growth variations among individuals and yields representative and reliable growth data from individuals that have successfully established in the canopy (Brienen and Zuidema 2006a). Such information cannot be obtained using radiocarbon dating or short-term increment measurements derived from permanent monitoring plots or dendrometers (Schöngart 2008; Rozendaal et al. 2010). Simulations of growth trajectories based on short-term measurements from monitoring trees of different diameter classes generally include slow-growers from small sizes classes as part of the modeling approach (e.g., Lieberman and Lieberman 1985; Nebel et al. 2001). However, many of these individuals do not reach mature size (Free et al. 2014), which can result in underestimates of juvenile growth and, consequently, overestimates of lifespans of tree species (Brienen and Zuidema 2006a; Rozendaal et al. 2010; Schöngart 2010).

The growing evidence for annual growth ring formation in tropical trees (Table 3.1, Fig. 3.1) has led to many dendroecological studies in recent decades that have contributed to the knowledge on tree growth dynamics by providing data for age-size variation within and among tropical tree species and ecosystems (Brienen and Zuidema 2006a, 2007; Schöngart 2008, 2010; Rozendaal et al. 2010; Rozendaal and Zuidema 2011; Jiménez and del Valle 2011; Giraldo and del Valle 2012; López et al. 2013; Free et al. 2014; Rosa et al. 2017; Locosselli et al. 2017), the role of growth autocorrelation and persistent growth (Brienen et al. 2006; Brienen and Zuidema 2007), and canopy accession patterns (Brienen and Zuidema 2006a; Brienen et al. 2010b; Schöngart et al. 2015). Age-size relationships are available for many tree species in different tropical regions of America such as the semi-deciduous forests of Venezuela (Worbes 1999) and Bolivia (Brienen and Zuidema 2006a; Rozendaal and Zuidema 2011), the Bolivian savannahs (López et al. 2013), the central Amazonian floodplains (Rosa 2008; Schöngart 2008, 2010; Leoni et al. 2011), and the terra firme forests of the Southern Amazon basin (Free et al. 2014). Recently published studies compare age-size relationships of *Calophyllum brasiliense* from several wetland types (Rosa et al. 2017) and *Hymenaea* spp. from different forest ecosystems (Locosselli et al. 2017) across different Brazilian ecoregions.

Mean age-size relationships of species growing in the same ecosystem show large variation as evidenced by characteristic species of different successional stages from the nutrient-rich and highly dynamic white-water floodplains (várzea) of central Amazonia (Fig. 3.3a). Mean ages of similarly sized trees varies by a factor of up to 20 between fast-growing pioneer species with low wood densities and short lifespans of not more than a few decades and slow-growing species from the late successional stages with high wood densities and lifespans of more than 300 years. Age-size relationships based on average growth data, however, do not consider growth variations among individual trees of a species. Brienen et al. (2016) found that for trees 10 cm in diameter of a *Cedrela odorata* population growing in Bolivian moist forests, ages ranged from 9 to 75 years. Free et al. (2014) described *Swietenia macrophylla* growing in terra firme forests of Southern Amazonia that had ages at 30 cm diameter that ranged from 8 to 88 years. Such broad variation among trees can be explained by differences between individuals in light and water availability, as well as soil fertility and genetic differences (Brienen and Zuidema 2006a; Brienen et al. 2010b; Rozendaal et al. 2010; Free et al. 2014). In the central Amazonian floodplain forests *Macrolobium acaciifolium* trees growing in the nutrient-rich várzea had ages ranging from 23 to 71 years for a diameter of 30 cm, contrary to the nutrient-poor black-water system (igapó) where trees at this size were much older (61–158 years) (Fig. 3.3b).

Individual tree growth is often autocorrelated, implying temporal correlations between growth rates in subsequent years (Brienen et al. 2006; Free et al. 2014). Juvenile trees, in particular, tend to have strongly autocorrelated growth which is likely due to different light conditions during their establishment. Such growth differences among individual trees can persist over long time periods leading to strong variations in growth trajectories and consequently age-size relationships within a population (Brienen et al. 2010b; Rozendaal et al. 2010). Analyses of growth suppressions and releases in growth trajectories during the phase of canopy establishment provide important information on the shade tolerance of a tree species, forest dynamics, and disturbance regimes. So far only a few tropical dendroecological studies have exploited the potential of tree-ring analysis to focus on these aspects of forest stand dynamics (Baker et al. 2005; Baker and Bunyavejchewin 2006, 2017; Brienen and Zuidema 2006a; Brienen et al. 2010b; Schöngart et al. 2015). Release and suppression patterns differ considerably among tree species and across sites during their establishment in the canopy mainly due to temporal variation in light, but also water and nutrient availability. In Bolivian moist forests growth trajectories of six non-pioneer tree species presented distinct

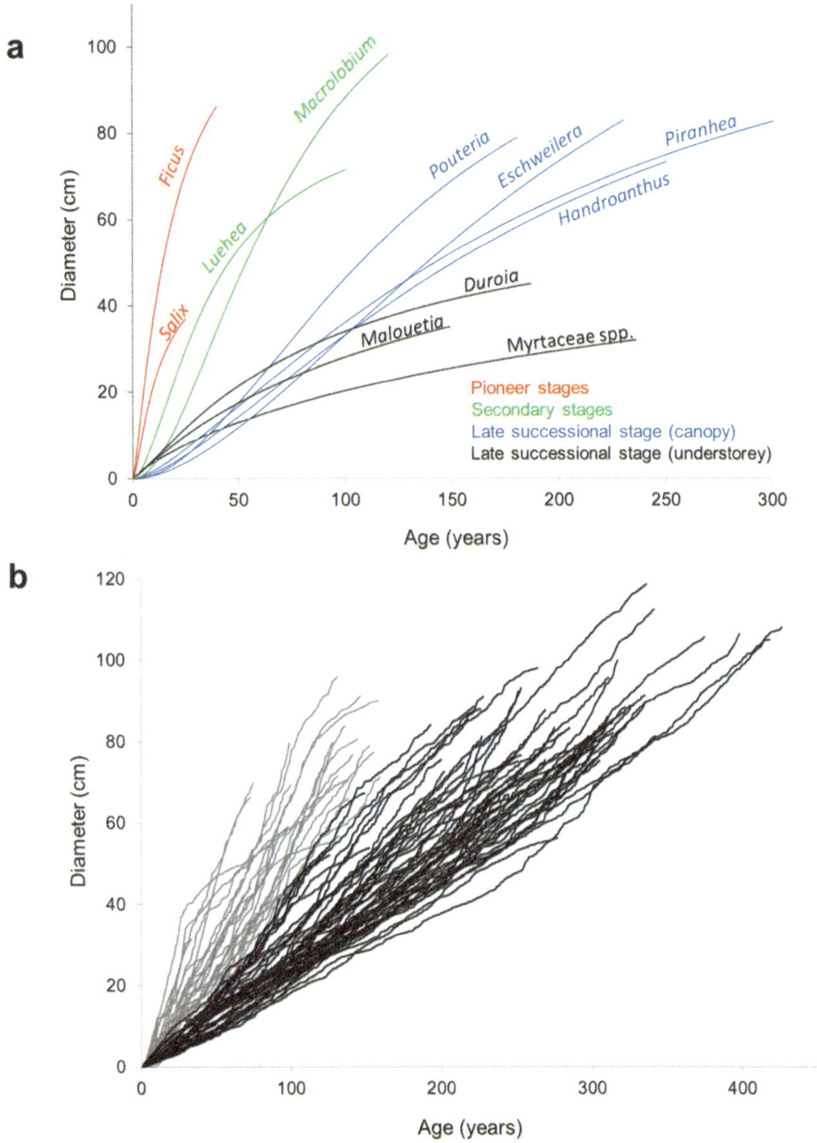

Fig. 3.3 (**a**) Mean age-size relationship of characteristic tree species from different successional stages of the central Amazonian white-water floodplains (várzea): *Ficus insipida* (Moraceae) and *Salix martiana* (Salicaceae) from pioneer stages; *Luehea cymulosa* (Malvaceae) and *Macrolobium acaciifolium* (Fabaceae) from secondary stages; canopy species *Piranhea trifoliata* (Picrodendraceae), *Eschweilera parvifolia* (Lecythidaceae), *Handroanthus barbatus* (Bignoniaceae) and understory species *Duroia duckei* (Rubiaceae), *Malouetia tamaquarina* (Apocynaceae) and Myrtaceae spp. from the late successional stage (data: Schöngart et al. 2007, Schöngart 2010, Leoni et al. 2011); (**b**) individual growth trajectories from *Macrolobium acaciifolium* (Fabaceae) growing in the nutrient-rich várzea floodplains (*grey curves*; n = 24) and the nutrient-poor black-water igapó floodplain forest (*black curves*; n = 55) of central Amazonia (data: Schöngart et al. 2005, Schöngart unpubl.)

release and suppression patterns in tree growth during canopy establishment and also in the duration of suppressed and released growth. This indicates different shade tolerances and capacities to respond to gap formation (Brienen and Zuidema 2006a). A comparison of growth trajectories from juvenile trees of *Cedrela odorata* growing in a Bolivian moist forest with low light levels in the understory and a Mexican dry forest with a lower and more open canopy suggests that light availability is the main limiting factor during canopy accession in moist forests, while in the dry forests the spatial heterogeneity in water (and nutrient) availability had a life-long impact on tree growth (Brienen et al. 2010b).

Due to frequently occurring tree-ring anomalies in many tropical tree species (wedging, double, false, and discontinuous growth rings) that can result in inconsistent ring-width patterns among radii and individuals, and which propagates to dating errors, ideally growth trajectories should be analyzed using crossdated time series. López and Villalba (2016b) evaluated lifetime growth rates from eight tropical tree species growing in the Bolivian Cerrado, comparing mean diameter growth rates from crossdated individuals and those based on estimates by ring counting based on wood anatomical features. They did not find significant differences between mean diameter growth rates and the dating errors were <5% suggesting that tree species with high-quality tree rings provide reliable estimates on tree age, growth rates, and growth trajectories based on ring counting (Baker and Bunyavejchewin 2017).

3.5 Dendroecological Applications for Forest Management and Conservation

The application of tree-ring data to develop sustainable forest management plans is not new. Sir Dietrich Brandis, a remarkable German botanist and forester who was sent to Burma (today Myanmar) and India during the second half of the nineteenth century by the British Empire to develop sustainable forest management, was one of the first forest scientists to systematically apply tree rings in the tropics. To develop management plans for the valuable teak timber resources (*Tectona grandis*) in Burma, Brandis (1856) performed inventories of natural teak forests and grouped the trees into four diameter classes. The upper diameter class with more than four cubits circumference (corresponding to a diameter at breast height (DBH) of about 58 cm) was destined for harvest. Brandis obtained diameter growth rates by simply counting the tree rings (assuming their annual formation) and estimated a felling cycle of 24 years based on the mean passage time through size classes. Later, Brandis (1879) confirmed the formation of annual tree rings for teak comparing the number of growth rings with the known age of plantations. Despite the early scientific evidence for annual tree-ring formation in the tropics of Southeast Asia during the end of the nineteenth century and the begin of the twentieth century (Gamble 1881; Ursprung 1904; Coster 1927, 1928) and first dendroecological applications (e.g., Geiger 1915; Berlage 1931), tree-ring analysis has rarely been applied as a tool for forest management and conservation since then due to the emerging and pervasive myth of absence of growth rings in tropical tree species. The high species richness and diversity that characterize tropical forests together with methodological problems in determining tree ages and lifetime growth rates contributed to the failure of tropical forest management for long periods (e.g., Dawkins and Philip 1998).

However, the growing evidence of annual tree-ring formation in the tropics and literature providing relevant information on varying growth patterns and age-size relationships within and among tree species and environments has begun to change this picture. Increasingly dendroecological data are used as a tool in tropical forest management to project timber harvests (Brienen and Zuidema 2006b, 2007; Rozendaal et al. 2010) and to evaluate and adjust current forest management practices to increase the level of sustainability (Schöngart 2008, 2010; López et al. 2013; Free et al. 2014). The development of sustainable management systems is a promising strategy to conserve tropical forests by providing the long-term use of natural resources (timber and non-wood forest products) and simultaneously maintaining the multiple ecological functions and services of the forests. Estimates indicate that about 400 million hectares of tropical forests are currently in a permanent timber estate and partially integrated in networks of protected areas (Edwards et al. 2014). In these areas polycyclic harvesting systems, such as the Selection Silvicultural Management System (Wyatt-Smith 1995; Bruenig 1998), are generally applied (Lamprecht 1989; Whitmore 1993; Dawkins and Philip 1998). These systems aim to maintain an uneven-aged structure, to minimize disturbances of the forest by selective tree harvests using a minimum logging diameter (MLD) and a felling cycle that defines the time period between successive timber harvests in the same coupe area. Selective harvests in combination with reduced-impact logging are promoted as a model for sustainable forest management (Dykstra and Heinrich 1996; Vidal et al. 1997; Gerwing 2002; Putz et al. 2008) to lower residual stand damage and soil disturbance as well as to reduce logging waste and losses and increase rates of economic return in the long term (Johns et al. 1996; Boltz et al. 2011; Holmes et al. 2002). However, the technical and logistical improvements of selective timber harvesting are not sufficient to guarantee sustainable management of timber resources unless they are integrated into a silvicultural management and conservation system that is based on empirical data on the ages, lifetime growth rates, regeneration processes, and life histories of the commercial and non-commercial tree species that structure and drive the forest ecosystem (Schöngart 2010). Consequently, there is much skepticism about growth rates being used as a basis for managing tropical forests, because the established time and diameter limitations are estimations or simply conform to legal restrictions and are not derived from sound scientific data.

Brazilian forest legislation defines a common MLD of 50 cm and felling cycles of 25–35 years, defined by the normative instruction IN n° 5 (11th December 2006) established by the Brazilian Institute for Environmental and Renewable Natural Resources (IBAMA) and Resolution n° 406 (2nd February 2009) of the National Council of Environment (CONAMA). In Bolivia forest legislation defines a minimum felling cycle as 20 years with MLDs varying from 40 to 70 cm depending on the tree species and forest type (Brienen and Zuidema 2006b; López et al. 2013). The mean age of 45 commercial tree species at these fixed diameter-cutting limits varies from 15 to 270 years among tree species and forest types in the tropical regions of South America (Table 3.2). Mean age at MLD of seven commercial species from the Bolivian savannahs (Cerrado) present ranges of 68–85 years (MLD of 40 cm) and 43–100 years (MLD of 50 cm) (López et al. 2013). An even higher variation of mean age (63–179 years for MLD of 50–60 cm) was observed for six commercial tree species from the Bolivian semi-deciduous forests (Brienen and Zuidema 2006b; Rozendaal et al. 2010). In the southern Amazon basin (terra firme) seven commercial tree species present mean ages at MLD between 31 and 117 years (Lobão 2011; Free et al. 2014; Oliveira 2014; Andrade 2015). In the central Amazonian floodplains mean age at a MLD of 50 cm differs tenfold (15–151 years) among 19 tree species from the nutrient-rich várzea, in contrast to the nutrient-poor black-water igapó where mean age of four species at this diameter limit tends to be much higher (162–270 years) (Schöngart et al. 2007; Rosa 2008; Schöngart 2008, 2010; Da Fonseca et al. 2009). These enormous variations of age at MLDs indicate already that felling cycles fixed by forest rules and legislation do not meet the biological reality of many of these species.

Based on mean age-size relationships of commercial tree species from the central Amazonian várzea floodplains derived from tree-ring data, growth models have been developed to project species-specific MLDs and felling cycles (Schöngart 2008). This modeling approach named GOL (Growth-Oriented Logging) integrates species-specific age-diameter and diameter-height relationships into allometric equations to simulate volume growth over the lifespan of a species. The MLD is estimated by the corresponding diameter at the age when the tree species attains its maximum annual volume increment. The felling cycle is derived from the mean passage time through 10-cm diameter classes until reaching the defined MLD. For commercial tree species in the várzea estimated species-specific MLDs ranged from 25 to almost 130 cm, which differ considerably from the commonly applied MLD of 50 cm (Rosa 2008; Schöngart 2008; Leoni et al. 2011). Estimated felling cycles of 12 commercial tree species in the várzea floodplains varied tenfold from 3 to 32 years (Schöngart 2008).

The GOL concept provided a framework to validate the established norms and regulations of Brazilian forest legislation in terms of felling cycles and MLDs based on the growth potential of commercial tree species. It demonstrated that forest management must consider species-specific growth variations to attain a higher level of sustainability (van Gardingen et al. 2006; Schöngart 2008; Sebbenn et al. 2008). Fortunately, the obtained results were considered in an innovative forest legislation (IN n° 009 from 12th November 2010) for the várzea floodplain forests implemented by the Ministry of Environment and Sustainable Development of the Amazonas State (SDS) defining felling cycles of 12 years for the fast-growing, low-density species (wood density below 0.6 g cm^{-3}) and 24 years for the slow-growing, high-density tree species (wood density above 0.6 g cm^{-3}) with species-specific MLDs.

Forest management must also take into account variation in tree growth in different environments. Tree-ring studies performed in the nutrient-poor black-water

Table 3.2 Mean ages at the minimum logging diameter (MLD) of commercial tree species from different forest types in tropical South America derived from tree-ring data

Tree species	Forest type, country	MLD (cm)	Mean age at MLD (years)	References
Albizia niopoides	Montane moist forest, Colombia	50	52	Giraldo and del Valle (2012)
Albizia subdimidiata	Várzea floodplain, central Amazonia, Brazil	50	56	Schöngart (2008)
Amburana cearensis	Savannah (Cerrado), Bolivia	40	68	López et al. (2013)
Amburana cearensis	Semi-deciduous forest, southwest Amazonia, Bolivia	50	95	Brienen and Zuidema (2006b)
Bertholletia excelsa	Semi-deciduous forest, southwest Amazonia, Bolivia	60	166	Brienen and Zuidema (2006a)
Bertholletia excelsa	Terra firme forest, central Amazonia, Brazil	50	107–121	Schöngart et al. (2015)
Calophyllum brasiliense	Igapó (black-water floodplain), central Amazonia, Brazil	50	261	Schöngart (2010)
Calophyllum brasiliense	Igapó (clear-water floodplain), south Amazonia, Brazil	50	105–117	Rosa et al. (2017)
Calophyllum brasiliense	Várzea floodplain, central Amazonia, Brazil	50	70	Rosa et al. (2017)
Calophyllum brasiliense	Wetland, Pantanal, Brazil	50	103–109	Rosa et al. (2017)
Calophyllum brasiliense	Wetland, Coastal Atlantic rainforest, Brazil	50	96	Rosa et al. (2017)
Cedrela fissilis	Savannah (Cerrado), Bolivia	50	100	López et al. (2013)
Cedrela odorata	Semi-deciduous forest, southwest Amazonia, Bolivia	60	95	Brienen and Zuidema (2006b)
Cedrela odorata	Várzea floodplain, central Amazonia, Brazil	50	72	Rosa (2008)
Cedrela odorata	Terra firme forest, south Amazonia, Brazil	50	84	Lobão (2011)
Cedrelinga catenaeformis	Semi-deciduous forest, southwest Amazonia, Bolivia	50	61	Brienen and Zuidema (2006b)
Cedrelinga catenaeformis	Semi-deciduous forest, southwest Amazonia, Bolivia	60	63	Rozendaal et al. (2010)
Centrolobium microchaete	Savannah (Cerrado), Bolivia	40	98–110	López et al. (2013)
Chrysophyllum argenteum	Várzea floodplain, central Amazonia, Brazil	50	131	Schöngart (2008)

(continued)

Table 3.2 (continued)

Clarisia racemosa	Semi-deciduous forest, southwest Amazonia, Bolivia	45	179	Rozendaal et al. (2010)
Copaifera langsdorffii	Plantation, Coastal Atlantic rainforest, Brazil	50	51	Costa et al. (2015)
Dalbergia nigra	Plantation, Coastal Atlantic rainforest, Brazil	50	61	Costa et al. (2015)
Eschweilera albiflora	Várzea floodplain, central Amazonia, Brazil	50	151	Schöngart (2008)
Ficus boliviana	Savannah (Cerrado), Bolivia	50	43	López et al. (2013)
Ficus insipida	Várzea floodplain, central Amazonia, Brazil	50	15	Schöngart et al. (2007)
Goupia glabra	Terra firme forest, south Amazonia, Brazil	50	107	Oliveira (2014)
Handroanthus aureus	Wetland, Pantanal, Brazil	50	156	Leite (2012)
Handroanthus barbatus	Igapó (black-water floodplain), central Amazonia, Brazil	50	270	Da Fonseca et al. (2009)
Handroanthus barbatus	Várzea floodplain, central Amazonia, Brazil	50	117	Da Fonseca et al. (2009)
Handroanthus serratifolius	Terra firme forest, south Amazonia, Brazil	50	114	Andrade (2015)
Hura crepitans	Várzea floodplain, central Amazonia, Brazil	50	41	Rosa (2008)
Hymenaea courbaril	Terra firme forest, south Amazonia, Brazil	50	117	Andrade (2015)
Ilex inundata	Várzea floodplain, central Amazonia, Brazil	50	55	Schöngart (2008)
Luehea cymulosa	Várzea floodplain, central Amazonia, Brazil	50	43	Schöngart (2008)
Machaerium scleroxylon	Dry tropical forest, Bolivia	40	122	Paredes-Villanueva et al. (2013)
Macrolobium acaciifolium	Igapó (black-water floodplain), central Amazonia, Brazil	50	171	Schöngart (2010)
Macrolobium acaciifolium	Várzea floodplain, central Amazonia, Brazil	50	54	Schöngart (2010)
Malouetia tamaquarina	Várzea floodplain, central Amazonia, Brazil	25	81	Leoni et al. (2011)
Manilkara huberi	Terra firme forest, central Amazonia, Brazil	50	223	Pinto (2012)
Ocotea cymbarum	Várzea floodplain, central Amazonia, Brazil	50	59	Rosa (2008)
Peltogyne cf. *heterophylla*	Semi-deciduous forest, southwest Amazonia, Bolivia	50	135	Brienen and Zuidema (2006b)
Peltogyne cf. *heterophylla*	Semi-deciduous forest, southwest Amazonia, Bolivia	50	150	Rozendaal et al. (2010)

(continued)

Table 3.2 (continued)

Tree species	Forest type, country	MLD (cm)	Mean age at MLD (years)	References
Piranhea trifoliata	Várzea floodplain, central Amazonia, Brazil	50	142	Schöngart (2008)
Platymiscium ulei	Savannah (Cerrado), Bolivia	40	85	López et al. (2013)
Pouteria elegans	Várzea floodplain, central Amazonia, Brazil	50	106	Schöngart (2008)
Prioria copaifera	Floodplain forest, Colombia	40	90	Jiménez and del Valle (2011)
Pseudobombax munguba	Várzea floodplain, central Amazonia, Brazil	50	33	Schöngart (2008)
Pterocarpus rohrii	Plantation, Coastal Atlantic rainforest, Brazil	50	51	Costa et al. (2015)
Qualea albiflora	Terra firme forest, south Amazonia, Brazil	50	81	Oliveira (2014)
Qualea ingens	Flooded riparian forest, Cerrado, Brazil	50	62	Mohr (2013)
Schizolobium parayba	Plantation, Coastal Atlantic rainforest, Brazil	50	57	Costa et al. (2015)
Schizolobium parayba var. *amazonicum*	Terra firme forest, south Amazonia, Brazil	50	31	Lobão (2011)
Senna multijuga	Plantation, Coastal Atlantic rainforest, Brazil	50	55	Costa et al. (2015)
Sloanea terniflora	Várzea floodplain, central Amazonia, Brazil	50	67	Schöngart (2008)
Sterculia apetala	Várzea floodplain, central Amazonia, Brazil	50	47	Rosa (2008)
Sterculia apetala	Wetland, Pantanal, Brazil	50	52–64	Santos (2006)
Swietenia macrophylla	Terra firme forest, south Amazonia, Brazil	60	77	Free et al. (2014)
Vatairea guianensis	Igapó (black-water floodplain), central Amazonia, Brazil	50	162	Da Fonseca et al. (2009)
Vatairea guianensis	Várzea floodplain, central Amazonia, Brazil	50	70	Da Fonseca et al. (2009)
Vochysia divergens	Wetland, Pantanal, Brazil	50	61	Schöngart et al. (2011)

igapó of the central Amazon region showed significantly lower diameter growth rates for commercial tree species compared to the nutrient-rich várzea (Schöngart et al. 2005; Da Fonseca et al. 2009; Scabin et al. 2012). Mean age of *Macrolobium acaciifolium* differed from 67 years (MLD: 62 cm) in the várzea to 326 years (MLD: 83 cm) in the igapó (Schöngart 2010). Similarly, *Calophyllum brasiliense* had a mean age of 91 years in the várzea (MLD: 64 cm) and 154–217 years (MLD: 35–63 cm) in the black-water igapó, while in the clear-water igapó, which is intermediate in nutrient status, the mean age was 149–177 years for MLDs of 64–81 cm (Rosa et al. 2017). These findings indicate the vulnerability of these oligotrophic ecosystems to unregulated selective logging and suggest a revision of forest management practices for the black-water floodplains in the Amazon region (Da Fonseca et al. 2009; Schöngart 2010; Scabin et al. 2012; Rosa et al. 2017).

Brienen and Zuidema (2006b) performed yield projections for four commercial tree species under current forest management practices in the Bolivian Amazon (felling cycle of 20 years, MLD of 50 cm, harvest intensity of 80% of trees above MLD) applying simple models for population growth. They suggested that the volume stocks of these species will recuperate 100% of the initially harvested volume only after a period of 32–84 years. For the valuable and threatened tree species *Swietenia macophylla* in the southern Brazilian Amazon sophisticated models based on tree-ring data (Free et al. 2014) and plot monitoring (Grogan et al. 2014) that incorporate demographic data project a steep population decline and volume depletion after 2–3 felling cycles (30 years).

However, estimated felling cycles and MLDs can only increase the sustainability of managing timber stocks if the exploited species will be able to regenerate and establish to maintain the capacity of the remaining population to replace the harvested volume within a felling cycle (Schöngart 2010). Therefore, it is of utmost importance to obtain further information about the influence of environmental factors on tree growth and the effects of light conditions, nutrient supply and water availability in the establishment and growth of tree species. Studies based on mean age-size relationships also do not account for growth variations among individual trees. Analyses of individual growth trajectories exploring growth autocorrelations within and among individuals as well as release and suppression patterns during canopy establishment provide essential insights into the ecology of a tree species. These analyses have strong applications in developing silvicultural treatments that promote fast growth through juvenile stages or exploit the capacity of tree species to respond to increasing light availability by crown liberation or liana cuttings (Brienen and Zuidema 2006a; Free et al. 2014; Cunha et al. 2016). Furthermore, there is an urgent need to incorporate demographic data of tree populations obtained from permanent monitoring plots, such as size structure, mortality rates, fruit production, seed germination, seedling establishment, and canopy disturbance rates, into yield projections to simulate and evaluate the impact of current management practices and to test projected MLDs and felling cycles (Grogan et al. 2014; Free et al. 2017).

3.6 Main Conclusions and Outlook

In the tropical regions of America, dendroecological studies started only about 50 years ago, but have advanced exponentially during recent decades. Over this period annual tree-ring formation has been demonstrated in 220 tree species, covering a variety of biomes and ecosystems across the Neotropics. Our main findings are:

- External factors leading to the formation of annual growth rings in tropical tree species are rainfall seasonality, annual long-term flooding (flood-pulse), soil water salinity (mangroves) and with increasing latitude and altitude also photoperiod and temperature.
- Tree-ring analyses indicate maximum longevity of about 400–600 years for angiosperms in the Neotropics, similar to ages reported for angiosperms in other climates and much lower than maximum ages indicated by radiocarbon dating.
- At marginal sites characterized by extreme environmental conditions where forest establishment is often in the form of monodominant populations with low competition, we can expect maximum longevities, as reported for other climate zones.
- The growing literature of dendroecological applications in the tropics of America demonstrates the large variation of age and diameter growth among individual trees, species, different ecosystems, and ecoregions. There is an urgent need to extend this information to more tree species and sites.
- Empirical estimates derived from tree rings for species- and site-specific growth data, tree ages, age-size relationships and growth variation among individuals are fundamental to developing sustainable forest management plans and conserving Neotropical forests.

Our review highlights the potential for tree-ring analyses and dendroecological applications in the Neotropics and is intended to encourage scientists to strengthen their collaborations between research groups, institutions and countries, including the training of young scholars and capacity building, sharing their experiences and knowledge. Dendrochronology is now firmly established within the tropics and will contribute towards a better understanding of tropical forest dynamics by delivering empirical data on tree age, forest productivity, climate-growth relationships, disturbance histories, and indicators for sustainable management and conservation. These are vital for projecting the fate of tropical forest ecosystems in a world increasingly dominated and threatened by human activity.

References

Abrams MD, Hock WK (2006) Annual growth rings and the impact of Benlate 50 DF fungicide on citrus trees in seasonally dry tropical plantations of northern Costa Rica. For Ecol Manage 227:96–101. doi:10.1016/j.foreco.2006.02.019

Alves ES, Angyalossy-Alfonso V (2000) Ecological trends in the wood anatomy of some Brazilian species. 1. Growth rings and vessels. IAWA J 21(1):3–30. doi:10.1163/22941932-90000233

Anchukaitis KJ, Evans MN, Wheelwright NT et al (2008) Stable isotope chronology and climate signal calibration in neotropical montane cloud forest trees. J Geophys Res 113:G03030. doi:10.1029/2007JG000613

Andrade ES, Garcia SSC, Albernaz ALKM et al (2017) Growth ring analysis of *Euxylophora paraensis* through X-ray microdensitometry. Ciência Rural 47(04):e20150895. doi:10.1590/0103-8478cr20150895

Andrade VHF (2015) Modelos de crescimento para *Hymenaea courbaril* L. and *Handroanthus serratifolius* (Vahl) S.O. Grosse em floresta de terra firme utilizando análise de anéis de crescimento. Dissertation, Universidade Federal do Paraná

Andreu-Hayles L, Santos GM, Herrera-Ramírez DA et al (2015) Matching dendrochronological dates with the southern hemisphere ^{14}C bomb curve to confirm annual tree rings in *Pseudolmedia rigida* from Bolivia. Radiocarbon 57(1):1–13. doi:10.2458/azu_rc.57.18192

Anholetto Jr CR (2013) Dendroecologia e composição isotópica (δ^{13}C) dos anéis de crescimento de *Cedrela odorata*, Meliaceae, na Caatinga e Mata Atlântica do Estado de Sergipe, Brasil. Dissertation, Escola Superior de Agricultura Luiz de Queiroz

Argollo J, Solíz C, Villalba R (2004) Potencialidad dendrocronológica de *Polylepis tarapacana* en los Andes Centrales de Bolivia. Ecología en Bolivia 39(1):5–24

Assahira C, Piedade MTF, Trumbore SE et al (2017) Tree mortality of a flood-adapted species in response of hydrographic changes caused by an Amazonian river dam. For Ecol Manag. doi:10.1016/j.foreco.2017.04.016

Baas P, Vetter RE (eds) (1995) Growth rings in tropical trees. IAWA Bull 10:95–174. doi:10.1002/fedr.19901011103

Baker PJ, Bunyavejchewin S (2006) Suppression release and canopy recruitment in five tree species from a seasonal tropical forest in western Thailand. J Trop Ecol 22:521–529. doi:10.1017/S0266467406003312

Baker PJ, Bunyavejchewin S (2017) Complex historical disturbance regimes shape forest dynamics across a seasonal tropical landscape in western Thailand. In: Amoroso MM, Daniels LD, Baker PJ, Camarero JJ (eds) Dendroecology: tree-ring analyses applied to ecological studies. Springer, Cham

Baker PJ, Bunyavejchewin S, Oliver CD et al (2005) Disturbance history and historical stand dynamics of a seasonal tropical forest in Western Thailand. Ecol Monogr 75(3):317–343. doi:10.1890/04-0488

Baker JCA, Hunt SFP, Clerici SJ et al (2015) Oxygen isotopes in tree rings show good coherence between species and sites in Bolivia. Glob Planet Chang 133:298–308. doi:10.1016/j.gloplacha.2015.09.008

Ballantyne AP, Baker PA, Chambers JQ et al (2011) Regional differences in South American monsoon precipitation inferred from the growth and isotopic composition of tropical trees. Earth Interact 15:1–35. doi:10.1175/2010EI277.1

Barbosa ACMC, Stahle DW, Martins TV et al (2013) Dendroclimatology on the São Francisco basin in Minas Gerais, Brazil. In: Abstracts of the International Workshop of Tropical Dendrochronology 2013, Universidade Federal de Lavras, 23–28 September 2013

Berlage HP (1931) Over het verband tusschen de dikte der jaarringen van djatiboomen (*Tectona grandis* L. f.) en den regenval op Java. Tectona 24:939–953

Biondi F (2001) A 400-year tree-ring chronology from the tropical treeline of North America. Ambio 30(3):162–166. doi:10.1639/0044-7447(2001)030[0162:aytrcf]2.0.co;2

Biondi F, Estrada IG, Ruiz JCG et al (2003) Tree growth response to the 1913 eruption of Volcán de Fuego de Colima, Mexico. Quat Res 59:293–299. doi:10.1016/S0033-5894(03)00034-6

Blagitz M, Botosso PC, Bianchini E et al (2016) Periodicidade do crescimento de espécies arbóreas da Floresta Estacional Semidecidual no Sul do Brasil. Sci For 44:163–173. doi:10.18671/scifor.v44n109.16

Boltz F, Carter DR, Holmes TP et al (2011) Financial returns under uncertainty for conventional and reduced impact logging in permanent production forests of the Brazilian Amazon. Ecol Econ 39:387–398. doi:10.1016/s0921-8009(01)00231-2

Borchert R (1999) Climatic periodicity, phenology, and cambium activity in tropical dry forest trees. IAWA J 20:239–247. doi:10.1163/22941932-90000687

Bormann FH, Berlyn G (eds) (1981) Age and Growth Rate of Tropical Trees. New Directions for Research. Yale Univ Sch For Eviron Stud Bull 94:1–137

Botosso PC, Vetter RE (1991) Alguns aspectos sobre a periodicidade e taxa de crescimento em 8 espécies arbóreas tropicais de floresta de terra firme (Amazônia). Rev Inst Flor 3(2):163–180

Botosso PC, Vetter RE, Tomazello Filho M (2000) Periodicidade e taxa de crescimento de árvores de cedro (*Cedrela odorata* L., Meliaceae), jacareúba (*Calophyllum angulare* A. C. Smith, Clusiaceae) e muirapiranga (*Eperua bijuga* Mart. Ex Benth, Leg. Caesalpinioideae) de floresta de Terra Firme, em Manaus-AM. In: Roig FA (ed) Dendrocronología en América Latina, EDIUNC, Mendoza, p 357-380

Brandes AFN, Albuquerque RP, Moraes LFD et al (2016) Annual tree rings in *Piptadenia gonoacantha* (Mart.) J.F.Macbr. in a restoration experiment in the Atlantic Forest: potential for dendroecological research. Acta Bot Bras 30(3):383–388. doi:10.1590/0102-33062016abb0101

Brandis D (1856) Report on the Teak Forests of Pegu. A Memorandum on the Teak in the Tharawaddy Forests. Printed by George Edward Eyre and William Spottiswoode, London

Brandis D (1879) Memorandum of the rate of growth of teak. Indian Forester IV:221–225

Bräuning A (2011) Editorial note for the special issue on 'Tropical Dendroecology'. Trees 25:1–2. doi:10.1007/s00468-010-0530-x

Bräuning A, Volland-Voigt F, Burchardt I et al (2009) Climatic control of radial growth of *Cedrela montana* in a humid mountain rainforest in southern Ecuador. Erdkunde 63(4):337–345. doi:10.3112/erdkunde.2009.04.04

Breitsprecher A, Bethel JS (1990) Stem-growth periodicity of trees in a tropical wet forests of Costa Rica. Ecology 71:1156–1164. doi:10.2307/1937383

Briceño A, Rangel J, Bogino S (2016) Estudio de los anillos de crecimiento de *Cordia alliodora* (Ruiz & Pav.) Oken en Colombia. Colombia Forestal 19(2):219–232. doi:10.14483/udistrital.jour.colomb.for.2016.2.a07

Brienen RJW, Zuidema PA (2005) Relating tree growth to rainfall in Bolivian rain forests: a test for six species using tree ring analysis. Oecologia 146:1–12. doi:10.1007/s00442-005-0160-y

Brienen RJW, Zuidema PA (2006a) Lifetime growth patterns and ages of Bolivian rain forest trees obtained by tree ring analysis. J Ecol 94:481–493. doi:10.1111/j.1365-2745.2005.01080.x

Brienen RJW, Zuidema PA (2006b) The use of tree rings in tropical forest management: projecting timber yields of four Bolivian tree species. For Ecol Manage 226:256–267. doi:10.1016/j.foreco.2006.01.038

Brienen RJW, Zuidema PA (2007) Incorporating persistent tree growth differences increases estimates of tropical timber yield. Front Ecol Environ 5:302–306. doi:10.1890/1540-9295(2007)5[302:rcptgd]2.0.co;2

Brienen RJW, Zuidema PA, During HJ (2006) Autocorrelated growth of tropical forest trees: unraveling patterns and quantifying consequences. For Ecol Manag 237:179–190. doi:10.1016/j.foreco.2006.09.042

Brienen RJW, Lebrija-Trejos E, van Breugel M et al (2009) The potential of tree rings for the study of forest succession in southern Mexico. Biotropica 41:186–195. doi:10.1111/j.1744-7429.2008.00462.x

Brienen RJW, Lebrija-Trejos E, Zuidema PA et al (2010a) Climate-growth analysis for a Mexican dry forest tree shows strong impact of sea surface temperatures and predicts future growth declines. Glob Chang Biol 16:2001–2012. doi:10.1111/j.1365-2486.2009.02059.x

Brienen RJW, Lebrija-Trejos E, Zuidema PA et al (2010b) Attaining the canopy in dry and moist tropical forests: strong differences in tree growth trajectories reflect variation in growing conditions. Oecologia 163:485–496. doi:10.1007/s00442-009-1540-5

Brienen RJW, Wanek W, Hietz P (2011) Stable carbon isotopes in tree rings indicate improved water use efficiency and drought responses of a tropical dry forest tree species. Trees 25:103–113. doi:10.1007/s00468-010-0474-1

Brienen RJW, Helle G, Pons TL et al (2012) Oxygen isotopes in tree rings are a good proxy for Amazon precipitation and El Niño-Southern Oscillation variability. PNAS 109(42):16957–16962. doi:10.1073/pnas.1205977109

Brienen RJW, Hietz P, Wanek W et al (2013) Oxygen isotopes in tree rings record variation in precipitation $\delta^{18}O$ and amount effects in the south of Mexico. J Geophys Res Biogeosci 118:1604–1615. doi:10.1002/2013jg002304

Brienen RJW, Schöngart J, Zuidema PA (2016) Tree rings in the tropics: insights into the ecology and climate sensitivity of tropical trees. In: Goldstein G, Santiago LS (eds) Tropical tree physiology. Adaptations and responses in a changing environment. Springer International Publishing, Switzerland, pp 439–461

Brown PM (1996) OLDLIST: a database of maximum tree ages. In: Dean JS, Meko DM, Swetnam YW (eds) Tree rings, environment, and humanity. Proceedings of the International Conference, Tucson, Arizona, 17–21 May, 1994. Radiocarbon 1996, pp 727–731

Bruenig EF (1998) Conservation and management of tropical rainforests: an integrated approach to sustainability, 2nd edn. CAB International, Wallingford

Bullock SH (1997) Effects of seasonal rainfall on radial growth in two tropical tree species. Int J Biometeorol 41:13–16. doi:10.1007/s004840050047

Callado CH, Neto SJS, Scarano FR et al (2001) Periodicity of growth rings in some flood-prone trees of the Atlantic Rain Forest in Rio de Janeiro, Brazil. Trees 15:492–497. doi:10.1007/s00468-001-0128-4

Callado CH, Roig FA, Tomazello Filho M et al (2013) Cambial growth periodicity studies of South American woody species—a review. IAWA J 34(3):213–230. doi:10.1163/22941932-00000019

Cardoso DS (2014) Caracterização anatômica da madeira e potencial dendrocronológico de *Schinopsis brasiliensis* Engl. (Anacardiaceae) na caatinga sergipana. Dissertation, Universidade Federal de Sergipe

Carmo JF (2016) Crescimento e propriedades da madeira de *Copaifera langsdorffii* Desf. sob regime de manejo florestal. Disseration, Universidade Federal Rural do Rio de Janeiro

Chagas MP (2009) Caracterização dos anéis de crescimento e dendrocronologia de árvores de *Grevillea robusta* A. Cunn, *Hovenia dulcis* Thunb., *Persea americana* Mill., *Tabebuia pentaphylla* Hemsl. e *Terminalia catappa* L. nos municípios de Piracicaba e Paulínia, SP. Dissertation, Escola Superior de Agricultura Luiz de Queiroz

Chambers JQ, Higuchi N, Schimmel JP (1998) Ancient trees in Amazonia. Nature 391:135–136. doi:10.1038/34325

Christie DA, Lara A, Barichivich J et al (2009) El Niño-Southern Oscillation signal in the world's highest-elevation tree-ring chronologies from the Altiplano, Central Andes. Palaeogeogr Palaeocl 281:309–319. doi:10.1016/j.palaeo.2007.11.013

Cintra BBL, Schietti J, Emillio T et al (2013) Soil physical restrictions and hydrology regulate stand age and wood biomass turnover rates of Purus–Madeira interfluvial wetlands in Amazonia. Biogeosciences 10:7759–7774. doi:10.5194/bg-10-7759-2013

Costa WS (2014) Anatomia da madeira e dendrocronologia de espécies arbóreas do gênero *Machaerium* Pers. (Leguminosae-Papilionoideae). Dissertation, Universidade do Estado do Rio De Janeiro

Costa MS, Ferreira KEB, Botosso PC et al (2015) Growth analysis of five Leguminosae native tree species from a seasonal semidecidual lowland forest in Brazil. Dendrochronologia 36:23–32. doi:10.1016/j.dendro.2015.08.004

Coster C (1927) Zur Anatomie und Physiologie der Zuwachszonen und Jahresringbildung in den Tropen. Ann Jard Bot Buitenzorg 37:49–160

Coster C (1928) Zur Anatomie und Physiologie der Zuwachszonen und Jahresringbildung in den Tropen. Ann Jard Bot Buitenzorg 38:1–114

Cunha TA, Finger CAG, Hasenauer H (2016) Tree basal area increment models for *Cedrela*, *Amburana*, *Copaifera* and *Swietenia* growing in the Amazon rain forests. For Ecol Manage 304:174–183. doi:10.1016/j.foreco.2015.12.031

Da Fonseca SF, Piedade MTF, Schöngart J (2009) Wood growth of *Tabebuia barbata* (E. Mey.) Sandwith (Bignoniaceae) and *Vatairea guianensis* Aubl. (Fabaceae) in Central Amazonian black-water (igapó) and white-water (várzea) floodplain forests. Trees 23:127–134. doi:10.1007/s00468-008-0261-4

Davidson EA, Araújo AC, Artaxo P et al (2012) The Amazon basin in transition. Nature 481:321–328. doi:10.1038/nature10717

Dawkins HC, Philip MS (1998) Tropical Moist forest silviculture and management: a history of success and failure. CAB International, Wallingford

De Ridder M, van den Bulcke J, van Acker J et al (2013) Tree-ring analysis of an African long-lived pioneer species as a tool for sustainable forest management. For Ecol Manage 304:417–426. doi:10.1016/j.foreco.2013.05.007

Détienne P (1995) Nature et périodicité des cernes dans quelques bois Guaynais. Bois et Forêts des Tropiques 243:65–75

Détienne P, Barbier C, Ayphassorho H et al (1988) Rythmes de croissance de quelques essences de Guayne Française. Bois et Forêts des Tropiques 217:63–76

Devall MS, Parresol BR, Wright SJ (1995) Dendroecological analysis of *Cordia alliodora*, *Pseudobombax septenatum* and *Annona spraguei* in central Panama. IAWA J 16(4):411–424. doi:10.1163/22941932-90001430

Dezzeo N, Worbes M, Ishii I et al (2003) Growth rings analysis of four tropical tree species in seasonally flooded forest of the Mapire River, a tributary of the lower Orinoko River, Venezuela. Plant Ecol 168:165–175. doi:10.1023/A:1024417610776

Di Filippo A, Pederson N, Baliva M et al (2015) The longevity of broadleaf deciduous trees in Northern Hemisphere temperate forests: insights from tree-ring series. Front Ecol Evol 3:46. doi:10.3389/fevo.2015.00046

Dünisch O, Bauch J, Sack M et al (1999) Growth dynamics in wood formation of plantation-grown *Swietenia macrophylla* King and *Carapa guianensis* Aubl. Mitt Bundesforschungsanstalt Forst- und Holzwirtschaft, Hamburg 193:79–96

Dünisch O, Montóia VR, Bauch J (2003) Dendroecological investigations on *Swietenia macrophylla* King and *Cedrela odorata* L. (Meliaceae) in the central Amazon. Trees 17(3):244–250. doi:10.1007/s00468-002-0230-2

Dykstra JA, Heinrich R (1996) FAO model code of forest harvesting practice. Food and Agriculture Organization of the United Nations (FAO), Rome

Eckstein D, Sass U, Baas P (eds) (1995) Growth periodicity in tropical trees: proceedings of an international meeting held in Kuala Lumpur, Malaysia, 16–18 November, 1994. IAWA J 16(4):323–442

Edwards DP, Tobias JA, Sheil D et al (2014) Maintaining ecosystem function and services in logged tropical forests. TREE 29(9):511–520. doi:10.1016/j.tree.2014.07.003

Enquist BJ, Leffler AJ (2001) Long-term tree ring chronologies from sympatric tropical dry-forest trees: individualistic responses to climatic variation. J Trop Ecol 17:41–60. doi:10.1017/s0266467401001031

Espinoza MJP, Guillen GJI, Morales MS et al (2014) Potencialidad de *Cedrela odorata* (Meliaceae) para estúdios dendrocronológicos en la selva central del Perú. Rev Biol Trop 62(2):783–793. doi:10.15517/rbt.v62i2.9835

Estrada GCD, Callado CH, Soares MLG et al (2008) Annual growth rings in the mangrove *Laguncularia racemosa* (Combretaceae). Trees 22:663–670. doi:10.1007/s00468-008-0224-9

Ferreira GB (2012) Análise dendroclimatológica do cedro (*Cedrela fissilis* L.) para reconstrução do cenário ambiental recente da cidade de São Paulo (SP). Dissertation, Univeridade de São Paulo

Ferreira-Fedele L, Tomazello Filho M, Botosso PC et al (2004) Periodicidade do crescimento de *Esenbeckia leiocarpa* Engl. (guarantã) em duas áreas da região Sudeste do Estado de São Paulo. Scientia Florestalis 65:141–149

Fichtler E, Clark DA, Worbes M (2003) Age and long-term growth of trees in an old-growth tropical rain forest, based on analyses of tree rings and ^{14}C. Biotropica 35:306–317. doi:10.1646/03027

Fichtler E, Helle G, Worbes M (2010) Stable-carbon isotope time series from tropical tree rings indicate a precipitation signal. Tree-Ring Res 66(1):35–49. doi:10.3959/2008-20.1

Fortes CF (2006) Estudos Dendrocronológicos da Espécie arbórea *Vochysia divergens* Pohl (Vochysiaceae) no Pantanal Norte Matogrossense. Dissertation Universidade Federal de Mato Grosso, Brasil

Free CM, Landis RM, Grogan J et al (2014) Management implications of long-term tree growth and mortality rates: a modeling study of big-leaf mahogany (*Swietenia macrophylla*) in the Brazilian Amazon. For Ecol Manag 330:46–54. doi:10.1016/j.foreco.2014.05.057

Free CM, Grogan J, Schulze MD et al (2017) Current Brazilian forest management guidelines are unsustainable for *Swietenia*, *Cedrela*, *Amburana*, and *Copaifera*: a response to da Cunha and colleagues. For Ecol Manag 386:81–83. doi:10.1016/j.foreco.2016.09.031

Fritts HC (1976) Tree rings and climate. Academic, London

Funck J (2004) Untersuchungen zur Wachstumsdynamik von *Cariniana micrantha* (Ducke) in der Nähe von Itacoatiara/Amazonien. Albert-Ludwigs Universität Freiburg, Dissertation

Gamble JS (1881) A Manual of Indian timbers. An account of the Structure, growth, distribution, and qualities of Indian woods, 1st edn. Office of the Superintendent of Government Printing, Calcutta

Geiger F (1915) Anatomische Untersuchungen über die Jahresbildung von *Tectona grandis*. Jahrbücher für wissenschaftliche Botanik 55:521–607

Gerwing JJ (2002) Degradation of forests through logging and fire in the eastern Brazilian Amazon. For Ecol Manag 157:131–141. doi:10.1016/s0378-1127(00)00644-7

Giraldo VD, del Valle JI (2012) Modelación del crecimiento de *Albizia niopoides* (Mimosaceae) por métodos dendrocronológicos. Rev Biol Trop 60(3):1117–1136. doi:10.15517/rbt.v60i3.1762

Grogan J, Schulze M (2011) The impact of annual and seasonal rainfall patterns on growth and phenology of emergent tree species in Southeastern Amazonia, Brazil. Biotropica 44:332–340. doi:10.1111/j.1744-7429.2011.00825.x

Grogan J, Landis RM, Free CM et al (2014) Bigleaf mahogany *Swietenia macrophylla* population dynamics and implications for sustainable management. J Appl Ecol 51(3):664–674. doi:10.1111/1365-2664.12210

Gutiérrez LAB, Ramos GMV (2013) Anatomía de anillos de crecimiento de 80 especies arbóreas potenciales para estudios dendrocronológicos en la Selva Central, Perú. Rev Biol Trop 61(3):1025–1037. doi:10.15517/rbt.v61i3.11778

García-Cervigón AI, Camarero JJ, Espinosa CI (2017) Intra-annual stem increment patterns and climatic responses in five tree species from an Ecuadorian tropical dry forest. Trees 31:1057–1067. doi:10.1007/s00468-017-1530-x

Hallé F, Oldeman RAA, Tomlinson PB (1978) Tropical trees and forests. An architectural analysis. Springer, New York

Hastenrath S (1963) Dendrochronologie in El Salvador. Meteorol Rundsch 4:110–113

Hayden B (2008) Annual growth rings in dry tropical forest trees. Dissertation, Concordia University

Herrera DA, del Valle JI (2011) Reconstrucción de los niveles del Río Atrato con anillos de crecimiento de *Prioria copaifera*. Dyna 78:121–130

Holmes TP, Blate GM, Zweede JC et al (2002) Financial and ecological indicators of reduced impact logging performance in the eastern Amazon. For Ecol Manag 163:93–110. doi:10.1016/s0378-1127(01)00530-8

Hua Q, Barbetti M, Worbes M et al (1999) Review of radiocarbon data from atmospheric and tree ring samples for the period 1950–1977 AD. IAWA J 20(3):261–284. doi:10.1163/22941932-90000690

Hua Q, Barbetti M, Rakowski AZ (2013) Atmospheric radiocarbon for the period 1950–2010. Radiocarbon 55:2059–2072. doi:10.2458/azu_js_rc.v55i2.16177

Huante P, Rincón E, Swetnam TW (1991) Dendrochronology of *Abies religiosa* in Michoaca, Mexico. Tree-Ring Bull 51:15–28

Inga JG, del Valle JI (2017) Log-relative growth: a new dendrochronological approach to study diameter growth in *Cedrela odorata* and *Juglans neotropica*, Central Forest, Peru. Dendrochronologia 44:117–129. doi:10.1016/j.dendro.2017.03.009

Ishii IH (1998) Estudos dendrocronológicos e determinação de idade de árvores das matas ciliares do Pantanal Sul Mato-grossense. Dissertation, Universidade de São Carlos

Jenkins HS (2009) Amazon climate reconstruction using growth rates and stable isotopes of tree ring cellulose from the Madre de Dios Basin, Peru. Dissertation, Duke University

Jiménez JAG, del Valle JIA (2011) Estudio del crecimiento de *Prioria copaifera* (Caesalpinaceae) mediante técnicas dendrocronológicas. Rev Biol Trop 59(4):1813–1831

Johns JS, Baretto P, Uhl C (1996) Logging damage during planned and unplanned logging operations in the eastern Amazon. For Ecol Manag 89:59–77. doi:10.1016/s0378-1127(96)03869-8
Jomelli V, Pavlova I, Guin O et al (2012) Analysis of the dendroclimatic potential of *Polylepis pepei*, *P. subsericans* and *P. rugulosa* In the tropical Andes (Peru-Bolivia). Tree-Ring Res 68(2):91–103. doi:10.3959/2011-10.1
Junk WJ, Barley PB, Sparks RE (1989) The flood-pulse concept in river-floodplain systems. Can Spec Publ Fish Aquat Sci 106:110–127
Junk WJ, Wittmann F, Schöngart J et al (2015) A classification of the major habitats of Amazonian black-water river floodplains and a comparison with their white-water counterparts. Wetl Ecol Manag 23:677–693. doi:10.1007/s11273-015-9412-8
Lamprecht H (1989) Silviculture in the tropics: tropical forest ecosystems and their tree species—possibilities and methods for their long-term utilization. GTZ, Eschborn
Lara A (1991) The dynamics and disturbance regimes of *Fitzroya cupressoides* forests in the south-central Andes of Chile. Dissertation, University of Colorado
Larson DW, Matthes U, Gerrath JA et al (2000) Evidence for the widespread occurrence of ancient forests on cliffs. J Biogeogr 27:319–331. doi:10.1046/j.1365-2699.2000.00401.x
Latorraca JVC, Souza MT, Silva LDSAB et al (2015) Dendrocronologia de árvores de *Schizolobium parahyba* (Vell.) S. F. Blake de ocorrência na Rebio de Tinguá-RJ. Rev Árvore 39(2):385–394. doi:10.1590/0100-67622015000200018
Laurance WF, Nascimento HEM, Laurance SG et al (2004) Inferred longevity of Amazonian rainforest trees based on a long-term demographic study. For Ecol Manage 190:131–143. doi:10.1016/j.foreco.2003.09.011
Leite PTM (2012) Dendroecologia de *Tabebuia aurea* (Manso) Benth & Hook e *Tabebuia heptaphylla* (Vell.) Toledo (Bignoniaceae) no Pantanal de Mato Grosso, Brasil. Dissertation, Universidade Federal do Mato Grosso
Leoni JM, da Fonseca Jr SF, Schöngart J (2011) Growth and population structure of the tree species *Malouetia tamaquarina* (Aubl.) (Apocynaceae) in the central Amazonian floodplain forests and their implication for management. For Ecol Manage 261:62–67. doi:10.1016/j.foreco.2010.09.025
Lieberman M, Lieberman D (1985) Simulation of growth curves from periodic increment data. Ecology 66:632–635. doi:10.2307/1940415
Lisi CL, Pessenda LCR, Filho T et al (2001) ^{14}C bomb effects in tree rings of tropical and subtropical species of Brazil. Tree-Ring Res 57(2):191–196
Lisi CL, Tomazello Filho M, Botosso PC et al (2008) Tree-ring formation, radial increment periodicity, and phenology of tree species from a seasonal semi-deciduous forest in Southeast Brazil. IAWA J 29(2):189–207. doi:10.1163/22941932-90000179
Lobão MS (2011) Dendrocronologia, fenologia, atividade cambial e qualidade do lenho de árvores de *Cedrela odorata* L., *Cedrela fissilis* Vell. e *Schizolobium parahyba* var. *amazonicum* Hub ex Ducke, no Estado do Acre, Brasil. Dissertation, Escola Superior de Agricultura Luiz de Queiroz
Locosselli GM, Buckeridge MS, Moreira MZ et al (2013) A multi-proxy dendroecological analysis of two tropical species (*Hymenaea* spp., Leguminosae) growing in a vegetation mosaic. Trees 27:25–36. doi:10.1007/s00468-012-0764-x
Locosselli GM, Cardim RH, Ceccantini G (2016a) Rock outcrops reduce temperature-induced stress for tropical conifer by decoupling regional climate in the semiarid environment. Int J Biometeorol 60:639–649. doi:10.1007/s00484-015-1058-y
Locosselli GM, Schöngart J, Ceccantini G (2016b) Climate/growth relations and teleconnections for a *Hymenaea courbaril* (Leguminosae) population inhabiting the dry forest on karst. Trees 30:1127–1136. doi:10.1007/s00468-015-1351-8
Locosselli GM, Krottenthaler S, Pitsch P et al (2017) Age and growth rate of congeneric tree species (*Hymenaea* spp.-Leguminosae) inhabiting different tropical biomes. Erdkunde 71:45–57. doi:10.3112/erdkunde.2017.01.03
Loehle C (1988) Tree life history strategies: the role of defenses. Can J For Res 18:209–222. doi:10.1139/x88-032

López L, Villalba R (2011) Climate influences on the radial growth of *Centrolobium microchaete*, a valuable timber species from the tropical dry forests in Bolivia. Biotropica 43:41–49. doi:10.1111/j.1744-7429.2010.00653.x

López L, Villalba R (2016a) An assessment of *Schinopsis brasiliensis* Engler (Anacardiaceae) for dendroclimatological applications in the tropical Cerrado and Chaco forests, Bolivia. Dendrochronologia 40:85–92. doi:10.1016/j.dendro.2016.07.002

López L, Villalba R (2016b) Reliable estimate of radial growth for eight tropical tree species based on wood anatomical patterns. J Trop For Sci 28(2):139–152

López BC, Sabaté S, Gracia CA et al (2005) Wood anatomy, description of annual rings, and responses to ENSO events of *Prosopis pallida* H.B.K., a wide-spread woody plant of arid and semi-arid lands of Latin America. J Arid Environ 61:541–554. doi:10.1016/j.jaridenv.2004.10.008

López BC, Rodríguez R, Gracia CA et al (2006) Climatic signals in growth and its relation to ENSO events of two *Prosopis* species following a latitudinal gradient in South America. Glob Chang Biol 12:897–906. doi:10.1111/j.1365-2486.2006.01138.x

López L, Villalba R, Peña-Claros M (2012) Determining the annual periodicity of growth rings in seven tree species of a tropical moist forest in Santa Cruz, Bolivia. For Syst 21(3):508–514. doi:10.5424/fs/2012213-02966

López L, Villalba R, Bravo F (2013) Cumulative diameter growth and biological rotation age for seven tree species in the Cerrado biogeographical province of Bolivia. For Ecol Manage 292:49–55. doi:10.1016/j.foreco.2012.12.011

Lusk CH (1999) Long-lived light-demanding emergents in southern temperate forests: the case of *Weinmannia trichosperma* (Cunoniaceae) in Chile. Plant Ecol 140(1):111–115. doi:10.1023/A:1009764705942

Marcati CR, Angyalossy V, Evert RF (2006) Seasonal variation in wood formation of *Cedrela fissilis* (Meliaceae). IAWA J 27(2):199–211. doi:10.1163/22941932-90000149

Marcati CR, Milanez CRD, Machado SR (2008) Seasonal development of secondary xylem and phloem in *Schizolobium parahyba* (Vell.) Blake (Leguminosae: Caesalpinioideae). Trees 22:3–12. doi:10.1007/s00468-007-0173-8

Marcati C, Machado SR, Podadera D et al (2016) Cambial activity in dry and rainy season on branches from woody species growing in Brazilian Cerrado. Flora 223:1–10. doi:10.1016/j.flora.2016.04.008

Maria VRP (2002) Estudo da periodicidade do crescimento, fenologia, e relação com a atividade cambial de espécies arbóreas tropicais de florestas estacionais semideciduais. Dissertation, Escola Superior de Agricultura Luiz de Queiroz

Mariaux A (1967) Les cernes dans les bois tropicaux africains, nature et périodicité. Bois et Forêts des Tropiques 113:3–14

Martínez-Ramos M, Alvarez-Buylla ER (1998) How old are tropical rainforest trees? Trends Plant Sci 3(10):400–405. doi:10.1016/S1360-1385(98)01313-2

Mattos PP (1998) Identificação de anéis anuais de crescimento e estimativa de idade e incremento anual em diâmetro de espécies nativas do Pantanal da Nhecolândia, MS. Dissertation, Universidade Federal do Paraná

Mattos PP, Braz EM, Domene VD et al (2015) Climate-tree growth relationships of *Mimosa tenuiflora* in seasonally dry tropical forests, Brazil. Cerne 21(1):141–149. doi:10.1590/01047760201521011460

Medeiros RS (2016) Estudo da anatomia do lenho e dendrocronologia de árvores de *Copaifera multijuga* Hayne na Amazonia brasileira e sua relacao com o manejo e extracao de oleorresina. Dissertation, Escola Superior de Agricultura Luiz de Queiroz

Mehlig U, Menezes MPM, Reise A et al (2010) Mangrove vegetation of the Caeté Estuary. In: Saint-Paul U, Schneider H (eds) Mangrove dynamics and management in North Brazil, Ecological Studies, vol 211. Springer, Heidelberg, Dordrecht, London, New York, pp 71–107

Mendivelso HA, Camarero JJ, Obregón OR et al (2013) Differential growth responses to water balance of coexisting deciduous tree species are linked to wood density in a Bolivian tropical dry forest. Plos One 8:e73855. doi:10.1371/journal.pone.0073855

Menezes M, Berger U, Worbes M (2003) Annual growth rings and long-term growth patterns of mangrove trees from the Braganca Peninsula, Northern Brazil. Wetl Ecol Manag 11:233–242. doi:10.1023/A:1025059315146

Mohr A (2013) Ecologia populacional de *Qualea ingens* Warm. e *Ruizterania wittrockii* (Malme) Marc.-Berti (Vochysiaceae) na Região do Médio Araguaia, Mato Grosso. Dissertation, Universidade do Estado de Mato Grosso

Montanher DR (2012) Aspectos dendrocronológicos de três espécies arbóreas nas áreas alagáveis do alto Rio Paraná. Dissertation, Universidade Estadual de Maringá

Morales MS, Villalba R, Grau HR et al (2004) Rainfall-controlled tree growth in high-elevation subtropical treelines. Ecology 85(11):3080–3089. doi:10.1890/04-0139

Morel H, Mangenet T, Beauchêne J et al (2015) Seasonal variations in phenological traits: leaf shedding and cambial activity in *Parkia nitida* Miq. and *Parkia velutina* Benoist (Fabaceae) in tropical rainforest. Trees 29:973–984. doi:10.1007/s00468-015-1177-4

Moya J, Lara A (2011) Cronologías de ancho de anillos de queñoa (*Polylepis tarapacana*) para los últimos 500 años en el Altiplano de la región de Arica y Parinacota, Chile. Bosque 32(2):165–173. doi:10.4067/s0717-92002011000200007

Nascimento MBG (2013) Anéis de crescimento, incremento em circunferência do tronco e anatomia da madeira de espécies arbóreas da Floresta Estacional Semidecidual do Sul do Brasil. Dissertation, Universidade Estadual de Londrina

Nebel G, Dragsted J, Simonsen TR et al (2001) The Amazon flood plain forest tree *Maquira coriacea* (Karsten) CC Berg: aspects of ecology and management. For Ecol Manag 150:103–113. doi:10.1016/s0378-1127(00)00684-8

Nogueira Jr FC (2011) Estrutura e composição de uma vegetação ripária. Relações dendrocronológicos e climáticos na Serra dos Macacos em Tobias Barreto, Sergipe-Brasil. Dissertation, Universidade Federal do Sergipe

Nydal R, Lövseth K (1983) Tracing bomb ^{14}C in the atmosphere 1962-1980. J Geophys Res 88:3621–3642. doi:10.1029/jc088ic06p03621

Ohashi S, Durgante FM, Kagawa A et al (2016) Seasonal variations in the stable oxygen isotope ratio of wood cellulose reveal annual rings of trees in a Central Amazon terra firme forest. Oecologia 180:685–696. doi:10.1007/s00442-015-3509-x

Oliveira MF (2014) Critérios para o manejo sustentável de duas espécies madeireiras das florestas tropicais do Mato Grosso. Dissertation, Universidade Federal do Paraná

Oliveira JM, Santarosa E, Pillar DP et al (2009) Seasonal cambium activity in the subtropical rain forest tree *Araucaria angustifolia*. Trees 23:107–115. doi:10.1007/s00468-008-0259-y

Oliveira JM, Roig FA, Pillar DP (2010) Climatic signals in tree-rings of *Araucaria angustifolia* in the southern Brazilian highlands. Austral Ecol 35:134–147. doi:10.1111/j.1442-9993.2009.02018.x

Pagotto MA (2015) A vegetação lenhosa da caatinga em assentamento do estado de Sergipe: aspectos fitossociológicos, anatômicos e dendrocronológicos. Dissertation, Universidade Federal do Sergipe

Pagotto MA, Roig FA, Ribeiro AS et al (2015) Influence of regional rainfall and Atlantic sea surface temperature on tree-ring growth of *Poincianella pyramidalis*, semiarid forest from Brazil. Dendrochronologia 35:14–23. doi:10.1016/j.dendro.2015.05.007

Paredes-Villanueva K, Sánchez-Salguero R, Manzanedo RD et al (2013) Growth rate and climatic response of *Machaerium scleroxylon* in a dry tropical forest in southeastern Santa Cruz, Bolivia. Tree-Ring Res 69:63–79. doi:10.3959/1536-1098-69.2.63

Paredes-Villanueva K, López L, Brookhouse M et al (2015) Rainfall and temperature variability in Bolivia derived from the tree-ring width of *Amburana cearensis* (Fr. Allem.) A.C. Smith. Dendrochronologia 35:80–86. doi:10.1016/j.dendro.2015.04.003

Paredes-Villanueva K, López L, Cerrillo RMN (2016) Regional chronologies of *Cedrela fissilis* and *Cedrela angustifolia* in three forest types and their relation to climate. Trees 30:1581–1593. doi:10.1007/s00468-016-1391-8

Pereira GA, Barbosa ACMC, Stahle DW et al (2016) Climate-sensitive chronology of *Cedrela fissilis* from middle São Francisco basin, eastern Brazil. In: Abstracts of the Third American Dendrochronology Conference 2016, 28 March–1 April 2016, Mendoza

Petrillo M, Cherubini P, Fravolini G et al (2016) Time since death and decay rate constants of Norway spuce and European larch deadwood in subalpine forests determined using dendrochronology and radiocarbon dating. Biogeosciences 13:1537–1552. doi:10.5194/bg-13-1537-2016

Pinto LAA (2012) Estudos dendrocronológicos da *Manilkara huberi* (Sapotaceae) em uma floresta de terra firme da Amazônia central. Dissertation, Instituto Nacional de Pesquisas da Amazônia/Universidade do Estado do Amazonas

Pucha-Cofrep D, Peters T, Bräuning A (2015) Wet season precipitation during the past century reconstructed from tree-rings of a tropical dry forest in Southern Ecuador. Glob Planet Chang 133:65–78. doi:10.1016/j.gloplacha.2015.08.003

Putz FE, Sist P, Fredericksen T et al (2008) Reduced-impact logging: challenges and opportunities. For Ecol Manag 256:1427–1433. doi:10.1016/j.foreco.2008.03.036

Ramírez Correa JA, Grajales ECM, Escobar MB (2010) Anillos anuales y clima en *Rhizophora mangle* L. de la Bahía de Cispatá, Colombia. Rev Fac Nac Agron Medellín 63(2):5639–5650

Ramírez JA, del Valle JI (2011) Paleoclima de La Guajira, Colombia; según los anillos de crecimiento de *Capparis odoratissima* (Capparidaceae). Rev Biol Trop 59(3):1389–1405. doi:10.15517/rbt.v0i0.3406

Ramírez JA, del Valle JI (2012) Local and global climate signals from tree rings of *Parkinsonia praecox* in La Guajira, Colombia. Int J Climatol 32(7):1077–1082. doi:10.1002/joc.2335

Rigozzo NR, Liso CS, Tomazello Filho M et al (2012) Solar-terrestrial signal record in tree ring width time series from Brazil. Pure Appl Geophys 169:2181–2191. doi:10.1007/s00024-012-0480-x

Robert EMR, Schmitz N, Okelle JA et al (2011) Mangrove growth rings: fact or fiction? Trees 25:49–58. doi:10.1007/s00468-010-0487-9

Rodriguez R, Mabres A, Luckman B et al (2005) "El Niño" events recorded in dry-forest species of the lowlands of northwest Peru. Dendrochronologia 22:181–186. doi:10.1016/j.dendro.2005.05.002

Roig FA, Jimenez Osornio JJ, Villanueva Diaz J et al (2005) Anatomy of growth rings at the Yucatán Peninsula. Dendrochronologia 22:187–193. doi:10.1016/j.dendro.2005.05.007

Rosa SA (2008) Modelos de crescimento de quatro espécies madeireiras de floresta de várzea alta da Amazônia Central por meio de métodos dendrocronológicos. Dissertation, Instituto Nacional de Pesquisas da Amazônia, Universidade Federal do Amazonas

Rosa SA, Barbosa ACMC, Junk WJ et al (2017) Growth models based on tree-ring data for the Neotropical tree species *Calophyllum brasiliense* across different Brazilian wetlands: implications for conservation and management. Trees 31:729–742. doi:10.1007/s00468-016-1503-5

Rosero Alvorado J (2009) Dendrocronologia de árvores de mogno, *Swietenia macrophylla* King., Meliaceae, ocorrentes na floresta tropical Amazônica do Departamento de Madre de Dios, Peru. Dissertation, Escola Superior de Agricultura Luiz de Queiroz

Rozendaal DMA, Zuidema PA (2011) Dendroecology in the tropics: a review. Trees 25:3–16. doi:10.1007/s00468-010-0480-3

Rozendaal DMA, Soliz-Gamboa CC, Zuidema PA (2010) Timber yield projections for tropical tree species: the influence of fast juvenile growth on timber volume recovery. For Ecol Manage 259:2292–2300. doi:10.1016/j.foreco.2010.02.030

Sallo FS, Sanches L, Dias VRM et al (2017) Stem water storage dynamics of *Vochysia divergens* in a seasonally flooded environment. Agric For Meteorol 232:566–575. doi:10.1016/j.agrformet.2016.10.015

Santarosa E, Oliveira JM, Roig FA et al (2007) Crescimento sazonal em *Araucaria angustifolia*: evidências anatômicas. Rev Bras Biociênc 5(1):618–620

Santos Jr A (2006) Aspectos populacionais de *Sterculia apetala* (Jacq.) Karst (Sterculiaceae) como subsídios ao plano de conservação da arara-azul no Sul do Pantanal, Mato Grosso do Sul, Brasil. Dissertation, Universidade Federal do Mato Grosso d Sul

Santos GM, Linares R, Lisi CS et al (2015) Annual growth rings in a sample of Paraná pine (*Araucaria angustifolia*): toward improving the ^{14}C calibration curve for the Southern Hemisphere. Quat Geochronol 25:96–103. doi:10.1016/j.quageo.2014.10.004

Scabin A, Costa F, Schöngart J (2012) The spatial distribution of illegal logging in the Anavilhanas Archipelago (Central Amazonia) and logging impacts on the primary timber species. Environ Conserv 39(1):111–121. doi:10.1017/s0376892911000610

Schmitz N, Verheyden A, Kairo JG et al (2007) Successive cambia development in *Avicennia marina* (Forssk.) Vierh. is not climatically driven in the seasonal climate at Gazi Bay, Kenya. Dendrochronologia 25:87–96. doi:10.1016/j.dendro.2006.08.001

Schöngart J (2008) Growth-Oriented Logging (GOL): a new concept towards sustainable forest management in Central Amazonian várzea floodplains. For Ecol Manage 256:46–58. doi:10.1016/j.foreco.2008.03.037

Schöngart J (2010) Growth-oriented logging (GOL): the use of species-specific growth informationfor forest management in in central Amazonian floodplain forests. In: Junk WJ, MTF P, Wittmann F et al (eds) Central Amazonian floodplain forests: ecophysiology, biodiversity and sustainable management, Ecological Studies, vol 210. Springer, Dordrecht, Heidelberg, London, New York, pp 437–462

Schöngart J (2013) Dendroecological studies in tropical forests. Dissertation, Albert-Ludwigs Universität Freiburg, Germany

Schöngart J, Piedade MTF, Ludwigshausen S et al (2002) Phenology and stem-growth periodicity of tree species in Amazonian floodplain forests. J Trop Ecol 18:581–597. doi:10.1017/s0266467402002389

Schöngart J, Junk WJ, Piedade MTF et al (2004) Teleconnection between tree growth in the Amazonian floodplains and the El Niño-Southern Oscillation effect. Glob Chang Biol 10:683–692. doi:10.1111/j.1529-8817.2003.00754.x

Schöngart J, Piedade MTF, Wittmann F et al (2005) Wood growth patterns of *Macrolobium acaciifolium* (Benth.) Benth. (Fabaceae) in Amazonian black-water and white-water floodplain forests. Oecologia 145:454–461. doi:10.1007/s00442-005-0147-8

Schöngart J, Wittmann F, Worbes M et al (2007) Management criteria for *Ficus insipida* Willd. (Moraceae) in Amazonian white-water floodplain forests defined by tree-ring analysis. Ann For Sci 64:657–664. doi:10.1051/forest:2007044

Schöngart J, Wittmann F, Worbes M (2010) Biomass and NPP of Central Amazonian floodplain forests. In: Junk WJ, MTF P, Wittmann F et al (eds) Central Amazonian floodplain forests: ecophysiology, biodiversity and sustainable management, Ecological Studies, vol 210. Springer, Dordrecht, Heidelberg, London, New York, pp 347–388

Schöngart J, Arieira J, Fortes CF et al (2011) Age-related and stand-wise estimates of carbon stocks and sequestration in the aboveground coarse wood biomass of wetland forests in the northern Pantanal, Brazil. Biogeosciences 8:3407–3421. doi:10.5194/bg-8-3407-2011

Schöngart J, Gribel R, Ferreira da Fonseca-Junior S et al (2015) Age and growth patterns of Brazil nut trees (*Bertholletia excelsa* Bonpl.) in Amazonia, Brazil. Biotropica 47(5):550–558. doi:10.1111/btp.12243

Schulman E (1944) Dendrochronology in Mexico, I. Tree-Ring Bull 10(3):18–24

Schulman E (1956) Dendroclimatic changes in semiarid America. University of Arizona Press, Tuscon

Schweingruber FH (1983) Der Jahrring. Standort, Methodik, Zeit und Klima in der Dendrochronologie. Verlag Paul Haupt, Stuttgart, Bern

Schweingruber FH, Wirth C (2009) Old trees and the meaning of 'old'. In: Wirth C, Gleixner G, Heimann M (eds) Old-growth forests. Springer, Berlin, pp 35–54

Sebbenn AM, Degen B, Azevedo VCR et al (2008) Modelling the long-term impacts of selective logging on genetic diversity and demographic structure of four tropical tree species in the Amazon forest. For Ecol Manage 254:335–349. doi:10.1016/j.foreco.2007.08.009

Seitz RA, Kanninen M (1989) Tree ring analysis of *Araucaria angustifolia* in Southern Brazil: preliminary results. IAWA Bull 10(2):170–174. doi:10.1163/22941932-90000485

Silva LB (2006) Variação na estrutura da madeira de quatro espécies da Caatinga nordestina e seu potencial para o desenvolvimento sustentável. Dissertation, Universidade Estadual de Feira de Santana

Silva MS, Santos FAR, Callado CH et al (2016) Growth rings in woody species of Ombrophilous Dense Forest: occurrence, anatomical features and ecological considerations. Braz J Bot. doi:10.1007/s40415-016-0313-8

Slik JWF, Arroyo-Rodríguez V, Aiba SI et al (2015) An estimate of the number of tropical tree species. PNAS 112:7472–7477. doi:10.1073/pnas.1423147112

Solíz C, Villalba R, Argollo J et al (2009) Spatio-temporal variations in *Polylepis tarapacana* radial growth across the Bolivian Altiplano during the 20th century. Palaeogeogr Palaeocl 281:296–308. doi:10.1016/j.palaeo.2008.07.025

Soliz-Gamboa CC, Rozendaal DMA, Ceccantini G et al (2011) Evaluating the annual nature of juvenile rings in Bolivian tropical rainforest trees. Trees 25:17–27. doi:10.1007/s00468-010-0468-z

Souza BT, Estrada GCD, Soares MLF et al (2016) Occurrence of annual growth rings in *Rhizophora mangle* in a region with low climate seasonality. An Acad Bras Cienc 88(1):517–525. doi:10.1590/0001-3765201620150009

Spannl S, Homeier J, Bräuning A (2016) Nutrient-induced modifications of wood anatomical traits of *Alchornea lojaensis* (Euphorbiaceae). Front Earth Sci 4:50. doi:10.3389/feart.2016.00050

Speer JH, Orvis KH, Grissino-Mayer HD et al (2004) Assessing the dendrochronological potential of *Pinus occidentalis* Swartz in the Cordillera Central of the Dominican Republic. The Holocene 14:563–569. doi:10.1191/0959683604hl732rp

Stahle DW (1996) Tree rings and ancient forest relics. Arnoldia 56:2–10

Stahle DW, Villanueva J, Cleaveland MK et al (2000) Recent tree-ring research in Mexico. In: Roig FA (ed) Dendrocronología en América Latina. EDIUNC, Mendoza, pp 285–306

Stahle DW, Villanueva Diaz J, Burnette DJ et al (2011) Major Mesoamerican droughts of the past millennium. Geophys Res Lett 38:L05703. doi:10.1029/2010gl046472

Stahle DW, Burnette DJ, Villanueva J et al (2012) Tree-ring analysis of ancient baldcypress trees and subfossil wood. Quat Sci Rev 34:1–15. doi:10.1016/j.quascirev.2011.11.005

Stahle DW, Griffin RD, Meko DM et al (2013) The ancient blue oak woodlands of California: longevity and hydroclimatic history. Earth Interact 17:12. doi:10.1175/2013ei000518.1

Stahle DW, Cook ER, Burnette DJ et al (2016) The Mexican Drought Atlas: tree-ring reconstructions of the soil moisture balance during the late pre-Hispanic, colonial, and modern eras. Quat Sci Rev 149:34–60. doi:10.1016/j.quascirev.2016.06.018

Steinbrenner MKA (1989) Altersbestimmung mit dendrochronologischen Methoden an *Araucaria angustifolia* (Bert.) O. Ktze. in einem südbrasiliansichen Naturwald. Dissertation, Albert-Ludwigs-Universität Freiburg

Studhalter RA, Glock WS, Agerter SA (1963) Tree growth, some historical chapters in the study of diameter growth. Bot Rev 29:245–365. doi:10.1007/bf02860823

Stuiver M, Rebello AL, White JC et al (1981) Isotopic indicators of age/growth in tropical trees. In: Bormann FH, Berlyn G (eds) Age and growth rate of tropical trees: new directions for research. Yale University Press, New Haven, Connecticut, pp 75–82

Therrell MD, Stahle DW, Cleaveland MK (2002) Warm season tree growth and precipitation over Mexico. J Geophys Res 107(D14):4205. doi:10.1029/2001jd000851

Tomazello Filho M, Lisi CS, Hansen N et al (2004) Anatomical features of increment zones in different tree species in the State of São Paulo, Brazil. Sci For 66:46–55

Tschinkel HM (1966) Annual growth rings in *Cordia alliodora*. Turrialba 16:73–80

Tsuchiya A (1994) Relationships between wood volume increments by stem analysis and water balance in the semi-arid interior of northeastern Brazil. Mem Fac Integrated Arts and Sci, Hiroshima Univ IV(20):71–80

Ursprung A (1904) Zur Periodicität des Dickenwachsthums in den Tropen. Botanische Zeitung X:189–210

Van der Sleen P, Groenendijk P, Vlam M et al (2014) No growth stimulation of tropical trees by 150 years of CO_2 fertilization but water-use efficiency increased. Nat Geosci 8:24–28. doi:10.1038/ngeo2313

Van Gardingen PR, Valle D, Thompson I (2006) Evaluation of yield regulation options for primary forest in Tapajós National Forest, Brazil. For Ecol Manag 231:184–195. doi:10.1016/j.foreco.2006.05.047

Vasconcellos TJ, Costa MS, Barros CF et al (2016) Growth dynamics of *Centrolobium robustum* (Vell.) Mart. ex Benth. (Leguminosae-Papilionoideae) in the Atlantic Forest. Braz J Bot 39:925–934. doi:10.1007/s40415-016-0292-9

Venter O, Sanderson EW, Magrach A et al (2016) Sixteen years of change in the global terrestrial human footprint and implications for biodiversity conservation. Nat Commun 7:12558. doi:10.1038/ncomms12558

Vetter RE (1995) Untersuchungen über Zuwachsrhythmen an tropischen Bäumen in Amazonien. Dissertation, Albert-Ludwigs Universität Freiburg

Vetter RE, Botosso PC (1989) Remarks on age and growth rate determination of Amazonian trees. IAWA Bull 10(2):133–145. doi:10.1163/22941932-90000481

Vidal E, Johns J, Gerwing JJ et al (1997) Vine management for reduced-impact logging in eastern Amazonia. For Ecol Manag 98:105–114. doi:10.1016/s0378-1127(97)00051-0

Vieira S, Trumbore S, Camargo PB et al (2005) Slow growth rate of Amazonian trees: consequences for carbon cycling. PNAS 102:18502–18507. doi:10.1073/pnas.0505966102

Villalba R, Boninsegna JA, Holmes RL (1985) *Cedrela angustifolia* and *Juglans australis*: two new tropical species useful in dendrochronology. Tree-Ring Bull 45:25–35

Villalba R, Grau HR, Boninsegna JA et al (1998) Tree-ring evidence for long term precipitation changes in subtropical South America. Int J Climatol 18:1463–1478. doi:10.1002/(sici)1097-0088(19981115)18:13<1463::aid-joc324>3.3.co;2-1

Vogado NO (2014) Fenologia e dendrocronologia de duas espécies de Fabaceae em uma área de cerrado no Sudeste do Brasil. Dissertation, Universidade Estadual Paulista

Volland F, Pucha D, Bräuning A (2016) Hydro-climatic variability in southern Ecuador reflected by tree-ring oxygen isotopes. Erdkunde 1:69–82. doi:10.3112/erdkunde.2016.01.05

Volland-Voigt F, Bräuning A, Ganzhi O et al (2011) Radial stem variations of *Tabebuia chrysantha* (Bignoniaceae) in different tropical forest ecosystems of southern Ecuador. Trees 25:39–48. doi:10.1007/s00468-010-0461-6

Westbrook JA, Guilderson TP, Colinvaux PA (2006) Annual growth rings in a sample of *Hymenaea courbaril*. IAWA J 27(2):193–197. doi:10.1163/22941932-90000148

Whitmore TC (1975) Tropical rain forests of the far east. Clarendon Press, Oxford

Whitmore TC (1993) Tropische Regenwälder: Eine Einführung. Spektrum Akad. Verlag, Heidelberg, Berlin, New York

Wilczek A, Miodek A, Gizińska A (2014) Terminal xylem and initial parenchyma in anatomical investigations–a review of definitions. Opole Sci Soc Nat J 47:31–44

Worbes M (1984) Periodische Zuwachszonen an Bäumen zentralamazonischer Überschwemmungswälder. Naturwissenschaften 71:157–158. doi:10.1007/bf01137783

Worbes M (1985) Structural and other adaptations to longterm flooding by trees in Central Amazonia. Amazoniana 9:459–484

Worbes M (1986) Lebensbedingungen und Holzwachstum in zentralamazonischen Überschwemmungswäldern. Erich Goltze, Göttingen. Scripta Geobot 17:1–112

Worbes M (1988) Variety in structure of annual growth zones in *Tabebuia barbata* (E. Mey.) Sandw., Bignoniaceae, a tropical tree species from Central Amazonian inundation forests. Dendrochronologia 6:71–89

Worbes M (1989) Growth rings, increment and age of tree in inundation forest, savannas and a mountain forest in the Neotropics. IAWA Bull 10(2):109–122. doi:10.1163/22941932-90000479

Worbes M (1995) How to measure growth dynamics in tropical trees—A review. IAWA J 16:337–351. doi:10.1163/22941932-90001424

Worbes M (1996) Rhythmisches Wachstum und anatomisch-morphologische Anpassungen an Lebensstrategien von Bäumen in zentralamazonischen Überschwemmungswäldern. Mitt Dtsch Dendrol Ges 82:155–172

Worbes M (1999) Annual growth rings, rainfall-dependent growth and long-term growth patterns of tropical trees from the Caparo Forest Reserve in Venezuela. J Ecol 87:391–403. doi:10.1046/j.1365-2745.1999.00361.x

Worbes M (2002) One hundred years of tree ring research in the tropics—a brief history and an outlook to future challenges. Dendrochronologia 20(1-2):217–231. doi:10.1078/1125-7865-00018

Worbes M (1994) Grundlagen und Anwendungen der Jahresringforschung in den Tropen Dissertation, Universität Hamburg

Worbes M, Fichtler E (2010) Wood anatomy and tree-ring structure and their importance for tropical dendrochronology. In: Junk WJ, MTF P, Wittmann F et al (eds) Central Amazonian floodplain forests: ecophysiology, biodiversity and sustainable management, Ecological Studies, vol 210. Springer, Dordrecht, Heidelberg, London, New York, pp 329–346

Worbes M, Junk WJ (1989) Dating tropical trees by means of ^{14}C from bomb tests. Ecology 70(2):503–507. doi:10.2307/1937554

Worbes M, Junk WJ (1999) How old are tropical trees? The persistence of a myth. IAWA J 20(3):255–260. doi:10.1163/22941932-90000689

Worbes M, Raschke N (2012) Carbon allocation in a Costa Rican dry forest derived from tree ring analysis. Dendrochronologia 30:231–238. doi:10.1016/j.dendro.2011.11.001

Worbes M, Klosa D, Lewark S (1995) Rohdichtestruktur von Jahresringen tropischer Hölzer aus zentralamazonischen Überschwemmungswäldern. Holz Roh Werkst 53:63–67. doi:10.1007/bf02716390

Wyatt-Smith J (1995) Manual of Malayan Silviculture for Inland Forests, Malayan Forest Record, vol 2, 2nd edn. Forest Research Institute Malaysia, Kuala Lumpur

Zuidema PA, Brienen RJW, Schöngart J (2012) Tropical forest warming: looking backwards for more insights. Trends Ecol Evol 27:193–194. doi:10.1016/j.tree.2011.12.007

Zumaeta LEC (2009) Dendrocronología en árboles de Tornillo, *Cedrelinga cateniformis* Ducke (Fabaceae), del Centro de Investigaciones Jenaro Herrera en el noreste de la Amazonia, Región Loreto–Perú. Dissertation, Universidad Nacional Agraria la Molina

Chapter 4
Complex Historical Disturbance Regimes Shape Forest Dynamics Across a Seasonal Tropical Landscape in Western Thailand

Patrick J. Baker and Sarayudh Bunyavejchewin

> *... all that mysterious life of the wilderness that stirs in the forest, in the jungles, ... He has to live in the midst of the incomprehensible, which is detestable. And it has a fascination, too, that goes to work upon him.*
>
> *Charles Marlow in Joseph Conrad's* ***Heart of Darkness***

Abstract The seasonally dry tropical forests of continental Southeast Asia occur in landscape-scale mosaics composed of several distinct forest types. The current understanding of the role of disturbances, such as fire and wind storms, in shaping these landscapes and the individual forest types within them, is limited by a lack of historical data. In this chapter we present the first dendroecological reconstruction of the historical dynamics of a mixed deciduous forest in western Thailand and compare it to an earlier study of the historical dynamics of an adjacent area of seasonal evergreen forest. The tree-ring data from the mixed deciduous forest demonstrate pulses of widespread gap recruitment since the 1850s for more than a dozen tree species. Recruitment and growth release data indicate complex patterns of disturbance both within and between forest types. However, there is strong evidence for extensive disturbances that synchronously impacted both the mixed deciduous and seasonal evergreen forests. Notably, several of the reconstructed landscape-scale disturbances are associated with intense regional drought events driven by ENSO variability in the Pacific Ocean. Interactions between the complex disturbance history of the landscape, which generate environmental conditions

P.J. Baker (✉)
School of Ecosystem and Forest Sciences, University of Melbourne, Melbourne, VIC 3121, Australia
e-mail: patrick.baker@unimelb.edu.au

S. Bunyavejchewin
Thai Long-term Forest Ecological Research, CTFS-Forest GEO, Bangkhen, Bangkok 10220, Thailand
e-mail: sarayudh_b@yahoo.com

M.M. Amoroso et al. (eds.), *Dendroecology*, Ecological Studies 231,
DOI 10.1007/978-3-319-61669-8_4

necessary for recruitment, and the wide range of reproductive phenologies, which generate the propagules for recruitment, provide a mechanism for coexistence for the many tree species that co-occur in these seasonal tropical forests.

Keywords Tropical dendrochronology • Disturbance history • Landscape dynamics • Huai Kha Khaeng Wildlife Sanctuary • ENSO

4.1 Introduction

Tropical forests are among the most biologically diverse and structurally complex terrestrial ecosystems in the world. They have fascinated, challenged, and mystified scientists for centuries. Over the past 50 years as ecological theory has both deepened and broadened, tropical forests have provided the ultimate test of the most fundamental question in ecology: "*How do so many species coexist?*" Theoretical and methodological advances have challenged long-held beliefs, pivoting away from earlier views that depicted tropical forest as ecosystems in ecological and evolutionary stasis to a more dynamic view in which stochasticity and drift shape patterns of community composition at local and regional scales (Bush and Colinvaux 1994; Carson and Schnitzer 2008; Hubbell 2008; Bush and McMichael 2016). However, much of the empirical evidence for tropical forest dynamics comes from either short-term observational studies (years to a few decades) or from palaeoecological studies (millennia) (Zuidema et al. 2013; Jackson and Blois 2015). Contemporary understanding of the recent dynamics of tropical forests has greatly increased due in large part to the establishment and ongoing monitoring of large-scale permanent forest dynamics research plots that track hundreds of thousands of individuals from hundreds of species within a single site. The largest network of such plots is the Smithsonian Institution's Center for Tropical Forest Science–Forest Global Earth Observatory (CTFS-ForestGEO), which includes over 60 forest dynamics plots and approximately 6 million trees from 10,000 species (Anderson-Teixeira et al. 2015). However, despite providing unprecedented details on the population and community dynamics of tree species in tropical forests, few of these plots have been monitored for more than 20 years. In contrast, palaeoecological studies over the past few decades have provided fascinating insights into millennial scale dynamics of tropical forest communities, highlighted the ephemeral nature of communities and species in tropical landscapes (Bush and McMichael 2016), and documented the influence of both humans and climate in driving broad patterns of vegetation change over long-time scales (Bush and Colinvaux 1994; Bishop et al. 2003; Maxwell 2004; Buckley et al. 2010). However, the temporal resolution of palaeoecological studies limits inference about dynamics of tropical forests at the scale of the individual components of the forest, namely the trees, and the potential consequences of specific climatic or ecological events.

Despite these new insights into tropical forest dynamics, the lifespan of trees and frequency of rare, high-intensity disturbances typically occur at decadal and centennial scales, which are not well described by the short time-scales of permanent plot networks nor the long time-scales of palaeoecological studies (Zuidema et al. 2013; Jackson and Blois 2015). In temperate forests dendroecological studies have been central to addressing this "scale gap" (e.g., Swetnam 1993; Swetnam and Betancourt 1998; Pederson et al. 2014) and have provided the foundations for contemporary understanding of forest stand dynamics (Oliver and Larson 1996). In tropical forests, there are few tree-ring studies that have explicitly examined long-term stand development patterns and disturbance histories. In large part this is due to the absence or inconsistency of annual growth rings in most tropical tree species. However, despite most tropical tree species lacking annual growth rings, it has long been recognised that some, particularly in seasonal or monsoonal tropical climates, do form reliably annual growth rings (e.g., Schöngart et al., this volume; Gamble 1902; Worbes 2002). Where annual growth rings are present, they have primarily been used to assess long-term growth rates for forest management. For example, in South and Southeast Asia foresters of the late 1800s and early 1900s used annual growth rings to estimate growth rates and yield of teak (*Tectona grandis*, Verbenaceae), a species of great economic importance regionally (Brandis 1856; Gamble 1902; Troup 1921). However, until recently no studies had used tree rings to study the ecology of tropical forests. As a result, the basic details of individual stand dynamics that are taken for granted in temperate forests—the age of the trees, the number of age cohorts, patterns of suppression and release, and the timing and frequency of past disturbances—remain largely unknown to tropical forest ecologists.

Over the past 15 years, an area of forests in the Huai Kha Khaeng Wildlife Sanctuary (HKK) in Thailand's Western Forest Complex has been subjected to an unprecedented level of dendroecological analyses due to the presence of a large, long-term forest dynamics plot and group of common canopy tree species that form reliable annual growth rings. In this chapter we review and synthesize the dendroecological research at HKK, most of which has been done in the seasonal evergreen forest type. We then present new results on dendroecological research from an area of mixed deciduous forest within the same landscape. We conclude by examining landscape-scale patterns of forest dynamics that emerge from these various studies and discuss the factors that may influence them.

4.2 Huai Kha Khaeng Wildlife Sanctuary: Conservation Importance, Climate, Landscape Context, and Forest Types

Thailand's Western Forest Complex (WFC), the largest area of protected land (~18,000 km^2) in continental Southeast Asia, lies on the border with Myanmar between Kanchanaburi Province to the south and Tak Province to the north. The Thung-Yai Naresuan and Huai Kha Khaeng (HKK) Wildlife Sanctuaries are the two largest protected areas accounting for one-third of the WFC and are both internationally recognised as UNESCO World Heritage sites for their significant regional and global conservation value. HKK supports viable populations of large mammal species, many of which are severely threatened across their natural ranges. These include tigers (*Panthera tigris*), leopards (*Panthera panthera*), elephants (*Elephus maximas*), Asiatic black bear (*Ursus thibetanus*), Malayan sun bear (*Helarctos malayanus*), banteng (*Bos javanicus*), and gaur (*Bos gaurus*) (Simcharoen et al. 2007, 2008).

The climate at HKK is strongly seasonal with intraannual rainfall periodicity driven by the Asian monsoon system. Mean monthly temperature ranges from 19°C in January to 27°C in July; mean annual rainfall is ~1500 mm. The rainy season generally occurs from May to October although the timing of the onset of the monsoons may vary by as much as two months. From the end of the monsoon in October until its return the following year mean monthly rainfall is rarely >100 mm. This 5–6 month dry season provides the necessary seasonality in water availability to drive phenological cycles of leaf exchange (Williams et al. 2008) as well as cambial dormancy and reactivation (Nobuchi et al. 1995) in many tree species.

One of the defining features of the seasonal tropical forests of continental Southeast Asia is the co-occurrence and interdigitation of multiple forest types in a complex patchwork mosaic across the landscape (Bunyavejchewin et al. 2011). The three main forest types in these landscapes are seasonal evergreen forest, mixed deciduous forest, and deciduous dipterocarp forest. They differ substantially in structure, composition, and diversity. The seasonal evergreen forest is the most structurally complex, has the most trees and tree species per unit area, and maintains most of the leaves in the forest canopy through the dry season (Bunyavejchewin et al. 2001, 2002). The canopy of the seasonal evergreen forest has a significant component of Dipterocarpaceae, but individual dipterocarp species vary from locally abundant to absent. Grasses and bamboos do not occur in the shaded understorey of the seasonal evergreen forest. The mixed deciduous forest is less diverse and generally shorter than the seasonal evergreen forest, and during the dry season about 70% of the canopy is leafless. Bamboos are relatively common in the mixed deciduous forest and grasses occur occasionally (Bunyavejchewin et al. 2011). The deciduous dipterocarp forest is the least complex and least diverse of the three forest types. The canopy is dominated almost exclusively by a small group of deciduous dipterocarp species; typically, the deciduous dipterocarp species account for two-thirds or more of the stand basal area (Bunyavejchewin 1983; Nguyen and Baker 2016). The ground layer of the deciduous dipterocarp forest is dominated by grasses and during the dry season the canopy is almost completely leafless. As a consequence, the deciduous dipterocarp forest is the most fire-prone of the three forest types. While these three forest types occur in close proximity across the landscape at HKK and elsewhere throughout continental Southeast Asia, it is unclear what factors determine their spatial distribution within any given landscape. Examination of growth patterns of common species from seasonal evergreen and mixed deciduous forests planted in common gardens in both forest types found no evidence of growth differences in seedlings between forest types (Baker 1997). In addition, analysis of soil chemistry and structural properties found no evidence of differences between the forest types (Baker 1997). As we discuss below, size distributions and species composition of seasonal evergreen forest in the CTFS-ForestGEO 50-ha forest dynamics plot at HKK in combination with dendroecological analyses suggest that these forest types are not stationary on the landscape and chance historical events may influence their occurrence and relative abundance.

4.3 Forest Stand Dynamics in the Seasonal Evergreen Forest at HKK

Over the past decade several studies in an area of ~1000 ha of seasonal evergreen forest at HKK have applied standard dendroecological tools to explore patterns of establishment and growth of various tree species, both spatially and temporally, since the early 1800s. Together these studies have detailed a dynamic, species-rich forest that has been subjected to a complex disturbance history. In the first of these studies, Baker et al. (2005) used a combination of tree-ring studies, age estimation techniques, and architectural analyses to demonstrate that the seasonal evergreen forest on and adjacent to the 50-ha plot retains a distinct structural and floristic legacy from a single, catastrophic disturbance event in the mid-1800s that led to the establishment of a single cohort of trees that dominate the current forest canopy. The analyses also revealed that subsequent disturbances of lower intensity had created discrete pulses of widespread gap formation throughout the forest, generating temporal pulses of establishment in the 1910s, 1940s, and 1980s. These were likely created by extensive, but low-intensity, fires that occur during periods of extreme drought (Baker et al. 2008; Baker and Bunyavejchewin 2009). Adding a further layer of complexity to the dynamics of the seasonal evergreen forest, there have been relatively continuous low levels of recruitment and growth release created by background mortality and random gap formation over the past two centuries. The abundant and varied recruitment opportunities over this period, combined with the diversity of the tree species and their associated life histories represent an important potential mechanism for the maintenance of species diversity within the seasonal evergreen forest.

Baker et al. (2005) also presented evidence based on diameter distributions and the architecture of individual trees that the current area of seasonal evergreen forest had been more akin to a mixed deciduous forest before the catastrophic disturbance of the mid-1800s. This is of particular interest because the forest dynamics plot in seasonal evergreen forest at HKK shares ~70% of tree species with a similar, but smaller (16 ha), forest dynamics plot in mixed deciduous forest, suggesting that rare disturbance events may dramatically change recruitment filters, altering the relative abundances of tree species, and creating state changes in forest types within the landscape. It also suggests that species of low relative abundance within species-rich forests may represent important reservoirs of diversity over broader temporal and spatial scales—a spatial storage effect operating at the landscape scale (*sensu* Warner and Chesson (1985)).

Subsequent dendroecological studies at HKK have focused on clarifying mechanisms of canopy species recruitment. Middendorp et al. (2013) used geospatial analyses and tree-rings to document spatiotemporal interactions in recruitment patterns of *Melia azederach* over several hundred hectares. *Melia* is a well known gap-dependent pioneer species, so spatial aggregation of recruitment dates would be expected. They found that although the age distribution of *Melia* implied relatively continuous recruitment since the 1950s, when the spatial pattern of recruitment was analysed, there were discrete pulses of strongly spatially aggregated recruitment—that is, gap formation was temporally random, but spatially clumped. Vlam et al. (2014) showed a similar dynamic for the long-lived canopy species *Afzelia xylocarpa*, which had discrete pulses of recruitment in the 1850s and 1950s. They linked these recruitment events to a combination of canopy disturbance and relatively wet climatic conditions and highlighted the failure of recruitment during drier periods or periods without significant canopy disturbance.

Both *Melia* and *Afzelia* are fast-growing, shade-intolerant species that typically establish in treefall gaps and grow directly to the forest canopy. Other, more shade-tolerant, species may establish in the low-light conditions of the forest understorey, but still require the high-light conditions created by a gap to eventually grow into the canopy. Indeed, all of the seasonal evergreen forest tree species that we have studied—ranging from extreme heliophiles to shade-tolerant sub-canopy species—showed evidence of gap-related growth patterns whether at establishment, during development, or both (Baker and Bunyavejchewin 2006). By comparing the relative frequency of establishment in high-light vs. low-light environments and canopy accession through growth release, we demonstrated that even shade-tolerant canopy species, such as *Neolitsea obtusifolia*, rarely reach the forest canopy without either establishing in the high light environment of a canopy gap or being released by a gap at some point in their development. Indeed, across the range of life histories sampled, only 2% of individuals showed neither fast early-growth nor release from a canopy disturbance (Baker and Bunyavejchewin 2006). In short, canopy tree species are highly dependent on disturbance to reach the forest canopy. Taken together these studies have highlighted the important role of disturbances for the establishment (or lack thereof) of seedlings and their subsequent recruitment into the canopy across a range of different life histories within the seasonal evergreen forests at HKK.

4.4 Reconstructing Forest Stand Dynamics of Mixed Deciduous Forest

The broad overlap in species composition and widespread interdigitation of the mixed deciduous and seasonal evergreen forests at HKK raise important questions about the mechanisms that control the landscape-scale distribution of these recognisably distinct forest types. Direct observations of landscape-scale fires at HKK (Baker and Bunyavejchewin 2009) suggest that the impacts of fire are widespread

Table 4.1 Tree species from the mixed deciduous forest at the Kapook Kapiang 16-ha forest dynamics plot that were used for dendroecological analyses. N_{plot} and N_{cored} are the number of individuals that occur in the plot and the number of individuals that we cored, respectively

Species	Species code	Family	N_{plot}	N_{cored}
Afzelia xylocarpa	AFZEXY	Fabaceae	9	1
Cassia fistula	CASSFI	Fabaceae	133	12
Chukrassia tabularis	CHUKTA	Meliaceae	9	4
Colona floribunda	COLOFL	Malvaceae	1633	14
Gmelina arborea	GMELAR	Lamiaceae	14	5
Lagerstroemia villosa	LAGEVI	Lythraceae	422	21
Melia azederach	MELIAZ	Meliaceae	9	2
Vitex canescens	VITECA	Lamiaceae	54	22
Vitex glabrata	VITEGL	Lamiaceae	77	17
Vitex limnofolia	VITELI	Lamiaceae	43	20
Vitex peduncularis	VITEPE	Lamiaceae	108	26
Vitex pinnata	VITEPI	Lamiaceae	57	3
Xylia xylocarpa	XYLIXY	Fabaceae	244	15
Total			2812	162

and that, at least in the 1997–98 fires, the mixed deciduous forest experienced the most severe damage. Reconstruction of past recruitment and growth release patterns in mixed deciduous forests may shed some light on variability, and potential synchronicity, in disturbance effects between the mixed deciduous and seasonal evergreen forest types. Here we present new dendroecological analyses from a 16-ha forest dynamics plot in mixed deciduous forest approximately 5 km northeast of the seasonal evergreen forest in the CTFS-ForestGEO 50-ha plot. We sampled trees from 13 species that form high-quality annual growth rings (Table 4.1). Of these, only three species (*Afzelia xylocarpa*, *Chukrassia tabularis*, and *Melia azederach*) were common to the dendroecological study of the seasonal evergreen forest. The majority of sampled trees (52%) are from the genus *Vitex* in the Lamiaceae family, which includes other tropical tree species with high-quality annual growth rings (e.g., *Tectona grandis*). Our sample also included three species in the family Fabaceae (i.e., *Afzelia*, *Cassia*, *Xylia*) and two species in the Meliaceae (i.e., *Chukrassia* and *Melia*). Overall, the study species ranged from fast-growing pioneers, such as *Colona*, *Gmelina*, and *Melia*, to long-lived canopy tree species, including *Afzelia* and all of the *Vitex* species. The remaining species—*Cassia fistula*, *Chukrassia tabularis*, *Xylia xylocarpa*, and *Lagerstroemia villosa*—are all relatively shade-intolerant trees that are found in the mid-canopy or canopy of the mixed deciduous forest. While these species span much of the life history variation in the mixed deciduous forest, it is important to recognise that they represent only a small subset of the 245 species found in the 50-ha plot.

We selected our trees from the 16-ha mixed deciduous forest plot database and followed the same field collection, sample preparation, and measurement procedures detailed for the seasonal evergreen tree species in Baker et al. (2005). As noted there, the growth rings in tropical trees are often anatomically complex and can present significant challenges for crossdating (Stahle 1999). Despite our best efforts to crossdate the tree rings, we expect that there remain some dating errors. Instead of ignoring this when constructing age distributions for individual species or the stand as a whole, we incorporated uncertainty by using probability distributions of ages instead of the tree-ring based point estimates of age. To do this, we estimated a discrete equivalent of the Guassian distribution with mean equal to the point estimate of age (i.e., observed number of growth rings) and a standard deviation scaled to age. We used a Guassian distribution as we had no reason to believe that false rings were more or less common than missing rings, although that can be easily changed if future information suggests otherwise. We scaled the standard deviation to age because we expected that the dating errors would be positively correlated with tree age—that is, the longer the tree-ring time series, the more errors they were likely to have accumulated. Based on our experience with crossdating trees from HKK, we used a standard deviation equal to the mean divided by 50, as this provided probability distributions consistent with our expectations of potential dating errors. As an example of this process the age of a tree with 100 growth rings would be represented by a probability mass function (the discrete equivalent of a probability density function) with mean 100 and 95% of the distribution in the 2σ interval of [96,104]. We made these calculations for all sampled individuals and then combined the probability mass functions of all trees in the mixed deciduous forest into a single age distribution.

To estimate patterns of growth release we used the radial growth-averaging method described in Nowacki and Abrams (1997) and implemented in the R package TRADER (Altman et al. 2014). We used the same criteria for identifying different intensities of growth release as we had for the seasonal evergreen forest (Baker et al. 2005). For each tree core we grouped growth releases into two bins: (1) mean growth rate over a decade was 50–100% greater than the preceding decade and (2) mean growth rate was more than 100% greater than the preceding decade. We then categorised the type of release for each tree based on all cores taken from that tree. *Minor* releases were those in which only one core in a tree showed evidence of a 50–100% growth increase and *moderate* releases were those in which two or more cores showed 50–100% growth increase in the same year. *Major* releases were those in which only one core in a tree showed evidence of a >100% growth and *extreme* releases were those in which two or more cores showed >100% growth increase in the same year.

4.4.1 Age Distribution of the Mixed Deciduous Forest

In all we sampled 162 trees from 13 species within the mixed deciduous forest plot (Table 4.1). The age distribution for the trees that we sampled in the mixed deciduous forest plot was distinctly multimodal (Fig. 4.1) with age peaks centred

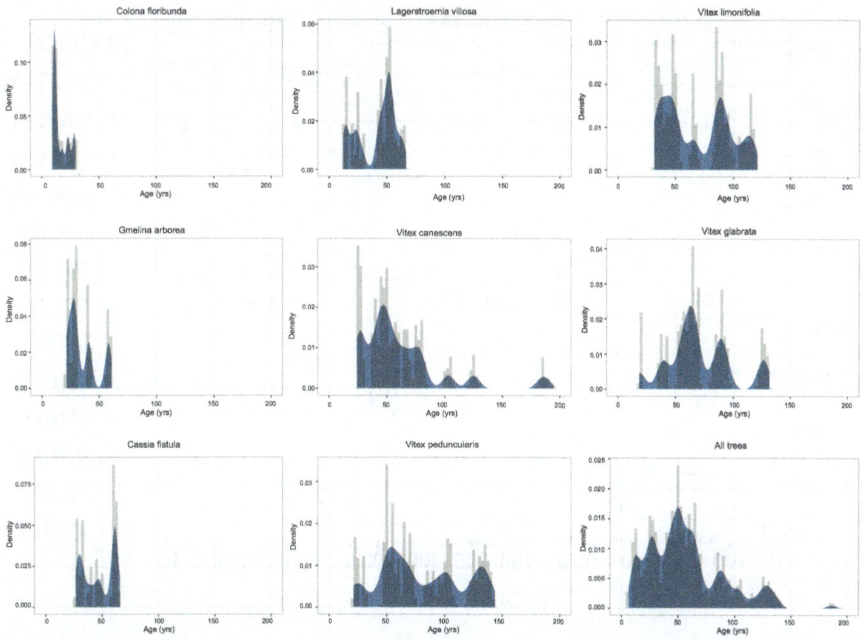

Fig. 4.1 Tree age distributions for individual species with >5 individuals and for all species pooled together. Age distributions and density profiles are estimated as Gaussian probability mass functions from point estimates of age derived from tree rings (see text for details)

on 1879, 1919, 1953, 1977, and 1993. There was, however, substantial variation in the age distributions of individual species. For example, all of the *Colona floribunda* was <30 years old, whereas only the *Vitex* species and *Afzelia* exceeded 80 years of age. The *Vitex* also had quite broad age distributions (>90 years) relative to the other species.

In addition to species-specific differences in the age distributions, there were also distinct spatial patterns among some, but not all, of the age cohorts. Figure 4.2 shows the spatial distribution of trees in age classes centred on the observed peaks in establishment dates. The most striking pattern is the clump of trees from the youngest cohort in the northeastern corner of the Kapook Kapiang plot. These trees are all *Colona floribunda* and are located near a parent tree that established in the previous pulse of regeneration. In the 19–38 year old cohort there are dense clumps of recruitment in the northwestern and southwestern corners of the plot. Unlike the younger *Colona floribunda* in the northeastern corner, both of these clumps of regeneration include multiple species. For example, regeneration in the southwestern corner of the plot includes *Cassia fistula*, *Colona floribunda*, *Gmelina arborea*, *Vitex peduncularis*, *Cassia fistula*, and *Xylia xylocarpa*. The 39–56 year old cohort also shows evidence of spatially aggregated regeneration in the southwestern and northeastern corners of the plot. Subsequent age cohorts show less evidence

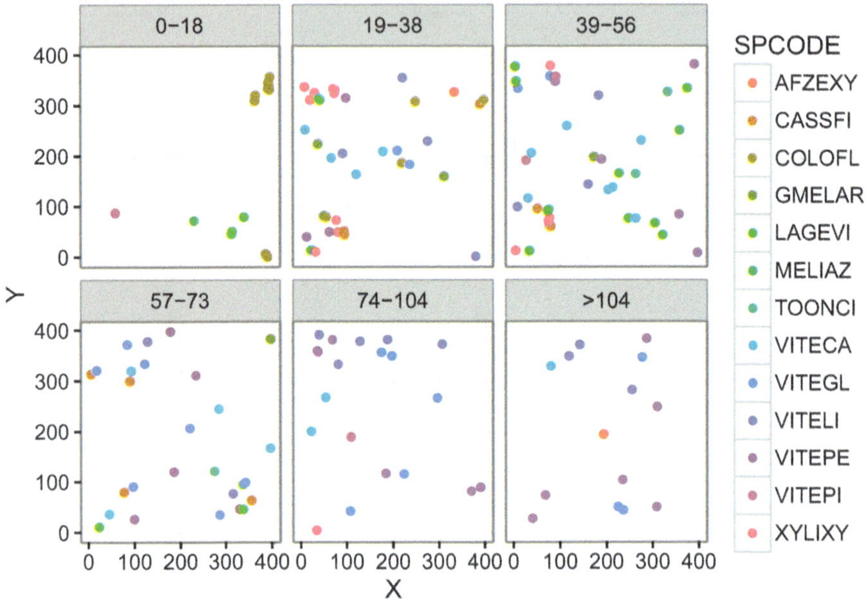

Fig. 4.2 Spatial patterns of tree ages in mixed deciduous forest at the Huai Kha Khaeng Wildlife Sanctuary in western Thailand. The ages were divided into six groups based on the age midpoints between peaks in the frequency distribution of ages shown in Fig. 4.1. Units are in metres and species codes are given in Table 4.1. The plots are oriented such that North is to the top and East is to the right

of spatial clumping, but that may be due to the broader age classes, smaller sample sizes, and/or density-dependent thinning that occurs in densely clumped regeneration over time. The spatial patterns of recruitment suggest that, at least for the species sampled, there are areas within the study plot that have had more active turnover than others. The south-central portion of the plot shows almost no evidence of recruitment of new individuals in the past century, whereas the southwestern-most hectare has had recruitment in every pulse of recruitment over the same period.

4.4.2 *Growth Releases in the Mixed Deciduous Forest*

The mixed deciduous forest trees that we sampled showed evidence of growth release in every 5-year period since 1905 (Fig. 4.3). The number of releases is much greater since 1950 due to the greater sample size of trees since then. The tree-ring data reveal substantial temporal variability in the frequency and intensity of growth releases among trees. The most pronounced peaks in the distribution of growth releases occur around 1983 and 1993 due to a large number of major

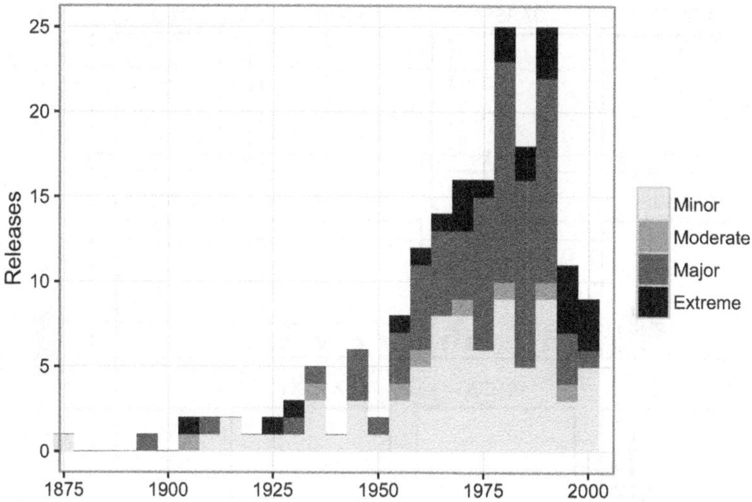

Fig. 4.3 Distribution of release events by intensity classes for all individuals sampled in the mixed deciduous forest study plot at the Huai Kha Khaeng Wildlife Sanctuary in western Thailand

and extreme releases. Species-specific patterns of growth release show substantial interspecific variation in the frequency and timing of growth releases (Fig. 4.4). Two species, *Colona floribunda* and *Melia azederach*, both fast-growing, short-lived pioneer species, did not experience any growth releases. However, the longer-lived, more shade-tolerant species, such as the *Vitex* spp., appeared to be sensitive to changing resource availability, but showed substantial variability in both the timing and intensity of growth releases. Several species, notably *Lagerstroemia villosa*, *Cassia fistula*, and *Vitex glabrata*, had a single, relatively brief, period of intense growth release. However, the timing of these species-specific pulses of sudden increased growth differed among these species. Several of the peaks in growth releases were closely aligned with periods of above-average tree recruitment. The two most notable examples are the 1977 peak in recruitment associated with the pulse of growth releases in *Lagerstroemia villosa* and the 1992/3 peak in recruitment (notably of *Colona floribunda*) and growth releases, particularly of *Cassia fistula* and *Vitex peduncularis*. The 1992/3 peak in recruitment and release is associated with a particularly dry year in which fires burned across much of the area in and around both seasonal evergreen and mixed deciduous forest dynamics plots (S. Bunyavejchewin, *personal observations*). It is worth noting, though, that over the past century there have been periods of recruitment that are not associated with distinct peaks in release, just as there have been periods with growth releases that are not associated with peaks in establishment. This suggests that at the spatial scale at which we sampled—tens of hectares within the mixed deciduous forest—there is a level of background mortality that leads to near continuous establishment and growth releases among the tree species that we sampled.

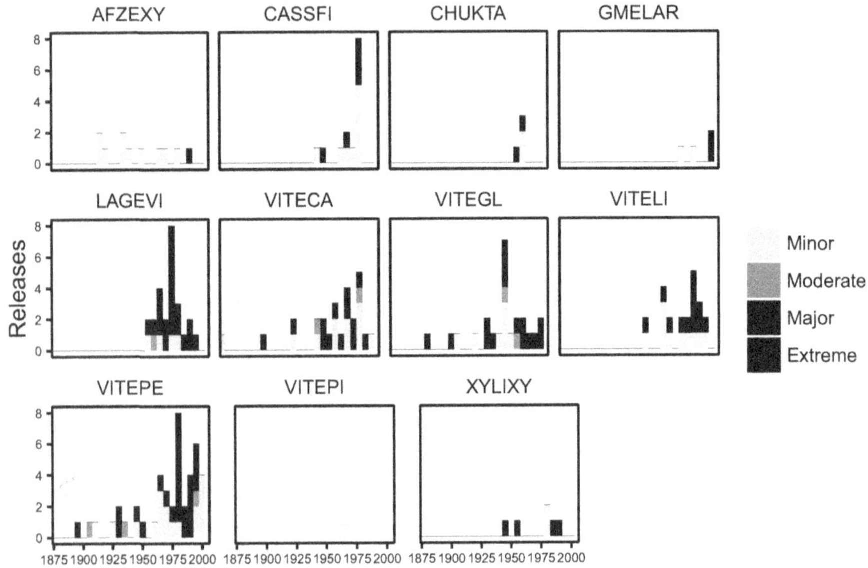

Fig. 4.4 Distribution of growth release events by species for trees cored at the mixed deciduous forest study plot in the Huai Kha Khaeng Wildlife Sanctuary in western Thailand. *Colona floribunda* and *Melia azederach* are not included as both species are fast-growing, short-lived pioneers that did not show any evidence of growth release. Species codes are given in Table 1

4.5 Discussion

4.5.1 Disturbance Regimes in Tropical Forests

All forested landscapes experience disturbances. Variation in the temporal and spatial scales of disturbances, and their intensity and type, determine how disturbances influence the structure, composition, and dynamics of forests (Oliver and Larson 1996). Frequent, small-scale disturbances of low intensity have received considerable attention, particularly in tropical forests where small canopy gaps created by one or a few trees falling are the most common form of disturbance (Denslow 1987; Brokaw 1987). Small canopy gaps directly influence above- and below-ground environmental conditions and resource availability, which impact recruitment patterns from the seed, seedling, and sapling banks of the forest understorey. Our dendroecological reconstructions of historical forest stand dynamics in both the seasonal evergreen and mixed deciduous forest types have documented the role of this background mortality in driving establishment and growth releases for almost every species we have sampled. Indeed, we have shown that in the absence of some form of canopy gap during ontogeny, the probability of successful recruitment to the forest canopy is close to zero (Baker and Bunyavejchewin 2006).

In recent decades, there has also been considerable interest in the role of low-frequency, large-scale, high-intensity disturbances in tropical forests (Whitmore and Burslem 1998; Vandermeer et al. 2000; Baker et al. 2005). Empirical evidence for rare, catastrophic disturbances in tropical forests is limited because of the low frequency with which they occur, the lack of long-term monitoring and reporting programs in tropical forests, and the lack of annual growth rings in most tropical tree species that can be used for dendroecological reconstructions. Consequently, most evidence for such disturbances is anecdotal, although direct observation of the impacts of hurricanes, fires, and earthquakes on tropical forests has occurred in some instances (e.g., Vandermeer et al. 2000; Yap et al. 2015, Leighton and Wirawan 1983). Remote-sensing technologies are beginning to change our perception of large forested landscapes in the tropics and have begun to record the impacts of large disturbances (Chambers et al. 2007). However, remote-sensing time series in the tropics are limited in their temporal reach and so provide little insight into historical patterns of disturbance. Dendroecological reconstructions of large disturbances in the past are commonplace in temperate forest regions, but, with the exception of our reconstruction of the impacts of a stand-replacing disturbance in the early 1800s on the dynamics of seasonal evergreen forest at HKK, have not been conducted in tropical forests.

There is a growing body of research, however, particularly in temperate forests (e.g., Amoroso et al. 2011; Trotsiuk et al. 2014), demonstrating that between these two extremes of the disturbance gradient (high-frequency, low-intensity, and small-scale vs. low-frequency, high-intensity, and large-scale) there are many other types of disturbances that contribute to the structural and compositional diversity of forests. In tropical forests, the paucity of information on intermediate-scale disturbances is a major source of uncertainty in understanding regional forest biomass dynamics and significantly limits our ability to represent forest dynamics in earth systems models (Chambers et al. 2009). The research we present here, in conjunction with our earlier work on the seasonal evergreen forests at HKK, is the first to provide empirical evidence for the role of intermediate-scale disturbances of mixed severity on historical forest dynamics across a tropical landscape.

Our dendroecological reconstructions of recruitment and release highlight the importance of these intermediate-scale disturbances on forest dynamics across a complex landscape mosaic of forest types in western Thailand. Our previous work in the seasonal evergreen forest described the repeated occurrence of discrete periods of widespread gap formation over the past 120 years. These pulses of recruitment and release are consistent with the canopy mortality patterns created by low-intensity wildfires like those that we observed during the 1997–98 ENSO event (Baker et al. 2008). During that period the intense drought conditions, which occurred across all of continental Southeast Asia, led to widespread reduction or loss of foliage among evergreen trees within the HKK landscape. The more open forest canopy led to greater insolation on the forest floor, which increased temperatures and decreased relative humidity of leaf litter layer, thereby increasing potential flammability. For the seasonal evergreen forests, the more open forest canopy and drier forest floor meant that the fires that were restricted to the deciduous forests during non-ENSO years could burn into the understorey. While these fires carry relatively little energy, even under extreme drought conditions, they have two distinct impacts on the seasonal evergreen forest. First, they kill the aboveground portion of most trees that are <2 cm DBH. A large proportion (>80%) of these individuals will resprout, but there are some species that are significantly more aggressive resprouters than others (Bunyavejchewin and Baker, *unpublished data*). Interspecific differences in post-fire resprouting ability may confer an advantage on some species if fires become more frequent. Second, the fires kill large canopy trees through smouldering debris piles composed of bark, fallen lianas, and other coarse woody debris that has accumulated at the base of individual trees. Our dendroecological reconstruction of recruitment and release dynamics in the mixed deciduous forest plot suggests that the widespread, but diffuse, canopy gap formation and subsequent recruitment that we documented in the seasonal evergreen forest also occurs in this forest type. Direct observation soon after the 1998 fires showed that the fire in the mixed deciduous forest created more and larger gaps than in the seasonal evergreen forest (Baker and Bunyavejchewin 2009), providing widespread opportunities for establishment of new seedlings and growth releases of individuals that survived the fires.

An obvious question is whether the pulses of recruitment from these widespread, but low-intensity, disturbances have been synchronous across both the evergreen and mixed deciduous forests and, if so, what might be driving these patterns. In the seasonal evergreen forest peaks of recruitment and release occurred in the early 1880s, the late 1910s/early 1920s, early 1950s and the late 1960s/early 1970s; in the mixed deciduous forest age peaks were centered on 1879, 1919, 1953, 1977, and 1993. These dates suggest the occurrence of synchronous disturbances across both forest types within the broader HKK landscape on at least three occasions in the past 125 years. The 1998 fires highlighted the potential role of regional drought and external climate forcing mechanisms such as ENSO in driving disturbances across large spatial scales in continental southeast Asia. The observed synchronicity of disturbance pulses implied by our regeneration and release data from the seasonal evergreen and mixed deciduous forest plots at HKK suggest that extreme regional climate events may drive landscape-scale forest dynamics by creating the necessary conditions for disturbances (in this case fire) and the subsequent patterns of recruitment and growth. To explore the potential role of regional climate drivers on the observed timing of disturbance and recruitment across our study sites at HKK, we compared the recruitment and release peaks to two palaeoclimate proxy reconstructions, specifically (1) the Monsoon Asia Drought Atlas, which estimates spatially gridded Palmer Drought Severity Index (PDSI) values across southeast Asia (Cook et al. 2010), and (2) the El Niño-Southern Oscillation (Li et al. 2013).

We found that three of the five periods of heightened recruitment and growth were associated with extreme regional drought conditions as evidenced by both PDSI and ENSO index values (Fig. 4.5). The 1879 peaks in recruitment in the mixed deciduous forest and the early 1880s peak in the seasonal evergreen forest closely follow the 1878 ENSO event, the most severe ENSO event in the past 200 years and the last year of an intense regional drought (The Great Drought, Cook et al. 2010).

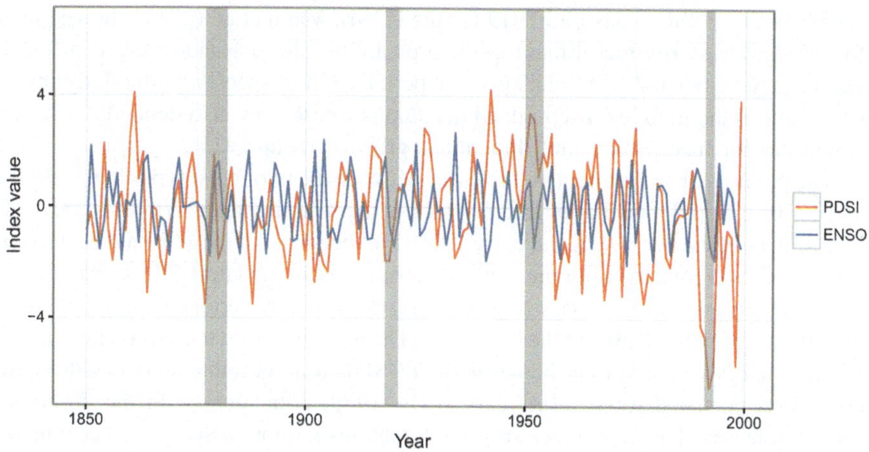

Fig. 4.5 Interannual variability in Palmer Drought Severity Index (PDSI) from the Monsoon Drought Atlas of Cook et al. (2010) and ENSO index values from Li et al. (2013). The values of the ENSO index have been multiplied by −1 to align them with the PDSI data. In the figure *positive values* are wet conditions and *negative values* are dry conditions. The *grey boxes* represent periods of recruitment common to both the seasonal evergreen forest and mixed deciduous forest study sites. Three of the four recruitment periods are associated with intense ENSO and drought conditions—the early 1950s are the exception

The peaks in recruitment and releases in the late 1910s and early 1920s are also associated with a severe ENSO and drought conditions from 1918 to 1920. The pulse of recruitment in 1993 at the mixed deciduous plot is associated with several years (1991–1993) of both intense ENSO and drought conditions and observed records of fires in 1992 at HKK. The seasonal evergreen forest reconstruction did not show evidence of recruitment in the early 1990s primarily because our sampling, conducted in 1998–99, was limited to trees >10 cm DBH, so any individuals establishing at that time would not have been large enough to core. However, the 1992 fires did burn into the northwestern margins of the plot, killing canopy trees and initiating recruitment in the newly formed canopy gaps (Baker and Bunyavejchewin, *personal observations*).

While these observations suggest a role for regional climate extremes in driving local forest dynamics, we also had one period of recruitment that was synchronous to both forests that was not associated with extreme climate and two peaks of recruitment that were not synchronous between the seasonal evergreen and mixed deciduous forests. The early 1950s peak, which is common in both plots and is the largest of the recruitment peaks in the mixed deciduous forest plots, occurred during a period of relatively moist regional conditions. The index values for both ENSO and the PDSI during the 1950s indicate wetter than average conditions across the region. Between the decade spanning 1948 and 1957, the period of peak recruitment, the ENSO index and PDSI values are only associated with dry conditions (i.e., positive and negative values, respectively) during a single year each

(1952 for the ENSO index and 1949 for the PDSI). Vlam et al. (2014) showed that for *Afzelia xylocarpa* recruitment in the area around the seasonal evergreen forest was largely associated with relatively wet periods. Other species, particularly those that occur in the more exposed mixed deciduous forest, may also depend on wetter conditions for successful establishment and growth (Troup 1921).

The patterns of recruitment and releases between the two forest types in the 1960s and 1970s differed from the others recorded in the tree-ring data. While the peak of recruitment in the late 1960s and early 1970s that occurred in the seasonal evergreen forest does not coincide with the peak of recruitment in the late 1970s in the mixed deciduous forest, the seasonal evergreen forest trees did experience a large number of extreme growth releases in the late 1970s. During this period there is no indication of extreme ENSO conditions; however, the PDSI data suggest that moderate drought conditions persisted for several years (1976–1979) coincident with the observed growth releases. The occurrence of growth releases without associated recruitment in the seasonal evergreen forest suggests that the multi-year drought conditions may have led to the mortality of some canopy trees. Gaps created by drought-induced mortality typically occur more slowly and cause less dramatic changes to the light conditions at the forest floor than those created by fire-induced mortality, where the base of the tree is burned out and the tree collapses to the forest floor.

4.5.2 Landscape-Scale Dynamics in a Seasonal Tropical Forest

Dendroecological reconstructions of past disturbances, episodic recruitment, and growth releases from the dominant forest types within a seasonal tropical landscape in western Thailand reveal a complex history of disturbances acting at multiple spatial scales and temporal frequencies, and influenced by a range of drivers, both internal and external to these ecosystems. The landscape of the Huai Kha Khaeng Wildlife Sanctuary has been impacted by catastrophic stand-replacing disturbances (most likely windstorms), broad-scale, but low-intensity, fires that cause widespread formation of small canopy gaps, and random gap formation associated with canopy tree mortality. The complex temporal and spatial interactions among the many disturbances over time have created a diverse mosaic of seasonal tropical forests in various stages of development. Within both forest types these low-intensity disturbances, whether widespread or random, have provided opportunities for establishment of new individuals of many of the tree species that occur in the community. These disturbances have not been sufficiently intense to overcome the compositional inertia of either forest type. Over the past century both of the forest types, despite repeated disturbances of varying intensity, have maintained their fundamental character. Only the catastrophic disturbance of the early 1800s appears to have led to a state change from one forest type to the other (Baker et al. 2005).

One important caveat to this picture of the landscape-scale dynamics within HKK is that we have no historical data on the stand dynamics of deciduous dipterocarp forest, one of the three common forest types within the Sanctuary and the most widespread forest type in continental Southeast Asia. At HKK the deciduous dipterocarp forest accounts for approximately 5% of the landscape, although it may have been more widespread in the past (Baker et al. 2005). None of the four deciduous dipterocarp species that dominate the canopy of the deciduous dipterocarp forest form annual growth rings, despite being deciduous for several months each year during the dry season. Absent dendrochronological data on the timing of establishment and patterns of growth release and canopy accession, we have a very limited understanding of the historical dynamics of the deciduous dipterocarp component of the landscape forest mosaic. A recent study in nearly identical deciduous dipterocarp forests in Central Vietnam has highlighted the role of historical variability in fire frequency as a potential factor limiting canopy recruitment of seedlings and saplings (e.g., Nguyen and Baker 2016). Zimmer and Baker (2009) demonstrated that prolonged fire-free intervals of sufficient length were necessary for the establishment of tropical pines in northern Thailand in forest systems that share many structural features with deciduous dipterocarp forests (i.e., open canopy, extremely low abundance of saplings, grassy understorey). Deciduous dipterocarp species often occur together with the tropical pines in northern Thailand suggesting that infrequent periods of benign climatic conditions in which fires are relatively rare may be a necessary condition for successful recruitment of seedlings into larger size classes and, ultimately, the canopy.

Understanding historical disturbance patterns—and their drivers and outcomes—within tropical forest landscapes is critical to forecasting future community states in response to external pressures such as increasing ignition sources and climate change (Jackson et al. 2009). Of equal importance, developing a better empirical understanding of past forest dynamics should contribute to better predictions of potential future shifts in the relative abundance of the different forest types across the landscapes of continental Southeast Asia.

Our results highlight the role of regional climate drivers such as the El Niño-Southern Oscillation in synchronising landscape-scale disturbances by creating intense drought conditions that lead to widespread canopy loss and rapid drying of the fine fuels on the forest floor. Given the relatively high human population levels throughout continental southeast Asia, fire occurrence is rarely ignition limited. Under normal climatic conditions the forest floor of the seasonal evergreen forests is relatively cool and moist due to the shading of the evergreen forest canopy, preventing egress of low-intensity fires from neighboring deciduous forest. However, during intense droughts, which are often associated with ENSO events, seasonal evergreen tree species can experience sufficient water stress to drop their leaves, thereby opening up the forest canopy and drying the forest floor. This provides the precursor fuel conditions required for the human-set fires originating on the margins of large protected areas such as HKK and Thung Yai-Naresuan to burn across the entire landscape, impacting all forest types over thousands of square kilometres of forest (Fig. 4.6, Bunyavejchewin et al. 2011).

The complex history of landscape-scale disturbances is also directly relevant to questions of community assembly and species coexistence in species-rich tropical

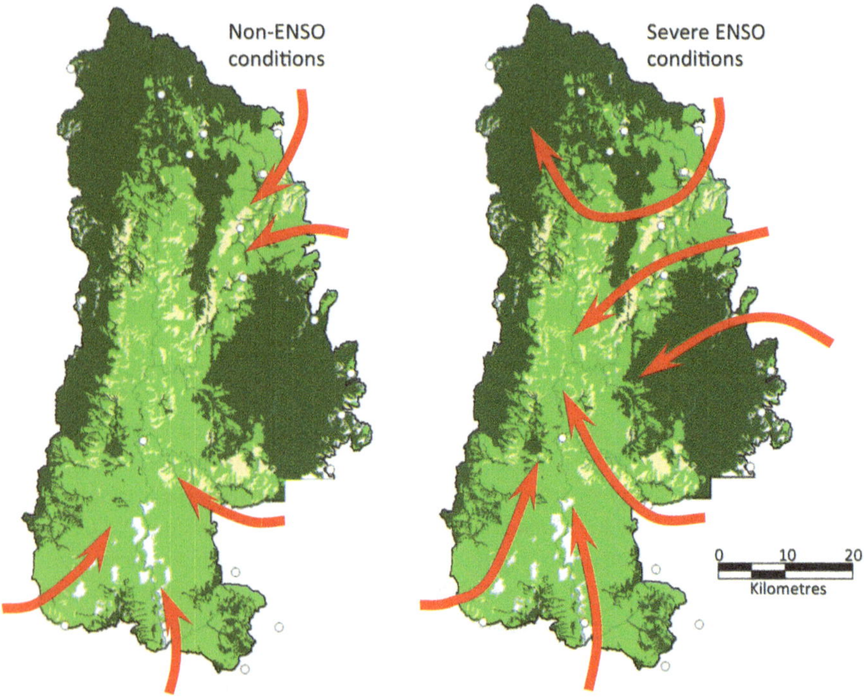

Fig. 4.6 Landscape-scale fire dynamics during non-ENSO and severe ENSO conditions at the Huai Kha Khaeng Wildlife Sanctuary in western Thailand. The *dark green* is seasonal evergreen forest, the *pale green* is mixed deciduous forest, and the *tan* is deciduous dipterocarp forest. During non-ENSO years, the fires come into the sanctuary from agricultural burning in the surrounding buffer zone, but cannot penetrate into the cool, moist understorey of the seasonal evergreen forest. During severe ENSO years partial or complete canopy loss from drought-induced deciduousness in typically evergreen trees leads to drier conditions at the forest floor, allowing fires from adjacent deciduous forest types to burn through the seasonal evergreen forest understorey

forest communities. To persist in a regional species pool, each species must be able to regenerate successfully. This requires reproductively mature individuals within the landscape and the appropriate environmental and ecological conditions for successful regeneration. Most models of species coexistence in tropical forest communities focus on equilibrating mechanisms such as Janzen-Connell dynamics that prevent common species from eliminating less common species by conferring an adaptive advantage to rarity (e.g., Hille Ris Lambers et al. 2002; Comita et al. 2010). Warner and Chesson (1985) proposed that species could coexist if the rarer species is able to persist within a community long enough to take advantage of infrequent opportunities for regeneration. While our results do not explicitly test for the storage effect in these tropical forest communities, they provide the mechanism by which individual species could benefit greatly from disturbances. While extreme drought has been seen as a key environmental filter in moist tropical forests where it greatly increases mortality of adult trees (e.g., Potts 2003; Slik 2004) and seedlings (e.g., Engelbrecht et al. 2007; Comita 2014), in the seasonally dry tropical forests that we studied drought causes widespread fires, which generate landscape-scale regeneration opportunities for many species. The widespread gap formation that accompanies landscape-scale fires creates thousands of gaps in the forest canopy. If a relatively rare species can persist long enough to take reproductive advantage of one of these disturbance events (i.e., storing its reproductive output for the appropriate conditions), it would potentially be able to establish across a wide range of sites, allowing it to maintain a presence in the community for many more decades. Importantly, the wide range of disturbance intensities—from catastrophic, stand-replacing events to individual tree mortality—provides a diversity of environmental conditions for regeneration that varies in time and space.

4.6 Conclusions

Our dendroecological research at HKK has shed important light on the complex dynamics of seasonal tropical forests at the stand and landscape scales and has generated new insights into fundamental ecological questions about forest dynamics and species coexistence in species-rich tropical forests. Our reconstruction of the stand history of a mixed deciduous forest in western Thailand revealed pulses of widespread gap recruitment over the past 150 years. When combined with our previous dendroecological research on the seasonal evergreen forest at HKK, these data show complex patterns of disturbance within and across forest types. However, there is also strong evidence for landscape-scale disturbances that impact multiple forest types synchronously. Importantly, we found that several of the reconstructed landscape-scale disturbances are associated with intense regional drought events driven by ENSO variability in the Pacific Ocean. Further studies in similar forests across continental southeast Asia would allow us to explore the spatial scale at which regional climate drivers influence forest disturbance and regeneration patterns. In addition, the interaction between the complex disturbance history of the landscape, which generates environmental conditions necessary for recruitment, and the wide range of reproductive phenologies, which generates the propagules for recruitment, provides a potentially important mechanism for coexistence of the many tree species that co-occur in these forests.

Dendroecological studies in continental southeast Asia have the potential to answer a range of fundamental questions about the scale of disturbance and dynamics in these remarkable forests. For instance, are the landscape-scale pulses of disturbance and subsequent recruitment observed at HKK occurring across the entire region? And, at what spatial scale are individual tree species able to take advantage of the regeneration opportunities created by regional-scale disturbances? Can regional-scale disturbances initiated by broad climate drivers such as ENSO provide opportunities for range shifts in a changing climate? These and other questions will benefit greatly from further dendroecological research in the seasonal tropical landscapes of continental southeast Asia.

Acknowledgements The dendroecological research on the mixed deciduous forest study plot at Huai Kha Khaeng was funded by a grant from the Center of Tropical Forest Science, Smithsonian Institution to PJB. We would like to thank the staff of the Kapook Kapiang Ranger Station, the Khlong Phuu Research Station, the Huai Kha Khaeng Wildlife Sanctuary, and the National Research Council of Thailand for supporting our research at HKK over the past 25 years. PJB was supported by an Australian Research Council Future Fellowship (FT120100715) during the writing of this chapter.

References

Altman J, Fibich P, Dolezal J, Aakala T (2014) TRADER: A package for Tree Ring Analysis of Disturbance Events in R. Dendrochronologia 32(2):107–112

Amoroso MM, Daniels LD, Bataineh M, Andison DW (2011) Evidence of mixed-severity fires in the foothills of the Rocky Mountains of west-central Alberta, Canada. For Ecol Manag 262(12):2240–2249

Anderson-Teixeira KJ, Davies SJ, Bennett AC, Gonzalez-Akre EB, Muller-Landau HC, Joseph Wright S, Abu Salim K, Almeyda Zambrano AM, Alonso A, Baltzer JL, Basset Y, Bourg NA, Broadbent EN, Brockelman WY, Bunyavejchewin S, Burslem DFRP, Butt N, Cao M, Cardenas D, Chuyong GB, Clay K, Cordell S, Dattaraja HS, Deng X, Detto M, Du X, Duque A, Erikson DL, Ewango CEN, Fischer GA, Fletcher C, Foster RB, Giardina CP, Gilbert GS, Gunatilleke N, Gunatilleke S, Hao Z, Hargrove WW, Hart TB, Hau BCH, He F, Hoffman FM, Howe RW, Hubbell SP, Inman Narahari FM, Jansen PA, Jiang M, Johnson DJ, Kanzaki M, Kassim AR, Kenfack D, Kibet S, Kinnaird MF, Korte L, Král K, Kumar J, Larson AJ, Li Y, Li X, Liu S, Lum SKY, Lutz JA, Ma K, Maddalena DM, Makana JR, Malhi Y, Marthews T, Mat Serudin R, McMahon SM, McShea WJ, Memiaghe HR, Mi X, Mizuno T, Morecroft M, Myers JA, Novotny V, Oliveira AA, Ong PS, Orwig DA, Ostertag R, Ouden J, Parker GG, Phillips RP, Sack L, Sainge MN, Sang W, Sri-Ngernyuang K, Sukumar R, Sun IF, Sungpalee W, Suresh HS, Tan S, Thomas SC, Thomas DW, Thompson J, Turner BL, Uriarte M, Valencia R, Vallejo MI, Vicentini A, Vrška T, Wang X, Wang X, Weiblen G, Wolf A, Xu H, Yap S, Zimmerman J (2015) CTFS-ForestGEO: a worldwide network monitoring forests in an era of global change. Glob Change Biol 21(2):528–549

Baker PJ (1997) Seedling establishment and growth across forest types in an evergreen/deciduous forest mosaic in western Thailand. Nat Hist Bull Siam Soc 45:17–41

Baker PJ, Bunyavejchewin S (2006) Suppression, release and canopy recruitment in five tree species from a seasonal tropical forest in western Thailand. J Trop Ecol 22(5):521

Baker PJ, Bunyavejchewin S (2009) Fire behavior and fire effects across the forest landscape of continental Southeast Asia. In: Cochrane MA (ed) Tropical fire ecology. Springer, Berlin/Heidelberg, pp 311–334

Baker PJ, Bunyavejchewin S, Oliver C, Ashton PS (2005) Disturbance history and historical stand dynamics of a seasonal tropical forest in western Thailand. Ecol Monogr 75(3):317–343

Baker PJ, Bunyavejchewin S, Robinson AP (2008) The impacts of large-scale, low-intensity fires on the forests of continental south-east Asia. Int J Wild Fire 17(6):782–792

Bishop P, Penny D, Stark M, Scott M (2003) A 3.5 ka record of paleoenvironments and human occupation at Angkor Borei, Mekong Delta, southern Cambodia. Geoarchaeology 18(3):359–393

Brandis D (1856) Report on the teak forests of Pegu. Tech. Rep. 28, CB Lewis, Baptis Mission Press, Calcutta

Brokaw N (1987) Gap-phase regeneration of three pioneer tree species in a tropical forest. J Ecol 75:9–19

Buckley B, Anchukaitis K, Penny D (2010) Climate as a contributing factor in the demise of Angkor, Cambodia. Proc Natl Acad Sci USA 107(15):6748–6752

Bunyavejchewin S (1983) Canopy structure of the dry dipterocarp forest of Thailand. Thai For Bull 14:1–132
Bunyavejchewin S, Baker PJ, LaFrankie JV, Ashton PS (2001) Stand structure of a seasonal evergreen forest at the Huai Kha Khaeng Wildlife Sanctuary, western Thailand. Nat Hist Bull Siam Soc 49:89–106
Bunyavejchewin S, Baker PJ, LaFrankie JV, Ashton PS (2002) Floristic composition of a seasonal evergreen forest, Huai Kha Khaeng Wildlife Sanctuary, western Thailand. Nat Hist Bull Siam Soc 50:125–134
Bunyavejchewin S, Baker PJ, Davies SJ (2011) Seasonally dry tropical forests in continental Southeast Asia: structure, function and dynamics. In: McShea WJ, Davies SJ (eds) Ecology and conservation of seasonally dry forests in Asia. Rowman & Littlefield Publishers, Lanham, MD, pp 9–35
Bush MB, Colinvaux PA (1994) Tropical forest disturbance: paleoecological records from Darien, Panama. Ecology 75(6):1761–1768
Bush MB, McMichael CNH (2016) Holocene variability of an Amazonian hyperdominant. J Ecol 104:1370–1378
Carson WP, Schnitzer SA (eds) (2008) Tropical forest community ecology. Oxford University Press, Oxford
Chambers JQ, Asner GP, Morton D, Anderson L, Saatchi S, Espírito-Santo F, Palace M, Souza C (2007) Regional ecosystem structure and function: ecological insights from remote sensing of tropical forests. Trends Ecol Evol 22(8):414–423
Chambers JQ, Negrón-Juárez RI, Hurtt GC, Marra DM, Higuchi N (2009) Lack of intermediate-scale disturbance data prevents robust extrapolation of plot-level tree mortality rates for old-growth tropical forests. Ecol Lett 12(12):E22–E25
Comita LS (2014) Drought as a driver of tropical tree species regeneration dynamics and distribution patterns. In: Coomes DA, Burslem DFRP, Simonson WD (eds) Forests and global change. Cambridge University Press, Cambridge, pp 261–308
Comita LS, Muller-Landau HC, Aguilar S, Hubbell SP (2010) Asymmetric density dependence shapes species abundances in a tropical tree community. Science 329:330–332
Cook E, Anchukaitis K, Buckley B, D'Arrigo R (2010) Asian monsoon failure and megadrought during the last millennium. Science 328:486–489
Denslow JS (1987) Tropical rainforest gaps and tree species diversity. Annu Rev Ecol Evol Syst 18:431–451
Engelbrecht BMJ, Comita LS, Condit RS, Kursar TA, Tyree MT, Turner BL, Hubbell SP (2007) Drought sensitivity shapes species distribution patterns in tropical forests. Nature 447(7140):80–82
Gamble JS (1902) A manual of Indian timbers: an account of the growth, distribution, and uses of the trees and shrubs of India and Ceylon with descriptions of their wood structure. Sampson, Low, Marston, & Co., London
Hille Ris Lambers J, Clark JS, Beckage B (2002) Density-dependent mortality and the latitudinal gradient in species diversity. Nature 417(6890):732–735
Hubbell SP (2008) Approaching Ecol. Complexity from the perspective of symmetric neutral theory. In: Carson WP, Schnitzer SA (eds) Tropical forest community ecology. Oxford University Press, Oxford, pp 143–159
Jackson ST, Blois JL (2015) Community ecology in a changing environment: perspectives from the quaternary. Proc Natl Acad Sci USA 112(16):4915–4921
Jackson ST, Betancourt JL, Booth RK, Gray ST (2009) Ecology and the ratchet of events: Climate variability, niche dimensions, and species distributions. Proc Natl Acad Sci USA 106:19,685–19,692
Li J, Xie SP, Cook ER, Morales MS, Christie DA, Johnson NC, Chen F, D'arrigo R, Fowler AM, Gou X, Fang K (2013) El Niño modulations over the past seven centuries. Nat Clim Change 3(9):822–826
Maxwell AL (2004) Fire regimes in north-eastern Cambodian monsoonal forests, with a 9300-year sediment charcoal record. J Biogeogr 31:225–239

Middendorp RS, Vlam M, Rebel KT, Baker PJ, Bunyavejchewin S, Zuidema PA (2013) Disturbance history of a seasonal tropical forest in western Thailand: a spatial dendroecological analysis. Biotropica 45(5):578–586

Nguyen TT, Baker PJ (2016) Structure and composition of deciduous dipterocarp forest in Central Vietnam: patterns of species dominance and regeneration failure. Plant Ecol Div 9:589–601

Nobuchi T, Ogata Y, Siripatanadilok S (1995) Seasonal characteristics of wood formation in *Hopea odorata* and *Shorea henryana*. IAWA J 16:361–369

Nowacki G, Abrams M (1997) Radial-growth averaging criteria for reconstructing disturbance histories from presettlement-origin oaks. Ecol Monogr 67(2):225–249

Oliver CD, Larson BC (1996) Forest stand dynamics. Wiley, New York

Pederson N, Dyer JM, Mcewan RW, Hessl AE, Mock CJ, Orwig DA, Rieder HE, Cook BI (2014) The legacy of episodic climatic events in shaping temperate, broadleaf forests. Ecol Monogr 84(4):599–620

Potts MD (2003) Drought in a Bornean everwet rain forest. J Ecol 91(3):467–474

Simcharoen S, Pattanavibool A, Karanth KU, Nichols JD, Kumar NS (2007) How many tigers (*Panthera tigris*) are there in Huai Kha Khaeng Wildlife Sanctuary, Thailand? An estimate using photographic capture-recapture sampling. Oryx 41(4):1–7

Simcharoen S, Barlow ACD, Simcharoen A, Smith JLD (2008) Home range size and daytime habitat selection of leopards in Huai Kha Khaeng Wildlife Sanctuary, Thailand. Biol Conserv 141(9):2242–2250

Slik JWF (2004) El Nino droughts and their effects on tree species composition and diversity in tropical rain forests. Oecologia 141(1):7

Stahle DW (1999) Useful strategies for the development of tropical tree-ring chronologies. IAWA J 20:249–253

Swetnam T (1993) Fire history and climate change in giant sequoia groves. Science 262(5135):885

Swetnam TW, Betancourt JL (1998) Mesoscale disturbance and ecological response to decadal climatic variability in the American Southwest. J Climate 11:3128–3147

Trotsiuk V, Svoboda M, Janda P, Mikolas M, Bače R, Rejzek J, Samonil P, Chaskovskyy O, Korol M, Myklush S (2014) A mixed severity disturbance regime in the primary *Picea abies* (L.) Karst. forests of the Ukrainian Carpathians. For Ecol Manage 334:144–153

Troup RS (1921) The silviculture of indian trees, Leguminosae (Caesalpinieae) to Verbenaceae, vol II. Clarendon Press, Oxford

Vandermeer J, de la Cerda IG, Boucher D, Perfecto I (2000) Hurricane disturbance and tropical tree species diversity. Science 290:788–791

Vlam M, Baker PJ, Bunyavejchewin S, Mohren GMJ, Zuidema PA (2014) Understanding recruitment failure in tropical tree species: insights from a tree-ring study. For Ecol Manage 312:108–116

Warner RR, Chesson PL (1985) Coexistence mediated by recruitment fluctuations: A field guide to the storage effect. Am Nat 125:769–787

Whitmore T, Burslem DFRP (1998) Major disturbances in tropical forests. In: Newbery DM, Brown ND, Prins HHT (eds) Population and community dynamics in tropical forests. Blackwell Scientific Press, pp 549–565

Williams LJ, Bunyavejchewin S, Baker PJ (2008) Deciduousness in a seasonal tropical forest in western Thailand: interannual and intraspecific variation in timing, duration and environmental cues. Oecologia 155(3):571–582

Worbes M (2002) One hundred years of tree-ring research in the tropics - a brief history and an outlook to future challenges. Dendrochronologia 20:217–231

Yap SL, Davies SJ, Condit R (2015) Dynamic response of a Philippine dipterocarp forest to typhoon disturbance. J Veg Sci 27(1):133–143

Zimmer H, Baker PJ (2009) Climate and historical stand dynamics in the tropical pine forests of northern Thailand. For Ecol Manage 257(1):190–198

Zuidema PA, Baker PJ, Groenendijk P, Schippers P, van der Sleen P, Vlam M, Sterck F (2013) Tropical forests and global change: filling knowledge gaps. Trends Plant Sci 18(8):413–419

Chapter 5
Low-Hanging DendroDynamic Fruits Regarding Disturbance in Temperate, Mesic Forests

Neil Pederson, Amanda B. Young, Amanda B. Stan, Uyanga Ariya, and Dario Martin-Benito

Abstract Temperate, mesic forests (TMFs) are generally viewed as being in a shifting-mosaic or a kind of dynamic equilibrium at broad spatial scales. Gaining insight to the potential dynamics of TMFs at large-scales is crucial because these species-rich, highly productive forests are important drivers of regional water and carbon cycles for approximately one billion people.

Here, we briefly review the disturbance dynamics of TMFs in eastern North America, Central Europe, and Japan, as well as the temperate rainforests of northwestern North America and western Eurasia. We find that current knowledge effectively describes the relatively frequent disturbance processes from stand to landscape scales. The few large-scale dendroecological networks in TMFs, however, suggest that large, infrequent disturbances (LIDs), those at regional scales or larger, are also important drivers of forest development. We also discuss the

N. Pederson (✉)
Harvard Forest, Harvard University, 324 North Main Street, Petersham, MA 01754, USA
e-mail: neilpederson@fas.harvard.edu

A.B. Young
Department of Geography, Pennsylvania State University, 302 Walker Building, University Park, PA 16802, USA
e-mail: amandabeayoung@gmail.com

A.B. Stan
Department of Geography, Planning and Recreation, Northern Arizona University, PO Box 15016, Flagstaff, AZ 86011, USA
e-mail: Amanda.Stan@nau.edu

U. Ariya
Department of Environment and Forest Engineering, School of Engineering and Applied Sciences, National University of Mongolia, Ulaanbaatar, Mongolia
e-mail: treering7@yahoo.com

D. Martin-Benito
Institute of Terrestrial Ecosystems Science ETH Zürich, Universitätstrasse 16 8092, Zürich, Switzerland

Department of Silviculture and Forest Management, Forest Research Centre, INIA, Avda. La Coruña km.7.5, 28040, Madrid, Spain
e-mail: dmartin@inia.es

M.M. Amoroso et al. (eds.), *Dendroecology*, Ecological Studies 231,
DOI 10.1007/978-3-319-61669-8_5

potential reasons why the spatiotemporal scales at which most ecological research is conducted—centennial-scale investigations (or longer) at small spatial scales or regional investigations (and larger) that only cover a few years to a decade or two—might bias our perception of TMF dynamics.

Our review leads us to hypothesize that the dynamic equilibrium concept in TMFs could be challenged by findings derived from broad-scale networks containing centuries of forest dynamics at high temporal resolution. Dendroecology is poised for this challenge. We provide dendroecological solutions to test the hypothesis that LIDs are important for long-term TMF dynamics. Our review also leads us to assert that the successional trajectories in TMFs are likely more diverse than what is commonly considered. Constraining the potential number of trajectories likely constrains ecological theory and forest management. To expand these perceptions, we present a conceptual model to aid comprehension of the potential n-dimensional developmental trajectories in diverse TMFs.

As we close in on a century of tree-ring research, new frontiers are being opened through the development of new methods and techniques that enhance investigations of temperate mesic forests. Applying these advances will help to address urgent questions on macroecological dynamics of temperate, mesic forests as climate change intensifies over the next century.

Keywords Disturbance ecology • Macroecology • Forest development • Tree rings • Dendrochronology • Mesoscale

5.1 Introduction

Increasing temperatures, reduced soil moisture, and an increase in the frequency of extreme climatic events over the next century are expected to significantly impact the processes driving structure and function in temperate, mesic forests (Iverson et al. 2008; Elkin et al. 2013; Seidl et al. 2014; see Duveneck et al. (2016) for an alternate scenario). Such changes will have profound ecological and social impacts. Temperate, mesic forests (TMFs) cover approximately 12 million km^2 and provide ecosystem services to >1 billion people. Diverse composition, function, and structure in TMFs create productive systems that are critical to regional carbon and water cycles. The loss of key species or compositional shifts could have large consequences to ecosystem functionality (Ellison et al. 2005; Nadrowski et al. 2010; Gamfeldt et al. 2013). Uncertainties regarding the response of TMFs to climate are large, in part, because high species diversity could lead to numerous developmental trajectories.

TMFs are characterized by abundant precipitation and moderate temperatures, although monthly ranges of both vary across regions (Fig. 5.1). Climate in TMF regions leads to relatively high productivity, which can be a factor of the broad array of intensive land-use in these regions (Kaplan et al. 2009; Munoz et al. 2014). Importantly, TMFs are species rich or are hotspots of diversity outside topical zones.

Fig. 5.1 Examples of dense, temperate mesic forests around the Northern Hemisphere. The inset represents the structure of climate at the center point of each region except the rainforests of the Pacific Northwest of North America (PNW), where data is from the southern end of the study area. Clockwise from *upper left*: (*1*) Great Smoky Mountains, eastern U.S.A.; (*2*) Colchic Temperate Rainforest, eastern end of the Black Sea; (*3*) rainforests of the PNW; (*4*) Central Europe; and (*5*) Central Japan

Common genera of these regions include *Abies*, *Acer*, *Betula, Fagus*, *Quercus*, *Picea,* and *Pinus*. Eastern Asian and eastern North American TMFs have the highest diversity within the biome and are dominated by broadleaf species, including *Carya*, *Liriodendron*, *Magnolia*, *Nyssa*, *Tilia*, and *Ulmus* (Latham and Ricklefs 1993). In contrast, the TMFs in the Pacific Northwest of North America (PNW) are dominated by conifers (*Abies, Callitropsis, Pseudotsuga, Thuja,* and *Tsuga*). The Colchic Temperate Rainforest of western Eurasia, dominated by *Fagus, Abies, Quercus,* and *Picea,* is comparatively less diverse but includes many Arcto-Tertiary relict species and adds many mesophytic species in a region that is more broadly semi-arid. Species diversity is important in TMFs as it might impart resistance to extreme climatic events (Isbell et al. 2015) or reduce annual variations in photosynthetic capacity (e.g. Morin and Chuine 2014; Musavi et al. 2017).

The current compositions of TMFs are influenced by the time since species pool formation (e.g. Davis 1983), contingencies of past environmental conditions (means and extremes) (Ordonez and Svenning 2016a, b), and centuries and millennia of human land-use (Tinner and Lotter 2006; Schwörer et al. 2015). Paleoecological studies have documented large-scale changes in temperate, mesic biome vegetation in response to climate change at continental and interhemispheric scales (e.g. Svenning 2003; Gajewski et al. 2006), suggesting that forecasts indicating significant impacts by future climate change on TMFs are possible. Some of these changes have impacted biodiversity. Since the Pleistocene, for example, glaciations in Europe triggered successive extinctions of genera that persist in eastern Asia and eastern North America, notably *Liriodendron*, *Magnolia*, and *Carya* (Latham and Ricklefs 1993). As a result, early-twenty-first century European forests have relatively low diversity and are dominated by more drought and cold tolerant species than other TMFs (Svenning 2003; Liebergesell et al. 2016). We posit that ecological research conducted at the spatial scales of dynamics documented by paleoecological research is needed to better understand the potential trajectories of temperate, mesic forests.

Our goals here are to: (1) review the main drivers of disturbance dynamics in four TMF regions, (2) synthesize common disturbance dynamics, and (3) discuss new dendroecological approaches and concepts for TMFs, such as the idea of a dynamic equilibrium at larger scales. One underlying theme is to foster discussion about the ecology of these dense, structurally complex, and diverse forests. We believe that meso- to macroecological scale research using dendrochronological techniques will aid ecological theory on long-term forest development and assist in the continued improvement of vegetation models.

5.1.1 Temperate, Mesic Forest Development Theory

Temperate, mesic forests were often considered to be in a kind of a dynamic equilibrium, or shifting-mosaic steady state, at large scales where the proportion of the species is relatively constant and the landscape is covered by a full range of successional stages, the "steady-state landscape or mosaic-cycle concept of ecosystems" (Botkin et al. 1981; Remmert 1991; White et al. 1999). The concept of a dynamic equilibrium appears to be especially applicable to TMFs because asynchronous, but low-impact, tree mortality over space occurring at relatively high frequency are important drivers of forest structure and function (Runkle 1982, 1985; Yamamoto 1992). Through this process, long-lived, shade-tolerant species replace short-lived pioneer species that dominate recruitment after canopy gap formation. A dynamic equilibrium is then achieved at large scales due to the spatially-asynchronous, high-frequency nature of these events.

The concept of the shifting-mosaic has been challenged for decades (Sprugel 1991; White and Jentsch 2001; Jentsch et al. 2002). Changes in climate, for example, can alter forest dynamics beyond known successional pathways by moving species or processes beyond their resilience thresholds (White and Jentsch 2001), either directly or through complex interactions with myriad disturbances agents. In this context, the spatial and temporal scales of research are critical to describing and understanding the ecological processes underlying forest development. The scales of research are also critical in the search for generalizations beyond one specific ecosystem, i.e., "macroecology" (Brown 1995; Gaston and Blackburn 2008).

One unsettled question around the concept of a dynamic equilibrium is whether disturbances create this equilibrium at larger scales (White et al. 1999). As a rule of thumb, Shugart (1984) suggested that a landscape equilibrium could be achieved if the landscape was 50 times larger than the average disturbance patch. A fundamental limitation in addressing this question, however, is that many studies in TMFs are conducted at stand to landscape scales or over short time-periods; most are likely smaller than 50 times the average patch size. As a result, the prevailing scales of research are more likely to reveal discrete events that highlight frequent, local, and small-scale dynamics, but not likely to be sufficient to answer the question of a dynamic equilibrium at large spatial scales.

Centennial- to millennial-scale paleoecological research indicates that large, infrequent disturbances (LID) are known to be important for TMFs (Foster et al. 1998; Romme et al. 1998). But, by their own nature, LIDs are rare over time and less-well documented than most disturbances in these systems. The occurrence and impact of LIDs suggest there is a need for a conceptual model that further disrupts the concept of a dynamic equilibrium as well as the expected forest development trajectories based upon the prevailing scales of research.

5.1.2 Definitions

Additional definitions regarding the scale and intensity of disturbance in temperate, mesic forests are needed before we move to region-specific reviews. Large, infrequent disturbance are known to affect forests at regional or larger scales (e.g., >50,000 ha) and are likely to disrupt quantitative equilibrium (Romme et al. 1998). Potential triggers of LIDs include intense, prolonged or repeated drought, heatwaves, or hot droughts (Allen and Breshears 1998; Ciais et al. 2005; Michaelian et al. 2011); increased fires linked to larger scale atmospheric teleconnections such as ENSO (Swetnam and Betancourt 1990); and warming that synchronizes insect outbreaks (Raffa et al. 2008).

A new term, megadisturbances, speaks to a potential increase in severity of disturbance at scales larger than most disturbances observed during the twentieth century (Millar and Stephenson 2015). A megadisturbance, then, would seem to have the ability to drive temperate forests over their threshold or beyond their resilience. Both LIDs and megadisturbances occur at large scales and can resonate at broad spatial scales for centuries, creating ecological legacies. In contrast to megadisturbances, however, LIDs can be of lower severity. For example, a strong drought affecting large areas of TMFs and inducing an increased rate of tree mortality, mainly impacting the most drought-vulnerable species, would constitute a LID. If the return interval of this LID is shorter than the time required to reach the hypothetical dynamic equilibrium, this equilibrium would not occur at very large spatial scales. LIDs and megadisturbances have been most frequently documented and studied in semi-arid, dry, or cool regions. In TMF regions, documentation of LIDs comes almost exclusively through sediment core analysis where temporal and taxonomic resolution can be low. New approaches in dendroecology can enhance current theory in the development of TMFs.

5.2 Ecological and Dendroecological Case Studies

5.2.1 Eastern North American Broadleaf-dominated Forest

The dynamics of the diverse eastern North American TMFs, composed primarily of *Abies, Acer, Betula, Carya, Fagus, Liriodendron, Quercus, Pinus, Picea,* and *Tsuga*, are driven by a wide range of disturbances. The predominant agents of natural disturbance include windstorms (hurricanes, tornadoes, derechos), fire, drought-induced mortality, and the fine-scale dynamics of background mortality (White 1979; Runkle 2000; Dietze and Moorcroft 2011; Vanderwel et al. 2013). Windstorms have substantial and long-lasting impacts on forests, especially when they lead to stand-replacing disturbances (Lorimer 1977; Canham and Loucks 1984; Foster and Boose 1992; Everham and Brokaw 1996; see Peterson et al. (2016) for a recent synthesis). Windstorms do not impact ecosystems equally. Topography, stand structure, and composition influence the impact of windstorms (Canham et al. 2001; Woods 2004; Pearson 2010; Stueve et al. 2011; Fahey et al. 2015). Forests on exposed sites, and some species in certain forest structures (large trees, pine species) are generally more vulnerable to windstorms (Putz and Sharitz 1991; Foster and Boose 1992; Canham et al. 2001). Windstorms with intermediate frequency, magnitude, and intensity have a substantial cumulative impact—roughly 25,000 ha of forest in the northern Great Lakes region experienced severe damage over a 25-year period (Stueve et al. 2011).

Dendroecological and sediment-core studies show that fire occurred regularly in eastern North America (Guyette et al. 2002, 2006; Power et al. 2008; Clifford and Booth 2015). Dendroecology studies have shown that fire strongly influences regeneration, composition, and stand structure in conifer-dominated ecosystems within the broadleaf forest matrix (Aldrich et al. 2010; Hessl et al. 2011; Flatley et al. 2013), especially on dry, low productivity sites (Grissino-Mayer 2016). Fire impacts in *Quercus*-dominated forests, however, are less resolved (Matlack 2015; Stambaugh et al. 2015). Fire in the *Quercus*-dominated forest often occurs as surface fires that primarily kill small trees and crown fires are rare (Abrams 1992; Grissino-Mayer 2016). Because fires are larger and more frequent during drought (Lynch and Hessl 2010; Lafon and Quiring 2012; Grissino-Mayer 2016), the extreme 2016 drought in the Southern Appalachian Mountains highlights that more intense fire can occur in extreme conditions (Akpan 2016). Although replication is relatively low, dendroecological research indicates that fire was relatively frequent in *Quercus*-dominated forests during the late 1600s (Guyette et al. 2006) and increased as Europeans immigrated across the region (McEwan et al. 2007; Grissino-Mayer 2016). Fire then declined early in the twentieth century, coinciding with a decrease in *Quercus* and rise of mesophytic species in forest understories (Nowacki and Abrams 2008; McEwan et al. 2011). An analysis of dendroclimatic reconstructions of the Palmer Drought Severity Index (PDSI) for this region reveals a long-term wetting trend, which could be one of many factors driving both the rise of mesophytic species and a reduction in fire (McEwan et al. 2011). Continued fire research in *Quercus*-dominated forests is needed to understand the full ecological impact of fire in this region (Varner et al. 2016).

Dendroecology and a range of other paleo-proxies show that moisture availability is an important driver of forest dynamics (Cook 1991; Clinton et al. 1993; Foster et al. 2006; Choat et al. 2012; Martin-Benito and Pederson 2015; Booth et al. 2016; Charney et al. 2016). Drought was observed to kill trees during the first part of the twentieth century across eastern North America (Millers et al. 1989). In some areas, however, drought-induced mortality has not been recorded over the last 30–50 years (Karnig and Lyford 1968; Stringer et al. 1989), which is consistent with increased moisture availability in recent decades over much of this region (Pederson et al. 2015). Recent tree mortality in the southeastern U.S. has been tied to increased moisture stress (Haavik et al. 2015; Berdanier and Clark 2016). Given these climate-forest dynamics, it is reasonable to surmise that a future increase in drought frequency and severity could lead to higher mortality levels and favor drought- and fire-adapted tree species like *Quercus* and *Carya* (Clark et al. 2016).

Dendroecology in old-growth forests has a long history in the eastern North American TMF (Marshall 1927; Henry and Swan 1974; Lorimer 1980; Frelich and Lorimer 1991; Abrams and Orwig 1996; Chokkalingam and White 2001; Orwig et al. 2001; Fraver et al. 2009; Hart et al. 2012a). Many of these studies have been conducted at small spatial scales (i.e., stand level) and generally support gap-phase dynamics. Over the last decade, dendroecological research began to expand to include larger areas up to subcontinental scales (Rentch 2003; D'Amato and Orwig 2008; Pederson et al. 2014). Some of these studies support hypotheses of episodic disturbance at large spatial scales (Oliver and Stephens 1977; Foster 1988; Cho and Boerner 1995; Woods 2004) and show evidence of such events from sediment core studies (Davis 1983; Grimm 1983; Webb 1988; Peteet et al. 1990; Shuman et al. 2002; Ordonez and Williams 2013; Clifford and Booth 2015). To refine large-scale change in vegetation temporally and taxonomically in this region, dendroecological studies should be conducted at broad spatial scales.

5.2.2 *Temperate Rainforest of Northwestern North America*

In the wetter portions of the coastal temperate region of the Pacific Northwest of North America, from the Olympic Peninsula, Washington, U.S. to southeastern Alaska, U.S., dense rainforests are composed primarily of *Abies, Callitropsis, Thuja, Psuedotsuga,* and *Tsuga*, with some trees exceeding 60 m in height and having diameters up to 350 cm. Natural disturbance processes in this region are strongly influenced by a climate with high annual rainfall and cool temperatures (Fig. 5.1; Alaback 1991, Meidinger and Pojar 1991, DellaSala et al. 2011). Diverse topography and site characteristics impact common disturbance agents, including fire, wind, geomorphic events (i.e. mass wasting, avalanches), climatic events (flooding, frost, drought), and mammal browsing (Dorner and Wong 2002; Franklin et al. 2002; Wong et al. 2003; Alaback et al. 2013). Though often generalized, there is substantial diversity in stand conditions, disturbance, and potential developmental trajectories of forests in this region.

Stand-replacing fire and windthrow are infrequent disturbances restricted to specific topographic settings in the wetter rainforests of the PNW. Fire mostly occurs on south-facing slopes, with mean return intervals sometimes greater than 1000 years and some sites not burning for thousands of years (Lertzman et al. 2002; Gavin et al. 2003a, b, 2013; Hallett et al. 2003). Indigenous peoples are hypothesized to have created spatial and temporal complexity by interrupting the long intervals between fires with low-severity fires (Hoffman et al. 2016). When stand-level windthrow occurs, it mostly takes place on exposed portions of the landscape and not in protected valleys (Kramer et al. 2001; Harcombe et al. 2004; Pearson 2010). During the long intervals between severe disturbance, fine-scale gap dynamics lead to structurally complex forests (Lertzman et al. 1996; Daniels and Gray 2006; Alaback et al. 2013). These disturbance dynamics account for most of the structure of unmanaged, productive, upland forests throughout the region.

Dendroecology has improved our understanding of gap dynamics in coastal temperate rainforests. In mid-elevation, old-growth stands on the South Coast of British Columbia, Canada, Daniels and Klinka (1996) used dendroecology and stem maps to estimate growth releases and delineate past canopy gaps. Findings suggested that growth releases vary among shade-tolerant trees moving into the canopy of these stands, implying different strategies among the study species that might contribute to their coexistence (Daniels and Klinka 1996). Pursuing this further, Stan and Daniels (2010a, b, 2014) found a suite of variables including, tree diameter, growth rate prior to release, and distance from the gap center, influenced the magnitude and duration of a release. Also combining stem maps and dendroecology, Parish and Antos (2004, 2006) found that fine-scale gaps are the dominant disturbance process among South Coast forests with very old, slow-growing, stress-tolerant trees. These examples represent the few dendroecological studies of gap dynamics in the region.

Projections of warmer and wetter conditions with a decrease in total precipitation as snow are expected to result in a series of ecosystem-level effects throughout the PNW by the end of the century (Wang et al. 2012; Shanley et al. 2015). The decline of *Callitropsis nootkatensis* in southeastern Alaska is one climate-driven phenomenon that has been altering ecosystems since the end of the Little Ice Age (Hennon et al. 1990). In this case, warming temperatures and reduced snowpack appears to have triggered a widespread death of this species (Hennon et al. 2012). If warming expands this decline into more productive, upland forests, the fine-scale processes that dominate stand dynamics could be greatly altered: progressive

Callitropsis nootkatensis decline and mortality would likely create conditions for a relatively synchronous canopy disturbance. Furthermore, an anticipated warmer and drier summer climate (Wang et al. 2012) might increase the rate of tree mortality and gap formation (van Mantgem et al. 2009). Undoubtedly, dendroecology can help us understand the impacts of altered disturbance regimes on PNW forest communities over time.

Special Highlight: The Colchic Temperate Rainforest of Western Asia
On the eastern Black Sea coast resides the Colchic temperate rainforest, a dense forest with trees up to 45 m in height (DellaSala 2011). At lower elevations, broadleaf forests are mainly comprised of *Fagus orientalis* and a mix of species including *Castanea sativa*. Coniferous forests (i.e., *Picea orientalis* and *Abies nordmanniana*) develop in somewhat dryer locations. Abundant *Rhododendron*, *Ilex*, and *Laurocerasus* characterize the subcanopy of these systems. The Colchic temperate rainforest contains one of the most important relicts of the Arcto-Tertiary forests in western Eurasia (Connor et al. 2007).

Because precipitation is abundant (>2000 mm) and evenly distributed throughout the year (Fig. 5.1), changes in natural vegetation are generally thought to be gradual; humans are seen as the main triggers of abrupt change (Connor et al. 2007). Precipitation, however, can vary significantly during critical parts of the growing season (Martin-Benito et al. 2016). With this insight on climatic variability, it is reasonable to hypothesize that extreme climate could initiate abrupt changes in vegetation. Paleoecological research shows evidence of these changes in the past (Connor et al. 2007), which make the possibility of significant alteration in the ecology of these forests in the future realistic. In light of modeling forecasts of a subtropical climate in this region, the Colchic temperate rainforest could expand into areas with low species diversity or low biomass and, as a result, increase the species diversity of these adjacent areas.

5.2.3 European Broadleaf-Dominated Forest

In the mixed temperate forests of European, comprised principally of *Acer*, *Fagus*, *Picea*, *Pinus*, *Larix*, and *Quercus*, windstorms and bark beetles (e.g., *Ips typographus* mainly affecting *Picea abies*) are the main disturbance agents at landscape scales. Windstorms primarily drive successional dynamics through gap creation and the loss of biomass (Pontailler et al. 1997; Ulanova 2000; Seidl et al. 2014). Bark beetle disturbance could increase tree mortality if future abiotic stress weakens trees and predisposes them to insect attacks (Seidl et al. 2014). Windthrow, however, can also facilitate bark beetle infestations (Temperli et al. 2013). Disease-induced mortality events are significant disturbances, but they occur either at the species level or in very specific situations (Robin and Heiniger 2001; Dobrowolska et al. 2011). Fire can have some of the strongest impacts through modification of stand developmental trajectories to earlier successional stages or limiting species composition, particularly at the southern border of the TMF or in the Southern Alps (Ascoli et al. 2013; Maringer et al. 2016); fire is less frequent in the central broadleaf forest (Niklasson et al. 2010). The most common types of disturbance observed in European TMFs are low-intensity, but high-frequency processes that make up the general background mortality rate.

Drought is increasingly considered an important disturbance agent at continental scales (Ciais et al. 2005). Although drought mortality events are rare (though see Siwkcki and Ufnalski 1998; Bréda et al. 2006), and mostly limited to the southern edge of the biome (Camarero et al. 2011; Cailleret et al. 2014), they are reported as far north as Norway (Solberg 2004). In contrast to prevailing expectations in European forests, a dominant species, *Fagus sylvatica*, appears to be more drought-sensitive at the core of its range than at its southern limit (Cavin and Jump 2017). Climate projections suggest that drought will increase, making European TMFs potentially more vulnerable to drought at large scales (Elkin et al. 2013; Temperli et al. 2013) and exacerbating the impacts of other disturbance agents (Seidl et al. 2017).

Important to long-term development of European temperate forests are the interactions among multiple disturbance agents. The combination of windthrow or drought with an outbreak of bark beetles can result in a spatially-extensive disturbance and, occasionally, extirpation (Temperli et al. 2013). Of all identified natural disturbances in the region, drought might be the strongest factor facilitating other agents (Siwkcki and Ufnalski 1998; Bréda et al. 2006). Forest decline often follows interactions among multiple stressors that compound the impact of a single disturbance agent (Manion 1981; Temperli et al. 2013). Some interactions between disturbance agents could have a negative feedback or a long-term decline in disturbance (Seidl et al. 2017). The combination of drought and bark beetle outbreaks, for example, might lead to the disappearance of the beetle's host (Temperli et al. 2013).

Evidence suggests that synchronization of stand dynamics and forest development across large scales could occur if severe droughts or heat waves (Ciais et al. 2005; Cook et al. 2014, 2015) exacerbate the impact of large windstorms, like storm Vivian in 1990 and Lothar in 1999. Drought, for example, can induce growth declines across species over wide areas. Paleoecological studies in central Europe show that forest composition was affected in part by drought over the Holocene (Tinner and Lotter 2001) and that the most drought sensitive species (e.g., *Fagus sylvatica* or *Abies alba*) could become locally extinct a few decades after extreme drought (Tinner and Lotter 2001; Camarero et al. 2011). Given that interactions among multiple disturbance agents can increase the severity and spatial extent of disturbance, we hypothesize that regional to continental-scale disturbances are important, but poorly understood, processes of long-term forest development in the region.

To date, dendroecological research in Europe has provided little evidence of regional to continental-scale disturbance and, in general, broadly documents gap-dynamics as the main process affecting forests (e.g. Trotsiuk et al. 2012). The long history of human land-use and small tracts of old-growth forest (Nagel et al. 2007) might be the reason that many tree-ring studies have been conducted in second-growth forests (Cherubini et al. 1996; Emborg et al. 2000; Rozas 2001, 2004; Rozas et al. 2008) or at small spatial scales in old-growth forests (Piovesan et al. 2005; Splechtna et al. 2005; Nagel et al. 2007; Svoboda et al. 2012; Trotsiuk et al. 2016). Dendroecological research at broad spatial scales will help in testing the hypothesis that regional to continental-scale disturbance is important in the long-term development of TMFs in Europe.

5.2.4 Broadleaf-Dominated Forests of Japan

Typhoons and *Sasa* spp. (Bamboo) are important factors of TMF dynamics in Japan, with typhoons creating canopy gaps and initiating tree regeneration. Japanese temperate forests are divided into mid-elevation forests dominated by broadleaf species (*Castanopsis, Fagus, Quercus*) and high-elevation forests of mixed broadleaf and conifer species (*Abies, Betula, Pinus, Picea*). Many studies have attempted to detect the occurrence of typhoon damage using ecological and dendroecological techniques (e.g., Franklin et al. 1979; Kanzaki and Yoda 1986; Kubota and Hara 1995; Namikawa 1996), but interestingly, intense typhoons do not appear to alter the rate of gap formation (Naka 1982; Kanzaki and Yoda 1986; Yamamoto 1996; Ida 2000; Altman et al. 2016; Ariya et al. 2016). Synchronous regeneration of trees has also been reported in this region over areas up to several square kilometers (Numata 1970; Nakashizuka 1987). If synchronous recruitment occurs at large scales and if severe typhoons do not change rates of gap formation, an important question is, "*What are the drivers of large-scale regeneration?*" One potential driver is the episodic dynamics of *Sasa*. Following blooming, culms die synchronously and create space for tree establishment (Numata 1970; Nakashizuka 1987; Taylor et al. 2004). A series of successional trajectories are then possible: trees fill the *Sasa* gap, *Sasa* regrows and "maintaining" its space, or there is a change in species composition. A dense *Sasa* canopy can lead to the creation of larger canopy gaps (Nakashizuka 1987). In this case, if gaps are not filled, canopy edges are at a greater risk of wind damage, potentially leading to expanded gaps. Therefore, interactions between *Sasa* dynamics and typhoon frequency could drive synchronous regeneration dynamics.

High-elevation forests in central Japan offer an interesting opportunity for dendroecological research in TMFs. Above the mixed subalpine forest, treeline is composed of the broadleaf *Betula ermanii*. There has been little to no *Betula ermanii* regeneration in the subalpine forests in recent decades (Kohyama 1984; Yamamoto 1993, 1996). *Betula ermanii* has been observed to establish on mineral soil created by uprooted gapmakers (Kohyama 1984; Nakashizuka 1987) and on elevated surfaces such as logs (Hiura et al. 1996) and in areas with continuous disturbance at small-scales (Kubota and Hara 1995). But, while one study found that 80% of dead *Betula ermanii* are uprooted, it seemed that the majority of these trees were too small to produce much mineral soil (Kanzaki and Yoda 1986). *Abies* can create important substrate for *Betula ermanii* regeneration in the form of logs, but they have often been observed to have remain as standing snags (Kanzaki and Yoda 1986; Ida 2000). The lack of *Betula* regeneration in these forests, despite frequent disturbance, point to alternative modes and processes of forest succession and development.

Dynamics of TMFs in Japan are also determined by interactions between elevation and forest composition. Canopy gaps in mid-elevation forests occupy 13–24% of the landscape (Yamamoto 1996) while canopy gaps in mixed coniferous subalpine forests occupy 4–20% of the landscape (Yamamoto 1996; Takahashi 1997; Narukawa and Yamamoto 2001). Curiously, gap sizes depend partly on species composition in mid-elevation forests but not in high-elevation forests. In mid-elevation forests dominated by *Quercus* and *Castanopsis*, gaps are typically ten times larger than those in *Fagus* dominated forests (ca. 700 m^2 vs. 70 m^2) (Naka 1982; Nakashizuka 1987; Hiura 1995; Ida 2000) while the average gap size in high-elevation forests is 100 m^2 regardless of species. Differences in gap-size between these forests appear independent of age and perhaps are related to the relatively-larger crowns of broadleaf trees.

Dendroecology has provided substantial insight into recruitment dynamics in mid-elevation TMFs in Japan. One study in a mid-elevation old-growth forest dominated by *Fagus* found the recruitment and initial radial growth of three species (Ariya et al. 2016) follows observed traits and behaviors of various tree species (Yamamoto 1995; Yoshida and Kamitani 1998; Abe et al. 2005). Shade-tolerant *Fagus* exhibits low radial-growth and recruits into the forest from advanced regeneration. In contrast, wide rings early in the life of shade-intolerant *Betula grossa* and *Magnolia obovata* indicate gap-origin trees. The coexistence of these species in the canopy is likely the result of severe disturbances or large gaps, although no severe disturbances were found after 1900 (Ariya et al. 2016).

Using multiple lines of evidence, dendroecology can likely help disentangle the spatiotemporal interactions between typhoons and *Sasa* and aid in understanding the puzzling issues of regeneration dynamics at high-elevations. For instance, shade-tolerant *Abies* recruits through advanced regeneration at high-elevation, (Mori and Komiyama 2008) while shade-intolerant *Betula* requires large gaps. Therefore, dendroecological research can determine the timing of gap creation following *Betula* recruitment (Yamamoto 1996) or the date of growth release in *Abies* (Abrams et al. 1999). As dendroecology becomes more frequently utilized in deciphering the ecology of TMFs in Japan, multiple lines of evidence from these studies can help determine mechanisms of long-term development in the region.

5.3 Synthesis of Temperate, Mesic Forest Literature

Constant, but low-impact and stochastic tree mortality ensures the perpetuity of fine-scale dynamics in TMFs (White 1979; Yamamoto 1992; Rebertus and Veblen 1993; Lertzman et al. 1996). Whether through the low rate of asynchronous gap dynamics (e.g., tree mortality of 1–3%/year in the eastern U.S. (Runkle 1998), turnover times of 350–950 years in the PNW (Lertzman et al. 1996)), the often asynchronous patterns of large-scale fire (Clark and Royall 1996, Parshall and Foster 2002, Guyette et al. 2006; though see McMurry et al. 2007), or intense, but spatially-limited windstorms (Everham and Brokaw 1996; Foster et al. 1998; DellaSala 2011; Vanderwel et al. 2013), these disturbances generally do not match the spatial scale or synchrony of disturbance anticipated by some model forecasts due to the intensification of anthropogenic climate change. The mismatch might be related to the prevalent scales of research and the rarity of extreme climate events during the short periods of observation that dominate most studies.

Large, infrequent disturbances are a component of disturbance theory in TMFs (Oliver and Larson 1990, Foster et al. 1998, White et al. 1999, Bormann and Likens 1994), and have been documented by paleoecology from regional to hemispheric scales (Svenning 2003; Gajewski et al. 2006). Dendroecology in TMFs has provided some evidence for LIDs, with a few studies suggesting they could be more frequent than what is commonly expected (Pederson et al. 2014; Rodríguez-Catón et al. 2015; Sánchez-Salguero et al. 2015). A lack of studies conducted at large spatial scales might overemphasize the asynchrony of disturbance in TMFs, which can lead to inaccuracies in forecasting long-term dynamics (*sensu* Wiens 1989).

Despite theory and despite evidence of LIDs, the preponderance of evidence for disturbance in TMFs centers on frequent and fine-scale processes. While there are many potential reasons for this, we focus on a few key factors. First, the small scale of most dendroecological studies likely limits the potential to observe the full spatial extent of past disturbance, though see (Rentch 2003; D'Amato and Orwig 2008; Trotsiuk et al. 2014, 2016). Second, there is a miniscule amount of remaining old-growth forest in TMFs, and the common longevity of trees in these forests, from 300–400 years (Di Filippo et al. 2015), likely limits the discovery of LIDs. Finally, of the studies in contemporary forests conducted at broad scales, most are time-limited to decades or less, which creates a mismatch between the time scale of research and the slow processes of forest development. Repeated surveys of old-growth forests, for example, have indicated: (1) a relatively stable canopy with essentially no change in composition and the lack of a severe, landscape-scale disturbance over three decades (e.g. Chapman and McEwan 2016) or (2) only one severe landscape-scale disturbance event over the course of seven decades (e.g. Woods 2004). A dendroecological investigation in the same forest studied by Chapman and McEwan (2016) indicated that: (1) two pulses of *Quercus* recruitment occurred ca. 200 years apart and (2) long periods of suppression before canopy accession (McEwan et al. 2014). Similarly, a dendroecological study in a *Pseudotsuga menziesii* stand in the PNW found three recruitment pulses 150–200 years apart over 500 years (Winter et al. 2002). We know of no repeated surveys that have observed processes where recruitment pulses effectively led to canopy trees centuries after a disturbance. Thus, the development of dendroecological networks in TMFs, like the fire-scar networks in the semi-arid US (Swetnam and Betancourt 1990; Falk et al. 2011), can aid in understanding the dynamics of mesic forests across spatial and temporal scales.

Finally, interactions between disturbances are important when considering forest dynamics. Repeated drought, interactions with other disturbance agents (frost, beetles), or heat-enhanced disturbance often elevate tree mortality (e.g. Long 1914; Hursh and Haasis 1931; Pedersen 1998; Ciais et al. 2005; Seidl et al. 2014; Allen et al. 2015; Haavik et al. 2015). Similarly, disturbances acting upon vegetation "substrate" (forest structure, composition) interact with historical land-use. In regions where protected forests are recovering from prior land-use (heavy cutting, agricultural abandonment), the accumulation of large trees and biomass has the potential to increase their vulnerability to disturbance and climatic extremes (e.g. Bennett et al. 2015; Kauppi et al. 2015).

5.4 Advancement in Disturbance Theory with Basic Dendroecological Techniques

New applications of well-established dendrochronological techniques have the potential to significantly advance existing theory of forest development. Dendrochronological crossdating provides the dating precision to analyze extreme events that would otherwise be effectively muted, masked, or shifted in time (Black et al. 2016). Crossdating, unfortunately, has not always been conducted in all tree-ring studies, which is understandable given the known difficulty of dating some species and particularly understory trees (e.g. Takaoka 1993; Lorimer et al. 1999; Grundmann et al. 2008). Notably, some understory diffuse-porous trees, have weak chronology interseries correlations ($R < 0.50$ with ≥ 20 series) and many locally-absent rings (LARs; Table 5.1). For example, 80 *Betula ermanii* trees produced 242 LARs compared to only 8 LARs from 100 *Abies mariesii* trees in central Japan (Young 2016). In the northeastern U.S., the number of LARs per radius in diffuse porous species varied, ranging from 10–16 rings in *Acer rubrum* and up to 18 LARs in one *Betula alleghaniensis* (Table 5.1). In one extreme case from Turkey, 24 LARs were found in a *Fagus orientalis* over only 32 years of growth (D. Martin Benito, unpublished data). For comparison, these rates of LARs in TMF collections are much greater than rates in most International Tree-Ring Databank collections (St George et al. 2013). Interestingly, we have found the rate of LARs

Table 5.1 Rate of missing rings in and site-level correlations of dendroecological collections in the northeastern U.S.

	% Missing rings (# measured rings)		% Missing rings after 1970 (# measured rings)		Site-level interseries correlations (Spearman)
					Understory, Overstory (size = # of series)
Species	Understory	Overstory	Understory	Overstory	0.3 | 0.4 | 0.5 | 0.6
Acer rubrum	2.1%	0.9%	4.1%	1.4%	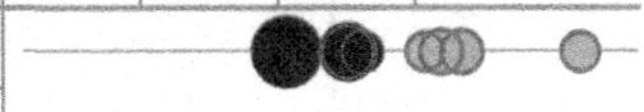
	(19047)	(10303)	(8625)	(4156)	
Acer saccharum	6.3%	1.3%	10.2%	0.7%	
	(4880)	(17937)	(2261)	(6276)	
Betula alleghaniensis	2.7%	0.4%	4.2%	0.9%	
	(5522)	(1380)	(3099)	(681)	
Fagus grandifolia	0.7%	0.4%	0.9%	0.1%	
	(16656)	(114047)	(8069)	(4125)	

Interseries correlations are derived from populations with ≥20 series. Sample size is represented by the size of each circle. Of the 18 populations, the range is 20–119 series with a median of 33. *Black circles* = understory trees, *grey circles* = overstory trees

in *Acer rubrum* in the northeastern U.S. has increased since the 1970s (Fig. 5.2a), which is a phenomenon that also appears in *Betula ermanii* trees in central Japan (Young 2016). High-rates of LARs in TMFs underscores the difficulty, as well as the necessity of crossdating all samples to the best of one's ability.

Crossdating matters in reconstructing stand history. When we do not account for LARs in *Acer rubrum* at the Harvard Forest in the U.S., it can appear that the 1938 hurricane was instrumental in recruitment of this species. When recruitment history is based on crossdated samples, however, it becomes evident that a disturbance in the 1920s triggered a spike in *Acer rubrum* recruitment (Fig. 5.2b). Similar findings were made in *Betula ermanii* living in a mixed TMF in northern Hokkaido, Japan, where the failure to account for LARs led to an incorrect reconstruction of age structure (Takaoka 1993). Despite suggestions from some early studies in TMFs, suppressed trees can be crossdated (Table 5.1). These and other dendrochronological studies make clear the importance of crossdating understory and overstory trees to better understand forest dynamics. Such work has revealed information not often noted in silvics manuals, such as centuries-long understory persistence of *Quercus* and other species, despite being classified as shade intolerant or mid-intolerant to overstory competition (Black et al. 2008; Pederson 2010).

Advancements in forest ecology theory can also be made through the *cautious* analysis of annually binned data. This kind of analysis increases the likelihood of correctly attributing drivers of disturbance while at the same time potentially

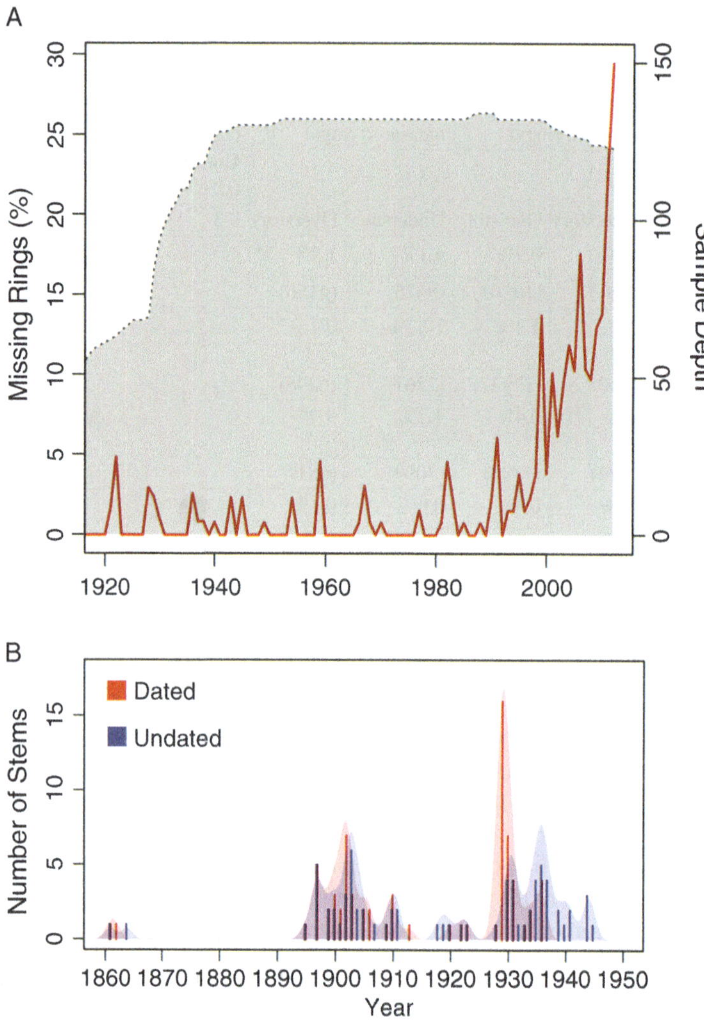

Fig. 5.2 The propensity of missing rings in *Acer rubrum* per year (**a**) and how it affects age structure analysis of trees estimated to be <5 years from pith or at pith (**b**). *Grey* in (a) represents the number of time series while the *red line* is the annual percentage of missing rings per time-series present. Recruitment in (**b**) is annually binned. Peaks are identified using a probability density function with a Guassian kernel with a smoothing bandwidth of one following the method of Trotsiuk et al. (submitted). A significant shift in age structure is observed with undated trees beginning in the mid- to late-1930s

revealing episodic dynamics that could be obscured with decadal binning, as shown in the *Acer rubrum* example above (Fig. 5.2b). Annual binning might not be completely necessary, but the combination of crossdating and binning data at bins smaller than decades can improve our understanding on the episodic (or not) nature of forests.

Finally, dendroecological analysis of historic timbers can recover long-forgotten forest dynamics (Trouet et al. 2017; de Graauw 2017). Such studies are potentially powerful in documenting past LIDs or where human land-use reduced most of the old-growth forest (Trouet et al. 2017). As TMF dendroecology follows the pioneering ecology of western U.S. forests at large spatial scales over hundreds of years (e.g. Brubaker 1980; Swetnam and Betancourt 1990), evidence of synchronous dynamics at large scales has increased (Pederson et al. 2014; Rodríguez-Catón et al. 2015; Sánchez-Salguero et al. 2015). These efforts need to continue. Precisely-dated records have revealed patterns of disturbance that challenge long-held theories on dynamic equilibriums in forested ecosystems.

5.5 A Temperate Mesic Forest Conundrum

We have noticed a bit of a conundrum in the concept of a dynamic equilibrium in broadleaf-dominated TMFs—some early successional species often make up substantial portions of old forests. This phenomenon is well documented in coniferous forests of the PNW where *Pseudotsuga menziesii* can live 400–700 years and in the forests of southwest Japan where *Cryptomeria japonica* can live 2000–3000 years (Veblen 1986; Suzuki and Tsukahara 1987). Increasingly, *Liriodendron tulipifera* trees 350–510 years old, beyond the presumed maximum age for the species, are being found in the southeastern U.S. (Pederson 2010). In central Japan, recent studies in mid- and high-elevation forests are finding *Betula* spp. trees that are still present in the canopy after 150–300+ years (Ariya et al. 2016; Young 2016). These ages and age structures do not fully reflect the paradigm of broadleaf-dominated TMFs. The presence of old, early-successional trees in TMFs show that they are meaningful to their long-term structure, function, and development.

The reasons as to why early-successional species compose a significant amount of old forest are not entirely clear. In general, longevity of shade-intolerant or pioneer species is lower than that of shade-tolerant or later-successional species. Despite often having relatively high rates of growth, some shade-intolerant species, however, are being found to reach very old ages even in closed canopy forests (e.g. in North America *Betula lenta*, *Liriodendron tulipifera* in the east or *Pseudotsuga menziesii* in the west). As such, there is a signal starting to arise above the noise in TMFs as well as in tropical forests (e.g. Baker et al. 2005; Baker and Bunyavejchewin 2017): recruitment pulses at large scales and early-successional species that occupy the forest canopy for centuries is evidence that severe disturbance is another key process of forest development. Put another way, centuries of fine- and intermediate-scale processes do not fully erase the legacies of LIDs or possible megadisturbances in TMFs.

5.5.1 New Evidence of Large-scale Disturbance in Temperate Mesic Forests

As the scale of dendroecological research in TMFs increases, there is more evidence that large-scale events occurred within the life span of living trees. A synthesis of tree recruitment dates from old-growth broadleaf-dominated forests in the eastern U.S. makes evident of what appears to be a recruitment pulse in the mid- to late-1600s over ca. 1.4 million km^2 (Pederson et al. 2014). An analysis of a collection created during the early-1900s that targeted old cross-sections from stave (barrel) mills (and similar sources) also finds a pulse of tree recruitment in the mid- to late-1600s in the western portion of the eastern U.S. (Trouet et al. 2017). Additionally, the age structure of dendroclimatological collections made during the late-1900s of the oldest-looking trees reveals what appears to be the same 1600s recruitment pulse. In fact, the age structure during the 1600s of the dendroclimatological collection was not significantly different from the collection made in the early-1900s (Trouet et al. 2017). These data provide more evidence of a large-scale disturbance in the eastern U.S. during the mid- to late-1600s.

Here, we present a new synthesis of age structures in ten forests collected over the last decade hinting at a similar recruitment event in the northeastern U.S. during and soon after the 1780s. Data in this synthesis only includes cores that have the pith of an individual or an inner ring that appears likely to be within a decade of the pith. We present annual recruitment dates from 217 trees that were collected through plot or targeted sampling. Recruitment dates of this synthesis ranges from 1571–1830 CE (Fig. 5.3a).

The first set of evidence that drew our attention to a possible recruitment event during the 1780s comes from Wachusett Mountain in central New England and includes the oldest-known *Quercus rubra* (Orwig et al. 2001). Dating to the 1660s, the oldest *Quercus rubra* trees on Wachusett Mountain uphold the hypothesis of subcontinental recruitment in the mid- to late-1600s. A second recruitment pulse of *Quercus rubra*, however, is seen between 1770–1799. Upon recognizing a potential regional-scale event, we synthesized recruitment dates from trees collected over an area of ca. 22,000 km^2. Together, these data indicate a significant increase in tree recruitment across forest types and species starting around the 1780s. A moving t-test technique of these data suggests an abrupt and significant increase in recruitment in 1777 ($p < 0.05$). Following a trough in recruitment, another peak appears in the late 1820s (Fig. 5.3b).

The synchrony of the ca. 1780s event in the northeastern U.S. appears stronger among *Quercus* species (Fig. 5.3a). *Quercus rubra* recruited between 1770–1799 in

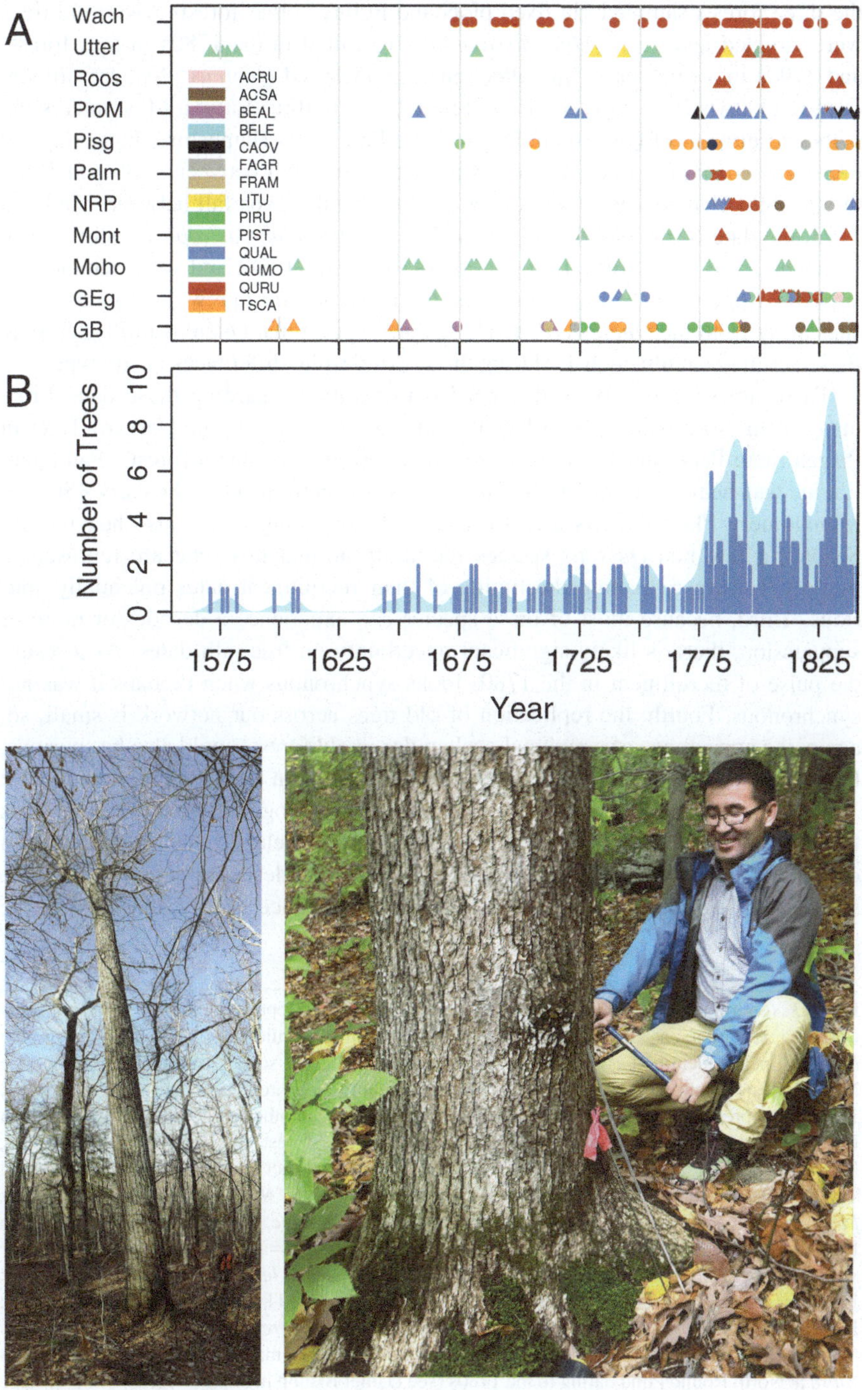

Fig. 5.3 (**a**) Recruitment of trees at the forest level over ca. 22,000 km^2 of the Northeastern U.S. from 1571–1830 based on inner pith dates (n = 217). Circles represent trees from plot sampling while triangles represent trees collected through targeted sampling, either oldest-appearing trees or

the three forests sampled via fixed plots and in three other forests where old trees were targeted and cored. *Quercus montana* recruited in the 1780s in four forests and 1790s in a fifth forest. In collections that included *Quercus alba*, recruitment occurred between the early-1780s and mid-1790s in three forests. Most interesting is the recruitment of *Quercus* in Pisgah State Park, New Hampshire. Inner rings of the six *Quercus alba* trees with dates at or near pith fall between 1780 and 1787. Of the eight *Quercus rubra* with dates at or near pith, seven fall between 1788 and 1793; the date of the eighth *Quercus rubra* is 1798. *Quercus alba* is estimated to be about 2% of the forest (Daley et al. 1930; Foster 1988) and were sampled on a narrow ridge with shallow soils while the *Quercus rubra* grew on slightly richer sites away from the *Quercus alba*. These data suggest that *Quercus* in Pisgah State Park recruited simultaneously during the 1780s despite differences in site type.

There are several caveats that need consideration regarding these data. First, many of the forests have been heavily cut; only Gill Brook, the Harvard Tract in Pisgah State Park, and Wachusett Mountain appear to be uncut forests. Palmghatt Ravine was heavily cut in 1870s and the pulse of recruitment in the early-1800s at Montgomery Place follows a real estate deal suggesting it was also heavily cut. Second, a few non-*Quercus* species recruiting around this time are represented by only 1–2 trees, making the timing of their recruitment dates potentially spurious. Third, because most of these species can experience a decades or more of suppression, there is likely significant uncertainty on true pith dates. As a result, the pulse of recruitment in the 1780s looks synchronous when perhaps it was not synchronous. Fourth, the replication of old trees across our network is small, so, again, the appearance of a regional-scale pulse in old forests could also be spurious. Finally, past cutting could have removed trees other than those that recruited in the 1780s and 1790s. It is difficult to see, however, how logging would remove trees from other decades except the late-1700s cohort. A selection against trees from outside the 1780s is possible if trees from other decades were roughly the same size during logging due to the same general disturbance history and grew at the

Fig. 5.3 (continued) a collection of canopy trees that appear to represent the range of age classes in the forest (Pederson 2010). (**b**) Regional summary of tree recruitment binned annually in *blue bars* with peaks identified as in section "New Evidence of Large-scale Disturbance in Temperate Mesic Forests", but with a smoothing bandwith of three. *Circles* represent data from fixed plots. *Triangles* represent data where trees were targeted for age or other ecological attributes. Each line represents a different forest or sampling type within a forest (*Wach* Wachusett Mountain, MA; *Utter* Uttertown, NJ; *Roos* Rooster Hill, NY; *ProM* Prospect Mountain, NY; *Pisg* Harvard Tract in Pisgah State Park, NH; *Palm* Palmaghatt Ravine, NY; *NRP* North Round Pond, NH; *Mont* Montgomery Place, NY; *Moho* Mohonk Nature Preserve, NY; *GEg* Goose Egg State Forest, NY; *GB* Gill Brook, NY. Species presented are: ACRU = *Acer rubrum*; ACSA = *Acer saccharum*; BEAL = *Betula alleghaniensis*; BELE = *Betula lenta*; CAOV = *Carya ovata*; FAGR = *Fagus grandifolia*; FRAM = *Fraxinus americana*; LITU = *Liriodendron tulipifera*; PIRU = *Picea rubens*; PIST = *Pinus strobus*; QUAL = *Quercus alba*; QUMO = *Quercus montana*; QURU = *Quercus rubra*; TSCA = *Tsuga canadensis*. Image *lower left*: a large *Quercus rubra* at North Round Pond dating to the 1780s (see Daniel Bishop in orange vest for scale). Image *lower right*: Oyunsanaa Byambasuren coring a *Quercus alba* at North Round Pond dating to the 1780s

same rate. Given the many kinds of disturbance, life-history traits, and biodiversity in these forests, however, it is hard to fathom how all trees but those in the 1780s and 1790s were roughly the same size during logging and experienced to the same general disturbance history. When we isolate the three forests that are likely uncut, a small pulse in recruitment occurs starting at about 1780 (figure not shown). Because of these considerations, we cannot yet exclude logging history as the reason for the pulse of recruitment in the late-1700s or that this recruitment event is simply spurious.

Regardless of the issues regarding the age-structure data, a synthesis of these data indicates the late-1700s to be important for tree recruitment in our study region (Fig. 5.3). Supporting this finding, research in northern Pennsylvania (Nowacki and Abrams 1994) and the southeastern U.S. (Lorimer 1980; Hart et al. 2012b; Pederson et al. 2014) suggest the post-1770s to be an important time in the dynamics in eastern U.S. TMFs. Not only do these data indicate ecologically-synchronous peaks in recruitment, but they indicate troughs in recruitment during the first half of the 1800s. Together, the peaks and valleys of recruitment could represent the multi-layered disturbance history of these forests due to various types and intensities of disturbance, from gap dynamics to, potentially, regional-scale disturbance. For now, we consider the data in Fig. 5.3 as a hypothesis that requires a significant amount of additional data collection and rigorous analysis to draw any kind of a conclusion.

5.6 The N-Dimensional, Forest Development Model

5.6.1 Theoretical Background

Despite the potential for a large number of trajectories through the tapestry of diverse disturbances types and tree species, temperate, mesic forests have often been considered to be in a shifting-mosaic or dynamic equilibrium for centuries or millennia (e.g. Botkin et al. 1981; Canham and Loucks 1984; Remmert 1991; Bormann and Likens 1994). As much theory for TMFs was developed in regions less diverse than the southeastern U.S. or temperate east Asia (e.g. Cowles 1899; Forcier 1975) and through the study of chronosequences, it is not too surprising that dynamic equilibrium came to the forefront in the theory of forest development. This and our limited understanding of how multiple drivers of disturbance interact with other factors and with past disturbances (e.g. tree-weakening droughts with bark beetles, fires with bark beetles, or windthrow-caused gaps with drought) might reinforce somewhat predictable changes over time and a limited number of successional trajectories. Forecasts, however, suggest the potential ecological impact of climate change could be so severe that future forests might become non-analogue communities (Williams and Jackson 2007).

Here we introduce the N-dimensional, Forest Development model (nFD), a conceptual model to aid in comprehending the potential trajectories of forest development, i.e. what could have been and what might be. Looking back in time, we can only detect the trajectory upon which systems have developed and not the potential pathways or contingencies for that system. This view could limit comprehension of contemporary TMFs. Although our knowledge of past structure and composition can make the short-term trajectory seem predictable (Oliver and Larson 1990), our observations arise mostly from a limited range of environmental conditions at any one site. Past, unobserved environmental conditions limit our understanding of the forest under conditions outside of the historical range (e.g., atmospheric CO_2 concentrations, temperature or precipitation and any possible interaction between them). Our objective with nFD is to visually illustrate expanded dimensions of potential forest succession, from small-scale dynamics to non-linear responses, with the intent to aid long-term management and ecological research that is not channeled by what has been realized in extant forests or those of the past.

Before describing nFD, several thoughts accompany our model. Most conceptual models frequently portray forest succession as linear, although theory contains nonlinear trajectories, or display a limited number of pathways (e.g. Holling 1986; Perera et al. 2007). While certainly not the intention, these conceptual models can often direct us towards expected destinations that unfortunately places diverse ecosystems like TMFs into a small number of narrow channels of development over time. If we continue that line of thinking, only imagining a single pathway with little to no other trajectories from the present into the future, it might lead us into a futile battle to direct ecosystem development towards conditions for which we only have tangible evidence of and not the full spectrum of possibilities for that system.

Our model has five main influences (Hutchinson 1944; Billings 1952; Holling 1986; Lavigne 1996, 1997; Jackson 2006). Hutchinson (1944) and Billings (1952) map out the myriad forces affecting plants and ecosystems. We extrapolate the multitude of forces for TMFs because of the high number of canopy species, interactions between this diversity, the many potential impacts of disturbance, and the historical contingencies of these interactions among the diversity in growth patterns and life history types. In addition to the diversity of these forces, repeated disturbance events drive TMF ecosystems indicating the order of drivers and the variation in severity or duration of drivers likely matters, too (Sippel et al. 2016). A repeated combination of a particular order or severity could conceivably push species-diverse systems in an n-dimensional set of pathways. In this light, while the North Atlantic Food Web (Lavigne 1996, 1997) and Billings' model (Billings 1952) likely overestimate the importance of some relationships, they do help us comprehend the idea that diverse ecosystems can branch in multiple directions through time.

Borrowing from the "Radical Gleasonian" concept of Jackson (2006) and alternative stable states framework (Beisner et al. 2003), ecosystems in nFD experience rather predictable sets of disturbance–frequent, gap dynamics and other low-intensity processes. A LID or a series of extreme events that culminate in rare and catastrophic events, Black Swan events (Taleb 2007), however, have the potential to create the substrate (structure, composition) upon which the aforementioned predicable disturbances act. The sequence, intensity, duration, and return interval between extreme events are highly stochastic and individualistic and, in forests with a dozen or more canopy species, have the capacity to either (1) incur little to no change if diversity begets resistance (Isbell et al. 2015), (2) move these systems linearly or non-linearly in one of the many n-dimensional sets of pathways, or (3) for Black Swans, potentially eliminate all but one trajectory for forest development.

Because humans and ecosystems are connected through space and time, historical events in forests leave legacies that impact and interact with future events. Considering current knowledge of these forests–their structure, composition, and disturbance regime–it may be possible to project the general trajectory of these systems due to predictable species-specific responses to disturbance and competition. This ability to correctly project the development of these systems, however, most likely decreases as conditions move away from the conditions under which observations and model development and validation have been performed. In no way do we mean to imply chaos regarding the potential trajectories of forest development. Composition, site type, and other factors will constrain the potential pathways. We wish to portray the directions and dimensions that forests can develop into or from so a more fluid and dynamic concept of forest ecosystems can be imagined.

5.6.2 *Working the N-Dimensional, Forest Development Model*

Each trajectory (pathway) in nFD represents a forest and each "ball" represents a disturbance (Fig. 5.4). When a forest is significantly altered by disturbance, the trajectory changes color. If the system is structurally altered, but not compositionally altered, a stasis in color represents cyclic succession (Forcier 1975). The common disturbance regime of temperate, mesic forests, of frequent, low-intensity, asynchronous, and small-scale disturbance, is represented by the mortal coils on each pathway (Fig. 5.4). That is, the curves in each trajectory, the mortal coils, represent death of canopy trees due to regular background processes. A disturbance with multiple colors represents potential trajectories. We limit the number of trajectories for clarity, but not for theory (as do most models).

For brevity, we describe two trajectories. The first trajectory precedes the blue disturbance labeled 'T', which is the current point in Time. The first trajectory represents the paleoecological history of our study system. Starting at the top (#1 on Fig. 5.4), the dark-yellow trajectory symbolizes a system dominated by *Quercus* pollen. There is some change in composition over time as the trajectory changes from brown to pink. A severe fire (the pink disturbance) firmly shifts the pollen into a *Quercus-Pinus* mixed system. The *Quercus-Pinus* mixed system then bends sharply as an unknown factor alters the pollen composition slightly more towards *Pinus* (purple), then back to the mixed *Quercus-Pinus* composition (pink). An unknown driver abruptly brings *Carya* into the mix (brick red trajectory). A

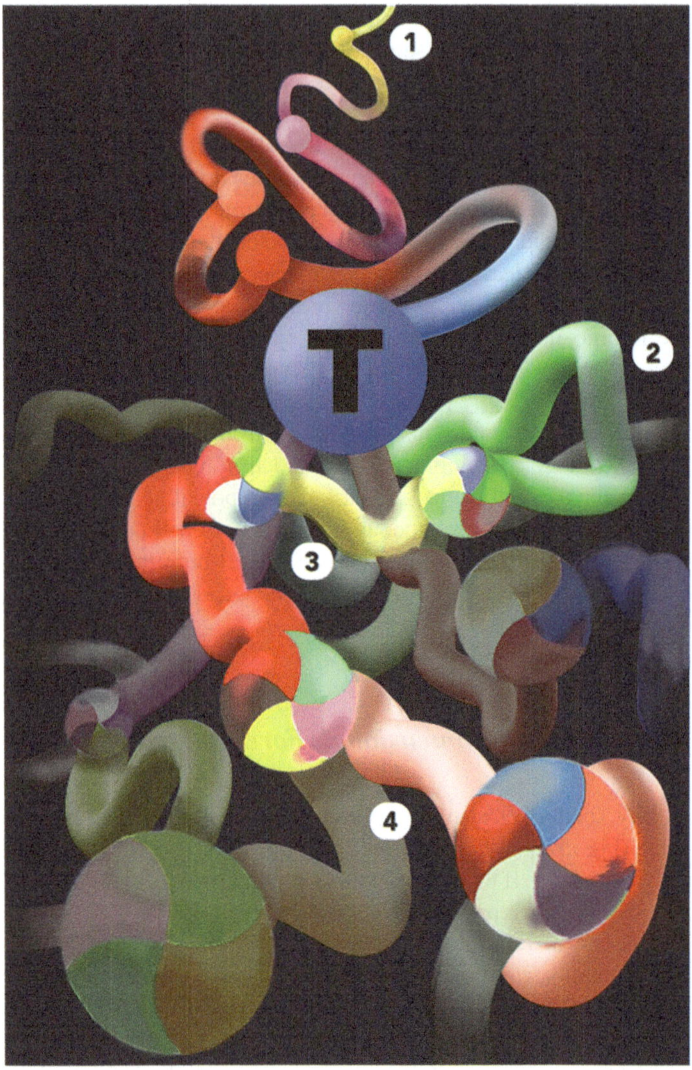

Fig. 5.4 The N-dimensional, Forest Development model. Potential forest development occurs on each trajectory (pathway). Disturbance is represented by a ball on the trajectory. The disturbance with 'T' represents the current forest. The trajectory starting with (*1*) is paleoecological history. The trajectory leading to (*2*) represents the future. Significant alteration in forest composition changes the color of the trajectory. Background disturbances that kill trees, but do not significantly alter TMFs, are represented by the mortal coils (*curves*) on each pathway. A disturbance with multiple colors represents potential trajectories. We limit the number of trajectories for clarity. The dimmed trajectories are paths not traveled. See text for more details

sharp bend and then the combination of two disturbances (moderate fire and a severe windstorm) and another unknown factor eventually reduces the amount of *Pinus* significantly enough that the trajectory becomes blue. Concurrent with the bluing of the trajectory is a long-term wetting trend. Note: as we look back in time, we have fewer details on all the drivers changing the system than we might when observing changes in today's forest. Similarly, we are not able to discern the potential trajectories looking back in time that the forests did not take.

To the right of 'T' is our second trajectory. Along this trajectory are the potential future scenarios of forest development. As the wetting trend of prior centuries continues, today's forest (T) is hit with a moderate windstorm, moving it along the green trajectory and a *Quercus-Carya* forest with mesophytic species in the understory (Nowacki and Abrams 2008; McEwan et al. 2011). The mesophytic species persist and continue to grow, but they are still in the understory (not altering the forest type). A disparity in functional traits, competitive abilities, and response to external drivers broaden the potential number of trajectories of this forest.

Moving further into the image, the trajectory oscillates as it experiences a series of intermediate disturbances, either the loss of an agent (fire), loss of foundational species (passenger pigeon, American chestnut), cultural impacts (fire suppression, increased road building), climate variations or shifts, or, more likely, a combination of these factors (McEwan et al. 2011). The trajectory bends sharply and becomes partially translucent. This represents a fire that is intense enough to reduce density and delay recovery post-disturbance that does not significantly alter composition (#2 on Fig. 5.4). The straightness of the trajectory following the point #2 symbolizes a forest with increased resistance to disturbance. One real-world example of this concept is the impact of the 1938 Hurricane on the Lyford Plot at the Harvard Forest. While the 1938 Hurricane knocked down 80% of the old-growth forest in the Harvard Tract of Pisgah State Park and altered its trajectory from a forest dominated by *Pinus strobus* to *Tsuga canadensis* (Foster 1988), the small trees in the Lyford Plot seem to have had some resistance to the hurricane (Rowlands 1941), thereby causing little to no change in its trajectory.

Back to the nDF exercise, the forest is then hit by a small, but intense disturbance that has the potential to move it in five directions. Only one potential developmental trajectory represents a significant departure from the current state, the dark red wedge, indicating that a significant change is less likely to occur. The other directions are pathways that are either closely related to the current composition of the forest (light green wedge), related to the last significant shift in forest composition (the blue wedge), represent where the forest cycles "back onto itself" (the dark green wedge), or a shift that is a mix of the blue and green forest states where a substantial component is fast-growing drought-deciduous *Liriodendron tulipifera* (yellow wedge).

Traveling along the yellow trajectory, the forest experiences a lower rate of background disturbance (#3 on Fig. 5.4) and is symbolized by having fewer mortal coils. Soon, the forest is hit with a LID or a combination of ecologically-simultaneous disturbances (drought, fire, ice storm, invasive species) that has the potential to shift the forest into five (or more) states. The current community assembly and developmental momentum constrains this forest to four closely related types (light green, blue, light grey, and yellow, where it cycles back onto itself). Before the development of a robust root system in the *Liriodendron tulipifera*, however, the forest experiences a frost that follows a megadrought (10 years of minor to extreme drought conditions), killing most of the canopy while favoring fire, *Quercus*, and *Carya* to send the forest along the red trajectory.

The *Quercus-Carya* dominated forest develops, flexing with the frequent mortal coils of low-intensity disturbance. While *Quercus* and *Carya* dominate the forest, there is still a substantial mixture of more mesophytic species. A wetter climate develops during this phase and a severe windstorm knocks down 25% of the *Quercus-Carya* canopy. The climatic shift allows the potential for suppressed, but fast-growing mesophytic species to rapidly move into the canopy as resources are relinquished by the dead and dying *Quercus* and *Carya* trees. A second LID or a combination of simultaneous disturbances hits the less resilient forest. Its resiliency is related to a canopy dominated by large *Quercus* trees; large trees are thought to be more vulnerable to rapid environmental change (Bennett et al. 2015; Kauppi et al. 2015). This disturbance can shift it into five (or more) states with four matching prior forest types (pink, red (cycling back onto itself), mint green, yellow, and brown, a significantly different state). The second LID is strong, but community assembly pushes onto a trajectory closely related to its current state, the pink phase (#4 on Fig. 5.4). So it goes.

Note: all along the future trajectories, we see the potential trajectories of each forest that were not taken (dimmed trajectories and disturbances) and limited space in this chapter precludes a fuller description of forest composition and alternative pathways in these diverse forests. We describe forest transitions that are closely related to the state at the time of disturbance, which are more likely to occur. Given the diversity of these forests, a greater range of forest types should be seen as possible.

5.6.3 Management Implications

As the forest trajectories described demonstrate, nFD creates a flexible framework for long-term forest management across local conditions and environmental change. Long-term management requires a great understanding of these systems, careful observation, and a flexible and patience philosophy that balances ecology and economics. Infusing management plans with this management ethic can sustain forests. Leon Neel et al. (2010) implied this when thinking about the ecological management of forests:

> " *... the natural world is resilient. An ecosystem, whatever condition it is in, is not a fixed, stable thing, and it cannot be preserved or managed as such. It is changing all the time.*"—
> Leon Neel

5.7 Conclusion

Our review of disturbance dynamics in the diverse and culturally important temperate, mesic forests of North America, Central Europe, and Japan indicates that constant, but low-impact and stochastic tree mortality is the best understood aspect of the dynamics in these systems. It seems likely that the common scales of dendroecological research reinforces this principle and, as a result, limits our understanding of other forest dynamics at large spatial scales. These common scales of research also likely limit our understanding of how multiple disturbance factors, especially the rare events of an extreme nature, interact with each other and past disturbances and forest conditions.

We believe that by expanding the common scales of dendroecological research we can better understand the flavors of disturbance and how various blends of disturbance shape temperate, mesic forests from decades to centuries. Greater insight into the complex interactions of disturbance in diverse systems will likely lead to an expansion of developmental theory in these forests. Our hypothesis is based on the few large-scale dendroecological studies in temperate, mesic forests that reveal a somewhat unexpected persistence of early-successional species in old forests, synchronous regional-scale canopy disturbance, punctuated, but infrequent tree recruitment at the stand- to landscape-scales, and, buried in fading records, evidence of recruitment (and the inferred synchronous opening of canopies) 200–350 years ago at regional to possibly subcontinental-scale. These findings suggest to us that the legacy of large, infrequent disturbance is not erased by centuries of fine- and intermediate-scale disturbance processes and integral to the development of temperate, mesic forests.

Dendroecology can open new frontiers in the ecology of temperate, mesic forests by conducting research at regional- to subcontinental-scales across a variety of regions that fully utilize basic dendrochronological methods and careful investigations of forest dynamics that takes advantage of dendrochronology's primary strength—annual temporal resolution. The fruits of such labors will be the reconstruction of centuries of more temporally precise and more complex histories of forest disturbance. These fruits can then be melded into sustainable long-term management plans based on the resilience and recovery of these forests in response to disturbance.

Acknowledgments Much thanks to Anna Guerrero for her thoughtful interpretation, creation, and revision of our conceptual diagram (Fig. 5.4). Thank you to Martin Brülhardt (ETH Zurich) for providing the picture of the central European forest in Fig. 5.1 and D. Bishop and S. Zhao for assistance in developing Figs. 5.2 and 5.3. Comments by P. Baker, A. Barker-Plotkin, E.P. Clark, L. Daniels, R. Delgado, D. Orwig, J. M. Varner, and X. Wang improved the manuscript. N. Pederson was partially supported by NSF EF-1241930, which supports the PalEON Project (paleonproject.org), as well as by the Harvard Forest. A. Young was supported by NSF DDRI (1434242), Society of Women Geographers and NASA Pennsylvania Space Grant Fellowship. D. Martin-Benito was supported by a Marie-Curie IEF Grant (EU-Grant 329935) and a Fulbright-Ministerio de Ciencia e Innovación (Spain) postdoctoral fellowship.

References

Abe M, Miguchi H, Honda A et al (2005) Short-term changes affecting regeneration of Fagus crenata after the simultaneous death of *Sasa kurilensis*. J Veg Sci 16(1):49–56

Abrams MD (1992) Fire and the development of oak forests. BioScience 42(5):346–353

Abrams MD, Orwig DA (1996) A 300-year history of disturbance and canopy recruitment for co-occurring white pine and hemlock on the Allegheny Plateau USA. J Ecol 84(3):353–363

Abrams MD, Copenheaver CA, Terazawa K et al (1999) A 370-year dendroecological history of an old-growth *Abies-Acer-Quercus* forest in Hokkaido northern Japan. Can J For Res 29(12):1891–1899

Akpan N (2016) How big droughts forest fires could be the new normal in Appalachia. PBS Newshour. http://www.pbs.org/newshour/updates/widespread-forest-fires-claims-may-signal-new-normal-appalachian-mountains/. Accessed 6 July 2017

Alaback PB (1991) Comparative ecology of temperate rainforests of the Americas along analogous climatic gradients. Rev Chil Hist Nat 64:399–412

Alaback P, Nowacki G, Saunders S (2013) Disturbance ecology of the temperate rainforests of Southeast Alaska and adjacent British Columbia. In: Orians G (ed) North Pacific temperate rainforests: ecology and conservation, 1st edn. University of Washington Press, Seattle, pp 73–88

Aldrich SR, Lafon CW, Grissino-Mayer HD et al (2010) Three centuries of fire in montane pine-oak stands on a temperate forest landscape. Appl Veg Sci 13:36–46

Allen CD, Breshears DD (1998) Drought-induced shift of a forest–woodland ecotone: rapid landscape response to climate variation. Proc Natl Acad Sci U S A 95(25):14839–14842

Allen CD, Breshears DD, McDowell NG (2015) On underestimation of global vulnerability to tree mortality and forest die-off from hotter drought in the Anthropocene. Ecosphere 6(8):1–55

Altman J, Fibich P, Leps J et al (2016) Linking spatiotemporal disturbance history with tree regeneration and diversity in an old-growth forest in northern Japan. Perspect Plant Ecol Evol Syst 21:1–13

Ariya U, Hamano K-Y, Makimoto T et al (2016) Temporal and spatial dynamics of an old-growth beech forest in western. Jap J For Res 21(2):73–83

Ascoli D, Castagneri D, Valsecchi C et al (2013) Post-fire restoration of beech stands in the Southern Alps by natural regeneration. Ecol Eng 54:210–217

Baker PJ, Bunyavejchewin S (2017) Complex historical disturbance regimes shape forest dynamics across a seasonal tropical landscape in western Thailand. In: Amoroso M, Daniels L, Baker PJ, Camarero JJ (eds) Dendroecology: tree-ring analyses applied to ecological studies. Springer, New York

Baker PJ, Bunyavejchewin S, Oliver CD et al (2005) Disturbance history and historical stand dynamics of a seasonal tropical forest in western Thailand. Ecol Monogr 75(3):317–343

Beisner BE, Haydon DT, Cuddington K (2003) Alternative stable states in ecology. Front Ecol Environ 1(7):376–382

Bennett AC, McDowell NG, Allen CD, Anderson-Teixeira KJ (2015) Larger trees suffer most during drought in forests worldwide. Nat Plants 1:15139

Berdanier AB, Clark JS (2016) Multiyear drought-induced morbidity preceding tree death in southeastern US forests. Ecol Appl 26(1):17–23

Billings WD (1952) The environmental complex in relation to plant growth and distribution. Q Rev Biol 27(3):251–265

Black BA, Colbert JJ, Pederson N (2008) Relationships between radial growth rates and lifespan within North American tree species. Ecoscience 15(3):349–357

Black BA, Griffin D, van der Sleen P et al (2016) The value of crossdating to retain high-frequency variability climate signals and extreme events in environmental proxies. Glob Chang Biol 22(7):2582–2595

Booth RK, Ireland AW, LeBoeuf K et al (2016) Late Holocene climate-induced forest transformation and peatland establishment in the central Appalachians. Quat Res 85(2):204–210

Bormann FH, Likens G (1994) Pattern and process in a forested ecosystem: disturbance development and the steady state based on the Hubbard Brook ecosystem study. Springer, New York
Botkin DB, Shugart HH, West D (1981) Forest succession: concepts and application. In: West DC, Shugart HH, Botkin DF (eds) Springer advanced texts in life sciences. Springer, New York
Bréda N, Huc R, Granier A et al (2006) Temperate forest trees and stands under severe drought: a review of ecophysiological responses adaptation processes and long-term consequences. Ann For Sci 63(6):625–644
Brown JH (1995) Macroecology. University of Chicago Press, Chicago
Brubaker LB (1980) Spatial patterns of tree growth anomalies in the Pacific Northwest. Ecology 61:798–807
Cailleret M, Nourtier M, Amm A et al (2014) Drought-induced decline and mortality of silver fir differ among three sites in Southern France. Ann For Sci 71(6):643–657
Camarero JJ, Bigler C, Linares JC et al (2011) Synergistic effects of past historical logging and drought on the decline of Pyrenean silver fir forests. For Ecol Manage 262(5):759–769
Canham CD, Loucks L (1984) Catastrophic windthrow in the presettlement forests of Wisconsin. Ecology 65(3):803–809
Canham CD, Papaik MJ, Latty EF (2001) Interspecific variation in susceptibility to windthrow as a function of tree size and storm severity for northern temperate tree species. Can J For Res 31(1):1–10
Cavin L, Jump A (2017) Highest drought sensitivity and lowest resistance to growth suppression is found in the range core of the tree *Fagus sylvatica* L not the equatorial range edge. Glob Chang Biol 23(1):362–379
Chapman JI, McEwan RW (2016) Thirty years of compositional change in an old-growth temperate forest: the role of topographic gradients in oak-maple dynamics. PLoS One 11(7):e0160238
Charney ND, Babst F, Poulter B et al (2016) Observed forest sensitivity to climate implies large changes in 21st century North American forest growth. Ecol Lett 19(9):1119–1128
Cherubini P, Schweingruber FH, Piussi P (1996) Spatiotenraporal growth dynamics and disturbances in a subalpine spruce forest in the Alps: a dendroecological reconstruction. Can J For Res 26(6):991–1001
Cho D-S, Boerner REJ (1995) Dendrochronological analysis of the canopy history of two Ohio old-growth forests. Vegetatio 120(2):173–183
Choat B, Jansen S, Brodribb TJ et al (2012) Global convergence in the vulnerability of forests to drought. Nature 491(7426):752–755
Chokkalingam U, White A (2001) Structure and spatial patterns of trees in old-growth northern hardwood and mixed forests of northern Maine. Plant Ecol 156(2):139–160
Ciais P, Reichstein M, Viovy N et al (2005) Europe-wide reduction in primary productivity caused by the heat and drought in 2003. Nature 437:529–533
Clark JS, Royall PD (1996) Local and regional sediment charcoal evidence for fire regimes in presettlement north-eastern North America. J Ecol 84(3):35–382
Clark JS, Iverson L, Woodall CW et al (2016) The impacts of increasing drought on forest dynamics structure and biodiversity in the United States. Glob Chang Biol 22(7):2329–2352
Clifford MJ, Booth RK (2015) Late-Holocene drought and fire drove a widespread change in forest community composition in eastern North America. The Holocene 25(7):1102–1110
Clinton BD, Boring LR, Swank WT (1993) Canopy gap characteristics and drought influences in oak forests of the Coweeta Basin. Ecology 74(5):1551–1558
Connor SE, Thomas I, Kvavadze EV (2007) A 5600-yr history of changing vegetation sea levels and human impacts from the Black Sea Coast of Georgia. The Holocene 17(1):25–36
Cook ER (1991) In: Wyman RI (ed) Tree rings as indicators of climate change and the potential response of forests to the greenhouse effect. Chapman and Hall, New York, pp 56–64
Cook BI, Smerdon JE, Seager R et al (2014) Global warming and 21st century drying. Clim Dyn 43:2607
Cook BI, Ault TR, Smerdon JE (2015) Unprecedented 21st century drought risk in the American Southwest and Central Plains. Sci Adv 1(1):e1400082

Cowles HC (1899) The ecological relations of the vegetation on the sand dunes of Lake Michigan. Bot Gaz 27(2):95–117
D'Amato AW, Orwig DA (2008) Stand- and landscape-level disturbance dynamics in old-growth forests in western Massachusett. Ecol Monogr 78(4):507–522
Daley R, Branch W, Lotti T (1930) Life history of the climax forest on the Pisgah Tract Winchester, New Hampshire. MFS Thesis, Harvard University, Cambridge, MA
Daniels LD, Gray RW (2006) Disturbance regimes in coastal British Columbia. J Ecosyst Manage 7(2):44–56
Daniels L, Klinka K (1996) The dynamics of old-growth *Thuja-Tsuga* forests near Vancouver British Columbia. In: Dean JS, Meko DM, Swetnam TW (eds) Tree rings, environment and humanity, Radiocarbon, pp 379–393
Davis MB (1983) Quaternary history of deciduous forests of eastern North America and Europe. Ann Mo Bot Gard 70(3):550–563
DellaSala DA (2011) Rainforests at the margins: regional profiles. In: Temperate and boreal rainforests of the world: ecology and conservation. Springer, New York, pp 213–242
DellaSala DA, Moola F, Alaback P et al (2011) Temperate and boreal rainforests of the Pacific Coast of North America. In: Temperate and boreal rainforests of the world: ecology and conservation. Springer, New York, pp 42–81
Di Filippo A, Pederson N, Baliva M et al (2015) The longevity of broadleaf deciduous trees in Northern Hemisphere temperate forests: insights from tree-ring series. Front Ecol Evol 3:46
Dietze MC, Moorcroft PR (2011) Tree mortality in the eastern and central United States: patterns and drivers. Glob Chang Biol 17(11):3312–3326
Dobrowolska D, Hein S, Oosterbaan A et al (2011) A review of European ash (*Fraxinus excelsior* L): implications for silviculture. Forestry 84(2):133–148
Dorner B, Wong C (2002) Natural disturbance dynamics on the North Coast. Background Report prepared for the North Coast LRMP. Prince Rupert, BC
Duveneck MJ, Thompson JR, Gustafson EJ et al (2016) Recovery dynamics and climate change effects to future New England forests. Landsc Ecol 33:1–13
Elkin C, Gutiérrez AG, Leuzinger S et al (2013) A 2 °C warmer world is not safe for ecosystem services in the European Alps. Glob Chang Biol 19(6):1827–1840
Ellison AM, Bank MS, Clinton BD et al (2005) Loss of foundation species: consequences for the structure and dynamics of forested ecosystems. Front Ecol Environ 3(9):479–486
Emborg J, Christensen M, Heilmann-Clausen J (2000) The structural dynamics of Suserup Skov a near-natural temperate deciduous forest in Denmark. For Ecol Manage 126(2):173–189
Everham EM, Brokaw NVL (1996) Forest damage and recovery from catastrophic wind. Bot Rev 62:113–185
Fahey RT, Fotis AT, Woods D (2015) Quantifying canopy complexity and effects on productivity and resilience in late-successional hemlock–hardwood forests. Ecol Appl 25(3):834–847
Falk DA, Heyerdahl EK, Brown PM et al (2011) Multi-scale controls of historical forest-fire regimes: new insights from fire-scar networks. Front Ecol Environ 9(8):446–454
Flatley WT, Lafon CW, Grissino-Mayer HD et al (2013) Fire history and its relation to climate and land use in three Southern Appalachian landscapes in the Eastern US. Ecol Appl 23:1250–1266
Forcier LK (1975) Reproductive strategies and the co-occurrence of climax tree species. Science 189(4205):808–810
Foster DR (1988) Disturbance history community organization and vegetation dynamics of the old-growth Pisgah Forest South-western New Hampshire USA. J Ecol 76(1):105–134
Foster DR, Boose ER (1992) Patterns of forest damage resulting from catastrophic Patterns of forest wind in central New England USA. J Ecol 80(1):79–98
Foster DR, Knight DH, Franklin JF (1998) Landscape patterns and legacies resulting from large infrequent forest disturbances. Ecosystems 1(6):497–510
Foster DR, Oswald WW, Faison E et al (2006) A climatic driver for abrupt mid-Holocene vegetation dynamics and the hemlock decline in New England. Ecology 87(12):2959–2966
Franklin JF, Maeda T, Ohsumi Y et al (1979) Subalpine coniferous forests of central Honshu Japan. Ecol Monogr 49(3):311–334

Franklin JF, Spies TA, Van Pelt R et al (2002) Disturbances and structural development of natural forest ecosystems with silvicultural implications using Douglas-fir forests as an example. For Ecol Manage 155(1):399–423
Fraver S, White AS, Seymour RS (2009) Natural disturbance in an old-growth landscape of northern Maine USA. J Ecol 97(2):289–298
Frelich LE, Lorimer CG (1991) Natural disturbance regimes in hemlock-hardwood forests of the Upper Great Lakes Region. Ecol Monogr 61(2):145–164
Gajewski K, Viau AE, Sawada M et al (2006) Synchronicity in climate and vegetation transitions between Europe and North America during the Holocene. Clim Chang 78(2):341–361
Gamfeldt L, Snäll T, Bagchi R et al (2013) Higher levels of multiple ecosystem services are found in forests with more tree species. Nat Commun 4:1340–1340
Gaston K, Blackburn T (2008) Pattern and process in macroecology. Wiley, New York
Gavin DG, Brubaker LB, Lertzman KP (2003a) An 1800-year record of the spatial and temporal distribution of fire from the west coast of Vancouver Island. Can J For Res 33(4):573–586
Gavin DG, Brubaker LB, Lertzman KP (2003b) Holocene fire history of a coastal temperate rain forest based on soil charcoal radiocarbon dates. Ecology 84(1):186–201
Gavin DG, Brubaker LB, Greenwald DN (2013) Postglacial climate and fire-mediated vegetation change on the western Olympic Peninsula Washington (USA). Ecol Monogr 83(4):471–489
de Graauw K (2017) Historic log structures as ecological archives: a case study from eastern North America. Dendrochronologia
Grimm EC (1983) Chronology and dynamics of vegetation change in the prairie-woodland region of southern Minnesota, USA. New Phytol 93:311–350
Grissino-Mayer HD (2016) Fire as a once-dominant disturbance process in the yellow pine and mixed pine-hardwood forests of the Appalachian Mountains. In: Greenberg CH, Collins BS (eds) Natural disturbances and historic range of variation, man for ecosystems, vol 32, pp 123–146
Grundmann BM, Bonn S, Roloff A (2008) Cross-dating of highly sensitive Common beech (*Fagus sylvatica* L) tree-ring series with numerous missing rings. Dendrochronologia 26(2):109–113
Guyette RP, Muzika RM, Dey DC (2002) Dynamics of an anthropogenic fire regime. Ecosystems 5:472–486
Guyette RP, Dey DC, Stambaugh MC et al (2006) Fire scars reveal variability and dynamics of eastern fire regimes. In: Dickinson MB (ed) US Dept of Agriculture Forest Service Northern Research Station General Technical Report NRS-P-1. Newtown Square, PA, pp 20–39
Haavik LJ, Billings SA, Guldin JM et al (2015) Emergent insects pathogens and drought shape changing patterns in oak decline in North America and Europe. For Ecol Manage 354:190–205
Hallett DJ, Lepofsky DS, Mathewes RW et al (2003) 11000 years of fire history and climate in the mountain hemlock rain forests of southwestern British Columbia based on sedimentary charcoal. Can J For Res 33(2):292–312
Harcombe P, Greene S, Kramer M et al (2004) The influence of fire and windthrow dynamics on a coastal spruce–hemlock forest in Oregon USA based on aerial photographs spanning 40 years. For Ecol Manage 194(1):71–82
Hart JL, Buchanan ML, Clark SL et al (2012a) Canopy accession strategies and climate-growth relationships in A*cer rubrum*. For Ecol Manage 282:124–132
Hart JL, Clark SL, Torreano SJ et al (2012b) Composition structure and dendroecology of an old-growth *Quercus* forest on the tablelands of the Cumberland Plateau USA. For Ecol Manage 266:11–24
Hennon P, Shaw I, Hansen E (1990) Dating decline and mortality of *Chamaecyparis nootkatensis* in southeast Alaska. For Sci 36(3):502–515
Hennon PE, D'Amore DV, Schaberg PG et al (2012) Shifting climate altered niche and a dynamic conservation strategy for yellow-cedar in the North Pacific coastal rainforest. BioScience 62(2):147–158
Henry JD, Swan JMA (1974) Reconstructing forest history from live and dead plant material—an approach to the study of forest succession in southwest New Hampshire. Ecology 55:772–783

Hessl AE, Saladyga T, Schuler T et al (2011) Fire history from three species on a central Appalachian ridgetop. Can J For Res 41(10):2031–2039

Hiura T (1995) Gap formation and species diversity in Japanese beech forests: a test of the intermediate disturbance hypothesis on a geographic scale. Oecologia 104(3):265–271

Hiura T, Sano J, Konno Y (1996) Age structure and response to fine-scale disturbances of *Abies sachalinensis*, *Picea jezoensis*, *Picea glehnii*, and *Betula ermanii* growing under the influence of a dwarf bamboo understory in northern Japan. Can J For Res 26(2):289–297

Hoffman KM, Gavin DG, Starzomski BM (2016) Seven hundred years of human-driven and climate-influenced fire activity in a British Columbia coastal temperate rainforest. R Soc Open Sci 3(10):160608

Holling CS (1986) The resilience of terrestrial ecosystems: local surprise and global change. In: Clark WC, Munn RE (eds) Sustainable development of the biosphere. Cambridge University Press, Cambridge, UK, pp 292–317

Hursh CR, Haasis FW (1931) Effects of 1925 summer drought on southern Appalachian hardwoods. Ecology 12(2):380–386

Hutchinson GE (1944) Limnological studies in Connecticut. VII. A critical examination of the supposed relationship between phytoplakton periodicity and chemical changes in lake waters. Ecology 25(1):3–26

Ida H (2000) Treefall gap disturbance in an old-growth beech forest in southwestern Japan by a catastrophic typhoon. J Veg Sci 11(6):825–832

Isbell F, Craven D, Connolly J et al (2015) Biodiversity increases the resistance of ecosystem productivity to climate extremes. Nature 526(7574):574–577

Iverson LR, Prasad AM, Matthews SN et al (2008) Estimating potential habitat for 134 eastern US tree species under six climate scenarios. For Ecol Manage 254(3):390–406

Jackson ST (2006) Vegetation environment and time: the origination and termination of ecosystems. J Veg Sci 17(5):549–557

Jentsch A, Beierkuhnlein C, White PS (2002) Scale the dynamic stability of forest ecosystems and the persistence of biodiversity. Silva Fenn 36(1):393–400

Kanzaki M, Yoda K (1986) Regeneration in subalpine coniferous forests. II. Mortality and the pattern of death of canopy trees. Shokub Zass 99(1053):37–51

Kaplan JO, Krumhardt KM, Zimmermann N (2009) The prehistoric and preindustrial deforestation of Europe. Quat Sci Rev 28(27):3016–3034

Karnig JJ, Lyford WH (1968) Oak mortality and drought in the Hudson Highlands. In: Harvard black rock forest, vol 29. Black Rock Forest Papers, Cornwall, NY, pp 1–13

Kauppi P, Birdsey R, Pan Y et al (2015) Effects of land management on large trees and carbon stocks. Biogeosciences 12(3):855–862

Kohyama T (1984) Regeneration and coexistence of two *Abies* species dominating subalpine forests in central Japan. Oecologia 62(2):156–161

Kramer MG, Hansen AJ, Taper ML et al (2001) Abiotic controls on long-term windthrow disturbance and temperate rain forest dynamics in southeast Alaska. Ecology 82(10):2749–2768

Kubota Y, Hara T (1995) Tree competition and species coexistence in a sub-boreal forest northern Japan. Ann Bot 76(5):503–512

Lafon CW, Quiring SM (2012) Relationships of fire and precipitation regimes in temperate forests of the eastern United States. Earth Interact 16(11):1–15

Latham RE, Ricklefs RE (1993) Global patterns of tree species richness in moist forests: energy-diversity theory does not account for variation in species richness. Oikos 67(2):325–333

Lavigne D (1996) Ecological interactions between marine mammals, commercial fisheries, and their prey: unravelling the tangled web. In: Montevecchi WA (ed) Studies of high-latitude seabirds. 4. Trophic relationships and energetics of endotherms in cold ocean systems. Can Wild Ser, Occasional Pap 91, Ottawa, Canada, pp 59–71

Lavigne DM (1997) Ecological interactions between marine mammals commercial fisheries and their prey: unravelling the tangled web. Oceanogr Lit Rev 3(44):229

Lertzman KP, Sutherland GD, Inselberg A et al (1996) Canopy gaps and the landscape mosaic in a coastal temperate rain forest. Ecology 77(4):1254–1270

Lertzman K, Gavin D, Hallett D et al (2002) Long-term fire regime estimated from soil charcoal in coastal temperate rainforests. Conserv Ecol 6(2):5

Liebergesell M, Reu B, Stahl U et al (2016) Functional resilience against climate-driven extinctions—comparing the functional diversity of European and North American tree floras. PLoS One 11(2):e0148607

Long WH (1914) The death of chestnuts and oaks due to *Armillaria mellea*. Bull US Dept Ag 89:1–9

Lorimer CG (1977) The presettlement forest and natural disturbance cycle of northeastern Maine. Ecology 58(1):139–148

Lorimer CG (1980) Age structure and disturbance history of a Southern Appalachian virgin forest. Ecology 61(5):1169–1184

Lorimer CG, Dahir SE, Singer MT (1999) Frequency of partial and missing rings in *Acer saccharum* in relation to canopy position and growth rate. Plant Ecol 143(2):189–202

Lynch C, Hessl A (2010) Climatic controls on historical wildfires in West Virginia, 1939-2008. Phys Geogr 31(3):254–269

Manion PD (1981) Tree disease concepts. Prentice-Hall, Englewood Cliffs

van Mantgem PJ, Stephenson NL, Byrne JC et al (2009) Widespread increase of tree mortality rates in the western United States. Science 323:521–524

Maringer J, Ascoli D, Küffer N et al (2016) What drives European beech (*Fagus sylvatica* L) mortality after forest fires of varying severity? For Ecol Manage 368:81–93

Marshall R (1927) The growth of hemlock before and after release from suppression, Harvard For Bull 11. Harvard Forest, Petersham, MA, pp 1–43

Martin-Benito D, Pederson N (2015) Convergence in drought stress but a divergence of climatic drivers across a latitudinal gradient in a temperate broadleaf forest. J Biogeogr 42(5):925–937

Martin-Benito D, Ummenhofer CC, Köse N et al (2016) Tree-ring reconstructed May–June precipitation in the Caucasus since 1752 CE. Clim Dyn 47(9–10):3011–3027

Matlack GR (2015) Managing fire in the mesic deciduous forest when fire history is unknown: response to Stambaugh et al. Conserv Biol 29(3):947–949

McEwan RW, Hutchinson TF, Long RP et al (2007) Temporal and spatial patterns in fire occurrence during the establishment of mixed-oak forests in eastern North America. J Veg Sci 18(5):655–664

McEwan RW, Dyer JM, Pederson N (2011) Multiple interacting ecosystem drivers: toward an encompassing hypothesis of oak forest dynamics across eastern North America. Ecography 34(2):244–256

McEwan RW, Pederson N, Cooper A (2014) Fire and gap dynamics over 300 years in an old-growth temperate forest. Appl Veg Sci 17(2):312–322

McMurry ER, Stambaugh MC, Guyette RP et al (2007) Fire scars reveal source of New England's 1780 Dark Day. Int J Wildland Fire 16:266–270

Meidinger D, Pojar J (1991) Ecosystems of British Columbia Special Report. Series 6. British Columbia Ministry of Forests, Victoria, British Columbia, Canada

Michaelian M, Hogg EH, Hall RJ et al (2011) Massive mortality of aspen following severe drought along the southern edge of the Canadian boreal forest. Glob Chang Biol 17(6):2084–2094

Millar CI, Stephenson NL (2015) Temperate forest health in an era of emerging megadisturbance. Science 349(6250):823–826

Millers I, Shriner DS, Rizzo D (1989) History of hardwood decline in the eastern United States. USDA Dept Ag, For Ser, NE Exp Sta Gen Tech Rep NE-126:1–75

Mori AS, Komiyama A (2008) Differential survival among life stages contributes to co-dominance of *Abies mariesii* and *Abies veitchii* in a sub-alpine old-growth forest. J Veg Sci 19(2):239–244

Morin X, Chuine I (2014) Will tree species experience increased frost damage due to climate change because of changes in leaf phenology? Can J For Res 44(12):1555–1565

Munoz SE, Mladenoff DJ, Schroeder S et al (2014) Defining the spatial patterns of historical land use associated with the indigenous societies of eastern North America. J Biogeogr 41(12):2195–2210

Musavi T, Migliavacca M, Reichstein M et al (2017) Stand age and species richness dampen interannual variation of ecosystem-level photosynthetic capacity. Nat Ecol Evol 1:0048

Nadrowski K, Wirth C, Scherer-Lorenzen M (2010) Is forest diversity driving ecosystem function and service? Curr Opin Environ Sustain 2(1):75–79

Nagel T, Levanic T, Diaci J (2007) A dendroecological reconstruction of disturbance in an old-growth *Fagus-Abies* forest in Slovenia. Ann For Sci 64(8):891–897

Naka K (1982) Community dynamics of evergreen broadleaf forests in southwestern Japan. I. Wind damaged trees and canopy gaps in an evergreen oak forest. Shokubutsu-gaku-zasshi 95(4):385–399

Nakashizuka T (1987) Regeneration dynamics of beech forests in Japan. Vegetatio 69(1–3):169–175

Namikawa K (1996) Stand dynamics during a 12-year period in an old-growth cool temperate forest in northern Japan. Ecol Res 11(1):23–33

Narukawa Y, Yamamoto SI (2001) Gap formation microsite variation and the conifer seedling occurrence in a subalpine old-growth forest central Japan. Ecol Res 16(4):617–625

Neel L, Sutter PS, Way AG (2010) The art of managing longleaf: a personal history of the Stoddard-Neel approach. University of Georgia Press, Athens, Georgia

Niklasson M, Zin E, Zielonka T et al (2010) A 350-year tree-ring fire record from Białowieża Primeval Forest Poland: implications for Central European lowland fire histor. J Ecol 98(6):1319–1329

Nowacki GJ, Abrams MD (1994) Forest composition structure and disturbance history of the Alan Seeger Natural Area, Huntington, Pennsylvania. Bull Torrey Bot Club 121(3):277–291

Nowacki GJ, Abrams MD (2008) The demise of fire and "mesophication" of forests in the eastern United States. BioScience 58(2):123–138

Numata M (1970) Conservation implications of bamboo flowering and death in Japan. Biodivers Conserv 2(3):227–229

Oliver CD, Larson BC (1990) Forest stand dynamics. McGraw-Hill, New York

Oliver CD, Stephens EP (1977) Reconstruction of a mixed-species forest in central New England. Ecology 58(3):562–572

Ordonez A, Svenning JC (2016a) Functional diversity of North American broad-leaved trees is codetermined by past and current environmental factors. Ecosphere 7(2):e01237

Ordonez A, Svenning JC (2016b) Strong paleoclimatic legacies in current plant functional diversity patterns across Europe. Ecol Evol 6(10):3405–3416

Ordonez A, Williams JW (2013) Climatic and biotic velocities for woody taxa distributions over the last 16 000 years in eastern North America. Ecol Lett 16(6):773–781

Orwig DA, Cogbill CV, Foster DR et al (2001) Variations in old-growth structure and definitions: forest dynamics on Wachusett Mountain, Massachusetts. Ecol Appl 11(2):437–452

Parish R, Antos JA (2004) Structure and dynamics of an ancient montane forest in coastal British Columbia. Oecologia 141(4):562–576

Parish R, Antos JA (2006) Slow growth long-lived trees and minimal disturbance characterize the dynamics of an ancient montane forest in coastal British Columbia. Can J For Res 36(11):2826–2838

Parshall T, Foster DR (2002) Fire on the New England landscape: regional and temporal variation cultural and environmental controls. J Biogeogr 29(10–11):1305–1317

Pearson AF (2010) Natural and logging disturbances in the temperate rain forests of the Central Coast, British Columbia. Can J For Res 40(10):1970–1984

Pedersen BS (1998) The role of stress in the mortality of midwestern oaks as indicated by growth prior to death. Ecology 79(1):79–93

Pederson N (2010) External characteristics of old trees in the Eastern Deciduous Forest. Nat Areas J 30(4):396–407

Pederson N, Dyer JM, McEwan RW et al (2014) The legacy of episodic climatic events in shaping broadleaf-dominated forests. Ecol Monogr 84(84):599–620

Pederson N, D'Amato AW, Dyer JM et al (2015) Climate remains an important driver of post-European vegetation change in the eastern United States. Glob Chang Biol 21(6):2105–2110

Perera AH, Buse LJ, Weber MG (2007) Emulating natural forest landscape disturbances: concepts and applications. Columbia University Press, New York

Peteet DM, Vogel JS, Nelson DE et al (1990) Younger dryas climatic reversal in northeastern USA? AMS ages for an old problem. Quat Res 33(2):219–230

Peterson CJ, Cannon JB, Godfrey CM (2016) First steps toward defining the wind disturbance regime in central hardwoods forests. In: Greenberg CH, Collins BS (eds) Natural disturbances and historic range of variation, man for ecosystems, vol 32, pp 89–122

Piovesan G, Biondi F, Bernabei M et al (2005) Spatial and altitudinal bioclimatic zones of the Italian peninsula identified from a beech (*Fagus sylvatica* L) tree-ring network. Acta Oecol 27(3):197–210

Pontailler J-Y, Faille A, Lemée G (1997) Storms drive successional dynamics in natural forests: a case study in Fontainebleau forest (France). For Ecol Manage 98(1):1–15

Power MJ, Marlon J, Ortiz N (2008) Changes in fire regimes since the Last Glacial Maximum: an assessment based on a global synthesis and analysis of charcoal data. Clim Dyn 30(7–8):887–907

Putz FE, Sharitz RR (1991) Hurricane damage to old-growth forest in Congaree Swamp National Monument, South Carolina, USA. Can J For Res 21(12):1765–1770

Raffa KF, Aukema BH, Bentz BJ et al (2008) Cross-scale drivers of natural disturbances prone to anthropogenic amplification: the dynamics of bark beetle. BioScience 58(6):501–517

Rebertus AJ, Veblen TT (1993) Structure and tree-fall gap dynamics of old-growth *Nothofagus* forests in Tierra del Fuego, Argentina. J Veg Sci 4:641–654

Remmert H (1991) The mosaic-cycle concept of ecosystems—an overview. The mosaic-cycle concept of ecosystems. Ecol Stud 85:1–21

Rentch J (2003) Oak establishment and canopy accession strategies in five old-growth stands in the central hardwood forest region. For Ecol Manage 184(1–3):285–297

Robin C, Heiniger U (2001) Chestnut blight in Europe: diversity of *Cryphonectria parasitica* hypovirulence and biocontrol. For Snow Land Res 76(3):361–367

Rodríguez-Catón M, Villalba R, Srur AM et al (2015) Long-term trends in radial growth associated with *Nothofagus pumilio* forest decline in Patagonia: integrating local-into regional-scale patterns. For Ecol Manage 339:44–56

Romme WH, Everham EH, Frelich LE et al (1998) Are large infrequent disturbances qualitatively different from small frequent disturbances? Ecosystems 1(6):524–534

Rowlands W (1941) Damage to even-aged stands in Petersham, Massachusetts by the 1938 hurricane as influenced by stand condition. MF thesis, Harvard University, Cambridge, MA

Rozas V (2001) Detecting the impact of climate and disturbances on tree-rings of *Fagus sylvatica* L and *Quercus robur* L in a lowland forest in Cantabria, Northern Spain. Ann For Sci 58(3):237–251

Rozas V (2004) A dendroecological reconstruction of age structure and past management in an old-growth pollarded parkland in northern Spain. For Ecol Manage 195(1–2):205–219

Rozas V, Olano J, DeSoto L et al (2008) Large-scale structural variation and long-term growth dynamics of *Juniperus thurifera* trees in a managed woodland in Soria, central Spain. Ann For Sci 65(8):809–809

Runkle JR (1982) Patterns of disturbance in some old-growth mesic forests of eastern North America. Ecology 63(5):1533–1546

Runkle JR (1985) Disturbance regimes in temperate forests. In: Pickett STA, White PS (eds) The ecology of natural disturbance and gap dynamics. Academic, New York, pp 17–33

Runkle JR (1998) Changes in southern Appalachian canopy tree gaps sampled thrice. Ecology 79(5):1768–1780

Runkle JR (2000) Canopy tree turnover in old-growth mesic forests of eastern North America. Ecology 81(2):554–567

Sánchez-Salguero R, Linares JC, Camarero JJ et al (2015) Disentangling the effects of competition and climate on individual tree growth: a retrospective and dynamic approach in Scots pine. For Ecol Manage 358:12–25
Schwörer C, Colombaroli D, Kaltenrieder P (2015) Early human impact (5000–3000 BC) affects mountain forest dynamics in the Alps. J Ecol 103(2):281–295
Seidl R, Schelhaas M-J, Rammer W et al (2014) Increasing forest disturbances in Europe and their impact on carbon storage. Nat Clim Chang 4(9):806–810
Seidl R, Thom D, Kautz M et al (2017) Forest disturbances under climate change. Nat Clim Chang 7:395–402
Shanley CS, Pyare S, Goldstein MI et al (2015) Climate change implications in the northern coastal temperate rainforest of North America. Clim Chang 130(2):155–170
Shugart HH (1984) A theory of forest dynamics: the ecological implications of forest succession models. Springer, New York
Shuman B, Webb IIIT, Bartlein P et al (2002) The anatomy of a climatic oscillation: vegetation change in eastern North America during the Younger Dryas chronozone. Quat Sci Rev 21(16):1777–1791
Sippel S, Zscheischler J, Reichstein M (2016) Ecosystem impacts of climate extremes crucially depend on the timing. Proc Natl Acad Sci U S A 113(21):5768–5770
Siwkcki R, Ufnalski K (1998) Review of oak stand decline with special reference to the role of drought in Poland. Eur J For Pathol 28(2):99–112
Solberg S (2004) Summer drought: a driver for crown condition and mortality of Norway spruce in Norway. For Pathol 34(2):93–104
Splechtna BE, Gratzer G, Black BA (2005) Disturbance history of a European old-growth mixed-species forest—A spatial dendro-ecological analysis. J Veg Sci 16(5):511–522
Sprugel DG (1991) Disturbance equilibrium and environmental variability: what is 'natural' vegetation in a changing environment? Biodivers Conserv 58:1–18
St George S, Ault TR, Torbenson MC (2013) The rarity of absent growth rings in Northern Hemisphere forests outside the American Southwest. Geophys Res Lett 40(14):3727–3731
Stambaugh MC, Varner JM, Noss RF (2015) Clarifying the role of fire in the deciduous forests of eastern North America: reply to Matlack. Conserv Biol 29(3):942–946
Stan AB, Daniels LD (2010a) Calibrating the radial-growth averaging method for detecting releases in old-growth forests of coastal British Columbia, Canada. Dendrochronologia 28(3):135–147
Stan AB, Daniels LD (2010b) Growth releases of three shade-tolerant species following canopy gap formation in old-growth forests. J Veg Sci 21(1):74–87
Stan AB, Daniels LD (2014) Growth releases across a natural canopy gap-forest gradient in old-growth forests. For Ecol Manage 313:98–103
Stringer JW, Kimmerer TW, Overstreet JC (1989) Oak mortality in eastern Kentucky. Southern J Appl Forestry 13:86–91
Stueve KM, Perry CH, Nelson MD (2011) Ecological importance of intermediate windstorms rivals large infrequent disturbances in the northern Great Lakes. Ecosphere 2(1):art2
Suzuki E, Tsukahara J (1987) Age structure and regeneration of old growth *Cryptomeria japonica* forests on Yakushima Island. Shokubutsu-gaku-zasshi 100(3):223–241
Svenning JC (2003) Deterministic Plio-Pleistocene extinctions in the European cool-temperate tree flora. Ecol Lett 6(7):646–653
Svoboda M, Janda P, Nagel TA et al (2012) Disturbance history of an old-growth sub-alpine *Picea abies* stand in the Bohemian Forest, Czech Republic. J Veg Sci 23(1):86–97
Swetnam TW, Betancourt JL (1990) Fire-southern oscillation relations in the southwestern United States. Science 249(4972):1017–1020
Takahashi K (1997) Regeneration and coexistence of two subalpine conifer species in relation to dwarf bamboo in the understorey. J Veg Sci 8(4):529–536
Takaoka S (1993) The effect of missing rings on stand-age estimation of even-aged forests in northern Hokkaido, Japan. Ecol Res 8(3):341–347
Taleb NN (2007) The black swan: the impact of the highly improbable. Random House, New York

Taylor AH, Jinyan H, ShiQiang Z (2004) Canopy tree development and undergrowth bamboo dynamics in old-growth *Abies–Betula* forests in southwestern China: a 12-year study. For Ecol Manage 200(1):347–360
Temperli C, Bugmann H, Elkin C (2013) Cross-scale interactions among bark beetles climate change and wind disturbances: a landscape modeling approach. Ecol Monogr 83(3):383–402
Tinner W, Lotter AF (2001) Central European vegetation response to abrupt climate change at 82 ka. Geology 29(6):551–554
Tinner W, Lotter AF (2006) Holocene expansions of *Fagus silvatica* and *Abies alba* in Central Europe: where are we after eight decades of debate? Quat Sci Rev 25(5):526–549
Trotsiuk V, Hobi ML, Commarmot B (2012) Age structure and disturbance dynamics of the relic virgin beech forest Uholka (Ukrainian Carpathians). For Ecol Manage 265:181–190
Trotsiuk V, Svoboda M, Janda P et al (2014) A mixed severity disturbance regime in the primary *Picea abies* (L) Karst forests of the Ukrainian Carpathians. For Ecol Manage 334:144–153
Trotsiuk V, Svoboda M, Weber P et al (2016) The legacy of disturbance on individual tree and stand-level aboveground biomass accumulation and stocks in primary mountain *Picea abies* forests. For Ecol Manage 373:108–115
Trotsiuk V, Druckenbrod D, Pederson N et al (submitted) Testing the efficacy of tree-ring methods for detecting past disturbance events using experimental data and known events. To be submitted to Methods Ecol Evol. (In Perp)
Trouet V, Domínguez-Delmás M, Pearson C et al (2017) "Dendro-archeo-ecology" in North America and Europe: re-purposing historical materials to study ancient human-environment interactions. In: Amoroso M, Daniels L, Baker PJ, Camarero JJ et al (eds) Dendroecology: tree-ring analyses applied to ecological studies. Springer, New York
Ulanova NG (2000) The effects of windthrow on forests at different spatial scales: a review. For Ecol Manage 135(1):155–167
Vanderwel MC, Coomes DA, Purves DW (2013) Quantifying variation in forest disturbance and its effects on aboveground biomass dynamics across the eastern United States. Glob Chang Biol 19:1504–1517
Varner JM, Arthur M, Clark S et al (2016) Fire in Eastern North American oak ecosystems: filling the gaps. Fire Ecol 12(2):1–6
Veblen TT (1986) Treefalls and the coexistence of conifers in subalpine forests of the central Rockies. Ecology 67(3):644–649
Wang T, Hamann A, Spittlehouse DL et al (2012) ClimateWNA-high-resolution spatial climate data for western North America. J Appl Meteorol Climatol 51(1):16–29
Webb T III (1988) Eastern North America. In: Huntley B, Webb T III (eds) Vegetation history. Springer, Netherlands, pp 385–414
White PS (1979) Pattern process and natural disturbance in vegetation. Bot Rev 45(3):229–299
White PS, Jentsch A (2001) The search for generality in studies of disturbance and ecosystem dynamics. In: Esser K, Lüttge U, Kadereit JW et al (eds) Progress in botany. Springer, New York, pp 399–450
White PS, Harrod J, Romme W et al (1999) The role of disturbance and temporal dynamics. In: Szaro RC, Johnson NC, Sexton WT et al (eds) . Elsevier Science, Oxford, pp 281–312
Wiens JA (1989) Spatial scaling in ecology. Funct Ecol 3(4):385–397
Williams JW, Jackson ST (2007) Novel climates no-analog communities and ecological surprises. Front Eco Eviron 5(9):475–482
Winter LE, Brubaker LB, Franklin JF et al (2002) Canopy disturbances over the five-century lifetime of an old-growth Douglas-fir stand in the Pacific Northwest. Can J For Res 32(6):1057–1070
Wong C, Dorner B, Sandmann H (2003) Historical variability of natural disturbances in British Columbia: a literature review. FORREX–Forest Research Extension Partnership, Kamloops, British Columbia
Woods KD (2004) Intermediate disturbance in a late-successional hemlock-northern hardwood forest. J Ecol 92(3):464–476
Yamamoto S-I (1992) The gap theory in forest dynamics. Bot Mag Tokyo 105(2):375–383

Yamamoto S-I (1993) Gap characteristics and gap regeneration in a subalpine coniferous forest on Mt Ontake, central Honshu, Japan. Ecol Res 8(3):277–285

Yamamoto S-I (1995) Gap characteristics and gap regeneration in subalpine old-growth coniferous forests central Japan. Ecol Res 10(1):31–39

Yamamoto S (1996) Gap regeneration of major tree species in different forest types of Japan. Vegetatio 127(2):203–213

Yoshida T, Kamitani T (1998) Effects of crown release on basal area growth rates of some broad-leaved tree species with different shade-tolerance. J For Res 3(3):181–184

Young AB (2016) Species coexistance in an alpine forest in central Japan. PhD dissertation, The Pennsylvania State University, University Park, PA

Chapter 6
Integrating Dendroecology with Other Disciplines Improves Understanding of Upper and Latitudinal Treelines

Andrea H. Lloyd, Patrick F. Sullivan, and Andrew G. Bunn

Abstract Despite a consistent global relationship between the position of alpine and arctic treeline and temperature, fine-scale variability in treeline response to climate is widespread. In this chapter, we describe two advances in the application of dendroecology to treeline environments. First, we show that obtaining detailed, spatially explicit measurements of the environment can provide a more complete picture of how tree growth responds to climate. Substantial topoclimatic and environmental heterogeneity can occur at a very fine scale, a few tens of meters, at treeline. Unlike traditional approaches that aggregate tree growth across relatively broad areas, applying the approaches of landscape ecology and quantifying this heterogeneity can allow dendroecologists to better understand fine-scale, within-population variation in climate response. Second, the integration of dendroecological approaches with physiological ecology and soil biogeochemical studies can provide a more holistic understanding of tree growth, one that approaches trees as not merely being the sum of their rings, but as complex organisms whose growth integrates the impact of multiple limiting factors filtered through distinct physiological processes. In regions with cold soils, for example, we show that nutrient limitation can strongly mediate the response of tree growth to climate warming. Together, these approaches allow us to understand the causes of fine-scale variation in treeline response to warming, reconcile that variation with global-scale correlations between treeline and temperature, and better predict future responses of treeline ecosystems to warming.

A.H. Lloyd (✉)
Department of Biology, Middlebury College, 9 Old Chapel Road, Middlebury, VT 05753, USA
e-mail: lloyd@middlebury.edu

P.F. Sullivan
Environment and Natural Resources Institute, University of Alaska Anchorage, 3151 Alumni Loop, Anchorage, AK 99508, USA
e-mail: pfsullivan@uaa.alaska.edu

A.G. Bunn
Environmental Sciences, Western Washington University, 516 High St, Mailstop 9181, Bellingham, WA 98225-9181, USA
e-mail: andy.bunn@wwu.edu

M.M. Amoroso et al. (eds.), *Dendroecology*, Ecological Studies 231,
DOI 10.1007/978-3-319-61669-8_6

Keywords Dendroecology • Soil biogeochemistry • Stable isotope • Subalpine • Subarctic • Treeline • Topoclimate

6.1 Introduction

6.1.1 Causes of Treeline

The treeline ecotone, which we will define here as the zone extending from subalpine or subarctic forests to the elevational or poleward limit of trees (e.g., Holtmeier and Broll 2005), presents an ecological paradox. At a global scale, the location of treeline exhibits a remarkably consistent correlation with broad-scale features of climate, occurring within an exceptionally narrow range of air and ground temperatures. Mean root-zone temperatures in treeline ecosystems, for example, vary less than 2.5 °C globally, from 5.4 to 7.8 °C (Körner and Paulsen 2004). This uniformity suggests a shared constraint on tree growth and, indeed, treeline position globally can be accurately predicted from thermal parameters (Paulsen and Körner 2014). Consistent with the hypothesis that thermal conditions are a primary determinant of treeline position, the position of treeline varies globally over time in a manner that is roughly, although imperfectly, correlated with climate variation over long time scales (e.g., Kullman 1995; Lloyd and Graumlich 1997; MacDonald et al. 2000). A number of hypotheses have been proposed to explain these patterns in space and time, including direct effects of temperature on growth (e.g., Körner 1998), direct effects of temperature on tissue dieback (e.g., Slatyer and Noble 1992; Harsch and Bader 2011), and the indirect effects of climate on nutrient uptake (e.g., Sveinbjörnsson et al. 2002).

Despite this highly consistent pattern of correlation with temperature, treeline structure and dynamics prove to be extraordinarily variable and complex at finer (regional to local) scales. For example, although there is ample evidence that temperature plays a crucial role in determining the position, structure, and function of treeline forests globally, treelines in some climatic regions are limited by both temperature and moisture (e.g., Lloyd and Graumlich 1997) or primarily by moisture (e.g., Lavergne et al. 2015; Liang et al. 2014; Wang et al. 2015). Complexity at local to regional scales can also be seen in the spatial configuration of treelines, which varies from very abrupt to very diffuse (Harsch and Bader 2011), in the spatial variability in the response of treeline to climate (e.g., Harsch et al. 2009), and in patterns of treeline tree growth response to climate that vary across space and time (e.g., Lloyd and Fastie 2002; Wilmking et al. 2004; Driscoll et al. 2005; Lloyd et al. 2011; Shrestha et al. 2015). Some of the variation among treeline sites can be attributed to the prevalence of non-climatic controls—land-use change (Ameztegui et al. 2016) or edaphic factors (Holtmeier and Broll 2005; Suarez et al. 1999), for example. But much of the fine-scale variability, including variability in growth response to climate over time, remains unexplained. Recent reviews have emphasized the scale-dependence of controls over treeline dynamics (Case and

Duncan 2014; Holtmeier and Broll 2005; Malanson et al. 2007, 2011) and the variability in climatic controls and hence climate response among different types of treeline ecotone (e.g., Harsch and Bader 2011) in an effort to better categorize and understand sources of variability in treeline dynamics and structure, but a single, unified understanding of the mechanisms by which declining temperature restricts the growth of trees has remained elusive.

The importance of arriving at a clear mechanistic understanding of treeline is heightened by the apparent sensitivity of treeline ecosystems to climate change, and hence the likelihood of an advance of treeline into adjacent tundra communities. Advances of treeline have been detected at a number of sites, but treeline advance is far from ubiquitous (e.g., Harsch et al. 2009), and there is important variability in the rate of response (MacDonald et al. 1998, 2000; Lloyd 2005). Accurate predictions of future patterns of change at treeline are important for anticipating conservation consequences (e.g., associated with the decline of species endemic to tundra habitats) and ecosystem level feedbacks (e.g., associated with reduced surface albedo; Chapin et al. 2005), and depend critically on a sufficiently robust functional understanding of the relationship between treeline dynamics and climate.

6.1.2 Dendroecology at Treeline

Dendroecological approaches have made important contributions to our understanding of treeline, providing crucial information on the decadal, centennial, and even millennial-scale dynamics that are important in ecosystems dominated by long-lived species. The earliest applications of dendrochronology to treeline focused on questions about climatic controls over the growth of adult trees (e.g., Giddings 1943). Although such studies have tended to confirm that growth at treeline is limited by temperature, tree rings have been instrumental in demonstrating the complexities of growth limitation. In the boreal forest, for example, tree-ring studies have demonstrated that warming may be associated with *reduced* tree growth, even at the very margins of the treeline ecotone (e.g., Wilmking et al. 2004; Lloyd and Bunn 2007; D'Arrigo et al. 2008). More recently, tree rings have been used as the basis for reconstructions of population age structures and patterns of recruitment and mortality that have contributed to our understanding of the dynamics of treeline ecosystems. Such demographic studies have been an important tool for reconstructing changes in the position of treeline over decadal to (rarely) millennial time scales (e.g., Lloyd and Graumlich 1997; MacDonald et al. 1998; Lloyd 2005; Batllori and Gutierrez 2008). Providing greater spatial and temporal resolution than other paleoecological proxies, reconstructions of treeline dynamics have revealed—in the aggregate—complex and spatially variable patterns of change.

In this chapter, we propose two key advances in the application of dendroecology to the task of achieving a robust functional understanding of treeline. First, we propose that dendroecological studies can more accurately describe the spatial and temporal variability in treeline dynamics by sampling in a manner that does not assume homogeneity within any individual treeline site. Second, we propose that hybrid studies, which combine dendroecological approaches with the methods of physiological ecology and/or biogeochemistry, are a promising way forward in developing a functional, mechanistic understanding of treeline. Part of the challenge of achieving a robust functional understanding of treeline is the relatively slow pace at which treeline ecosystems change in response to shifts in the environment. This slow rate of change presents challenges to efforts to conduct controlled experiments, for example. The longer temporal perspective afforded by dendroecological approaches is critical, but the correlational foundation of tree-ring studies is limiting: the application of physiological methods, in combination with tree-ring studies, will likely provide the best of both worlds.

6.2 Case Study 1: A Topoclimatic Approach to Understanding Treeline

Traditionally, dendroclimatologists and dendroecologists have approached the concept of what constitutes a study site, and which trees are included in the sampling of any particular site, quite differently from one another. Indeed, the International Tree Ring Data Bank typically lists a single latitude and longitude for each ring-width file no matter the number of trees or area sampled. A dendroclimatologist often has an eye for the oldest trees (e.g., Kaufmann 1996; Stahle 1996; Stahle and Chaney 1994; Swetnam and Brown 1992; Huckaby et al. 2003) at a given location in order to extract the longest possible record. A dendroecologist, on the other hand, is more likely to lay out transects or delineate plots and sample trees that are more representative of the entire population in order to increase replicability or to make inference about a larger area. The advantage of the first approach is that a long record can be obtained, which can be crucial both for adequate calibration and verification of tree-ring response to climate and for developing a suitably long record for climate reconstructions. Sampling only the oldest trees also allows a dendroclimatologist to obtain a large sample over a reasonably large area, in which case small-scale variability (e.g., from topography) in growth may be drowned out by the large sample size. The advantage of the second approach is that stratified sampling methods can be used and more nuanced and complicated experimental designs can be employed: the fine-scale variability in growth across space that a dendroclimatologist might endeavor to eliminate with a large, widespread sample can become a useful part of the study design. We argue that if we explicitly consider spatial heterogeneity as a driver of tree growth in areas of complex topography we can marry these two approaches as a way of leveraging climate gradients over short geographic distances.

Ancient bristlecone pine (*Pinus longaeva* D.K. Bailey; Fig. 6.1a) is the charismatic tree species long of interest to dendrochronology because the trees are extremely long lived, with the potential to reach nearly 5000 years old, and the ring

a

b

Fig. 6.1 The two case studies presented here address two distinct treeline ecosystems, shown here. (**a**) Bristlecone pine (*Pinus longaeva*) at treeline in the Snake Range, Nevada. (**b**) White spruce (*Picea glauca*) treeline along the Agashashok River in northwestern Alaska

widths vary with climate (see canonical works such as Currey 1965; Ferguson 1968; Lamarche 1969; Lamarche and Stockton 1974). Although the records are of interest to dendroclimatologists because of their sheer length, bristlecone pine presents a challenge because the growth signal can be a mixture of different limiting variables with some trees responding to energy (temperature) limitation and some reflecting soil moisture limitations that, in turn, are driven by temperature in combination with precipitation. This variability in limiting factors can occur even among trees growing on the same mountain slope (Bunn et al. 2011; Salzer et al. 2009, 2013, 2014).

We present here a case study, based on unpublished (Bruening 2016; Tran 2016) and published (Salzer et al. 2009, 2014; Bunn et al. 2011) results, of 240 bristlecone pine samples growing within 100 vertical meters of treeline over an area of about 4 km^2 at Mount Washington in the Snake Range of the Great Basin (Nevada, USA, 38.91° N, 114.31° W, treeline elevation 3400 m.a.s.l.). This data set is well replicated to 500 CE with median segment length of 350 years and a median first-order autocorrelation of 0.69.

At this location and others like it, bristlecone pine at the lower forest border shows moisture limitations to growth (Lamarche 1974b; Hughes and Funkhouser 1998, 2003) while the upper forest trees are more typically considered as temperature limited (LaMarche 1974a, b; Lamarche and Stockton 1974; Salzer et al. 2009, 2013). However, even at the upper forest border, complex mountain topography can reverse this general pattern: in some topographic settings, trees growing near the treeline show growth patterns more similar to those found in trees at the lower forest border than the upper forest border (Salzer et al. 2009, 2014; Bunn et al. 2011). Previous work at bristlecone pine treeline has found that topographic setting can mediate growing season temperatures by several degrees C at a spatial scale of tens of meters and that, in turn, can affect tree growth (Bruening 2016; Tran 2016). If individual trees at a single site—say, a particular mountain slope—can be limited by different factors because of topoclimate, aggregating those trees into a single site chronology might well confound contrasting growth signals.

A field campaign at Mount Washington modeled topoclimate using dozens of temperature sensors arrayed across topographic gradients and resulted in temperature data interpolated to 10-m resolution as a function of topography (described by Bruening 2016). Inspired by Paulsen and Körner (2014), Bruening (2016) and Tran (2016) found that treeline position and growth patterns can be modeled well by topoclimate variables including the length of growing season (in days) and seasonal mean air temperature. The length of the growing season is a sum of days with daily mean temperature above 0.9 °C, and seasonal mean temperature is an average of the daily mean temperatures for the days included in the growing season length calculations (Paulsen and Körner 2014). For the area considered in this study, entirely within 100 m of treeline, there was a median seasonal mean temperature of 8.2 °C (range: 7.4–9.0 °C) and a median growing season length of 164 days (range: 143–188 days).

A hierarchical cluster analysis of the ring-width data in this case study indicated two dominant growth signals. The difference in cluster membership was most cleanly modeled by seasonal mean temperature, with trees growing in topoclimate settings colder than 8 °C displaying a qualitatively and quantitatively different pattern than trees growing in topoclimate settings with seasonal mean temperature warmer than 8 °C. Because of a robust sample depth ($n = 240$ trees) and the lack of typical "juvenile" growth patterns, we followed Salzer et al. (2009) and used raw ring widths for the chronologies. Mean value chronologies of average ring widths are shown in Fig. 6.2 for all samples (Fig. 6.2a) as well as trees in the colder settings with seasonal mean temperature less than 8 °C (Fig. 6.2b) and warmer settings with samples in settings with seasonal mean temperature greater than 8 °C (Fig. 6.2c).

In a qualitative sense, the "cold" (b) and "warm" (c) chronologies are mixtures of the ensemble (a) indicated by similar features among all three chronologies. In particular, low growth episodes such as the 1450s tend to occur in common, while growth patterns during warm periods like those near 1150 CE and after 1900 differ. The "cold" chronology, in those periods, shows increased growth while the "warm" chronology does not. However, the temporal frequency at which tree growth variation is concentrated differs between the two chronologies. Wavelet spectra for the "cold" chronology indicate that variation in this chronology is concentrated at lower frequencies—multi-decadal to centennial—while the warmer chronology is more likely to vary at higher frequencies. Correspondingly, the range of variability

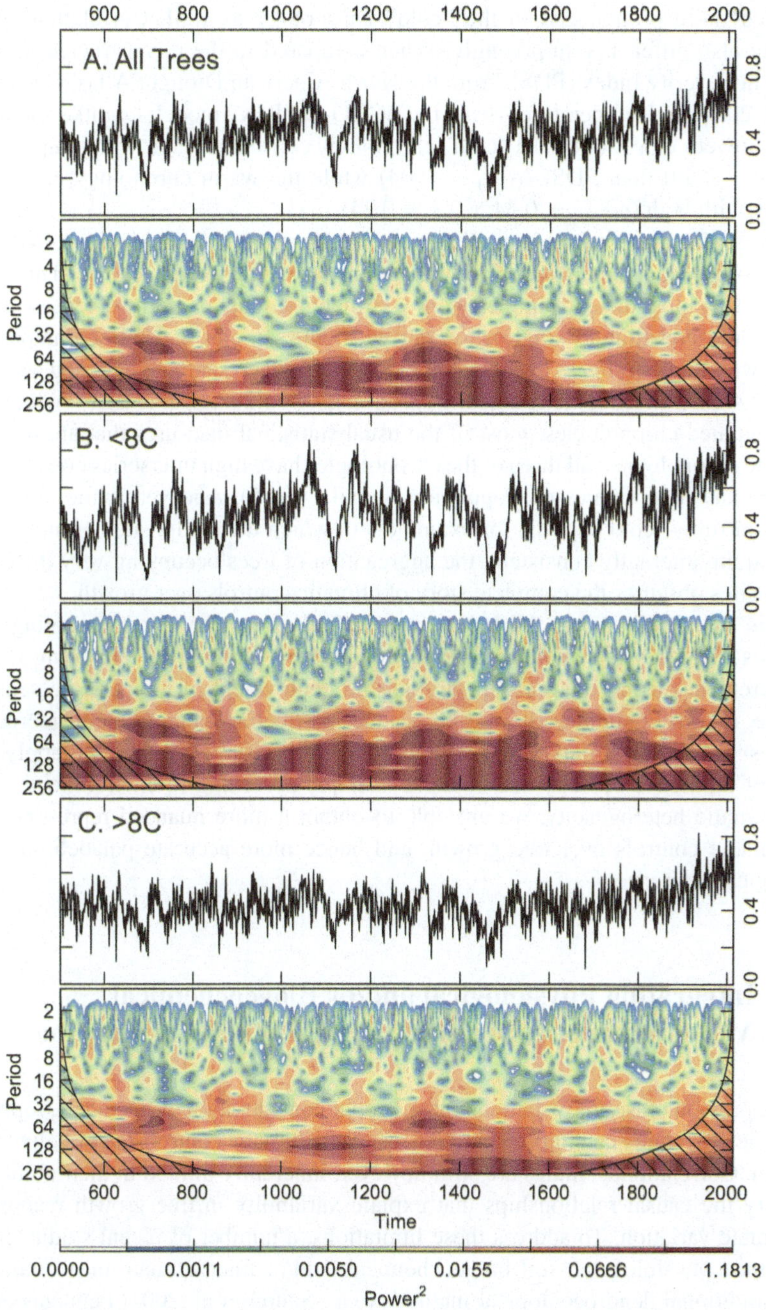

Fig. 6.2 Raw ring width chronologies and continuous wavelet spectra are sown for all 240 samples (**a**) as well as the 128 samples in topoclimate settings with seasonal mean temperatures <8 °C (**b**) and the 112 samples in topoclimate settings with seasonal mean temperatures >8 °C (**c**) The wavelet powers are scaled by quantile (0–10%, 10–20%, etc.) and period is in years

is much more pronounced in the "cold" chronology as well. Correlations with climate also differ, not surprisingly. When compared to the reconstructed Palmer Drought Severity Index (PDSI; from the North American Drought Atlas, Cook and Krusic 2004) and temperatures (from the ECHO-G ERIK2 model; Legutke and Voss 1999, Stevens et al. 2008), the "cold" chronology correlates better with temperature ($r_{TEMP} = 0.50$) than PDSI ($r_{PDSI} = 0.04$) while the warm chronology correlates weakly with both ($r_{TEMP} = 0.34$, $r_{PDSI} = 0.24$).

The implications of these findings are significant for dendroecology. All 240 bristlecone pine samples described here were collected within 100 m of treeline and the two most distant samples were less than 4 km apart. In most senses, these data would constitute a single site, and with 240 samples a dendroclimatologist might well think of this as a single collection with a deep sample depth and a single climate signal. Indeed, the composite chronology would pass, at least as well as the differentiated chronologies, most of the usual statistical measures that are used to validate chronologies: all three of the chronologies have high interseries correlations (>0.56) and high Expressed Population Signal ($\geq$0.85). And yet, while there are points in time—cold periods, for example—in which the composite chronology is likely quite internally consistent, the aggregation of trees occupying very different micro-sites obscures the complete story of climatic controls over growth.

This case study shows that when we explicitly consider in our sampling and analysis the ways in which climate is mediated by topography, including direct measurement of that relationship through the deployment of temperature sensors on site, relatively simple criteria like distance to treeline cannot be considered as the best way to differentiate the bounds for a collection. If instead we apply the tools of landscape ecology and explicitly sample according to environmental and topoclimatic heterogeneity, we are able to obtain a more nuanced representation of climatic controls over tree growth, and hence more accurate paleoclimatic or ecological inferences.

6.3 Integrating Physiological and/or Biogeochemical Measurements in Dendroecological Studies at Treeline

As the previous case study demonstrates, proper consideration of fine-scale, topoclimatic heterogeneity can improve inferences about climatic controls over treeline tree growth. Correlational studies are still, however, inherently limited in their ability to identify the causal relationships that explain variability in tree growth responses to climate variation. To address these limitations, a number of recent studies have combined physiological, soil biogeochemical, and/or microclimate measurements with traditional dendroecological methods (e.g., Saurer et al. 2004; Peñuelas et al. 2011; Sullivan et al. 2015; Brownlee et al. 2016). Results of these hybrid studies have helped to improve understanding of mechanisms of tree growth responses to rising atmospheric CO_2 concentration and associated changes in climate. In this

Table 6.1 Potential variables for inclusion in a dendroecology study that incorporates physiological, biogeochemical and microenvironment measurements

Measurement	Type	Scale	Resolution
$\delta^{13}C$ in alpha-cellulose	Integrative	Tree	Seasonal-decadal
$\delta^{18}O$ in alpha-cellulose	Integrative	Tree	Seasonal-decadal
Foliar gas exchange	Point	Tree	Minutes
Remotely sensed leaf area	Point	Stand	Seasonal-annual
Foliar nutrition	Integrative	Tree	Weekly-annual
Sap flux	Continuous	Tree	Hourly
Xylem water potential	Continuous	Tree	Hourly
Eddy covariance	Continuous	Stand	Hourly
Soil nutrient availability	Integrative	Tree	Annual
Air temperature	Continuous	Tree-stand	Hourly
Relative humidity	Continuous	Tree-stand	Hourly
Precipitation	Continuous	Stand	Hourly
Photosynthetically active radiation	Continuous	Tree-stand	Hourly
Soil temperature	Continuous	Tree	Hourly
Soil moisture	Continuous	Tree	Hourly
Slope	Point	Tree-stand	NA
Aspect	Point	Tree-stand	NA
Topographic position	Point	Tree-stand	NA
Stand density	Point	Tree-stand	Annual-decadal
Basal area	Point	Tree-stand	Annual-decadal
Ground cover type/depth	Point	Tree-stand	Annual-decadal
Litter depth	Point	Tree-stand	Annual-decadal
Moss depth	Point	Tree-stand	Annual-decadal
Organic soil depth	Point	Tree-stand	Annual-decadal

Resolution defines the typical temporal resolution of measurements of this variable

section, we review three types of measurements that have been fruitfully combined with tree-ring measurements at treeline (Table 6.1), and discuss the methodological implications of conducting a hybrid study.

6.3.1 Overview of Hybrid Approaches

6.3.1.1 Stable Isotopes in Dendroecology

The most common method of integrating physiological measurements into dendroecological studies is to examine the stable isotopic composition of alpha-cellulose extracted from the annual rings. It is important to extract cellulose, rather than measuring the isotopic composition of the whole wood, because the components of wood (e.g., alpha-cellulose, hemicellulose, lignin) have different isotopic compositions and their relative abundance may vary over time in the annual rings (Wilson and Grinsted 1977). The stable isotopes of carbon and oxygen are the most common isotopes of interest.

Carbon isotope discrimination ($\Delta^{13}C$), which is approximately the difference between $\delta^{13}C$ of the plant tissue and $\delta^{13}C$ of atmospheric CO_2, has been related to the ratio of leaf intercellular to atmospheric CO_2 concentration (C_i/C_a) as follows (Farquhar et al. 1982):

$$\Delta^{13}C = a + (b-a)^* (C_i/C_a),$$

where a (4.4‰) is fractionation associated with diffusion of CO_2 through the stomata and b (27‰) is fractionation that results from carboxylation. C_i/C_a is reflective of the balance between photosynthesis and stomatal conductance and provides insights into water use efficiency (WUE) over the lifespan of the tree. The information that can be gleaned from $\Delta^{13}C$ is limited, however, in the sense that an increase in $\Delta^{13}C$ could reflect an increase in stomatal conductance or a decrease in assimilation, while a decrease in $\Delta^{13}C$ could result from a decrease in stomatal conductance or an increase in assimilation. In the context of rising atmospheric CO_2 concentration, stable $\Delta^{13}C$ over time generally indicates stomatal closure to maintain constant C_i/C_a, which is associated with increasing C_i and increasing WUE (Ehleringer and Cerling 1995).

One potential method of overcoming the limitations of $\Delta^{13}C$ is to measure $\delta^{18}O$ in tree-ring alpha-cellulose. The enzyme carbonic anhydrase catalyzes the near complete exchange of $^{18}O/^{16}O$ between H_2O and CO_2 prior to photosynthesis, leading to photosynthate that carries the isotopic fingerprint of the H_2O with which it was formed (DeNiro and Epstein 1979). The isotopic composition of leaf water is determined by the isotopic composition of the source water and by evaporation during transpiration. Source water $\delta^{18}O$ varies with temperature during precipitation (e.g., summer rain is enriched, while winter snow is depleted), storm track, evaporation from the soil and depth of water uptake from the soil (e.g., Dawson and Ehleringer 1991). Meanwhile, leaf water enrichment varies with the leaf to atmosphere vapor pressure difference, $\delta^{18}O$ of atmospheric water vapor, leaf structure, and transpiration (Craig and Gordon 1965; Farquhar and Lloyd 1993). Under some circumstances (Roden and Siegwolf 2012), $\delta^{18}O$ may be a useful tool to aid interpretation of $\Delta^{13}C$ in tree-ring alpha-cellulose. For example, enrichment of $\delta^{18}O$ over time in tree-ring alpha-cellulose could be indicative of increasing evaporative demand and might suggest that a corresponding decline in $\Delta^{13}C$ is indicative of drought-induced stomatal closure, rather than an increase in assimilation (Scheidegger et al. 2000). However, enrichment of $\delta^{18}O$ over time in tree-ring α-cellulose could also reflect increasing use of summer rain rather than winter snow, with no change in moisture availability (Saurer et al. 2002). The complexity of the controls on $\delta^{18}O$ in tree-ring cellulose and the difficulty of acquiring historical estimates of source water $\delta^{18}O$ have limited most studies to using $\delta^{18}O$ in tree-ring alpha-cellulose as a general indicator of relative changes in aridity (e.g., Saurer et al. 2008).

In treeline environments, stable isotopes have provided greater clarity on the mechanisms underlying the deterioration of historically strong positive correlations between temperature and tree growth (e.g., D'Arrigo et al. 2008). This phenomenon, known as "divergence", is potentially attributable to a number of factors (reviewed in D'Arrigo et al. 2008), including methodological issues related to the method by which tree ring series are detrended (e.g., Esper and Frank 2009; Briffa and Melvin 2011; Sullivan et al. 2016). To the extent that divergence reflects real changes in the relationship between growth and climate, however, it casts doubt on predictions that climate warming will lead to widespread, sustained increases in growth of treeline trees. Because changes in temperature and water availability are often confounded at any single site, it is challenging, if not impossible, to differentiate their effects using correlative analyses alone. For instance, a "divergent" growth trend could be attributable to either the direct effects of above optimal temperatures or greater evapotranspiration/reduced water availability. Stable isotopes in tree-ring alpha-cellulose have been used in a limited number of studies to examine whether or not divergent growth of treeline trees is associated with drought-induced stomatal closure. A study of larch in northern Siberia showed that warm and dry conditions during June and July were associated with reduced growth and lower carbon isotope discrimination (Sidorova et al. 2009). Meanwhile, a study of white spruce near the arctic treeline in Alaska showed that divergence was not associated with a change in $\Delta^{13}C$ (Brownlee et al. 2016). In this case, divergent trees exhibited lower photosynthesis and lower WUE than trees that increased growth during a period of strong climate warming, suggesting that divergence is probably not associated with increasing water deficit in this case. Incorporating stable isotope measurements into future studies of divergence at high latitudes would be a valuable step toward improving understanding of variation in the causes of divergence across trees, sites, regions, and species.

6.3.1.2 Incorporating Gas Exchange Measurements in Dendroecology

While stable isotopes in tree-ring alpha-cellulose integrate physiological conditions throughout each growing season over the lifespan of the tree, they rarely allow for isolation of individual physiological variables, such as photosynthesis (assimilation) and stomatal conductance. Measurements of gas exchange at the scale of the leaf, canopy, and ecosystem may aid interpretation of inter-annual variability in tree growth. In general, the number of physiological variables that can be measured or accurately estimated decreases as the scale of measurement increases. The widest range of physiological information can be obtained at the leaf- or shoot-level, where climate-controlled gas exchange cuvettes enable accurate estimation of variables such as photosynthesis, stomatal conductance, evapotranspiration, respiration and chlorophyll fluorescence (e.g., LI-COR Environmental, Lincoln, NE, USA). Cuvette measurements also allow for examination of the form of physiological responses to changing light, CO_2 concentration, temperature and humidity. One of the most useful measurements of gas exchange that can be made at the scale of the whole tree canopy is sap flux density, which can be used to estimate both whole tree canopy transpiration and canopy stomatal conductance (Granier 1987). One of the great benefits of sap flux measurements is that they are automated and can be made at sub-hourly intervals over long periods of time. Similarly, automated sensors for xylem water potential have recently become available and would be a valuable compliment to a suite of sap flux sensors (ICT International, Armidale, NSW, Australia).

In recent years, an extensive network of eddy covariance (EC) towers has been established in forested ecosystems around the globe. The EC method provides sub-hourly ecosystem-scale estimates of net ecosystem production, which are commonly broken down into the component fluxes: gross primary productivity and ecosystem respiration. All three of these variables may be useful in explaining the cause(s) of inter-annual variability in the tree-ring record, yet surprisingly few studies have taken advantage of this valuable resource (Rocha et al. 2006; Babst et al. 2014). As EC records continue to grow in length, we anticipate that combining tree-ring and long-term ecosystem-scale gas exchange measurements will lead to valuable insights regarding climate effects on forest ecophysiology and wood production. Unfortunately, because trees often reach their upper elevation and northern limits in complex topography, there are few EC towers currently operating at treeline. Overcoming this limitation to ecosystem-scale gas exchange measurements in complex terrain could open the door to valuable insights into the causes and consequences of changes in treeline.

6.3.1.3 Integrating Soil Biogeochemistry and Dendroecology

Soil nutrient availability is generally thought to be the most important limiting factor for forest productivity at the global scale (Vitousek and Horwath 1991; LeBauer and Treseder 2008). Nitrogen (N) and phosphorus (P) are the most common limiting nutrients, largely because of their importance in the light and light-independent reactions of photosynthesis. Nutrients may limit tree growth as a result of parent material type, soil age and/or soil microclimate effects on root, microbial and/or mycorrhizal activity. Ecosystems with cold and/or wet soils, such as the northern limits of the boreal forest, may be particularly prone to nutrient limitation (Sveinbjörnsson et al. 1992, Schulze et al. 1994, Loomis et al. 2006, McNown and Sullivan 2013).

One of the simplest methods of assessing tree and/or plot nutrient availability is to examine nutrient concentrations and element ratios in foliar tissue (Aerts and Chapin 2000). Foliar nutrient concentrations are not perfect indicators of nutrient acquisition. For instance, the rapid growth of trees on relatively fertile sites may dilute foliar nutrient concentrations. It is also important to recognize that foliar nutrient concentrations (e.g., mg of nutrient per g of dry foliage) are affected by changes in both the numerator and the denominator. Thus, changes in the content of other foliar components can change nutrient concentrations (mg/g) in the absence of changes in nutrient content (mg/leaf). For this reason, it is important to record the surface area or the number of leaves analyzed, so data may also be expressed on a per leaf or per m^2 basis. Despite these potential shortcomings, foliar nutrient concentrations have proven to be excellent predictors of tree growth across vast areas near the northern limits of the boreal forest (Ellison 2015).

While foliar nutrient concentrations represent an imperfect measure of soil nutrient availability, a relatively straightforward and more direct measurement of soil nutrient availability involves the use of ion exchange resin membranes (Qian and Schoenau 2002). Ion exchange membranes are typically installed in the soil as soon as possible after thaw and recovered just before soils freeze in the fall. As ions are released from soil organic matter or microbial biomass, they become adsorbed to membranes, where they accumulate over time and, in theory, provide a time-integrative measure of soil nutrient availability over the course of the growing season. Drawbacks to ion exchange membranes include the possibility that some ions may become desorbed over time (Giblin et al. 1994), the difficulty of installing membranes well below the surface without disturbing the soil and the fact that many trees acquire a large percentage of their nutrient requirement through mycorrhizae (Aerts and Chapin 2000), which is a flux of nutrients that would not be captured by exchange membranes.

6.3.1.4 Methodological Considerations in Hybrid Studies

There are a number of methodological considerations when designing a hybrid study arising from differences among approaches in the scale, effort, and cost of analysis. We highlight four factors here: the number of trees to sample, the timing of sampling, the question of where within a tree to sample, and the selection of which trees to sample. In a dendroecological study, a large number of trees can typically be sampled comparatively quickly at relatively low cost. The cost and effort of isotopic, gas exchange, and biogeochemical analyses is generally greater. In isotopic studies, for example, the high cost of analysis necessitates consideration of how many trees to analyze for how many years, whether to pool samples across years or across trees and whether to separate earlywood and latewood. The best approach will depend on the objectives of the study. With that said, the very small numbers of trees sampled in early isotope dendroclimatology studies (e.g., Leavitt and Long 1984) will probably not be a good fit for most dendroecology studies, which often have at least a secondary focus on variation across trees and study sites. At treeline, for example, the focus on understanding the spatially complex patterns of change both within and among sites requires a fairly large sample of trees, which might necessitate pooling samples in some fashion to reduce the overall cost of the analysis. Nested approaches to sampling—in which more intensive measurements are made on a subset of trees sampled for dendroecological purposes—may therefore be necessary in many hybrid studies.

Hybrid studies benefit from coordinating diverse measurements on the same trees—measuring gas exchange, for example, on all or some of the trees that were cored. In such cases, it is best to make the physiological or soil biogeochemical measurements over several growing seasons prior to collecting increment cores for the tree-ring measurements. This ensures that injury from coring will not affect the gas exchange measurements and that inter-annual variability in the physiological or soil biogeochemical measurements can be directly related to the tree-ring record.

Issues arise in hybrid studies in terms of where on a tree sampling is conducted. At least some of the trees sampled for tree-ring work are likely to be large old trees, with most of their canopy above heights that are convenient to make gas exchange measurements. Trees are well known to exhibit vertical gradients in leaf- or shoot-level physiology (e.g., Schulze et al. 1977). Thus, to ensure data are representative of the whole canopy, it is best to make measurements at multiple heights and depths within the canopy. Resources like the Wind River Canopy Crane are extremely valuable in this context, but they are also rare, particularly at treeline. In some species and for certain gas exchange variables, it may be possible to make gas exchange measurements on very recently detached foliage sampled using a pole saw or pruner. This challenge highlights the value of sap flux measurements, which are typically made at breast height, but provide data that integrate the whole tree canopy, regardless of tree size.

Finally, perhaps the greatest methodological question of all in dendrochronology, which becomes even more acute in hybrid studies, is the issue of which trees to sample at a given site. Dendroecologists most often aim for a sample that is representative of the tree population in terms of the distribution of sizes/ages and likely sensitivities to climate (Sullivan and Csank 2016). One approach to improve the likelihood that sampled trees are representative of the site is to first inventory and map trees, then randomly select trees from the map for sampling (Nehrbass-Ahles et al. 2014). Alternatively, a stratified random sampling design, reflecting important topographic or environmental heterogeneity, could produce an unbiased sample of trees for this kind of study. If a sufficient number of trees are selected for measurement, this approach should ensure that trees of different sizes/ages are selected, in proportion to their abundance at the site. This approach would also facilitate scaling the data up from the level of the tree to that of the site, which would be a great advantage if the study were conducted within the footprint of an EC tower, a satellite pixel, or if a goal of the study is to examine changes in stand productivity over time. Cost and labor will likely dictate that only a subset of the trees selected for increment core sampling can be monitored for gas exchange, soil biogeochemistry and/or tree microclimate. These trees could be randomly selected from those identified for increment core sampling and then checked against the larger random sample to ensure that trees in each size class are going to be sampled in proportion to their abundance.

Regardless of the suite of measurements selected (Table 6.1), we anticipate that studies integrating physiological and/or soil biogeochemical measurements into more traditional tree-ring studies will make valuable contributions to our understanding of factors limiting growth of trees at treeline and in forests around the globe. In the case study that follows, we provide an example of how a hybrid study can provide important insights into patterns of and controls over tree growth at treeline.

6.4 Case Study 2: Integrating Physiological, Biogeochemical, and Dendroecological Methods to Study the Growth of Treeline Trees

The integration of physiological, biogeochemical, and dendroecological methods has proven useful in recent studies near the Arctic treeline in northwest Alaska (Fig. 6.1b), which is a region where white spruce growth has generally increased and many treelines have advanced into the tundra in recent decades (Suarez et al. 1999). The central aim of research in this region has been to improve understanding of the causes of variation in growth rate and growth response to climate across habitats. This line of inquiry was motivated by several earlier studies that showed contrasting growth responses to twentieth century climate warming over small spatial scales at treeline in Alaska (Lloyd and Fastie 2002; Wilmking et al. 2004). In a comparison of three habitats (terrace, forest, and treeline) with similar aboveground climates, but strongly different soil conditions, McNown and Sullivan (2013) combined measurements of soil nutrient availability with weekly measurements of foliar nutrition and needle gas exchange. Across these three habitats, relatively warm soils of the terrace give way to intermediate soil conditions in the forest and finally to the cold soils, underlain by discontinuous permafrost, at treeline. Results showed limited evidence of drought-induced stomatal closure, limited seasonality in light-saturated photosynthesis, and strong declines in soil nutrient availability, foliar nutrient concentrations, and photosynthesis along the gradient from the terrace to treeline habitats. Sullivan et al. (2015) extended the study to include examination of seasonal variation in growth of all major organs (branch extension, xylem production, and fine root growth), along with historical growth preserved in the tree-ring record. These analyses revealed the surprising finding that seasonal peaks in growth of all organs preceded the warmest times of the season, both above- and below-ground. This pattern, which was particularly pronounced at treeline, suggests that treeline trees were not taking advantage of the full thermal growing season and points to a limiting factor other than temperature.

That limiting factor appears to be nutrient availability. Examination of the tree-ring record showed that trees growing on the terrace, where soil nutrient availability is relatively high, showed strong positive growth responses to twentieth century climate warming. Meanwhile, trees growing in the forest and at treeline, where soil nutrient availability is lower, showed progressively smaller positive growth responses (Fig. 6.3). Temperature effects on growth of treeline trees may therefore be mediated by soil nutrient availability at these study sites. Despite a common pattern of warming across the three habitats, significant increases in growth occurred only at the terrace site, where microbial activity in the comparatively warmer soils may have been sufficiently stimulated to alleviate nutrient limitation. Factors like soil biogeochemistry, which can mediate the effect of warming, complicate our ability to explain historical and predict future growth trends and highlight the importance of study designs that situate dendroecological responses within a broader understanding of physiological and ecosystem-level dynamics.

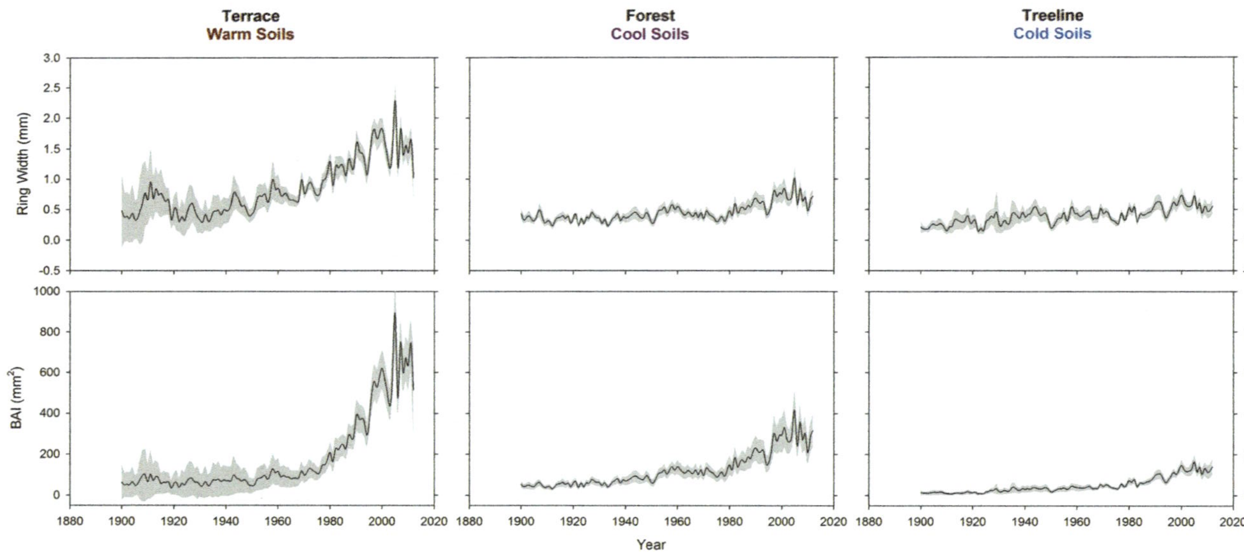

Fig. 6.3 Varying patterns of ring width and basal area increment (BAI) change over time in white spruce growing in three habitats near Arctic treeline in northwestern Alaska

6.5 Conclusions

These two case studies, drawn from very different treeline ecosystems, indicate that fine-scale environmental variability, properly quantified and incorporated into sampling design and data analysis, can yield key insights into treeline dynamics and shed light on the surprisingly complex responses of treeline ecosystems to climate warming. Both case studies employ dendroecology within a broader ecological context, integrating dendroecological approaches with those from landscape ecology, physiological ecology, and ecosystem ecology to yield a functional understanding of treeline ecosystems and treeline tree growth. In both cases, fine-scale spatial complexity in controls over tree growth challenges traditional dendroecological approaches; the insights gleaned from application of landscape, physiological, and ecosystem ecological approaches allowed that complexity to be used, however, as the basis for important insights about treeline dynamics.

We draw two important conclusions, which are perhaps best understood as recommendations for future studies, from these examples (Fig. 6.4). First, detailed, spatially explicit measurements of environmental variability—including topography, microclimate, and soil physical and biogeochemical properties—are a necessity, allowing researchers to understand the underlying environmental gradients to which trees are responding. As our case study on bristlecone pine in the Great Basin demonstrates, substantial topoclimatic heterogeneity can occur on a very fine spatial scale, and this becomes a significant factor explaining patterns of tree growth (Fig. 6.4a). In northwestern Alaska, substantial variation in soil microclimate and hence soil nutrient availability occurs on a relatively fine spatial scale and this, too, is a critical factor driving divergent patterns of tree growth (Fig. 6.4b). Absent information about fine-scale topoclimatic and/or soil biophysical parameters, the tree-ring records from these sites would either obscure relevant variability by grouping trees from divergent microclimates together, as shown in the Great Basin, or would yield a puzzling difference in response between topographic locations, as shown near the Agashashok River in northwestern Alaska. In both examples, quantifying the local-scale, within-site variability provides a critical basis for understanding why a uniform regional driver, rising air temperature, produces different responses within a site, with one subpopulation of trees in each instance responding with significantly increased growth to the warming and another subpopulation showing little or no response. Armed with finely resolved environmental data, we see that these differences reflect fine-scale variation in limiting factors, with the response to warming in some trees limited by low moisture availability and in others by access to soil nutrients. The case studies presented here, and the differences between them, also hint at broader global patterns in treeline response to warming. In the Great Basin, which is located at the warm end of the continuum of global treeline climates (Körner and Paulsen 2004), response to temperature is mediated by moisture availability, and the most responsive trees occur in the cooler microsites, where moisture limitation is lowest. In northwestern Alaska, located at the cool end of the continuum of global treeline climates, the most

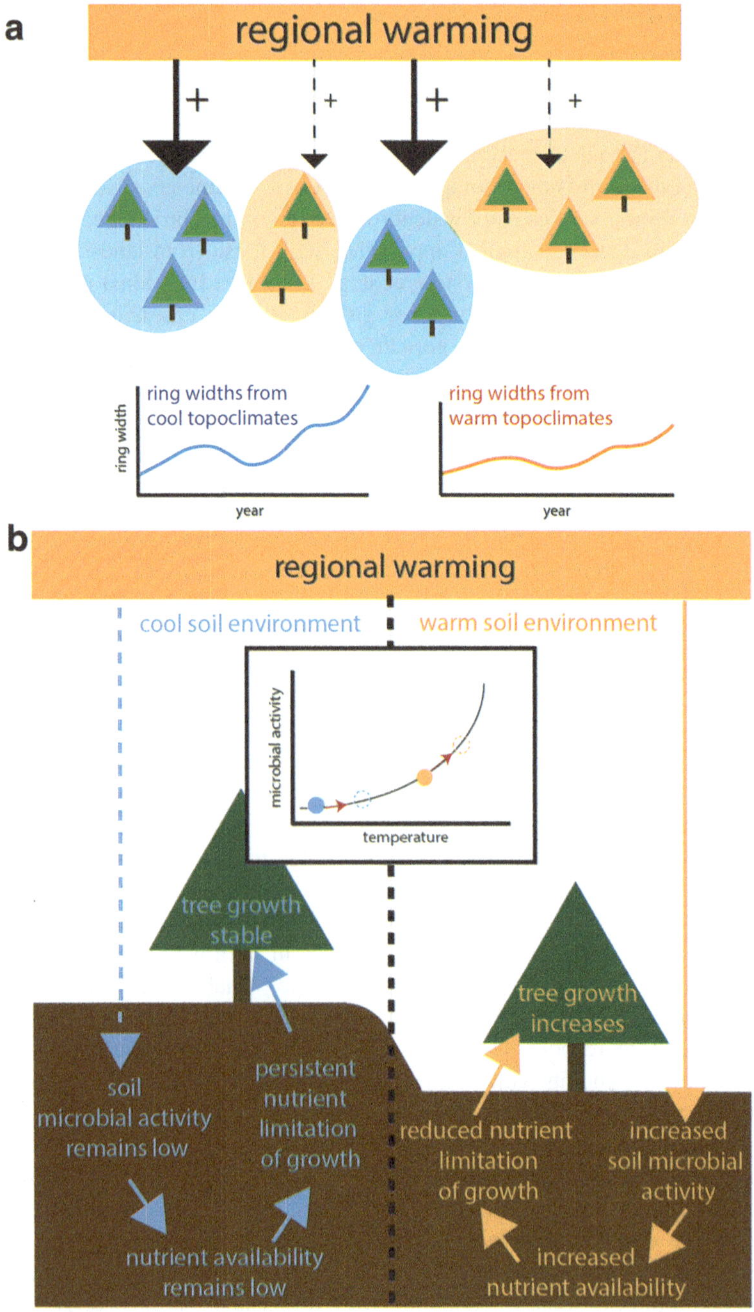

Fig. 6.4 (**a**) Quantifying fine-scale variation in topo-climate can allow better differentiation of within-population variation in response to temperature and, as a result, more robust inferences about the nature of climatic controls over tree growth Trees occupying warm topoclimates near treeline show a less pronounced response to warming (e.g., Fig. 6.2c) than those occupying cool topoclimates near treeline (e.g., Fig. 6.2b). (**b**) Hybrid studies that pair dendroecology with

responsive trees occur in the warmer microsites, where soil nutrient availability is greater. Although two study sites is an insufficient sample size from which to draw broad conclusions, this difference in how temperature effects are mediated across the landscape is a promising direction for future research.

Second, the integration of dendroecology with other ecological approaches allows us to understand trees not simply as the sum of their rings, but as complex organisms whose overall growth reflects the impact of multiple limiting factors filtered through a number of simultaneously occurring physiological processes. The widespread observation that limited growth of treeline trees is associated with low temperatures is, in the end, supported by both of the case studies described here. However, by expanding our inquiry beyond wood production to incorporate other physiological processes, we arrive at a more nuanced understanding of how temperature affects trees—understanding, in the case of northwestern Alaska, that the most significant impact of temperature may be indirect, expressed through soil microbial processes. Distinct soil microclimates in that area mediate the effect of temperature (Fig. 6.4b). Because soil microbial activity shows a non-linear response to rising temperature (Lloyd and Taylor 1994), a given amount of warming may result in greater stimulation of microbial activity in warm sites compared with cool sites (compare filled and unfilled points at each temperature range in Fig. 6.4b, inset). In sites with relatively warm soils and high nutrient availability, trees are able to capitalize on rising air temperatures with greater growth. Trees in these sites show a significant increase in ring-width over time (e.g., Fig. 6.3, Terrace sites). In sites with relatively cool soils, in contrast, rising air temperature is insufficient to stimulate microbial activity. Growth remains nutrient limited, and ring-width does not increase over time (e.g., Fig. 6.3, Treeline sites). Information on soils and tree physiological parameters are crucial in revealing that the differences in growth response to warming reflect variation across the landscape in soil conditions and microbial activity which, in turn, mediate tree response to climate warming.

The high degree of local and regional scale variability in how treeline trees and treeline forests respond to regional warming has been challenging to reconcile with global-scale evidence that suggests temperature limitation is a key driver of global treeline dynamics. The case studies presented here suggest that the causes of local- and regional-scale variability in treeline dynamics can be understood through greater attention to quantifying fine-scale variation in environment (topoclimate, soil biophysical parameters) and by integrating dendroecological approaches with plant physiological measurements to more fully understand the limitations on plant growth. As we have shown, variation within any single site in response to climate

Fig. 6.4 (continued) measurements of tree physiology and soil biogeochemistry can reveal limiting factors that mediate tree response to a regional-scale driver such as air temperature, and thus allow more robust explanations for differences in growth response to that broader-scale change Trees occupying habitats with cool soils remain nutrient-limited as climate warms, while those in habitats with warmer soils experience an increase in nutrient availability and thus increased growth *Inset*: Microbial activity increases exponentially with rising soil temperature, so a rise in soil temperatures has a comparatively greater effect in habitats with warm soils than those with cool soils

warming does not necessarily contradict broad-scale correlations with temperature but, instead, can be a consequence of fine-scale variation in thermal environment or in factors mediating response to temperature.

References

Aerts R, Chapin FS III (2000) The mineral nutrition of wild plants revisited: a re-evaluation of processes and patterns. Adv Ecol Res 30:1–67

Ameztegui A, Coll L, Brotons L, Ninot JM (2016) Land-use legacies rather than climate change are driving the recent upward shift of the mountain tree line in the Pyrenees. Glob Ecol Biogeogr 25:263–273. doi:10.1111/geb.12407

Babst F, Bouriaud O, Papale D et al (2014) Above-ground woody carbon sequestration measured from tree-rings is coherent with net ecosystem productivity at five eddy-covariance sites. New Phytol 201:1289–1303

Batllori E, Gutierrez E (2008) Regional tree line dynamics in response to global change in the Pyrenees. J Ecol 96:1275–1288

Briffa KR, Melvin TM (2011) A closer look at regional curve standardization of tree-ring records: justification of the need, a warning of some pitfalls, and suggested improvements in its application. In: Hughes MK, Diaz HF, Swetnam TW (eds) Dendroclimatology: progress and prospects, Developments in paleoenvironmental research, vol 11. Springer, Netherlands, pp 113–146

Brownlee AD, Sullivan PF, Csank AD et al (2016) Drought-induced stomatal closure probably cannot explain divergent white spruce growth in the Brooks Range, Alaska, USA. Ecology 97:145–159

Bruening JM (2016) Fine-scale topoclimate modeling and climatic treeline prediction of great basin bristlecone pine (*Pinus Longaeva*) in the American Southwest. Masters Thesis collection, Western Washington University, Bellingham WA

Bunn AG, Hughes MK, Salzer MW (2011) Topographically modified tree-ring chronologies as a potential means to improve paleoclimate inference. Clim Chang 8:627–634

Case BS, Duncan RP (2014) A novel framework for disentangling the scale-dependent influences of abiotic factors on alpine treeline position. Ecography 37:838–851

Cook ER, Krusic PJ (2004) The North American Drought Atlas. http://iridl.ldeo.columbia.edu/SOURCES/.LDEO/.TRL/.NADA2004/.pdsi-atlas.html

Craig H, Gordon LI (1965) Deuterium and oxygen-18 variations in the ocean and marine atmosphere. In: Tongiorgi E (ed) Proceedings of a conference on stable isotopes in oceanographic studies and paleotemperatures. Consiglio Nazionale Delle Ricerche, Pisa, pp 9–130

Currey DR (1965) An ancient bristlecone pine stand in eastern Nevada. Ecology 46:564–566

Chapin FS, Sturm M, Serreze MC, McFadden JP, Key JR, Lloyd AH, McGuire AD, Rupp TS, Lynch AH, Schimel JP, Beringer J, Chapman WL, Epstein HE, Euskirchen ES, Hinzman LD, Jia G, Ping CL, Tape KD, Thompson CD, Walker DA, Welker JM (2005) Role of land-surface changes in arctic summer warming. Science 310:657–660

D'Arrigo R, Wilson R, Liepert B, Cherubini P (2008) On the 'Divergence Problem' in Northern Forests: a review of the tree-ring evidence and possible causes. Glob Planet Chang 60:289–305

Dawson TE, Ehleringer JR (1991) Streamside trees that do not use stream water. Nature 350:335–337

DeNiro MJ, Epstein S (1979) Relationship between the organic isotope ratios of terrestrial plant cellulose, carbon dioxide and water. Science 204:51–53

Driscoll WW, Wiles GC, D'Arrigo RD, Wilmking M (2005) Divergent tree growth response to recent climatic warming, Lake Clark National Park and Preserve, Alaska. Geophys Res Lett. doi:10.1029/2005GL024258

Ehleringer JR, Cerling TE (1995) Atmospheric CO_2 and the ratio of intercellular to ambient CO_2 concentrations in plants. Tree Physiol 15:105–111

Ellison SBZ (2015) Can low soil nutrient availability explain divergent white spruce growth in the eastern Brooks Range? MS Thesis, University of Alaska Anchorage, Anchorage
Esper J, Frank D (2009) Divergence pitfalls in tree-ring research. Clim Chang 94:261–266
Farquhar GD, Lloyd J (1993) Carbon and oxygen isotope effects in the exchange of carbon dioxide between terrestrial plants and the atmosphere. In: Ehleringer JR, Hall AE, Farquhar GD (eds) Stable isotopes and plant carbon-water relations. Academic, San Diego, pp 47–70
Farquhar GD, O'Leary MH, Berry JA (1982) On the relationship between carbon isotope discrimination and the inter-cellular carbon-dioxide concentration in leaves. Aust J Plant Physiol 9:121–137
Ferguson CW (1968) Bristlecone Pine Science and Esthetics: A 7100-year tree-Ring chronology aids scientists; old trees draw visitors to California mountains. Science 159(3817):839–846
Giblin AE, Laundre JA, Nadelhoffer KJ, Shaver GR (1994) Measuring nutrient availability in arctic soils using ion exchange resins: a field test. Soil Sci Soc Am J 58:1154–1162
Giddings JL (1943) Some climatic aspects of tree growth in Alaska. Tree-Ring Bull 9:26–32
Granier A (1987) Evaluation of transpiration in a Douglas-fir stand by means of sap flow measurements. Tree Physiol 3:309–320
Harsch MA, Bader MY (2011) Treeline form—a potential key to understanding treeline dynamics. Global Ecol Biogeogr 20:582–596
Harsch MA, Hulme PE, McGlone MS, Duncan RP (2009) Are treelines advancing? A global meta-analysis of treeline response to climate warming. Ecol Lett 12:1040–1049
Holtmeier FK, Broll G (2005) Sensitivity and response of northern hemisphere altitudinal and polar treelines to environmental change at landscape and local scales. Global Ecol Biogeogr 14(3):95–410
Huckaby LS, Kaufmann MR, Fornwalt PJ et al (2003) Field guide to old ponderosa pines in the Colorado front range United States, Department of Agriculture, General Technical Report RMRS-GTR-109
Hughes MK, Funkhouser G (1998) Extremes of moisture availability reconstructed from tree rings for recent millennia in the Great Basin of Western North America. In: Beniston M, Innes J (eds) The impacts of climate variability on forests. Springer, Berlin
Hughes MK, Funkhouser G (2003) Frequency-dependent climate signal in upper and lower forest border tree rings in the mountains of the Great Basin. Clim Chang 59:233–244
Kaufmann MR (1996) To live fast or not: growth, vigor and longevity of old-growth ponderosa pine and lodgepole pine trees. Tree Physiol 16:139–144
Körner C (1998) A re-assessment of high elevation treeline positions and their explanation. Oecologia 115:445–459
Körner C, Paulsen J (2004) A world-wide study of high altitude treeline temperatures. J Biogeogr 31:713–732
Kullman L (1995) Holocene tree-limit and climate history from the Scandes Mountains, Sweden. Ecology 76:2490–2502
LaMarche VC (1969) Environment in Relation to Age of Bristlecone Pines. Ecology 50:53–59
LaMarche VC (1974a) Paleoclimatic inferences from long tree-ring records. Science 183:1043–1048
LaMarche VC (1974b) Frequency-dependent relationships between tree-ring series along an ecological gradient and some dendroclimatic implications. Tree-Ring Bull 34:1–20
LaMarche VC, Stockton CW (1974) Chronologies from temperature-sensitive bristlecone pines at upper treeline in Western United States. Tree-Ring Bull 34:21–45
Lavergne A, Daux V, Villalba R, Barichivich J (2015) Temporal changes in climatic limitation of tree-growth at upper treeline forests: contrasted responses along the west-to-east humidity gradient in Northern Patagonia. Dendrochronologia 36:49–59
Leavitt SW, Long A (1984) A Sampling strategy for stable carbon isotope analysis of tree rings in pine. Nature 311:145–147
LeBauer DS, Treseder KK (2008) Nitrogen limitation of net primary productivity in terrestrial ecosystems is globally distributed. Ecology 89:371–379

Legutke S, Voss R (1999) The Hamburg Atmosphere-Ocean Couple Circulation Model ECHO-G DKRZ technical report 18. DKRZ, Hamburg, Germany
Liang E, Dawadi B, Pederson N, Eckstein D (2014) Is the growth of birch at the upper timberline in the Himalays limited by moisture or by temperature? Ecology 95:2453–2465. doi:10.1890/13-1904.1
Lloyd AH (2005) Ecological histories from Alaskan tree lines provide insight into future change. Ecology 86:1687–1695
Lloyd AH, Bunn AG (2007) Responses of the circumpolar boreal forest to 20th century climate variability. Environ Res Lett 2:045013. doi:10.1088/1748-9326/2/4/045013
Lloyd AH, Fastie CL (2002) Spatial variability in the growth and climate response of treeline trees in Alaska. Clim Chang 52:481–509
Lloyd AH, Graumlich LJ (1997) Holocene dynamics of treeline forests in the Sierra Nevada. Ecology 78:1199–1210
Lloyd J, Taylor A (1994) On the temperature dependence of soil respiration. Funct Ecol 8:315–323
Lloyd AH, Bunn AG, Berner L (2011) A latitudinal gradient in tree growth response to climate warming in the Siberian taiga. Glob Chang Biol 17:1935–1945
Loomis PF, Ruess RW, Sveinbjörnsson B, Kielland K (2006) Nitrogen cycling at treeline: latitudinal and elevational patterns across a boreal landscape. Ecoscience 13:544–556
MacDonald GM, Szeicz JM, Claricoates J, Dale KA (1998) Response of the central Canadian treeline to recent climate changes. Ann Assoc Am Geogr 88:183–208
MacDonald GM, Velichko AA, Kremenetski CV et al (2000) Holocene treeline history and climate change across northern Eurasia. Quat Res 53:302–311
Malanson GP, Butler DR, Fagre DB et al (2007) Alpine treeline of western North America: linking organism-to-landscape dynamics. Phys Geogr 28:378–396. doi:10.2747/0272-3646.28.5.378
Malanson GP, Resler LM, Bader MY et al (2011) Mountain treelines: a roadmap for research orientation. Arct Antarct Alp Res 43:167–177. doi:10.1657/1938-4246-43.2.167
McNown RW, Sullivan PF (2013) Low photosynthesis of treeline white spruce is associated with limited soil nitrogen availability in the Western Brooks Range, Alaska. Funct Ecol 27:672–683
Nehrbass-Ahles C, Babst F, Klesse S et al (2014) The influence of sampling design on tree-ring based quantification of forest growth. Glob Chang Biol 20:2867–2885
Paulsen J, Körner C (2014) A climate-based model to predict potential treeline position around the globe. Alp Bot 124:1–12
Peñuelas J, Canadell JG, Ogaya R (2011) Increased water-use efficiency during the 20th century did not translate into enhanced tree growth. Glob Ecol Biogeogr 20:597–608
Qian P, Schoenau JJ (2002) Practical applications of ion exchange resins in agricultural and environmental soil research. Can J Soil Sci 82:9–21
Rocha AV, Goulden ML, Dunn AL, Wofsy SC (2006) On linking interannual tree ring variability with observations of whole-forest CO_2 flux. Glob Chang Biol 12:1378–1389
Roden J, Siegwolf R (2012) Is the dual-isotope conceptual model fully operational? Tree Physiol 32:1179–1182
Salzer MW, Hughes MK, Bunn AG, Kipfmueller KF (2009) Recent unprecedented tree-ring growth in bristlecone pine at the highest elevations and possible causes. Proc Natl Acad Sci U S A 106(48):20348–20353
Salzer MW, Bunn AG, Graham NE, Hughes MK (2013) Five millennia of paleotemperature from tree rings in the Great Basin, USA. Clim Dyn 42:1517–1526
Salzer MW, Larson ER, Bunn AG, Hughes MK (2014) Changing climate response in near-treeline bristlecone pine with elevation and aspect. Environ Res Lett 9:114007
Saurer M, Schweingruber F, Vaganov E et al (2002) Spatial and temporal oxygen isotope trends at the northern tree-line in Eurasia. Geophys Res Lett 29:1296
Saurer M, Siegwolf RTW, Schweingruber FH (2004) Carbon isotope discrimination indicates improving water-use efficiency of trees in northern Eurasia over the last 100 years. Glob Chang Biol 10:2109–2120
Saurer M, Cherubini P, Reynolds-Henne CE et al (2008) An investigation of the common signal in tree ring stable isotope chronologies at temperate sites. J Geophys Res 113:G04035

Scheidegger Y, Saurer M, Bahn M, Siegwolf R (2000) Linking stable oxygen and carbon isotopes with stomatal conductance and photosynthetic capacity: a conceptual model. Oecologia 125:350–357

Schulze ED, Fuchs MI, Fuchs M (1977) Spatial distribution of photosynthetic capacity and performance in a mountain spruce forest of northern Germany I. Biomass distribution and daily CO_2 uptake in different crown layers. Oecologia 29:43–61

Schulze ED, Chapin FS III, Gebauer G (1994) Nitrogen nutrition and isotope differences among life forms at the northern treeline of Alaska. Oecologia 100:406–412

Shrestha KB, Hofgaard A, Vandik V (2015) Tree-growth response to climatic variability in two climatically contrasting treeline ecotone areas, central Himalaya, Nepal. Can J For Res 45:1643–1653

Sidorova OV, Siegwolf RTW, Saurer M et al (2009) Do centennial tree-ring and stable isotope trends of Larix gmelinii (Rupr.) Rupr. indicate increasing water shortage in the Siberian north? Oecologia 161:825–835

Slatyer RO, Noble IR (1992) Dynamics of montane treelines. In: Hansen AJ, di Caestri F (eds) Landscape boundaries. Springer, New York, pp 346–359

Stahle DW (1996) Tree rings and ancient forest history. In: Davis MB (ed) Eastern old-growth. Forests Island Press, Washington, DC, pp 321–343

Stahle DW, Chaney PL (1994) A predictive model for the location of ancient forests. Nat Areas J 14:151–158

Stevens MB, Gonzalez-Rouco JF, Beltrami H (2008) North American climate of the last millennium: underground temperatures and model comparison. J Geophys Res 113(F1)

Suarez F, Binkley D, Kaye MW, Stottlemyer R (1999) Expansion of forest stands into tundra in the Noatak National Preserve, northwest Alaska. Ecoscience 6:465–470

Sullivan PF, Csank AZ (2016) Contrasting sampling designs among archived datasets: implications for synthesis efforts. Tree Physiol 36:1057–1059

Sullivan PF, Ellison SBZ, McNown RW et al (2015) Evidence of soil nutrient availability as the proximate constraint on growth of treeline trees in northwest Alaska. Ecology 96:716–727

Sullivan PF, Pattison RR, Brownlee AH et al (2016) Effect of tree-ring detrending method on apparent growth trends of black and white spruce in interior Alaska. Environ Res Lett 11:114007

Sveinbjörnsson B, Nordell O, Kauhanen H (1992) Nutrient relations of mountain birch growth at and below the elevational tree-line in Swedish Lapland. Funct Ecol 6:213–220

Sveinbjörnsson B, Hofgaard A, Lloyd AH (2002) Natural causes of the tundra-taiga boundary. Ambio Spec No 12:23–29

Swetnam TW, Brown PM (1992) Oldest known confiers in the Southwestern United States: temporal and spatial patterns of maximum age. In: Kaufman MR, Moir WH, Bassett RL (eds) Old growth forests in the Southwest and Rocky Mountain Regions USDA Forest Service General Technical Report RM-213. pp 24–38

Tran TJ (2016) Cluster analysis as a means of examining topographically-mediated bristlecone pine (*Pinus Longaeva*) Growth in the American Southwest. Masters Thesis Collection, WWU, Bellingham, WA

Vitousek PM, Horwath RW (1991) Nitrogen limitation on land and in the sea—how can it occur? Biogeochemistry 13:87–115

Wang Y, Liang E, Ellison AM et al (2015) Acilitation stabilizes moisture-controlled alpine juniper shrublines in the central Tibetan Plateau. Glob Plan Chang 132:20–30

Wilmking M, Juday GP, Barber VA, Zald HS (2004) Recent climate warming forces contrasting growth responses of white spruce at treeline in Alaska through temperature thresholds. Glob Chang Biol 10:1724–1736

Wilson AT, Grinsted MJ (1977) $^{12}C/^{13}C$ in cellulose and lignin as paleothermometers. Nature 265:133–135

Chapter 7
Dendroecological Applications to Coarse Woody Debris Dynamics

Shawn Fraver, Tuomas Aakala, and Anthony W. D'Amato

Abstract Coarse woody debris plays a crucial role in forest ecosystems. The current amount of woody debris on a given site represents a balance between additions (tree mortality) and depletions (wood decomposition, combustion, transport). Understanding woody debris dynamics has recently gained much attention, primarily because of the need to improve forest carbon accounting and modelling. Woody debris itself also holds great potential for use in dendrochronological studies, including those aimed at revealing forest stand dynamics. As such, tree-ring data from woody debris is at times used to make inferences about past stand dynamics; however, at other times tree-ring data from samples in close proximity to woody debris are used to make inferences about the dynamics of woody debris itself. Our case study provides an example of the latter application by addressing woody debris dynamics in three old-growth *Picea rubens* stands in Maine, USA. Our findings show striking fluctuations in woody debris mass over a 100-year period (1900–2000), with pulses in woody debris inputs corresponding to reconstructed forest disturbances. These fluctuations highlight the need to characterize woody debris dynamics for refining modelling efforts and developing restoration prescriptions in ecosystems with disturbance regimes dominated by gap- and meso-scale disturbances.

Keywords Forest carbon cycle • *Picea rubens* • Stand dynamics • Tree mortality • Woody detritus • Wood decay rates • Wood decomposition

---

S. Fraver (✉)
School of Forest Resources, University of Maine, 5755 Nutting Hall, Orono, ME 04469, USA
e-mail: shawn.fraver@maine.edu

T. Aakala
Department of Forest Sciences, University of Helsinki, P.O. Box 27, Helsinki, FI 00014, Finland
e-mail: tuomas.aakala@helsinki.fi

A.W. D'Amato
Rubenstein School of Environment and Natural Resources, University of Vermont, 204E Aiken Center, Burlington, VT 05405, USA
e-mail: awdamato@uvm.edu

M.M. Amoroso et al. (eds.), *Dendroecology*, Ecological Studies 231,
DOI 10.1007/978-3-319-61669-8_7

7.1 Introduction

Coarse woody debris (here dead standing or downed trees >10 cm diameter) plays a crucial role in forest ecosystems, where it influences biological diversity, nutrient cycling, soil development, geomorphological processes, and the spread and severity of wildfire (Harmon et al. 1986; Schoennagel et al. 2004; Stokland et al. 2012; Ruiz-Villanueva et al. 2016). The current amount of woody debris on a given site represents a balance between additions (tree mortality) and depletions (wood decomposition, combustion, transport). Understanding woody debris dynamics has recently gained much attention given the growing need to improve forest carbon accounting and modelling (Kurz et al. 2009; Woodall et al. 2015), to model changes in habitat suitability for deadwood-dependent species (Tikkanen et al. 2007), and to assess the longevity of forest fuels (He et al. 2004).

Another important—but often overlooked—attribute of woody debris is its potential use for dendroecological purposes. When applied to forest dynamics, tree-ring data from woody debris is used to make inferences about past stand dynamics; however, at other times tree-ring data from samples in close proximity to woody debris, often combined with samples from debris itself, are used to make inferences about the dynamics of woody debris. We review both applications in this chapter. Our case study below provides an example of the latter application. We focus the following discussion on terrestrial systems, acknowledging that a large body of literature addresses samples obtained from riverine or lake environments (e.g., Guyette et al. 2008; Jones et al. 2011; Gennaretti et al. 2014).

7.2 Dendroecological Data *Obtained from* Coarse Woody Debris

Samples collected from dead stems are used for many of the same purposes as using live-tree samples in dendroecological studies: obtaining tree recruitment dates, growth patterns, disturbance histories, and mortality dates. Benefits of sampling woody debris include extending the temporal coverage of tree-ring series, or bolstering sample sizes obtained from living trees. Also, when larger samples (i.e., stem cross-sections or partial cross-sections) are needed in sensitive areas such as nature reserves, woody debris sampling can be used instead of sampling live trees, as it causes less aesthetic and ecological disturbance. Obtaining samples from well-preserved dead trees is commonly employed in other dendrochronological fields; in fact, it is the hallmark of long-term climate reconstructions (Lara and Villalba 1993; Linderholm et al. 2010; Cook and Kairiukstis 2013; Edvardsson et al. 2016), fire history reconstructions from fire scars (Niklasson and Granström 2000; Swetnam and Betancourt 2010), and archeological studies involving wood dating (Speer 2010). Similarly, preserved subfossil wood can extend chronologies back in time to assess past geomorphic events (Stoffel and Corona 2014; Rădoane et al. 2015), tree-line and species distribution changes (Pregitzer et al. 2000; Nicolussi et al. 2005), and lake level changes (Gunnarson 2001). One particularly interesting recent use of subfossil wood is to assess anthropogenic changes to forests by comparing average tree longevities, as well as tree sizes, between ancient and modern samples (Lindbladh et al. 2013). Despite these diverse applications, the use of samples from partially decayed woody debris, particularly decaying logs on the forest floor, has received far less attention, despite the potential applications in dendroecological studies.

One common task in various dendroecological studies of forest dynamics is to estimate mortality years, based on the crossdated outermost rings in samples (either cores, wedges, or cross-sections) taken from the dead trees. This information is used to reconstruct the timing of past beetle outbreaks (Mast and Veblen 1994; Khakimulina et al. 2016), to date the arrival of non-native pests (Siegert et al. 2014), to evaluate spatial patterns in tree mortality (Aakala et al. 2007; Bigler et al. 2007), to determine tree growth rates prior to death (Cherubini et al. 2002; Lännenpää et al. 2008; Amoroso et al. 2015), to date past canopy disturbances (Henry and Swan 1974; Caron et al. 2009), to test the relationship between tree mortality pulses and past droughts (Bigler et al. 2007; Aakala et al. 2011), and to determine the time needed for a fallen log to become a suitable substrate for seedling establishment (i.e., a nurse log; Zielonka 2006). Crossdating the mortality year can also improve the temporal resolution in permanent sample plots with census intervals longer than a year (Mast and Veblen 1994; Jones and Daniels 2012); such information is crucial when attributing forest dynamics to specific causes such as climatic variability. Considering recent changes in tree mortality reported from different parts of the world (e.g., van Mantgem et al. 2009; Allen et al. 2010), annually-resolved time series of tree mortality dates are becoming increasingly important.

However, crossdating and interpretation of the outermost rings on dead samples creates challenges, as pointed out by nearly every dendrochronological study employing this method. Assigning the year of death is especially challenging when the tree had experienced slow growth prior to death, which may be common in late-successional or old-growth forests (Fraver et al. 2008). That is, during periods of stress or senescence prior to death, annual rings may not form, particularly lower on the stem where samples are typically obtained. These missing outer rings can be quite numerous on poor vigor trees: Cherubini et al. (2002) report 31 such missing rings in *Pinus mugo*, and we have identified 36 in *Picea abies* (unpublished work from Sweden). As a further example from our own work from Finland, in a sample of recently fallen or broken *Picea abies* trees with green foliage still attached, we crossdated the outermost ring from cores and recorded the number of missing rings in the end of the series. While most appeared to have zero or only a few missing rings, some had 17 missing rings (Fig. 7.1). Further, in many cases where the stem appears to be sound, outermost rings may have been lost to weathering (Jemison 1937; Aakala et al. 2007), thereby precluding an accurate estimate of year of death. Finally, uprooted trees that are assumed to be dead may in fact continue to produce annual rings for several years after they fall (Shorohova and Kapitsa 2014), further confounding the dating.

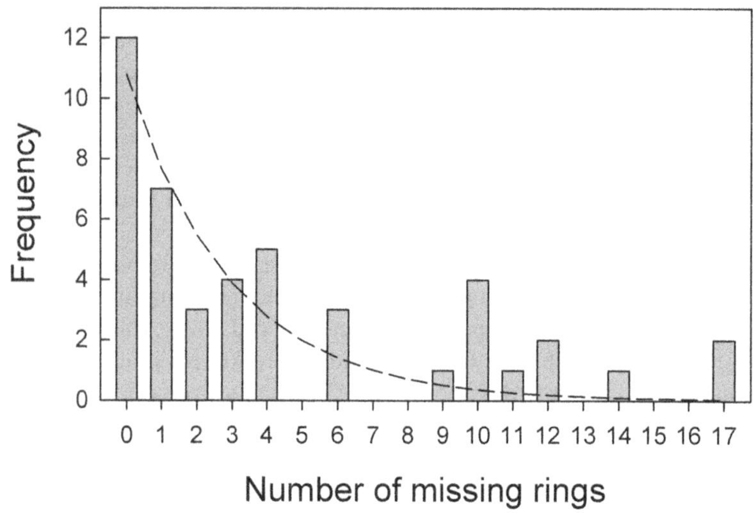

Fig. 7.1 The frequency of missing rings at the end of tree-ring series, from recently fallen *Picea abies* trees in old-growth forests in northwestern Finland and the Murmansk Oblast in northwestern Russia. Missing rings at the end of the series confound the assessment of mortality years. The dashed line represents a negative exponential fit to the data (data from Lännenpää et al. 2008)

Samples are often collected from woody debris to estimate tree establishment or germination dates. For shade-intolerant species in particular, these dates may suggest the occurrence of past disturbances that fostered tree establishment. The use of woody debris for this purpose can bolster the information obtained from living trees regarding establishment dates, particularly for the earlier periods (Fraver et al. 2008), or may provide information unavailable from living trees (Henry and Swan 1974). We have found this application very useful in our studies of stand dynamics, where we have included samples carefully collected from seemingly well-decayed woody debris on the forest floor (Fraver et al. 2008; Fraver and Palik 2012). However, we point out that the date of the innermost ring (from these or any samples), unless obtained from the true root collar, underestimates the time since germination (emergence of the hypocotyl; Telewski 1993). The discrepancy between the innermost ring, even when obtained near the apparent root collar, and the germination date can be substantial, at times exceeding 26 years (Niklasson 2002) or 35 years (Parent et al. 2002). Several authors have suggested methods for reducing this discrepancy, based on coring height above root collar and early growth rates (Villalba and Veblen 1997; Fraver et al. 2011) or by presenting age structures in terms of recruitment as opposed to establishment dates (Fraver and White 2005a; D'Amato and Orwig 2008). The latter approach is more useful for shade-tolerant trees that can persist in the understory for decades awaiting canopy opening. For these trees, the year of establishment is often less informative from the forest-dynamics point of view.

Growth patterns evident in samples from woody debris, along with those of living trees, have been used to make inferences about past forest disturbance. Evidence is derived from growth releases, that is, abrupt and sustained increases in radial growth that suggest the loss of a formerly overtopping canopy (Black and Abrams 2003; Fraver and White 2005b). Henry and Swans' (1974) classic paper exemplifies the utility of examining growth releases in woody debris samples, along with those of living trees, to reconstruct more than three centuries of disturbance history of an old-growth mixed *Tsuga canadensis-Pinus strobus* stand in New Hampshire, USA. Their methods also include estimated mortality and recruitment dates from woody debris. Despite the somewhat rudimentary dendroecological methods by today's standards, their paper remains a 'must-read' for those interested stand dynamics. More recently, Winter et al. (2002) applied the growth-release method to snag samples to date canopy disturbances in an old-growth *Pseudotsuga menziesii* forest in Washington, USA. Caron et al. (2009) used the method applied to cross-sections from snags and downed woody debris to date canopy gap formation in a *Picea abies* forest of northwestern Finland. Although the growth-release method is very commonly used with samples from living trees, it appears to be underutilized with regard to woody debris samples, particularly well-decayed pieces.

Finally, we note that as debris advances through the decay process, it becomes increasingly unsuitable for dendroecological purposes. Decay rates vary considerably among species and across regional climatic gradients (Russell et al. 2014). However, at a local scale many factors interact to determine decay rates, which together create considerable variability. One of the more influential factors, albeit challenging to quantify, is the abundance and composition of the fungal community (Lindner et al. 2011; Bradford et al. 2014). The fungal community also influences the suitability of woody debris for dendrochronological purposes by determining if brown- *vs.* white-rot is present. Brown-rot fungi primarily decompose cellulose and hemicellulose (leaving lignin), often preserving the ring structure even in highly decomposed wood, allowing it to be used if carefully collected and prepared. In contrast, white rot fungi decompose cellulose, hemicellulose, and lignin, resulting in wood structure unsuitable for dendrochronological purposes. Other biotic factors include (1) species-specific chemical and anatomical traits (Cornwell et al. 2009); (2) anomalous anatomical features, such as reaction wood (Blanchette et al. 1994) or resin-rich, fire-affected wood (Verrall 1938) which is quite decay-resistant, a fact well known to the dendrochronologist working with fire scars; (3) tree growth rates prior to death, as slow-growing conifers tend to decay more slowly (Edman et al. 2006); (4) the position of the stem relative to the forest floor, as ground contact hastens decay (Næsset 1999); (5) stem size because larger stems typically decay more slowly (Beets et al. 2008); and (6) overgrowth of logs by ground-layer vegetation, which retards decomposition (Hagemann et al. 2010). Because snags typically remain drier than downed woody debris (Lambert et al. 1980), they generally decay more slowly (Storaunet and Rolstad 2002; Vanderwel et al. 2006; Aakala 2010), given that the activity of wood-decay fungi slows dramatically when wood dries below mass moisture contents of 25–30% (Panshin and de Zeeuw 1980).

7.3 Dendroecological Methods *Applied to* Coarse Woody Debris Dynamics

One of the most valuable applications of dendroecology in the context of woody debris is quantifying the dynamics of the woody debris itself. Methods may include analyses of growth releases, reaction wood, impact scars, or mortality dates of samples taken *adjacent to* woody debris—which in many cases is itself too decayed for sampling—to make inferences about woody debris input and decay rates. Such information is invaluable for elucidating the factors affecting woody debris condition for a given forest type and/or developmental stage, given the integral link between disturbance history and the nature and abundance of woody debris (D'Amato et al. 2008; Harmon 2009).

Authors have used dendroecological methods to determine when particular logs entered the woody debris pool. Several early studies are worth highlighting in this regard. For example, Dynesius and Jonsson (1991) applied growth releases, impact scar dates, and reaction wood formation on samples adjacent to uprooted trees to determine the date of the uprootings in *Picea abies* forests of northern Sweden, concluding that estimates are dramatically improved through the simultaneous use of these and other methods. Daniels et al. (1997) combined sampling of the logs themselves (to determine mortality dates) with samples taken from adjacent trees (to assess growth releases, scars, and reaction wood) to determine how long ago woody debris pieces from *Thuja plicata* had died, with results showing some pieces had died nearly 300 years before.

A number of dendroecological studies have emphasized the distinction between time since tree death and time since tree fall. This distinction has important implications for modelling snag longevity, as well as decomposition studies based on chronosequences, because decomposition progresses slowly during the standing dead phase, then increases after falling and coming in contact with the forest floor (Storaunet and Rolstad 2002; Aakala 2010). Storaunet and Rolstad (2002) used outermost rings on fallen *Picea abies* trees to estimate the year of death, followed by dating the impact scars and reaction wood formation on neighboring trees damaged by the fall. They report a remarkable difference between the two events, with a range of 0–91 years (mean 22 years). As the fall of standing dead trees is usually the result of structural degradation near the stem base, fall rates are to a large extent determined by decay progression at the base and are thus governed by the same factors controlling wood decay in general (see above). It follows then that the amount of time a snag stands before falling is likewise highly variable even within a forest type or tree species.

Times since death or fall are often obtained to determine wood decay rates using a chronosequence, that is, a set of logs with various mortality dates spanning long time periods. Dendroecological methods lend themselves quite well to such chronosequences, because they provide annually resolved mortality and/or tree-fall dates that mark the starting point of the chronosequence. Though similar data could potentially be obtained from long-term permanent plots, such data is seldom annually resolved, plots are few in number, and even the longest-established plots have limited longevity relative to the potential of dendroecology. As above, time since death or fall when used specifically for constructing chronosequences may be obtained by crossdating the outermost ring in well-preserved samples (Brown et al. 1998; Holeksa et al. 2008; Aakala 2010, 2011), dating impact scars caused when the tree in question fell against a neighbor (Means et al. 1985; Yatskov et al. 2003; Shorohova and Kapitsa 2014), aging post-disturbance regeneration (Carmona et al. 2002), dating growth releases evident in adjacent trees (Shorohova and Kapitsa 2014), and dating the onset of reaction wood of saplings displaced by the fallen tree (Fraver et al. 2013). From each piece of woody debris forming the chronosequence, a sample of wood is collected and returned to the lab for density or mass determination. By combining data from samples spanning a range of years, decay rates can thus be estimated. The metric most commonly evaluated in decay-rate studies is wood density, not mass or volume (Laiho and Prescott 2004; Rock et al. 2008), simply because initial wood densities can be obtained from the literature, while initial masses or volumes are rarely known. However, we argue that mass loss, not density reduction, is more useful for most purposes, including modelling forest carbon. Because mass is obtained by multiplying density by volume, Fraver et al. (2013) demonstrated that density reduction could be combined with volume reduction (changes in log shape, *via* gradual collapse over time) in a chronosequence to arrive at mass-loss equations. This concept was elaborated by Russell et al. (2014) and linked with repeated forest inventory data to produce woody debris mass-loss equations for all common tree species in the eastern United States.

Several authors have combined dendroecological analyses with decay-class transition models to obtain decay rates and to characterize woody debris dynamics over time. Decay classes are subjectively assigned to each sampled woody debris piece during field inventories using visual and tactile characteristics, and are meant to describe the general stage of decay (Sollins 1982). By knowing the mortality year as well as the decay class for each sample, one can estimate the time spent in each progressive class and thus develop depletion curves that depict the probability of samples persisting over time. Aakala et al. (2008) employed this approach to model snag longevity (time remaining standing before fall) in conifer forests of Quebec. They found distinct differences in longevities for the two species under study, with approximately a 10-year difference in the 'half-lives' of *Picea mariana* and *Abies balsamea*. Aakala (2010) later expanded this approach to include the progression through snag as well as woody debris decay classes for *Picea abies* forests in northern Europe, showing a latitudinal gradient in residence times for both snags and woody debris. Using time-since-death data from both snags and downed debris (but without relying on decay-class transition models), Angers et al. (2010) developed species-specific 'survival' curves for snags, which depict the probability of snags remaining standing as time since death progresses. Their study nicely shows distinct curves for the four species studied.

Such studies also allow assessment of the utility of using decay class, which is easy to obtain, as a surrogate for time since death or time since fall, which is labor intensive to determine. As it turns out, studies have repeatedly demonstrated the relationship to be weak at best (Daniels et al. 1997; Storaunet and Rolstad 2002; Lombardi et al. 2008; Aakala 2010). The poor relationship results primarily from the enormous variability in decay rates (as above), as well as the subjectivity in assigning decay classes (Larjavaara and Muller-Landau 2010). Nevertheless, even without dendroecological data, repeated measurements of permanent plots, even at lengthy (e.g., 5 year) intervals, allow the calculation of transition probabilities from one decay class to the next, which in turn can be used estimate woody debris depletion curves (Kruys et al. 2002; Vanderwel et al. 2006) and residence times (Russell et al. 2014). A major weakness in using the chronosequence approach, whether applying decay rates or decay-class transitions, is that slow decaying woody debris pieces have a higher probability of being sampled, thereby potentially overestimating woody debris longevity. Kruys et al. (2002) demonstrated how, using a Horwitz-Thompson estimator, this effect can be taken into account, but this source of uncertainty is rarely considered.

Once woody debris input rates and decay rates have been determined, we can consider combining them to show how woody debris abundance changes through time, given that the amount of woody debris present on a site at any time represents a balance between additions and depletions. Woody debris mass or volume on a given site is often assumed to follow a U-shape pattern (high–low–relatively high) through recovery following disturbance. The pattern results from high initial disturbance-caused inputs ('legacy wood,' *sensu* Harmon 2009), followed by a nadir where the initial input has decayed but the recovering stand has not yet reached a stage in which mortality is contributing substantially to the woody debris pool. This intermediate stage is followed by increased mortality in the developing stand, adding significantly to the pool ('de novo wood,' *sensu* Harmon 2009). This conceptual model has proliferated in the literature since its introduction by Harmon et al. (1986), in part because it so nicely demonstrates the balance between inputs from disturbance and losses from decay. Nevertheless, it best represents forest types or regions where stand-replacing disturbances are a common feature of the prevailing disturbance regime. In many unmanaged forests such disturbances are rare, with return intervals four to five times the average longevity of late-successional canopy species (Lorimer and White 2003). In such regions, natural disturbances occur relatively frequently and cause only partial canopy tree mortality (see regional reviews by Seymour et al. 2002, Kuuluvainen and Aakala 2011, Pederson et al. 2017), resulting in somewhat continuous to episodic woody debris inputs at lower levels relative to those of the stand-replacing model. Despite the prevalence of partial disturbance regimes, where *de novo* wood is the primary woody debris input for centuries to millennia, few attempts have been made to characterize temporal patterns of woody debris abundance over time, particularly using empirical data, under these conditions (but see Janisch and Harmon 2002, Aakala 2011, D'Amato et al. 2017). Our case study below is one of the few attempts to do so.

7.4 Obtaining and Preparing Woody Debris Samples: Practical Matters

Much of our work has taken place in protected areas where destructive sampling, such as collecting radial cross-sections from woody debris, is prohibited primarily for aesthetic reasons. In these settings, it is necessary to use a large-diameter increment borer (such as the Haglöf 12-mm diameter increment borer) for sample collection (Fig. 7.2). Though the bit diameter is only 2.3 times larger than the standard 5.15-mm bit, it collects 5.4 times the volume of material for a given core length. This extra volume dramatically increases the likelihood of obtaining a usable sample of what is often punky, gallery-ridden, or otherwise partially decayed material. Straws of this size are available for field storage of these cores; however, we find it more convenient to place them in sleeves made of rigid *Rite in the Rain*™ paper, stapled and labeled as needed.

When using a 12-mm borer on samples where the year of death (outermost ring) is not of interest or unobtainable due to rot, we find it beneficial to first drill a shallow pilot hole (ca. 0.5–1 cm deep into sound wood) with a hand brace and a

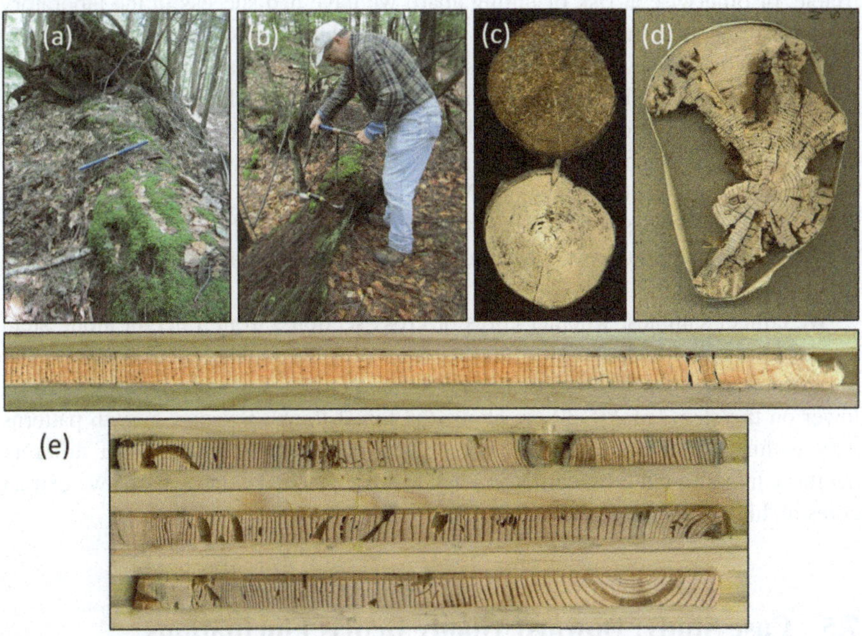

Fig. 7.2 (**a**, **b**) Extracting large diameter (12 mm) increment cores from decayed *Pinus strobus* logs windthrown in a 1938 hurricane (Pisgah Tract, New Hampshire, USA). (**c**) Decayed cross-section before (*upper*) and after (*lower*) stabilization by glue emulsion. (**d**) Decayed cross-section after glue-mounting, sanding, and polishing. (**e**) Examples of large-diameter (12 mm) increment cores mounted and prepared for crossdating

standard 12-mm (or 0.5-inch) auger bit. The pilot hole greatly facilitates starting the increment borer, which can be a challenge due to its two-thread construction. The pilot-hole approach is especially handy when sample trees are *case hardened*, a condition in which differential drying between the stem interior and exterior produces an outer shell that is unusually hard to penetrate. When outermost rings are needed and available, it may be useful to first remove bark, if present, and cover the area to be cored with adhesive tape to help maintain the surface material intact.

Wooden mounts for 12-mm increment cores are commercially available in the USA; however, these can easily be made on a table saw. We make ours using ca. 2.5 cm wide conifer stock, cutting a 12 mm (width) × 5 mm (depth) dado, which leaves adequate space for glue. Cores can be glued into the dado and held in place with masking tape or binder clips until glue dries.

Where the collection of cross-sections from dead trees is permitted, we recommend doing so. As in many dendrochronological studies, we stabilize cross-sections with plastic wrap and/or duct tape in the field to secure them during transport to the laboratory. For very decayed or otherwise fragile stems, wrapping duct tape around the stem prior to cutting the sample makes it possible to obtain useable samples. Once air-dried, relatively sound cross-sections can be mounted to a board for security using hot glue (Angers et al. 2010). When cross-sections are punky, friable, or otherwise at risk of falling apart, we have had success in the laboratory by first submerging them for several days in a solution of seven parts of water to one part inexpensive water-soluble school/craft glue (for a detailed description, see Krusic and Hornbeck 1989). Once removed from solution and thoroughly dried, samples can often be machined and sanded to produce a surface suitable for dendrochronological purposes. A commercial wood hardening agent can be used for this same purpose (Pedersen and McCune 2002). Others have had success with freezing moist wood samples and preparing the surface while still frozen with a scalpel or by sanding to reveal ring boundaries (Herman et al. 1972; Storaunet and Rolstad 2002).

We collect samples, whether cores or cross-sections, from a height along stem similar to that at which samples forming the dating chronology were obtained. For our purposes this is often breast height, although some research requires samples lower on the stem (ca. 20–30 cm above the forest floor). Because growth patterns vary within the stem (Kerhoulas and Kane 2012), samples taken from different heights can confound crossdating. As a guide, for downed woody debris we extract cores no higher than 2 m above the base.

7.5 Case Study: Downed Woody Debris Fluctuations Through Time in Old-Growth *Picea rubens* Stands

In this case study, we illustrate how dendroecological data can be used to represent fluctuations in *de novo* downed woody debris mass (the right-hand side of the

U-shape model, see above) in a late-developmental-stage forest far removed from stand-replacing disturbances, yet having experienced repeated partial canopy losses over extended time periods. This approach has rarely been attempted using empirical data (but see Aakala 2011, D'Amato et al. 2017). Indeed, the challenges of doing so are formidable, as it requires that we (1) estimate woody debris mass that would have been present at the starting point of the temporal sequence, (2) estimate annual woody debris inputs over the time period in question using dendroecological approaches, and (3) assign a suitable mass-loss function to allow the gradual decay of those inputs, each explained below. Our objective is to reconstruct woody debris mass fluctuations through time for a forest type in which dynamics are driven by gap- and meso-scale disturbances. We recognize is not possible to reconstruct a definitive record of woody debris mass over time, given the variation and uncertainties surrounding each of these three challenges. Instead, we illustrate how dendroecological data allows reasonable estimates of the temporal dynamics of coarse woody debris.

7.5.1 Methods

We here capitalize on existing detailed dendroecological, structural, and spatial data to reconstruct woody debris abundance over a 100-year period, from 1900 to 2000, in three old-growth red spruce (*Picea rubens*) stands located in a 2000-ha old-growth reserve in northern Maine, USA. These data were initially collected (in 2000 and 2001) to reconstruct the spatial and temporal aspects of canopy disturbance for this forest type. For that purpose, one 50 × 50 m plot was established in each stand; for details see Fraver and White (2005a). Data sets resulting from that work include (1) tree-ring series from all living trees (≥10 cm diameter) and dead trees, when not precluded by decay, (2) diameters, heights, and spatial locations (X, Y coordinates) of all trees and saplings, (3) mapped locations of current (2000) canopy gaps, and (4) complete inventory of all coarse woody debris, recording end diameters, lengths, species, decay classes (as above) and spatial locations of intact pieces. Results from that previous work reveal periodic pulses of moderate-severity disturbance by wind storms and insect outbreaks interposed upon a background of scattered fine-scale canopy disturbances (Fraver and White 2005a; Fig. 7.3).

From these same data, we estimated the timing and volume of woody debris inputs on an annual basis; however, this required two distinct approaches depending on the timing of tree death. For the earlier decades (1900 to ca. 1980), we relied on growth releases in surviving trees as evidence of the timing and location of past canopy gaps using methods and thresholds described in Fraver and White (2005b). To estimate the volume of woody debris added during each canopy gap disturbance, we assumed the size of the tree that ultimately filled the canopy gap (based on our 2000 inventory) approximated the size of the tree that fell and formed the gap (i.e., gap makers), as justified and outlined in Lorimer and Frelich (1989). That is, diameters of current gap-filling trees served as surrogates for those of the gap

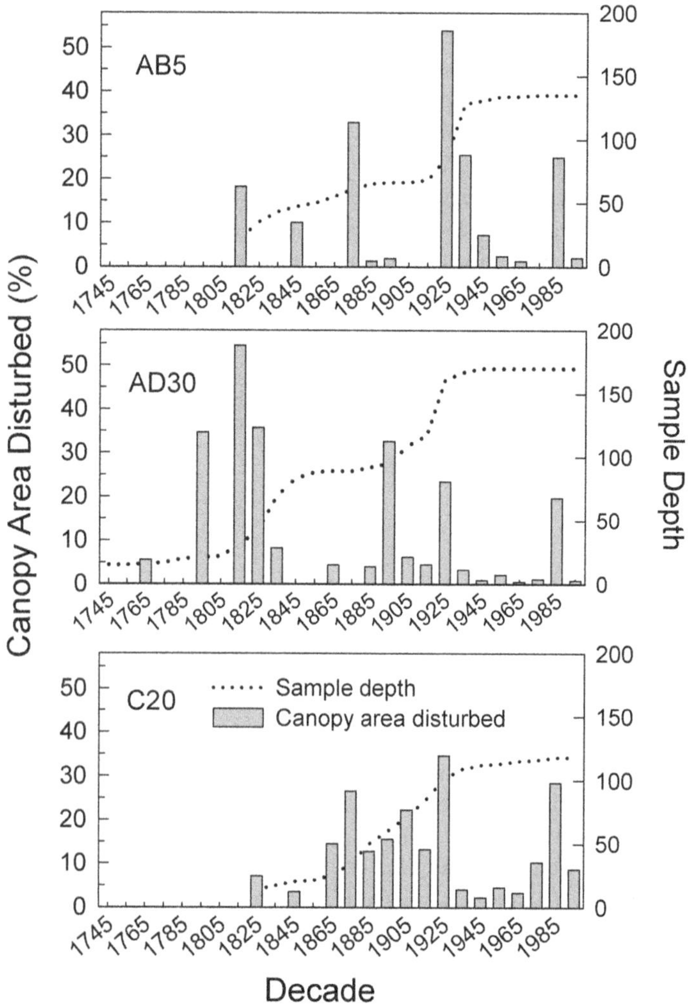

Fig. 7.3 Disturbance chronologies for three old-growth *Picea rubens* stands, showing peaks and hiatuses in canopy disturbance. Sample depth refers to the number of canopy trees used in the chronology. Figure modified from Fraver and White (2005a); printed with permission from *Journal of Vegetation Science*

makers. When combined with regional allometric equations (Honer 1965; Li et al. 2012), the current diameters allowed us to estimate stem volumes of gap makers, that is, volumes of coarse woody debris added. For modelling purpose we chose to convert our volume estimates to mass, given the availability of mass-loss equations (Russell et al. 2015) and the absence of volume-loss equations. Volumes were thus converted to dry mass using specific gravities presented in Miles and Smith (2009).

The growth-release method, however, cannot be applied to recent decades because the requisite post-disturbance time period is insufficient for assessing growth response. For recent decades (ca. 1980–2000), we relied on multiple lines of dendroecological, spatial, and structural evidence to estimate the timing of woody debris inputs. These methods include (1) crossdated outermost rings of cored woody debris, (2) onset of compression wood in living trees displaced from vertical by the disturbance, (3) increased height growth recorded on mapped saplings, (4) the vertical arrangement of stacked fallen logs (most recent above previously fallen), and (5) the location of woody debris relative to current canopy gaps. Taken together, these methods allowed us to date the input of individual pieces of woody debris for these recent decades. Their volumes were calculated using the conic-paraboloid formula (Fraver et al. 2007), based on end diameters and length recorded on each in 2001. As above, woody debris volumes were converted to dry mass for modelling purposes using specific gravities presented in Miles and Smith (2009).

These methods produced estimated annual woody debris mass inputs for each year between 1900 and 2000. The mass of these inputs was then subjected to a decay function, also calculated annually, assuming mass loss followed a negative exponential decline, using species-specific mass-loss k-constants (Russell et al. 2015). For the earlier decades, we assumed woody debris inputs were entirely *Picea rubens* and used the associated k-constant. For recent decades, where species was known, we used species-specific k-constants. We note that the negative exponential rate loss, despite its widespread acceptance, may not best describe mass loss (Freschet et al. 2012; Fraver et al. 2013); however, these were the only mass-loss equations available.

The final challenge in creating the temporal sequence was to estimate the woody debris mass that was present in 1900, the starting point of our sequence. To this end, we relied on the growth-release method to estimate woody debris input rates for the 60-year period prior to 1900, and applied our volume estimations, mass conversions, and mass-loss rates (as above) to these inputs. We chose a 60-year period because decaying *Picea rubens* has an average residence time of ca. 60 years (Russell et al. 2015), meaning that woody debris generated in 1840 would have been largely decayed by 1900, which represents a stable starting point.

7.5.2 Results and Discussion

Our temporal sequences represent woody debris mass fluctuations under a non-stand-replacing disturbance regime. The sequences from all three sites show striking variability in woody debris mass through the 100-year period (Fig. 7.4). Mass ranged from ca. 10–31 Mg/ha on plot AB5; from 14–46 Mg/ha on plot AD30, and from 17–42 Mg/ha on plot C20. These values *roughly* correspond to woody debris volumes of 39–116, 52–173, and 65–158 m^3/ha, respectively, based on *Picea rubens* densities averaged across decay classes (Harmon et al. 2008). The

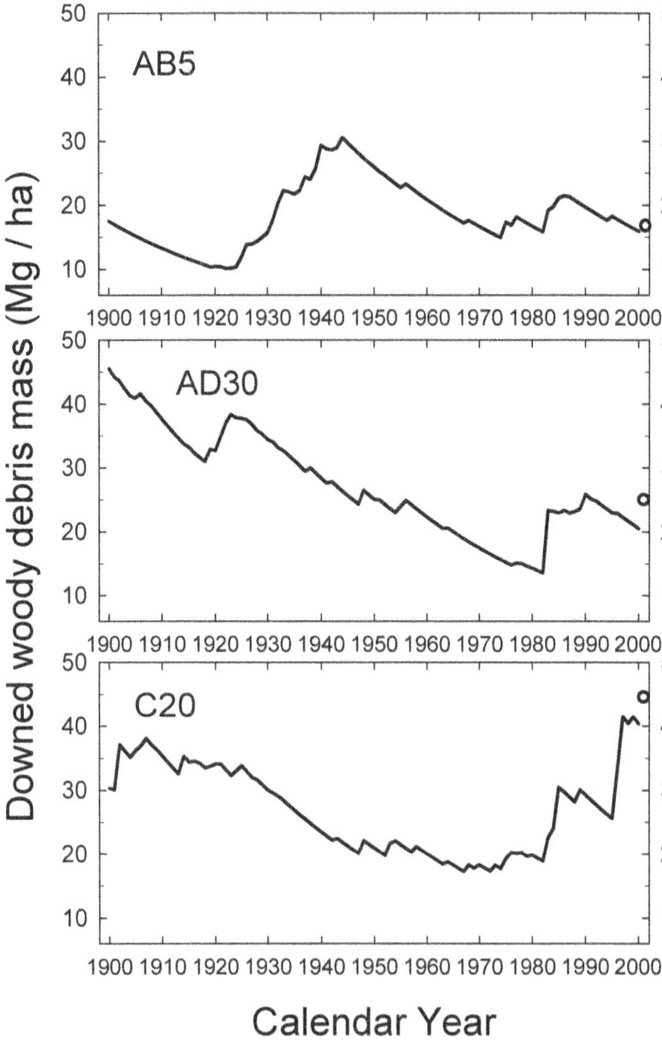

Fig. 7.4 Fluctuations in downed woody debris mass for a 100-year period in three old-growth *Picea rubens* forest stands. Peaks in debris input correspond to past disturbances, also reconstructed from tree-ring data. The small circles in 2001 depict the debris volume calculated from a field inventory conducted in that year. The "skewed" shape of the curves nicely illustrates the fast-in-slow-out nature of woody debris dynamics

common patterns among the plots are consistent with documented landscape-wide disturbances (Fig. 7.3). For example, two spruce budworm outbreaks in the late 1910s and early 1980s and a 1983 wind storm caused pulses in woody debris input evident on all three plots. The relatively low woody debris masses from roughly the 1960s through the early 1980s reflect decades of relative quiescence evident on all three plots. Nevertheless, individual plots also show unique patterns (Fig. 7.3), the result of past disturbances that have differentially affected the plots, as described in Fraver and White (2005a). For example, the broad peak in woody debris mass spanning the 1930s through the 1950s on plot AB5 reflects the peaks in canopy disturbance at that time. The high initial mass on plot AD30 reflects the residual mass from a disturbance peak in the 1880s. The relatively high initial mass and an early broad peak on plot C20 reflect moderate rates of disturbance that began in the 1860s and persisted through the 1920s. The unique signature of disturbance history on woody debris mass for each plot highlights the potential pitfalls of using metrics such as canopy tree age to model patterns in woody debris abundance in these and other systems (cf. Tyrrell and Crow 1994a, b).

We acknowledge the uncertainty in our reconstructed temporal sequences. Uncertainty enters in at each step in our reconstruction, from estimating the initial (1900) woody debris mass, estimating input mass from growth-release and other data, and using mass-loss decay functions to gradually deplete input mass. In addition, we have not accounted for snag-fall contributions to the woody debris pool, because the slow death of trees that remain standing is not captured by the growth-release method, which assumes the sudden loss (not slow demise) of canopy trees to which formerly suppressed individuals respond. Despite these uncertainties and limitations, the endpoints of the modelled temporal sequences (year 2000) closely approximate the masses calculated from 2001 woody debris inventory (Fig. 7.4), suggesting that the sequences developed using dendroecological data reasonably represent past fluctuations.

Together, our reconstructed temporal patterns depict what we believe to be typical fluctuations for this forest type, and perhaps representative of other forest types that experience partial canopy loss over extended periods. Aakala's (2011) simulations based on empirical data show somewhat similar fluctuations and ranges, although based on woody debris volume, in *Picea abies* forests of boreal Finland and Russia. Similarly, Jönsson et al. (2011) showed fluctuations, with broad landscape-level similarity in woody debris input (without including decay) among *Picea abies* stands in Sweden; however, individual stands had input patterns unique to their own disturbance histories.

Given the large number of studies that document disturbance regimes characterized by partial canopy loss (regional reviews by Seymour et al. 2002, Kuuluvainen and Aakala 2011), it seems likely that the type of woody debris fluctuations shown here would be found in many other forest types worldwide. Our empirical results shed light on the patterns underlying the right-hand portion of the popular U-shaped model of woody debris abundance over time. Specifically, they depict a continuously fluctuating pattern that may persists for centuries to millennia in these systems until an infrequent stand-replacing disturbance reverts this pattern to a U-shaped trend. In addition, these results, as well as those presented in Aakala (2011) and D'Amato et al. (2017), call into question the assumption that old-growth stands reach an equilibrium with respect to woody debris abundance (e.g., Tyrrell and Crow 1994b).

These fluctuations in woody debris abundance have a number of important ecological implications. First, they regulate population dynamics of saproxylic (i.e., deadwood-dependent) organisms by creating a 'boom-or-bust' situation regarding available substrate. The periods of low abundance, particularly in isolated stands, may create a risk of extirpation for red-listed saproxylics that depend on a rather continuous supply of woody debris of particular tree species or decay classes (Jönsson et al. 2011; Stokland et al. 2012). In addition, forest management approaches for conserving these organisms need to account for these dynamics by creating woody debris at each harvest entry to maintain suitable substrates in various stages of decay over time (D'Amato et al. 2015). Fluctuations also have important implications and create significant challenges for modelling carbon dynamics in forest types typified by partial disturbance regimes. Finally, when viewed from the perspective of fuel loads, these fluctuations show how fuels and associated fire risks can vary through time, even more so when we consider that combustion properties change through decay (Hyde et al. 2011).

In summary, our case study exemplifies the application of detailed dendroecological data, combined with spatial data, to better understand the dynamics of coarse woody debris. It also highlights the challenges involved in doing so, even with high-quality data. Our temporal sequences demonstrate the fluctuations and ranges of woody debris masses in old-growth *Picea rubens* stands; however, we suspect other forest types subject to non-stand replacing disturbance regimes may show these same types of fluctuations. These fluctuations highlight the need to characterize fine-scale woody debris dynamics to guide modeling and restoration efforts for ecosystems with disturbance regimes dominated by gap- and meso-scale disturbances.

7.6 Future Directions

Given the critical role coarse woody debris plays in forest ecosystems—carbon storage, nutrient cycling, soil development, and maintenance of biodiversity—it is imperative that we understand how its abundance changes through time. In particular, the recent need to better characterize the forest carbon cycle in the context of climate change further highlights the importance of understanding these temporal changes. The field of dendroecology lends itself perfectly to coarse woody debris dynamics because it provides annual resolution that can be applied to woody debris additions (i.e., tree mortality or tree-fall dates) as well as decay rates (using chronosequences or decay-class transition models). Nevertheless, several of the limitations addressed throughout this chapter, as well as findings from recent studies, highlight knowledge gaps and hence directions for future research, each addressed briefly below.

7.6.1 Disturbance Agents and Woody Debris Dynamics

The manner in which trees are killed (i.e., the disturbance agent) strongly influences woody debris dynamics, particularly the length of time trees remain in the snag phase (potentially decades for fire-killed trees, yet instantaneous transition to downed woody debris for windthrow), as well as the decay rates of downed woody debris. Thus the disturbance agents themselves create considerable uncertainty in our estimates of snag and downed woody debris longevities; however, little work has specifically addressed this gap in our understanding. Carefully designed dendrochronology studies including a range of disturbance agents and forest types could dramatically improve our predictions of long-term woody debris dynamics.

7.6.2 'Buried' Woody Debris

Recent work suggests that under certain conditions downed woody debris accumulates gradually as it becomes buried by litter and overgrown by ground vegetation (Moroni et al. 2015). Once buried, its decomposition appears to slow markedly (Arsenault et al. 2007; Hagemann et al. 2010). Under such conditions, the volume of this partially buried wood can exceed the volume of all other live and deadwood volumes combined (Moroni et al. 2015). Clearly, work is needed to quantify both the abundance (mass, volume) and decay rates of this often-overlooked feature. Its purported low decay rates suggest buried wood may provide material suitable for dendrochronological studies aimed at quantifying mass loss over time. Results of such work would improve carbon accounting and modelling, and they may clarify our interpretation of ecosystem processes where this missing carbon pool may have been overlooked.

7.6.3 Tropical Woody Debris Dynamics

As dendrochronology capabilities continue to improve in the tropics (Pumijumnong 2013), so do opportunities to quantify woody debris dynamics in these regions. Currently, our knowledge of woody debris dynamics from the tropics is quite poor relative to the temperate and boreal regions (Russell et al. 2015; Magnússon et al. 2016), despite the importance of tropical forests in the global carbon cycle and global biodiversity.

7.6.4 Model Improvements

Most existing gap-dynamics, carbon-accounting, and terrestrial ecosystem models treat woody debris and woody debris dynamics somewhat superficially. Many of these models currently rely on calibration from forest inventory data, which span several decades at most, are rarely annually resolved, and are disproportionately from second-growth or managed stands. This lack of temporal depth and annual resolution means that ecological processes are poorly represented and insufficiently constrained, making the models reliant on assumptions that can be difficult to justify and creating considerable uncertainty in predictions. Improving the predictive confidence of these models requires the use of dendroecological data, which when applied to this purpose may span a century or more (as in the case study above) and can provide annual resolution. High-quality data such as these allow us to better estimate woody debris dynamics—including inherent year-to-year variability—which is essential for improving the predictive confidence of ecosystem-dynamics models.

Acknowledgments We thank Al White for guidance in the early portions of the case study, David Foster and Dave Orwig for permitting and facilitating the sampling at Harvard Forest's Pisgah Tract (Fig. 7.2), and The Nature Conservancy, Maine Chapter, for permitting the sampling at the Big Reed Forest Reserve. We thank Lori Daniels and Markus Stoffel for providing editorial suggestions that greatly improved the chapter.

References

Aakala T (2010) Coarse woody debris in late-successional *Picea abies* forests in northern Europe: variability in quantities and models of decay class dynamics. For Ecol Manag 260(5):770–779

Aakala T (2011) Temporal variability of deadwood volume and quality in boreal old-growth forests. Silva Fenn 45:969–981

Aakala T, Kuuluvainen T, Grandpré LD, Gauthier S (2007) Trees dying standing in the northeastern boreal old-growth forests of Quebec: spatial patterns, rates, and temporal variation. Can J For Res 37(1):50–61

Aakala T, Kuuluvainen T, Gauthier S, De Grandpré L (2008) Standing dead trees and their decay-class dynamics in the northeastern boreal old-growth forests of Quebec. For Ecol Manag 255(3):410–420

Aakala T, Kuuluvainen T, Wallenius T, Kauhanen H (2011) Tree mortality episodes in the intact Picea abies-dominated taiga in the Arkhangelsk region of northern European Russia. J Veg Sci 22(2):322–333

Allen CD, Macalady AK, Chenchouni H et al (2010) A global overview of drought and heat-induced tree mortality reveals emerging climate change risks for forests. For Ecol Manag 259(4):660–684

Amoroso MM, Daniels LD, Villalba R, Cherubin P (2015) Does drought incite tree decline and death in *Austrocedrus chilensis* forests? J Veg Sci 26(6):1171–1183

Angers VA, Drapeau P, Bergeron Y (2010) Snag degradation pathways of four North American boreal tree species. For Ecol Manag 259(3):246–256

Arsenault D, Boucher É, Bouchon É (2007) Asynchronous forest–stream coupling in a fire-prone boreal landscape: insights from woody debris. J Ecol 95:789–801

Beets PN, Hood IA, Kimberley MO et al (2008) Coarse woody debris decay rates for seven indigenous tree species in the central North Island of New Zealand. For Ecol Manag 256:548–557
Bigler C, Gavin DG, Gunning C, Veblen TT (2007) Drought induces lagged tree mortality in a subalpine forest in the Rocky Mountains. Oikos 116(12):1983–1994
Black BA, Abrams M (2003) Use of boundary-line growth patterns as a basis for dendroecological release criteria. Ecol Appl 13(6):1733–1749
Blanchette RA, Obst JR, Timell TE (1994) Biodegradation of compression wood and tension wood by white and brown rot fungi. Holzforschung 48:34–42
Bradford MA, Warren RJ II, Baldrian P et al (2014) Climate fails to predict wood decomposition at regional scales. Nat Clim Chang 4:625–630
Brown PM, Shepperd WD, Mata SA, McClain DL (1998) Longevity of windthrown logs in a subalpine forest of central Colorado. Can J For Res 28(6):932–936
Carmona MR, Armesto JJ, Aravena JC, Pérez CA (2002) Coarse woody debris biomass in successional and primary temperate forests in Chiloé Island, Chile. For Ecol Manag 164(1):265–275
Caron MN, Kneeshaw DD, Grandpré LD et al (2009) Canopy gap characteristics and disturbance dynamics in old-growth *Picea abies* stands in northern Fennoscandia: Is the forest in quasi-equilibrium? Ann Bot Fenn 46(4):251–262
Cherubini P, Fontana G, Rigling D et al (2002) Tree-life history prior to death: two fungal root pathogens affect tree-ring growth differently. J Ecol 90(5):839–850
Cook ER, Kairiukstis LA (eds) (2013) Methods of dendrochronology: applications in the environmental sciences. Kluwer Academic Publishers, Dordrecht
Cornwell WK, Cornelissen JHC, Allison SD et al (2009) Plant traits and wood fates across the globe: rotted, burned, or consumed? Glob Chang Biol 15:2431–2449
D'Amato AW, Orwig DA (2008) Stand and landscape-level disturbance dynamics in old-growth forests in western Massachusetts. Ecol Monogr 78(4):507–522
D'Amato AW, Orwig DA, Foster DR (2008) The influence of successional processes and disturbance on the structure of *Tsuga canadensis* forests. Ecol Appl 18:1182–1199
D'Amato AW, Catanzaro P, Fletcher LS (2015) Early regeneration and structural responses to patch selection and structural retention within second-growth northern hardwoods. For Sci 61:183–189
D'Amato AW, Orwig DA, Foster DR et al (2017) Long-term structural and biomass dynamics of virgin *Tsuga canadensis-Pinus strobus* forests after hurricane disturbance. Ecology 98(3):721–733
Daniels LD, Dobry J, Klinka K, Feller MC (1997) Determining year of death of logs and snags of *Thuja plicata* in southwestern coastal British Columbia. Can J For Res 27(7):1132–1141
Dynesius M, Jonsson BG (1991) Dating uprooted trees: comparison and application of eight methods in a boreal forest. Can J For Res 21:655–665
Edman M, Möller R, Ericson L (2006) Effects of enhanced tree growth rate on the decay capacities of three saprotrophic wood-fungi. For Ecol Manag 232:12–18
Edvardsson J, Stoffel M, Corona C et al (2016) Subfossil peatland trees as proxies for palaeohydrology and climate reconstruction during the Holocene. Earth-Sci Rev 163:118–140
Fraver S, Palik BJ (2012) Stand and cohort structures of old-growth *Pinus resinosa*-dominated forests of northern Minnesota, USA. J Veg Sci 23:249–259
Fraver S, White AS (2005a) Disturbance dynamics of old-growth *Picea rubens* forests of northern Maine. J Veg Sci 16:597–610
Fraver S, White AS (2005b) Identifying growth releases in dendrochronological studies of forest disturbance. Can J For Res 35(7):1648–1656
Fraver S, Ringvall A, Jonsson BG (2007) Refining volume estimates of down woody debris. Can J For Res 37:627–633
Fraver S, Jonsson BG, Jönsson M, Esseen P-A (2008) Demographics and disturbance history of a boreal old-growth *Picea abies* forest. J Veg Sci 19:789–798
Fraver S, Bradford JB, Palik BJ (2011) Improving tree age estimates derived from increment cores: a case study of *Pinus resinosa*. For Sci 57:164–170

Fraver S, Milo AM, Bradford JB et al (2013) Woody debris volume depletion through decay: Implications for biomass and carbon accounting. Ecosystems 16(7):1262–1272
Freschet GT, Weedon JT, Aerts R et al (2012) Interspecific differences in wood decay rates: insights from a new short-term method to study long-term wood decomposition. J Ecol 100:161–170
Gennaretti F, Arseneault D, Bégin Y (2014) Millennial stocks and fluxes of large woody debris in lakes of the North American taiga. J Ecol 102(2):367–380
Gunnarson BE (2001) Lake level changes indicated by dendrochronology on subfossil pine, Jämtland, central Scandinavian Mountains, Sweden. Arct Antarct Alp Res 33(3):274–281
Guyette RP, Dey DC, Stambaugh MC (2008) The temporal distribution and carbon storage of large oak wood in streams and floodplain deposits. Ecosystems 11:643–653
Hagemann U, Moroni MT, Gleibner J, Makeschin F (2010) Accumulation and preservation of dead wood upon burial by bryophytes. Ecosystems 13:600–611
Harmon ME (2009) Woody detritus its contribution to carbon dynamics of old-growth forests: the temporal context. In: Wirth C, Gleixner G, Heimann M (eds) Old-growth forests. Springer, Berlin, pp 159–190
Harmon ME, Franklin JF, Swanson FJ et al (1986) Ecology of coarse woody debris in temperate ecosystems. Adv Ecol Res 15:133–302
Harmon ME, Woodall CW, Fasth B, Sexton J (2008) Woody detritus density and density reduction factors for tree species in the United States: A synthesis. GTR-NRS-29, US Department of Agriculture, Newtown Square, PA
He HS, Shang BZ, Crow TR et al (2004) Simulating forest fuel and fire risk dynamics across landscapes—LANDIS fuel module design. Ecol Model 180:135–151
Henry JD, Swan JM (1974) Reconstructing forest history from live and dead plant material—and approach to the study of forest succession in southwest New Hampshire. Ecology 55:772–783
Herman FR, Smith CE, Firth JE (1972) Ring count of decayed wood made easier through freezing. J For 70:743
Holeksa J, Zielonka T, Zywiec M (2008) Modeling the decay of coarse woody debris in a subalpine Norway spruce forest of the West Carpathians, Poland. Can J For Res 38:415–428
Honer TG (1965) A new total cubic foot volume function. Forest Chron 41(4):476–493
Hyde JC, Smith AM, Ottmar RD et al (2011) The combustion of sound and rotten coarse woody debris: a review. Int J Wildland Fire 20(2):163–174
Janisch JE, Harmon ME (2002) Successional changes in live and dead wood carbon stores: implications for net ecosystem productivity. Tree Physiol 22(2–3):77–89
Jemison GM (1937) Loss of weight of wood due to weathering. J For 35(5):460–462
Jones EL, Daniels LD (2012) Assessment of dendrochronological year-of-death estimates using permanent sample plot data. Tree-Ring Res 68:3–16
Jones TA, Daniels LD, Powell SRE (2011) Abundance and function of large woody debris in small, headwater streams in the foothills of the Rocky Mountains, Alberta, Canada. River Res Appl 27:297–311
Jönsson M, Fraver S, Jonsson BG (2011) Spatio-temporal variation of coarse woody debris input in woodland key habitats in central Sweden. Silva Fenn 45(5):957–967
Kerhoulas LP, Kane JM (2012) Sensitivity of ring growth and carbon allocation to climatic variation vary within ponderosa pine trees. Tree Physiol 32(1):14–23
Khakimulina T, Fraver S, Drobyshev I (2016) Mixed-severity natural disturbance regime dominates in an old-growth Norway spruce forest of northwest Russia. J Veg Sci 27(2):400–413
Krusic PJ Jr, Hornbeck JW (1989) Preserving decayed wood samples for tree-ring measurement. Tree-Ring Bull 49:23–27
Kruys N, Jonsson BG, Ståhl G (2002) A stage-based matrix model for decay-class dynamics of woody debris. Ecol Appl 12(3):773–781
Kurz WA, Dymond CC, White TM et al (2009) CBM-CFS3: a model of carbon-dynamics in forestry and land-use change implementing IPCC standards. Ecol Model 220:480–504
Kuuluvainen T, Aakala T (2011) Natural forest dynamics in boreal Fennoscandia: a review and classification. Silva Fenn 45(5):823–841

Laiho R, Prescott CE (2004) Decay and nutrient dynamics of coarse woody debris in northern coniferous forests: a synthesis. Can J For Res 34:763–777

Lambert RC, Lang GE, Reiners WA (1980) Loss of mass and chemical change in decaying boles of a subalpine balsam fir forest. Ecology 61:1460–1473

Lännenpää A, Aakala T, Kauhanen H, Kuuluvainen T (2008) Tree mortality agents in pristine Norway spruce forests in northern Fennoscandia. Silva Fenn 42(2):151–163

Lara A, Villalba R (1993) A 3620-year temperature record from *Fitzroya cupressoides* tree rings in southern South America. Science 260(5111):1104–1106

Larjavaara M, Muller-Landau HC (2010) Comparison of decay classification, knife test, and two penetrometers for estimating wood density of coarse woody debris. Can J For Res 40(12):2313–2321

Li R, Weiskittel A, Dick AR et al (2012) Regional stem taper equations for eleven conifer species in the Acadian region of North America: development and assessment. N J App For 29(1):5–14

Lindbladh M, Fraver S, Edvardsson J, Felton A (2013) Past forest composition, structures and processes—How paleoecology can contribute to forest conservation. Biol Conserv 168:116–127

Linderholm HW, Björklund JA, Seftigen K et al (2010) Dendroclimatology in Fennoscandia–from past accomplishments to future potential. Clim Past 6:93–114

Lindner DL, Vasaitis R, Kubartova A et al (2011) Initial fungal colonizer affects mass loss and fungal community development in *Picea abies* logs 6 yr after inoculation. Fungal Ecol 4:449–460

Lombardi F, Cherubini P, Lasserre B et al (2008) Tree rings used to assess time since death of deadwood of different decay classes in beech and silver fir forests in the central Apennines (Molise, Italy). Can J For Res 38(4):821–833

Lorimer CG, Frelich LE (1989) A methodology for estimating canopy disturbance frequency and intensity in dense temperate forests. Can J For Res 19:651–663

Lorimer CG, White AS (2003) Scale and frequency of natural disturbances in the northeastern US: implications for early successional forest habitats and regional age distributions. For Ecol Manag 185:41–64

Magnússon RÍ, Tietema A, Cornelissen JH et al (2016) Tamm review: sequestration of carbon from coarse woody debris in forest soils. For Ecol Manag 377:1–15

Mast JN, Veblen TT (1994) A dendrochronological method of studying tree mortality patterns. Phys Geogr 15:529–542

Means JE, Cromack K Jr, MacMillan PC (1985) Comparison of decomposition models using wood density of Douglas-fir logs. Can J For Res 15:1092–1098

Miles PD, Smith WB (2009) Specific gravity and other properties of wood and bark for 156 tree species found in North America (Vol. 38). US Department of Agriculture, Forest Service, Northern Research Station

Moroni MT, Morris DM, Shaw C et al (2015) Buried wood: A common yet poorly documented form of deadwood. Ecosystems 18(4):605–628

Næsset E (1999) Decomposition rate constants of *Picea abies* logs in southeastern Norway. Can J For Res 29:372–381

Nicolussi K, Kaufmann M, Patzelt G, Thurner A (2005) Holocene tree-line variability in the Kauner Valley, Central Eastern Alps, indicated by dendrochronological analysis of living trees and subfossil logs. Veg Hist Archaeobotany 14(3):221–234

Niklasson M (2002) A comparison of three age determination methods for suppressed Norway spruce: implications for age structure analysis. For Ecol Manag 161:279–288

Niklasson M, Granström A (2000) Numbers and sizes of fires: long-term spatially explicit fire history in a Swedish boreal landscape. Ecology 81(6):1484–1499

Panshin AJ, de Zeeuw C (1980) Textbook of wood technology. McGraw-Hill, Inc, New York, p722

Parent S, Morin H, Messier C (2002) Missing growth rings at the trunk base in suppressed balsam fir saplings. Can J For Res 32:1776–1783

Pedersen BS, McCune B (2002) A non-invasive method for reconstructing the relative mortality rates of trees in mixed-age, mixed-species forests. For Ecol Manag 155(1):303–314

Pederson N, Young A, Stan AB, Ariya U, Martin-Benito D (2017) Low-hanging dendrodynamic ecological fruits regarding disturbance in temperate, mesic forests. In: Amoroso MM, Daniels LD, Baker PJ, Julio Camarero J (eds) Dendroecology: tree-ring analyses applied to ecological studies. Springer, Cham

Pregitzer KS, Reed DD, Bornhorst TJ et al (2000) A buried spruce forest provides evidence at the stand and landscape scale for the effects of environment on vegetation at the Pleistocene/Holocene boundary. J Ecol 88:45–53

Pumijumnong N (2013) Dendrochronology in Southeast Asia. Trees 27:343–358

Rădoane M, Nechita C, Chiriloaei F et al (2015) Late Holocene fluvial activity and correlations with dendrochronology of subfossil trunks: Case studies of northeastern Romania. Geomorphology 239:142–159

Rock J, Badeck F-W, Harmon ME (2008) Estimating decomposition rate constants for European tree species from literature sources. Eur J For Res 127:301–313

Ruiz-Villanueva V, Gurnell A, Piégay H et al (2016) Recent advances quantifying the large-wood cycle in forested river basins: new methods, remaining challenges. Rev Geophys 54:611–652

Russell M, Woodall CW, Fraver S et al (2014) Residence times and decay rates for downed woody debris biomass/carbon in eastern US forests. Ecosystems 17:765–777

Russell M, Fraver S, Aakala T et al (2015) Tamm review: quantifying carbon stores and decomposition in dead wood. For Ecol Manag 350:107–128

Schoennagel T, Veblen TT, Romme WH (2004) The interaction of fire, fuels, and climate across Rocky Mountain Forests. Bioscience 54:661–676

Seymour RS, White AS, deMaynadier PG (2002) Natural disturbance regimes in northeastern North America—evaluating silvicultural systems using natural scales and frequencies. For Ecol Manag 155:357–367

Shorohova E, Kapitsa E (2014) Influence of the substrate and ecosystem attributes on the decomposition rates of coarse woody debris in European boreal forests. For Ecol Manag 315:173–184

Siegert NW, Mccullough DG, Liebhold AM, Telewski FW (2014) Dendrochronological reconstruction of the epicentre and early spread of emerald ash borer in North America. Divers Distrib 20:847–858

Sollins P (1982) Input and decay of coarse woody debris in coniferous stands in western Oregon and Wahington. Can J For Res 12:18–28

Speer JH (2010) Fundamentals of tree-ring research. University of Arizona Press, Tucson, Arizona, USA

Stoffel M, Corona C (2014) Dendroecological dating of geomorphic disturbance in trees. Tree-Ring Res 70:3–20

Stokland JN, Siitonen J, Jonsson BG (2012) Biodiversity in dead wood. Cambridge University Press, Cambridge, UK, p 509

Storaunet KO, Rolstad J (2002) Time since death and fall of Norway spruce logs in old-growth and selectively cut boreal forest. Can J For Res 32(10):1801–1812

Swetnam TW, Betancourt JL (2010) Mesoscale disturbance and ecological response to decadal climatic variability in the American Southwest. J Clim 11(12):3128–3147

Telewski FW (1993) Determining the germination date of woody plants: a proposed method for locating the root/shoot interface. Tree-Ring Bull 53:13–16

Tikkanen O-P, Heinonen T, Kouki J, Matero J (2007) Habitat suitability models of saproxylic red-listed boreal forest species in long-term matrix management: Cost-effective measures for multi-species conservation. Biol Conserv 140:359–372

Tyrrell LE, Crow TR (1994a) Structural characteristics of old-growth hemlock-hardwood forests in relation to age. Ecology 75:370–386

Tyrrell LE, Crow TR (1994b) Dynamics of dead wood in old-growth hemlock-hardwood forests of northern Wisconsin and northern Michigan. Can J For Res 24(8):1672–1683

Van Mantgem PJ, Stephenson NL, Byrne JC et al (2009) Widespread increase of tree mortality rates in the western United States. Science 323(5913):521–524

Vanderwel MC, Malcolm JR, Smith SM (2006) An integrated model for snag and downed woody debris decay class transitions. For Ecol Manag 234(1):48–59
Verrall AF (1938) The probable mechanism of the protective action of resin in fire wounds on red pine. J Forest 12:1231–1233
Villalba R, Veblen TT (1997) Improving estimates of total tree ages based on increment cores. Ecoscience 4:535–542
Winter LE, Brubaker LB, Franklin JF et al (2002) Canopy disturbances over the five-century lifetime of an old-growth Douglas-fir stand in the Pacific Northwest. Can J For Res 32(6):1057–1070
Woodall CW, Russell M, Walters BF et al (2015) Net carbon flux of dead wood in forests of the Eastern US. Oecologia 177:861–874
Yatskov M, Harmon ME, Krankina ON (2003) A chronosequence of wood decomposition in the boreal forests of Russia. Can J For Res 33:1211–1226
Zielonka T (2006) When does dead wood turn into a substrate for spruce replacement? J Veg Sci 17(6):739–746

Part II
Disturbance Regimes

Credit: J. Meyers, L. D. Daniels

Chapter 8
Deciphering the Complexity of Historical Fire Regimes: Diversity Among Forests of Western North America

Lori D. Daniels, Larissa L. Yocom Kent, Rosemary L. Sherriff, and Emily K. Heyerdahl

Abstract Wildfire is a key disturbance agent in forests worldwide, but recent large and costly fires have raised urgent questions about how different current fire regimes are from those of the past. Dendroecological reconstructions of historical fire frequency, severity, spatial variability, and extent, corroborated by other lines of evidence, are essential in addressing these questions. Existing methods can infer the severity of individual fires and stand-level fire regimes. However, novel research designs combining evidence of stand-level fire severity with fire extent are now being used to reconstruct spatial variability in historical fire regimes and to quantify the relative abundance of fire severity classes across landscapes, thereby facilitating comparison with modern fire regimes. Here we review how these new approaches build on traditional analyses of fire scars and forest age structures by presenting four case studies from the western United States and Canada. Collectively they demonstrate the importance of ecosystem-specific research that can guide management aiming to safeguard human, cultural and biological values in fire-prone forests and enhance forest resilience to the cumulative effects of global environmental change. Dendroecological reconstructions, combined with multiple lines of corroborating evidence, are key for achieving this goal.

L.D. Daniels (✉)
Department of Forest and Conservation Sciences, University of British Columbia, Vancouver, BC, Canada
e-mail: lori.daniels@ubc.ca

L.L. Yocom Kent
School of Forestry, Northern Arizona University, Flagstaff, AZ, USA
e-mail: larissa.yocom@nau.edu

R.L. Sherriff
Department of Geography, Humboldt State University, Arcata, CA, USA
e-mail: sherriff@humboldt.edu

E.K. Heyerdahl
Rocky Mountain Research Station, Fire Sciences Laboratory, USDA Forest Service, Missoula, MT, USA
e-mail: eheyerdahl@fs.fed.us

M.M. Amoroso et al. (eds.), *Dendroecology*, Ecological Studies 231,
DOI 10.1007/978-3-319-61669-8_8

Keywords Fire history • Fire scars • Post-fire cohorts • Simulation modeling • Alberta • Arizona • British Columbia • Colorado • Oregon

8.1 Introduction

Wildfire is a key agent of ecological disturbance in forests worldwide (Bowman et al. 2009). In western North America, fire drives vegetation dynamics, underlies many ecological patterns, and explains ecosystem heterogeneity at a range of spatio-temporal scales (Turner 2010). Subtle changes to fire regimes in western forests can have major impacts on forest composition, structure, biodiversity and productivity (Turner 2010). Changes to fire frequency and severity driven by land use and fire suppression, exacerbated by climate change, have the potential to exceed the historical range of variation of environmental conditions in which species evolved and ecosystems have functioned (Landres et al. 1999; Swetnam et al. 1999). Thus, understanding historical fire regimes and their drivers can help identify vulnerable ecosystems in which active management can enhance forest resilience to the cumulative impacts of environmental change (Stephens et al. 2013). Equally important, ecosystem-specific knowledge can identify forests in which fire regimes have not departed from the historical range of variation, and where management investments are not needed or may compromise ecosystem function or services (Schoennagel and Nelson 2011; Stephens and Fulé 2005; Odion et al. 2014).

Dendroecological research has been instrumental for reconstructing historical fire regimes and quantifying their key attributes. A fire regime is the pattern of the temporal (seasonality, frequency and predictability) and spatial (location, extent and spatial complexity) occurrences of fire, the magnitude (type, intensity and severity) of fire effects, and interactions of fire with other disturbance agents (Agee 1993; Turner 2010). Describing fire regime attributes requires accurate documentation of individual fires. To extend modern fire records back decades or centuries, reconstructions of past fires can be achieved by dendroecological analyses corroborated by multiple lines of evidence (Swetnam et al. 1999). This research framework has facilitated significant advances in our understanding of fire regimes over recent decades, shifting the focus from fire frequency to more nuanced analyses of fire severity, extent and spatial complexity.

In this review, we show how contemporary research focused on mixed-severity fire regimes builds on traditional analyses of fire scars and forest age structures. Novel research designs facilitating analyses across spatial scales and analytical methods for quantifying fire severity and spatial variability allow more nuanced characterization of the spatio-temporal variation in fire regimes. We present four case studies illustrating a variety of research approaches and the complexity of historical fire regimes across forest types in the western United States and Canada. Collectively, they demonstrate the importance of ecosystem-specific research to guide management that aims to safeguard human, cultural and biological values in fire-prone forests and enhance forest resilience to the cumulative effects of global environmental change.

8.2 Fire Regime Reconstructions

Traditionally, fire history research focused on two contrasting fire regime types simplified as high- *versus* low-frequency. The research methods, including dendrochronological evidence, used to reconstruct fire frequency of these contrasting regimes reflect the differences in impacts and physical evidence left by presumably surface *versus* crown fires.

8.2.1 High-frequency, Low-severity Fire Regimes

High-frequency fire regimes are characterized by relatively low-severity surface fires that burn at short intervals, causing low levels of canopy tree mortality (e.g., stand-maintaining fires; Schoennagel et al. 2004). Fire scars on individual live trees, stumps, snags or logs provide point-specific, direct evidence of low-severity fire that damages, but does not kill the tree (Swetnam and Baisan 1996, Swetnam et al. 1999, Yocom Kent 2014; Table 8.1; Fig. 8.1). Scars caused by fire are differentiated from other disturbances (e.g. bark beetles, animals, tree falls or cultural modification by humans) based on their triangular shape at the base of the tree and the presence of charred bark, healing lobes or exposed wood on the face of the scar (McBride 1983; Smith and Sutherland 2001). When properly crossdated, scars provide the exact year of fire occurrence and often season can be determined from the position of the scar within an annual growth ring (Dieterich and Swetnam 1984; Caprio and Swetnam 1995). The presence of resin ducts or micro rings provides secondary dendroecological evidence to corroborate fire scar dates (Brown and Swetnam 1994; Smith et al. 2016). Often large and presumably old trees with multiple scar lobes are targeted to maximize the fire record (Swetnam and Baisan 1996; Swetnam et al. 1999). Fire-scar records from individual trees can include multiple fires over several centuries, depending on the lifespan of the species, rates of wood decomposition, and occurrence of subsequent fires that consume fire-scarred trees (Swetnam and Baisan 1996; Swetnam et al. 1999).

At the site level, sampling multiple fire-scarred trees serves several purposes (Swetnam and Baisan 1996, Swetnam et al. 1999; Fig. 8.1). Replicate samples corroborate fire dates among trees and differentiate spreading fires that scar multiple trees from small isolated fires that scar single trees, although not all burned trees form a scar. Composite fire-scar records from multiple trees are used to calculate site-level fire frequency metrics, such as mean, median, minimum and maximum intervals and time since last fire. Because fire frequency metrics are sensitive to number of samples and the area sampled (Falk et al. 2007), consistency among sites reduces potential bias (Amoroso et al. 2011). Alternately, many fire history studies report fire intervals for the subset of years in which fires scarred at least two trees or at least 10% or 25% of recorder trees (Swetnam and Baisan 1996). Trees are considered recorders after an initial scar exposes the cambium and

Table 8.1 Comparison of fire scars and forest demography (tree age, growth and death data) for reconstructing past fires and interpreting historical fire regimes

Criterion	Fire scars	Forest demography
Advantages	Crossdated samples precise to year (and sometimes season) of fire; most utility for reconstructing low- and mixed-severity fire regimes	Most utility for reconstructing mixed- and high-severity fire regimes; corroborating evidence for low-severity fires or confirming fire as the agent of disturbance
Disadvantages	Time period limited by oldest wood; older scars lost in subsequent fires; limited to forest types where trees form scars; not all fires scar trees; research and sampling designs can introduce bias; mean intervals vary with sample size and area sampled	Reconstructions are strongest for the period following the last mixed- or high-severity fire; older evidence lost in subsequent fires; fire dates not precise due to post-fire regeneration lags; fine-scale fires difficult to reconstruct
Amount of uncertainty	None at the point of a scarred tree; uncertainty about fire extent between scarred trees	Causal agent and fire dates uncertain unless accompanied by fire scar
Direct/indirect evidence	Direct at point of a scarred tree; indirect when inferring spread between scarred trees	Direct for year of death of fire-killed trees; indirect when inferring fire from even-aged cohort or growth releases
Assumptions	Multiple fire-scarred trees in a study site in the same year are the result of a spreading surface fire, not individual fires at each tree	Even-aged cohorts, tree releases or deaths caused by fire and not another disturbance agent
Length of reconstruction	Decades to centuries depending on species and forest types; limited by loss in subsequent fires or decay	Decades to centuries depending on species and forest types; limited by loss in subsequent fires or decay
Spatial resolution	Individual tree to stand or landscape depending on research design	Stand to landscape depending on research design; small fires difficult to detect
Temporal resolution	Precise to year of fire, sometimes to season	Precise for year of death of fire-killed trees; not precise for tree ages or post-fire cohorts unless corroborated by a fire scar
Reconstruction of fire regime attributes	Direct evidence of fire type, severity, seasonality and frequency; indirect evidence of fire size	Direct or indirect evidence of fire type and severity; indirect evidence of frequency and size; no evidence of seasonality

Adapted from Yocom Kent 2014

increases susceptibility to subsequent fires (Romme 1980). The resulting "filtered" fire interval statistics are relatively immune to changes in sample size or area in some ecosystems (Van Horne and Fulé 2006; Farris et al. 2013), allowing comparison among sites or studies with differing sampling designs (Kitzberger et al. 2007; Falk et al. 2011), but not necessarily for all ecosystems or fire regime types (Stretch et al. 2016).

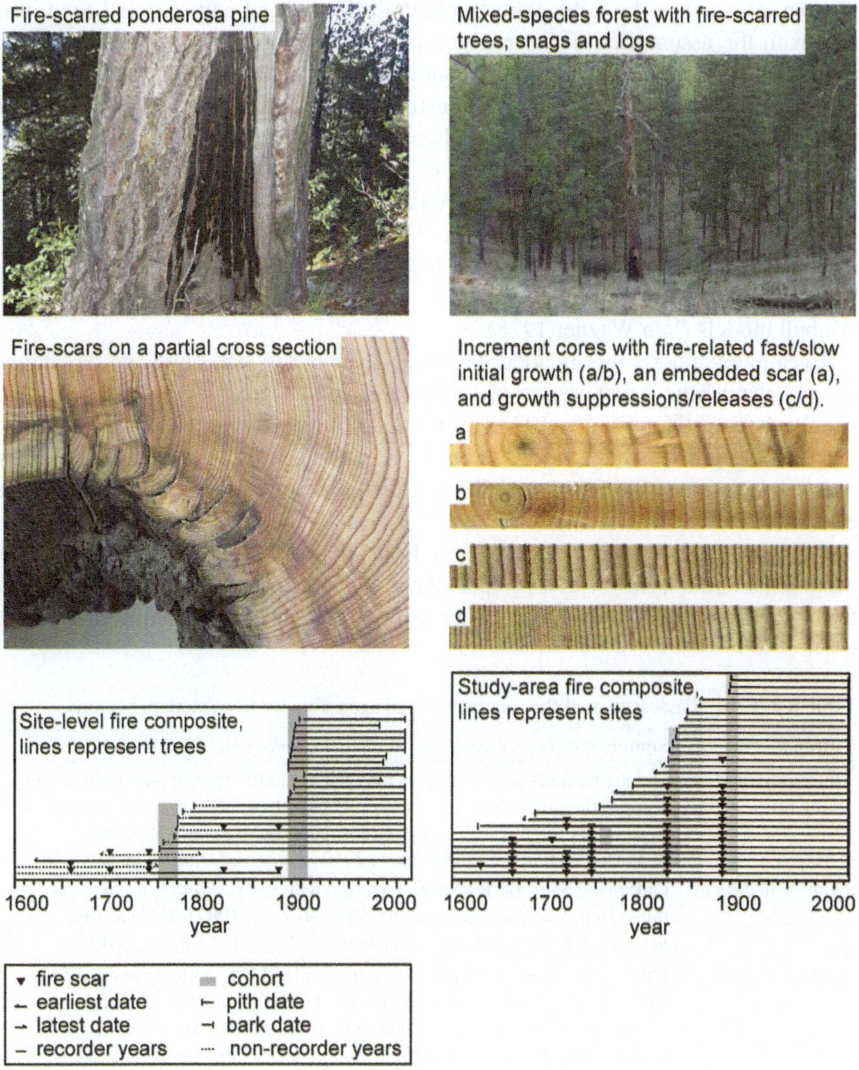

Fig. 8.1 Examples of fire scar, tree growth and forest demography evidence used to reconstruct fire history at tree, site and study area scales (Graphics by R.D. Chavardès)

8.2.2 Low-frequency, High-severity Fire Regimes

In contrast, low-frequency fire regimes are characterized by high-intensity crown fires burning at long intervals causing high levels of tree mortality (e.g., high-severity or stand-replacing fires; Schoennagel et al. 2004). These fires result in forest patches with relatively distinct boundaries that are distinguishable on air photographs and in the field. Forest polygons are mapped and sampled for forest age, with the assumption that even-aged patches of forest were initiated by high-severity fire. Age sampling is often combined with analysis of fire scars formed at polygon boundaries to refine these estimates to an annual level (Johnson and Gutsell 1994; Sibold et al. 2006). Landscape-scale fire history is represented by a stand-origin map (e.g., time since the last stand-replacing fire; Heinselman 1973) or a time-since-fire map (e.g., time since last fire of any severity; Johnson and Van Wagner 1985). Patch sizes and ages are summarized as cumulative time-since-fire (survival) distributions from which the fire cycle, the number of years required to burn an area equal to the study area, is calculated by fitting negative exponential or Weibull models (Van Wagner 1978).

Even-aged cohorts that include the oldest trees in the stand indicate past high-severity disturbances that created openings large enough for many trees to establish simultaneously (Heinselman 1973). Cohorts are indirect evidence of fire because the ages assigned to trees in a cohort are less precise than crossdated fire scars due to lags in tree establishment following disturbance, errors inherent to tree age estimates, and the facilitation of tree establishment by disturbances other than fire or during periods of suitable climate (Yocom Kent 2014; Table 8.1). The importance of these considerations also varies by species composition and environment. Therefore, tree age estimates need to be as accurate as possible (Table 8.2), sample sizes must

Table 8.2 Critical assessment of the accuracy of tree age estimates from increment cores

Attribute	Source of error and *dendroecological method to increase accuracy*
Outer-ring date of living trees	Outer ring dates of living trees may not equal the year of sampling due to competition, disturbance or environmental stress. *Crossdating cores from living trees verifies the calendar year of the outer ring (Cherubini et al.* 2002; *Amoroso and Daniels* 2010; *Jones and Daniels* 2012*)*
Outer-ring date of dead trees	Outer ring dates of dead trees are unknown. *Crossdating estimates the outer ring date and year of death (Daniels et al.* 1997*); but, inaccuracies may result from wood decay and erosion (Jones and Daniels* 2012*)*
Inner-ring date	False or missing rings cause inaccurate calendar years to be assigned to rings causing errors in estimated year of establishment and tree age (outer-ring date—inner-ring date +1). *Crossdating ensures an accurate date is assigned to the inner ring when determining tree age (Fritts and Swetnam* 1989*)*
Pith date	Cores may not intercept the pith, even after multiple cores are extracted from a tree. *The number of missing rings can be estimated from visual assessments (Applequist* 1958*) or geometric measurements of curved inner rings (Duncan* 1989*) or from the length of the core relative to the average tree radius inside the bark at coring height (Norton et al.* 1987*)*
Establishment date	Cores extracted above the root-shoot interface overestimate the establishment date and underestimate age. *Species-specific regressions of age-on-height correct for the number of years to coring height (Villalba and Veblen* 1997; *Wong and Lertzman* 2001*). Height growth rates may vary with resource availability at the time of establishment, requiring different corrections for trees with fast (wide rings) and slow (narrow rings) initial growth*

be sufficient to detect cohorts, and corroborating evidence should support inferences of past high-severity fires. Optimal sample sizes vary among forest types and age. For example, sampling ≥10 of the largest trees in even-aged post-fire lodgepole pine forests is likely to detect the oldest tree to estimate the fire year, but samples sizes of ≥15 large trees are needed in uneven-aged mixed-conifer forests (Kipfmueller and Baker 1998). Multiple lines of evidence differentiate post-fire cohorts from those initiated by insects, pathogens, wind or periods of suitable climate (Ehle and Baker 2003; Brown and Wu 2005; Sherriff and Veblen 2006). Dendroecological evidence corroborating high-severity fire includes (Sibold et al. 2006; Margolis et al. 2007, 2011): (a) a cohort comprised of fire- and shade-intolerant, fast-growing trees indicating establishment in open conditions, (b) fire scars on the boundary of even-aged stands concurrent with cohort initiation, (c) year of death of fire-killed trees (e.g. small or thin-barked trees, trees with charred bark) immediately preceding the cohort, although such evidence decays rapidly in many mesic forests, and (d) few or no remnant trees older than the cohort.

8.2.3 Mixed-Severity Fire Regimes

It is now widely recognized that classifying fire regimes as either high- or low-frequency oversimplifies the inherent variability in fires across a diversity of forests (Perry et al. 2011). The term "mixed-severity fire regime" describes the complex patterns and effects of heterogeneous fires across a range of spatial and temporal scales (Agee 1998; Lertzman et al. 1998; Schoennagel et al. 2004). Variable intensity within individual fires results in patches with low, moderate and high levels of tree mortality (Agee 1993). Within a fire, thin-barked species, small regenerating trees, and even large trees with thick bark are killed if exposed to torching, crowning or surface fire for sufficient duration to damage the cambium. Post-fire cohorts may colonize the resulting gaps, so forests are uneven-aged with multiple cohorts and remnant trees that survive fire, some of which form single to multiple fire scars. Successive fires at one location may burn with a range of severities resulting in compositionally and structurally diverse stands with complex dynamics (e.g., Heyerdahl et al. 2012; Marcoux et al. 2015). At the landscape scale, fires burn with a range of severities simultaneously and through time yielding heterogeneous patch sizes, shapes and spatial patterns (Hessburg et al. 2007; Halofsky et al. 2011; Perry et al. 2011).

Given the complexities of mixed-severity fire regimes, they are particularly challenging to characterize and reconstruct. Unlike high- and low-severity regimes, there is no *a priori* assumption about the impacts of individual fires. Instead, physical evidence is required to infer the severity, as well as the timing, of each fire. Scaling up from individual fires, the severity of successive fires at a site indicates fire history through time and comparing multiple sites across a landscape characterizes the historical fire regime. To simultaneously characterize the timing and severity of individual fires and spatio-temporal variations among sites requires multiple types

Table 8.3 Example criteria for classifying fire severity across spatio-temporal scales

Spatial scale	Temporal scale (source)	Criteria for severity classes		
		Low	Moderate	High
Individual fires	Contemporary (Agee 1993)	≤20% basal area of live trees killed	21–70% basal area of live trees killed	>70% basal area of live trees killed
	Reconstructed (Sherriff and Veblen 2006)	≥80% of living trees are remnants and <20% established post-fire; fire scar(s) present	<80% of living trees are remnants and 21–80% established post-fire	≤20% of living trees are remnants and >80% established post-fire
		Low	Mixed	High
Individual sites	Reconstructed through time (Sherriff and Veblen 2006)	MFI ≤ 30 years; trees with multiple fire scars; low-severity fires only	MFI > 30 years; low-, moderate- or high-severity fires through time	MFI >30 years; moderate- or high- severity fires through time
	Reconstructed through time (Heyerdahl et al. 2012)	≥1 fire scar year(s); no cohorts	≥1 fire scar year(s) and ≥1 cohort OR post-fire cohort with remnant trees	no fire scars; 1 post-fire cohort; no remnants
Study area	Reconstructed through time at individual sites (Marcoux et al. 2015)	≥81% low-severity and <20% mixed- or high-severity	<80% low severity and 21–80% mixed- or high-severity	≤20% low-severity and ≥81% mixed-or high-severity

of dendroecological evidence. The methods typically used to reconstruct high- or low-severity fires are best used in combination. In concert with fire scars, living and dead trees are sampled to represent tree population age structures and to identify cohorts, remnant survivors of past fires, and fire-killed trees (Ehle and Baker 2003; Baker et al. 2007). Independent, corroborating evidence is also sought (Swetnam et al. 1999; Baker et al. 2007).

Classifying the severity of historical fires has proven to be difficult in all but the most homogeneous systems. The severity of contemporary fires is often measured as the mortality rates of overstory trees (e.g. Agee 1993; Table 8.3). When reconstructing fire history, pre-fire forest densities are unknown (Schoennagel et al. 2011); therefore, severity is inferred from fire scars, forest demography and other corroborating evidence. Sherriff and Veblen (2006) introduced a method to classify the severity of individual fires as low, moderate or high based on the proportion of remnant trees (e.g. trees that established before fire) and those establishing in the 40 years after fire (Table 8.3; Case Study 1). Of these two metrics, remnant trees provide a more direct measure of fire severity than post-fire establishment, as tree colonization rates vary (e.g. <20 to >40 years; Amoroso et al. 2011, Chavardès and Daniels 2016) due to the availability of local seed sources and dispersion rates, availability of suitable microsites and variation in climate facilitating germination and survival (Sherriff and Veblen 2006). Both pre-fire remnant trees and post-fire tree establishment have uncertainties for evaluating fire severity and are not a perfect complement, but when used in conjunction with fire-scar dates, they represent a robust way to reconstruct historical fire severity in montane forests. Other corroborating evidence includes the presence and spatial distribution of fire scars, tree mortality synchronous with a fire date, and presence of resin ducts, micro-rings and radial-growth suppressions or releases in remnant trees (Brown and Swetnam 1994; Sherriff and Veblen 2006; Smith et al. 2016). The classification of historical fire severity works best for recent fires (Yocom Kent et al. 2015; Chavardès and Daniels 2016). Estimating the effect of fires that burned prior to a moderate- or high-severity fire is less reliable because of loss of evidence during subsequent fires, mortality and decay of older trees with time (Sherriff and Veblen 2006). To overcome this limitation, Tepley and Veblen (2015) recently introduced a metric to quantify variation in fire severity through time. In their models, tree density through time is a function of the cumulative number of fires that have burned at a site. The severity of an individual fire is a relative measure of the change in tree density before and after the nth fire at the site compared to landscape-level mean tree densities associated with the nth fire at that site.

There is no standard definition for low-, mixed- or high-severity fire regimes and different authors have classified them using different criteria. For example, at the site level, Sherriff and Veblen (2006) classified the predominant fire regime based on fire frequency, the severity class of individual fires and their cumulative effects over time (Table 8.3; Case Study 1). Similarly, Heyerdahl et al. (2012) classified site-level fire severity through time, but they did not attempt to classify individual fires within the sites (Table 8.3; Case Studies 2 and 3). In their approach, all site-level evidence was assessed simultaneously. Fire scars and remnant trees indicated past low- and moderate-severity fires. Post-fire cohorts indicated past moderate- and high-severity fires. Cohorts were identified based on the proportion of sampled trees within a defined establishment period (e.g., ≥25% of sampled trees established in a 30-year period) and were considered a post-fire cohort if trees were absent from the age classes immediately preceding a cohort (e.g., a moderate- or high-severity fire had killed young, small trees). Combinations of scars, remnant trees and cohorts were used to classify sites as having a low-, mixed- or high-severity fire history through time.

The relative abundance and distribution of forests with low-, mixed- or high-severity fire histories has emerged as a critical factor in determining if fire regimes have been altered during the twentieth century. Networks of research sites are used to reconstruct and classify fire regimes at landscape to regional scales (Falk et al. 2011; Table 8.3; Case Studies 1 and 2). Stratifying study areas according to biophysical attributes, then comparing fire regime attributes across strata is one way to quantify spatial variability and test for drivers of that variation (Heyerdahl et al.

2001; Taylor and Skinner 2003; Sherriff and Veblen 2007). Representative sampling allows inferences to be extrapolated from the site-level data to the study area (Sherriff et al. 2014; Marcoux et al. 2013, 2015). Current classification systems for fire regimes lack quantitative criteria to differentiate low-, mixed- and high-severity regimes (Brown et al. 2008; Perry et al. 2011); however, universal criteria may not be ecologically meaningful or relevant for management on all landscapes. Since fire regimes vary along a continuum rather than forming discrete classes, research on the relative importance of fires of different severities may more effectively guide spatial and temporal strategies to achieve management and conservation goals.

Most research on historical low- and mixed-severity fire regimes has focused on magnitude and temporal attributes, rather than spatial attributes. Combining dendroecological evidence with spatial analysis has yielded new ways to reconstruct severity within a fire and quantify the spatial extent of past fires. Research designs that include sample plots distributed on a systematic grid are conducive to such spatial analyses (Heyerdahl et al. 2001). Simulation modeling can also be used to infer spatial attributes of fire (e.g., Brown et al. 2008, Case Study 3). Such models can be corroborated by dendroecological evidence, lending confidence to their use in extrapolating fire behavior across landscapes and into the future to infer changes in fire regimes through time and to test the relative importance of fuels and weather as drivers of fire severity (Heyerdahl et al. 2014). Another quantitative approach for estimating spatial fire regime attributes is the spatial mean fire interval method (Hessl et al. 2007; Kernan and Hessl 2010). This method uses the fire evidence (e.g., fire-scar dates or tree ages representing stand-replacing fires; Yocom Kent et al. 2015, Greene and Daniels 2017; Case Study 4) at individual sample plots in a geographic information system (GIS) and inverse distance weighting to interpolate between plots and estimate the boundaries and size of individual fires (Hessl et al. 2007; Kernan and Hessl 2010; Swetnam et al. 2011). Similarities between reconstructions of fire extent and corresponding documentary fire records demonstrates the efficacy of this approach (Farris et al. 2010). The spatial extent of multiple fires through time is composited in a spatial mean fire interval map, to estimate point-specific fire frequencies and landscape-level fire rotation (Kernan and Hessl 2010). These maps enable qualitative and quantitative analysis of relationships between fire and topography (Kernan and Hessl 2010; O'Connor et al. 2014).

8.3 Mixed-Severity Fire Regimes Reconstructed Using Dendroecology

Below, we present four case studies from across the western United States and Canada to illustrate how dendroecological evidence is used to decipher the complexity of historical fire regimes (Fig. 8.2). We focus on mixed-severity fire in these case studies because it requires the use of various methods and multiple lines of evidence to reconstruct fire regimes. The first two case studies show how detailed reconstructions of past fire frequency and severity at representative sites can be used

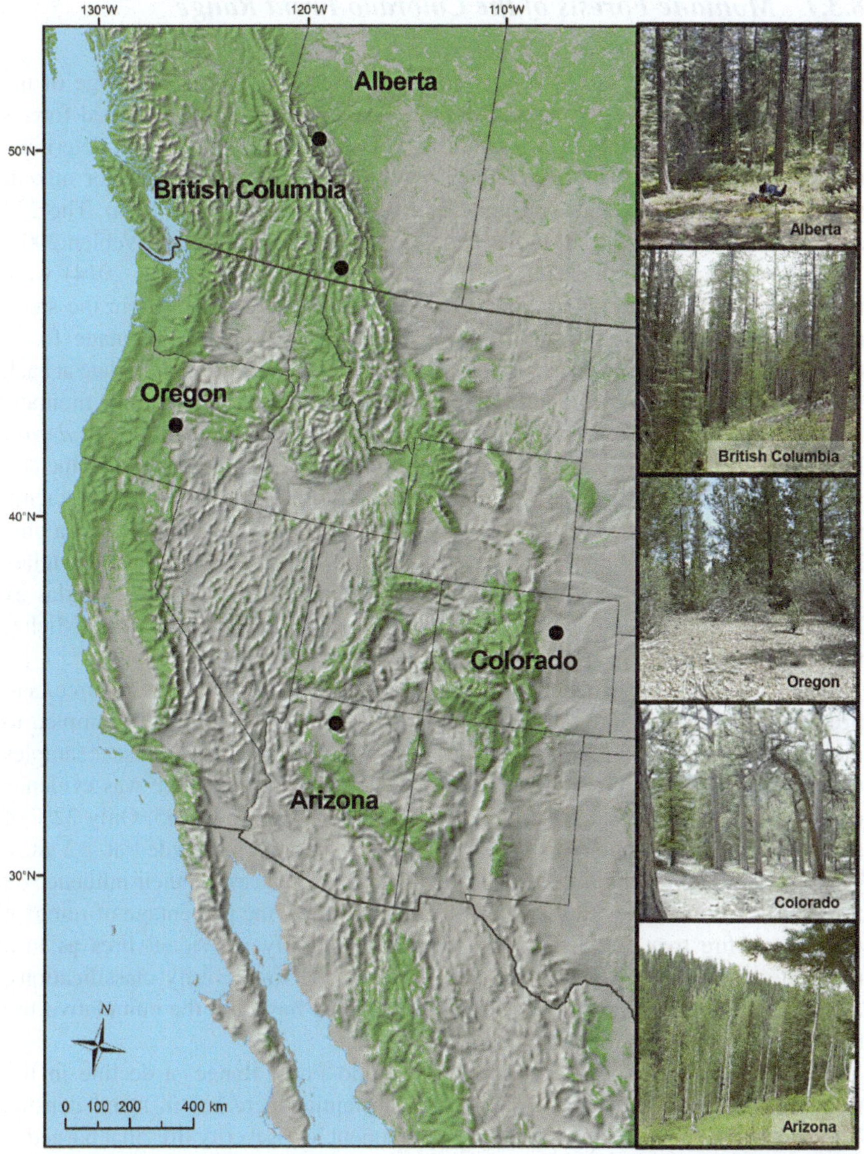

Fig. 8.2 Four case studies on mixed-severity fire regimes represent research in the Canadian Cordillera, central Oregon, Colorado Front Range and Grand Canyon in Arizona (Graphics by D. Snow and R.D. Chavardès)

to illustrate landscape-scale variation in fire regimes. The latter two case studies demonstrate a suite of methods to quantify spatio-temporal variation in fire severity within a landscape.

8.3.1 Montane Forests of the Colorado Front Range

In ponderosa pine and mixed-conifer forests of the Colorado Front Range of the Rocky Mountains, fire-scar records and tree age structures from unlogged forests were used to evaluate the historical fire regime (e.g., before 1920) and determine the parts of the landscape characterized by predominantly low-severity or mixed-severity (including moderate- or high-severity) fires (Figs. 8.2 and 8.3a). The 232 sites compiled from multiple studies (Veblen et al. 2000; Sherriff and Veblen 2006, 2007; Schoennagel et al. 2011; Gartner et al. 2012; Sherriff et al. 2014) were representative of the relative proportion of the dominant forest types in the study area, allowing inferences from site-level data to 564,413 ha of montane forest (Sherriff et al. 2014; an example of fire-scar dates and the last fire-cohort date at each site are compiled for 92 sites in Fig. 8.3a). In the study region, the lower montane zone (*c.* 1800–2200 m) comprises primarily pure ponderosa pine (*Pinus ponderosa* Lawson & C. Lawson) on south-facing slopes and a mixture of ponderosa pine and Douglas-fir (*Pseudotsuga menziesii var. glauca* (Beissn.) Franco) on north-facing slopes. The upper montane zone (*c.* 2200–3000 m) is comprised of ponderosa pine stands on south-facing slopes and denser stands of ponderosa pine and Douglas-fir on north-facing slopes along with lodgepole pine (*Pinus contorta* Douglas ex Loudon var. *latifolia* Engelm. ex S. Watson), aspen (*Populus tremuloides* Michx) and dispersed limber pine (*Pinus flexilis* James) trees at higher elevations.

At each site, fire scars (number of fire-scarred trees and fire scars per tree, and fire dates), forest structure, and tree age structure were systematically sampled to identify the fire-severity regime. In total, 7680 tree cores and 1262 fire-scar samples were used to delineate fire dates and severity across all sites. There was evidence of 322 spreading fires that scarred ≥2 trees between 1597 and 1995. Only 22% of spreading fires were unique to one site; all other fires were recorded at ≥2 sites. The severity of spreading fires at each site was classified based on their influence on forest structure. Along with the dates of spreading fires, the percentage of remnant *versus* post-fire tree establishment was used to classify individual fires as low, moderate or high severity (Table 8.3). Based on the fire-severity classifications, a predominant fire regime at each site was assigned based on the cumulative fire effects over time (Table 8.3).

In the lower montane forests in the Colorado Front Range, a decline in fire frequency over the past 100 years led to substantial increases in stand density. These forests were characterized mainly by frequent low-severity fires that burned at average intervals <30 years and maintained open forests by killing mostly juvenile trees, resulting in low densities of mature trees. The cessation of these fires coincides with increased stand densities. This pattern also occurred at some higher elevations on less steep sites where montane grasslands most likely occurred. Overall this change to the fire regime and forest structure represents a relatively small proportion of the montane forest of the Colorado Front Range as only 27.8% of the study area is mapped with the historical low-severity fire regime (Sherriff et al. 2014).

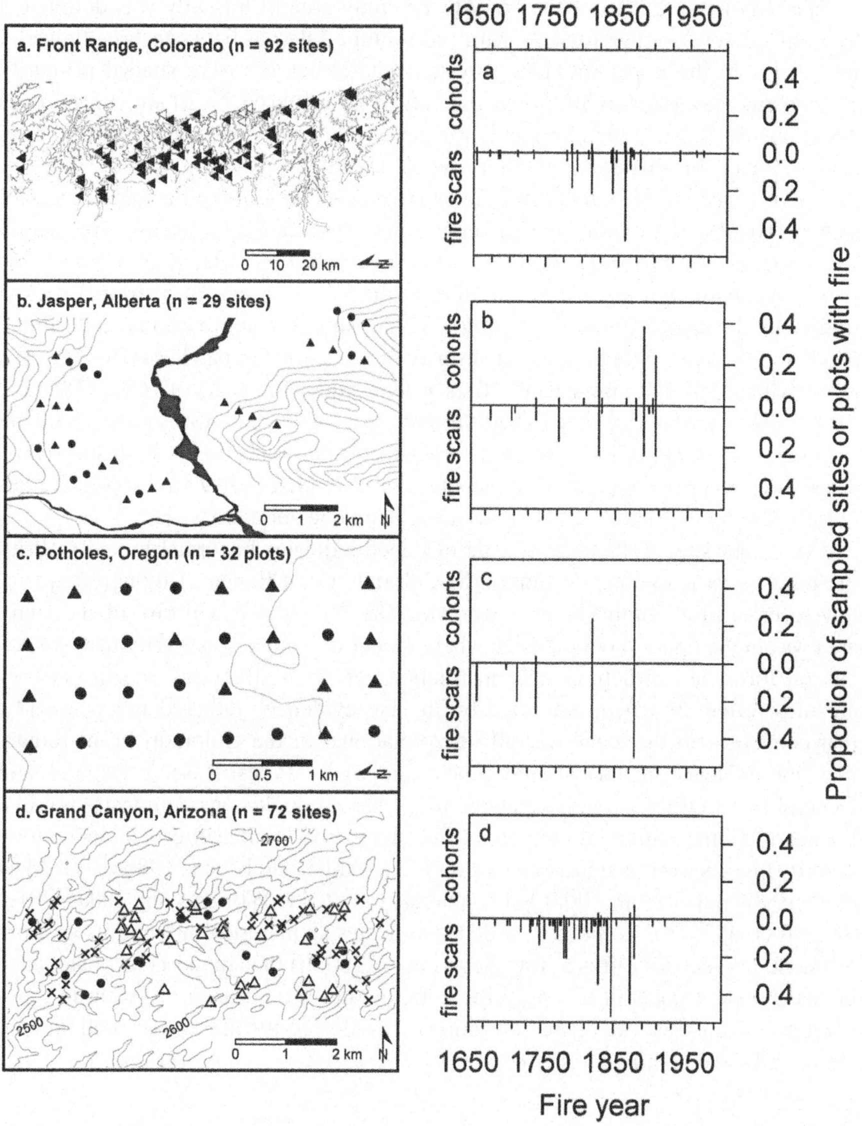

Fig. 8.3 Research designs and fire records for the four case studies on mixed-severity fire regimes in (**a**) Colorado (564,413 ha), (**b**) Alberta (2400 ha), (**c**) Oregon (783 ha), and (**d**) Arizona (1400 ha). In the maps, *solid circles* are sites/plots with cohorts only, *solid triangles* have cohorts and fire scars, *open triangles* have fire scars only, and x indicates no scars or cohorts

The majority (72.2%) of the 564,413 ha study area historically was dominated by a mixed-severity fire regime, with the average interval between fires typically >30 years. In the upper montane forests, stand structures were shaped primarily by moderate-severity (46.5% of sites) and high-severity (45.7% of sites) fires; only 7.8% of sites recorded predominantly low-severity fires. Evidence of mixed-severity fires occurred in all dominant forest types, including stands in which ≥80% of the canopy trees were ponderosa pine or lodgepole pine, as well as mixed-conifer stands. Fire effects varied from non-stand-replacing to canopy-replacing fires within sites and across broad landscapes, with severity often related to topographic variability. Although the ability to interpret the spatial extent of historical fires is limited by the sampling unit size (up to 232 ha), subsequent fire events, and sample depth, the evidence indicates mixed-severity fires occurred in patches up to 200 ha or more, with *c.* 50 ha of high-severity fires at some sites. Fire years of 1654, 1786, and 1859–1860 were particularly extensive with 36%, 43%, and 48% of the available recorder sites (with ≥2 fire-scarred trees) recording each fire year, respectively. These fires were recorded at multiple sites that extended over 9 km (1654), 7.5 km (1786), and up to 30 km (1859–1860) away from one another.

Consistent site- to landscape-scale evidence indicates historical mixed-severity fire regimes varied with elevation in the Colorado Front Range. Although frequent, low-severity fires dominated in lower montane forests, the majority of the study area was in the upper montane zone where moderate- and high-severity fires shaped current forest age structures. Reconstructions of stand structures and fire history are most effective when supported by diverse evidence, gathered independently, that converges to the same overall interpretations. In the Colorado Front Range, tree-ring evidence, historical landscape photographs, Forest Reserve Reports, and General Land Office surveys converge to the same conclusions demonstrating that the historical fire regime of ponderosa pine and mixed-conifer forests included low-severity fires as well as high-severity fires (e.g., Veblen and Lorenz 1986, 1991; Mast et al. 1998; Baker et al. 2007; Schoennagel et al. 2011; Williams and Baker 2012; Sherriff et al. 2014). Combined evidence indicates an historical fire regime of more frequent, low-severity fires at low elevation (<2260 m) that supports a convergence of management goals of ecological restoration and fire hazard mitigation in those habitats. In contrast, at higher elevations mixed-severity fires were predominant historically and continue to be so today.

8.3.2 Montane Forests of the Canadian Cordillera

Dendroecological evidence shows that fires of a range of severities historically burned in the Canadian Cordillera, despite the northerly latitude and continental climate (Fig. 8.2). Landscape studies conducted west of the Rockies in southeastern British Columbia (Marcoux et al. 2013, 2015; Greene and Daniels 2017) and east of the Rockies in Jasper National Park, Alberta (Chavardès and Daniels 2016) form a 500 km south-north gradient. Collectively, 102 sites in mesic mixed-conifer montane and lower subalpine forests were selected using a stratified-random research design to represent the environmental gradients in each landscape. Fire-scarred trees sampled in 1-ha plots provided evidence of low-to-moderate-severity fires. Forest composition, size- and age-structure were sampled at the center of fire-scar plots using an n-tree design. The resulting evidence indicated historical fires (1439–1966) burned at a wide range of frequencies and severities at site and landscape scales. Even-aged cohorts determined from stand age-structure analyses and an absence of fire scars or remnant trees provided evidence of high-severity fires at 43 of 102 sites. Crossdated, annually-resolved fire records were used to quantify return intervals of low-to-moderate severity fires at the other 59 sites. In southeastern British Columbia (n = 73 sites), fire-scars were most common on thick-barked western larch (*Larix occidentalis* Nutt.), Douglas-fir and ponderosa pine; in Jasper (n = 29) fire scars were most common on lodgepole pine and Douglas-fir.

Crossdated fire scars combined with age structure analyses allowed detailed site-level reconstructions of fire history and forest dynamics through time (Table 8.3). In southeast British Columbia, 37 fire years between 1600 and 2009 were identified based on fire scars (n = 31) or cohorts only (n = 6; Marcoux et al. 2013, 2015). Eleven of 20 sites included fire scars and cohorts indicating mixed-severity fire histories. These sites last burned between 1910 and 1953, yielding times since fire of 56–99 years (at time of sampling) which exceeded the maximum recorded fire return intervals at five sites and the mean interval at all sites. The remaining sites had cohorts only, indicating high-severity fires burned <150 years ago at four sites or >270 years ago at five sites. In this landscape, tree species composition varied with disturbance history. Mixed-severity sites were dominated by Douglas-fir and western larch that regenerated after frequent low- or moderate-severity fires indicated by scars. Periodic moderate-severity fires generated even-aged cohorts with surviving remnant trees. At higher-elevations, severe fires <150 years ago generated cohorts dominated by lodgepole pine and high-severity sites that last burned >270 years ago were dominated by subalpine fir (*Abies lasiocarpa* (Hook.) Nutt.). Tree size attributes did not distinguish mixed- from high-severity sites. Across fire history classes, canopy tree densities were 70–500 ha^{-1}. Subcanopy trees were up to 5600 ha^{-1} having established then persisted following the last fire at most sites. In this landscape, selective harvesting and fire suppression during the twentieth century have reduced fire frequency in montane forests and homogenized contemporary forest structures across elevations.

In Jasper, 13 fire years between 1646 and 1905 were identified based on fire scars (Chavardès and Daniels 2016; Fig. 8.3b). All 29 sites had at least one post-fire cohort (Fig. 8.3b). At 18 of 29 sites fire-scars and remnant trees combined with post-fire cohorts were legacies of past fires of mixed severity through time. Single even-aged cohorts provided evidence of high-severity fires at 11 sites. The most recent fires initiated post-fire cohorts at 24 sites, providing evidence of moderate-to-high severity fires between 1889 and 1905. No fire scars were detected after 1905, although young thin-barked trees could have recorded fire at all sites. The simultaneous, long fire-free intervals during the twentieth century are unprecedented in the 250-year fire-scar record. The dendroecological evidence showed that lodgepole pine, hybrid spruce and Douglas-fir establish simultaneously after low-, moderate- and high-severity fires. Thus, forest canopies were mixed in composition regardless of fire history and subtle differences such as the presence of remnant trees and discontinuous age structures distinguished sites with mixed-severity fire histories. In general, subcanopies were strongly dominated by shade-tolerant spruce, which are similar in age to the canopy trees despite their small size. Subcanopy spruce do not represent recent recruitment and stand infilling; rather, species-specific growth rates and adaptations to shade resulted in size stratification among species and canopy layers. These subcanopy trees have survived in the absence of surface fires during the twentieth century so that forests with mixed- and high-severity fire histories have developed similarly, homogenizing the landscape. Current dense subcanopies (up to 3100 ha^{-1}) provide ladder fuels, increase fire hazard and decrease forest resistance to high-severity fire, especially during droughts.

Dendroecological evidence indicates mixed-severity fire regimes dominate in many forests of the Canadian Cordillera. This interpretation contrasts with the traditional view that high-severity fire regimes dominated, a perception that strongly influences forest and fire management. Instead, low-to-moderate severity fires were common historically but have been effectively eliminated during the twentieth century. Altering these fire regimes can homogenize forests, reducing stand- and landscape-level diversity and resilience to environmental change.

8.3.3 Lodgepole Pine Forests of Central Oregon

Lodgepole pine forests are widely distributed in western North America and historically sustained a range of fire regimes (Schoennagel et al. 2008; Amoroso et al. 2011). While these fire regimes have been well documented at the high-severity end of this range (e.g., in portions of the Greater Yellowstone Area and parts of the U.S. northern Rocky Mountains), they are not as well documented in mixed-severity systems, such as central Oregon's Pumice Plateau Ecoregion (Heyerdahl et al. 2014; Fig. 8.2). This region covers over a million hectares and is characterized by thick deposits of pumice and ash that combine with generally flat topography to favor lodgepole pine, but restrict the establishment of other overstory and understory species (Geist and Cochran 1991; Simpson 2007). On parts of the Plateau, coarse-textured pumice substrates limit forest composition to low-density lodgepole pine with scattered ponderosa pine and a shrub understory dominated by antelope bitterbrush (*Purshia tridentata* (Pursh) DC.; hereafter bitterbrush). This woody shrub is intolerant of shade and highly flammable. It acts as a ladder fuel and facilitates passive crown fire (e.g., torching of individual trees or small patches of trees; Busse and Riegel 2009). It is the primary understory fuel in these forests because the coarse-textured, nutrient-poor pumice substrate limits the growth of grass and herbaceous fuels (Geist and Cochran 1991). Although fire initially reduces the abundance and biomass of bitterbrush, it also stimulates regeneration and populations can recover to pre-fire levels, especially where fire creates canopy gaps (Ruha et al. 1996; Busse and Riegel 2009).

The historical fire regime and its spatial complexity were reconstructed at 30 2-ha plots distributed over 783 ha of forest dominated by lodgepole pine with scattered ponderosa pine at an elevation of 1485 m (Heyerdahl et al. 2014; Fig. 8.3c). The occurrence of low-severity fires was reconstructed from 56 fire-scarred trees (37 ponderosa and 19 lodgepole pine). The occurrence of high- and moderate-severity fires was reconstructed from the establishment dates of 752 live or dead trees. From these establishment dates, a cohort was assumed to have initiated when ≥5 trees recruited within 20 years at a plot, preceded by at least 30 years without recruitment. Death cohorts were identified when ≥5 trees died in the same year.

During the analysis period (1650–1900), 6 fires were reconstructed from 129 fire scars and 19 of the cohort initiation dates (Fig. 8.3c). All but two of the cohorts satisfied the criteria for assignment to fire-scar dates in 1750, 1819, or 1877. From these 6 fires, 34 plot-composite fire intervals of variable length (26–82 years) were computed. Fire scars from five of these same six fires (all but 1700) were crossdated on trees sampled between plots. Twenty-one cohort initiation dates were identified. Although trees established between 1624 and 1962, all but one cohort initiated between 1822 and 1888. The remaining cohort initiated in 1752. Cohorts occurred in two-thirds of the plots, most of which had a single cohort except for one plot with two. Two death cohorts were identified in 1988, during a widespread outbreak of mountain pine beetles in central Oregon (Preisler et al. 2012).

For centuries (1650–1900), extensive mixed-severity fires occurred every 26–82 years, creating a multi-aged forest and shrub mosaic. Although widespread synchrony in fire-scar dates during several years suggests extensive low-severity fires, these scars were also synchronous with cohorts of tree recruitment, suggesting that individual fires included patches of both high- and low-severity fire. The inference of historical mixed-severity fire is also consistent with the general lack of serotinous lodgepole cones in central Oregon (Mowat 1960) because extensive high-severity fires select for cone serotiny in pines (Keeley and Zedler 1998). Fire intervals of 26–82 years are long enough for bitterbrush to regain sufficient cover and height to facilitate fire spread across the site and into the canopy in a mosaic pattern. In turn, this mosaic pattern would have allowed for post-fire regeneration of bitterbrush by creating canopy gaps while maintaining some unburned plants as seed sources and stimulating vigorous sprouting from undamaged portions of surviving plants (Ruha et al. 1996; Busse and Riegel 2009). This work supports findings about the drivers of mixed-severity fire regimes elsewhere in the region (Halofsky et al. 2011).

The effect of fire exclusion on the fire regime at this site is unusual among mixed-conifer forests in the interior Pacific Northwest (Hagmann et al. 2013). While forest composition is topoedaphically limited primarily to lodgepole, contemporary, low shrub fuel loads at the site are likely insufficient to spread fire to the canopy. In contrast, the tree-ring reconstructed fire history indicates that patches of high-severity fire occurred periodically, generating multi-aged stands that may have been more resilient to beetle attacks. Because topographic relief at the site is low, spreading fires were likely wind driven, and would have required sufficient surface fuel loads for horizontal and vertical spread. However, fuel loads, in particular the abundance and cover of bitterbrush, the primary understory species, has likely decreased since the exclusion of fire 130 years ago, reducing the ability of the site to support a mixed-severity fire regime. Because bitterbrush is both sensitive to and stimulated by fire, continued lack of fire within the ecosystem is likely to promote a negative feedback cycle. Without fire, canopy gaps are not created and bitterbrush sprouting is not stimulated, thereby restricting shrub growth. In turn, limited shrub abundance and cover restricts horizontal and vertical spread of fire, thus eliminating some opportunities for creation of canopy gaps and perpetuating conditions that limit fire.

It is challenging to extrapolate future fire regimes from tree-ring reconstructions of past fire alone because more than a century of fire exclusion has likely changed forest structure, hence fire behavior, and climate is projected to continue changing in the future. However, if simulation models corroborate past fire behavior reconstructed from tree rings, such models can be used to simulate potential future fire behavior. At Potholes, past, present, and future fire behavior was simulated with a landscape-scale program (FlamMap; Finney 2006) to infer whether modern fire behavior has changed since fire exclusion and to assess the relative importance of fuels and weather as drivers of fire severity. Simulations of past fire behavior were consistent with tree-ring reconstructions of patchy mixed-severity fire and suggested that fuel loadings and wind speed were the primary drivers of fire behavior at Potholes. In contrast, simulations of modern fire behavior suggested that a century of fire exclusion has reduced the potential for the high-severity patches of fire that were common historically. This occurred because the loadings of modern understory fuels (i.e., bitterbrush cover, the primary ladder fuel) have declined and are now insufficient to spread the mix of surface and torching fire that occurred at Potholes in the past. Simulations with historical fuels included torching, especially under extreme wind speeds, but simulations with modern fuels were dominated by surface fire except under extreme wind speeds. Active crown fire was very rare regardless of the scenario. In the absence of the abundant shrub fuels of the past, flame lengths of sufficient height to carry fire into the canopy occurred only with extreme winds. Coupling simulation modeling with tree-ring reconstructions of fire at Potholes provided a more complete picture of the consequences of changes in the forest due to fire exclusion.

8.3.4 The North Rim of Grand Canyon National Park, Arizona

Grand Canyon National Park in northern Arizona has a progressive and active fire management program, making frequent use of prescribed fire and managed wildfire. Park managers have a goal of restoring fire as a natural process throughout the park, including the mixed-conifer zone. The fire scar record provides strong evidence of historical low-severity fires (Fulé et al. 2003; Fig. 8.2). In this study, three complementary analyses using dendroecological evidence were applied to estimate the size of historical high-severity patches in the mixed-conifer forests (Yocom Kent et al. 2015).

The 1400-ha study area, at *c.* 2600 m in elevation, is located on the Kaibab Plateau in the North Rim area of the Park. The forest in the study area is highly stratified by aspect. The vegetation was mapped and polygons were classified the into four major forest types by dominant overstory tree species: white fir (*Abies concolor* (Gord. & Glend.) Lindl. ex Hildebr.), blue spruce (*Picea pungens* Engelm.), subalpine fir and aspen. Other common tree species are ponderosa pine, Douglas-fir and Engelmann spruce (*Picea engelmannii* Parry ex Engelm.). Using a stratified-random research design, plots were randomly placed in 72 mapped vegetation polygons, 18 in each forest type (Fig. 8.3d). At each plot, the *n*-tree method (Jonsson et al. 1992) was used to identify and core the ten large overstory trees closest to plot center. To quantify fire-related tree establishment, species-specific minimum diameters were used to target "large" trees that most likely established before 1879, the last widespread fire in the study area (Fulé et al. 2003). The cores from 647 trees were crossdated using a local chronology (Fulé et al. 2003) and inner-ring dates were determined. Fire dates were obtained from crossdated fire-scarred trees sampled in the area previously (Fulé et al. 2003; Fig. 8.3d).

High-severity fire patches were reconstructed using three different methods, each using dendroecological data. First, similar to the other case studies, even-aged cohorts establishing after high severity were identified based on the following conditions: (a) trees formed one even-aged cohort with an initiation date that corresponded to a nearby fire scar or a widespread fire date, (b) ≥2 cohorts were present, but the oldest cohort was even-aged with an initiation date that corresponded to a nearby fire scar or a widespread fire date, or (c) fire-sensitive species (aspen, white fir, subalpine fir, or spruce) were even-aged but there were some older remnant ponderosa pines or Douglas-firs. Plots met these criteria if ≥2 aspen pith dates followed fire by ≤5 years or if ≥4 conifer pith dates followed fire by ≤20 years. The size of each patch represented by a fire-initiated plot was determined using GIS.

The second method for reconstructing high-severity patches was a dichotomous classification based on the regeneration dynamics of aspen. Aspen regenerates quickly after high-severity fire by suckering from clonal root stock (Margolis et al. 2007). Thus, mapped patches dominated by aspen were classified as initiated by high-severity fire; all other patches were classified as not initiated by high-severity fire.

The third method used GIS and the inverse distance weighting method to interpolate high-severity fire patch boundaries and sizes (Hessl et al. 2007). The pith date of the oldest fire-sensitive tree (aspen, white fir, subalpine fir or spruce), refined by fire-scar dates, estimated the minimum time since the last high-severity fire. Fire boundaries and sizes for the seven high-severity fires (1785, 1813, 1829, 1841, 1847, 1873, and 1879) were estimated using 3 different thresholds of interpolated values to define high-severity *versus* low-severity or unburned forests.

Of these three methods, the aspen method yielded the smallest estimates of high-severity fire patches, while the interpolation method resulted in the largest estimates that varied depending on the threshold value for fire boundaries (Table 8.4). The aspen method and the fire-initiated polygon method may be more valuable

Table 8.4 Comparison of estimated metrics of high-severity fire patches reconstructed using three different methods (Yocom Kent et al. 2015)

Metrics of high-severity patches	Aspen method (N = 49)	Fire-initiated patch method (N = 18)	Interpolation method (N = 7) minimum threshold		
			0.5	0.4	0.3
Minimum size (ha)	0.1	0.5	1.9	3.9	9.0
Maximum size (ha)	10.8	35.6	200.2	309.1	555.9
Average size (ha)	1.1	4.2	30.2	53.3	102.4
Average area burned (ha) per fire year			129.3	182.8	263.2
Average number per fire year			4.3	3.4	2.6

for understanding estimates of minimum high-severity patch size, whereas the interpolation method may be more valuable for understanding estimates of the total area burned in high-severity fire in a given year.

Consensus among the three different methods indicates high-severity fire was a regular component of the historical fire regime along with low-severity fire, throughout the nineteenth century. In some fire years, such as 1847, high-severity fire was widespread. Patch size of high-severity fire during the 1800s likely ranged from small patches (0.1 ha) that allowed a few trees to establish to large patches (100 ha) that initiated multiple stands across the landscape. However, the forest in the study area was quite young, with the majority of trees having pith dates in the 1800s or later. A large stand-replacing fire during or prior to the mid-1700s cannot be ruled out.

The mixed-conifer forests on the North Rim of Grand Canyon National Park are highly stratified by aspect, and likely experienced a very complex historical fire regime with mixed low- and high-severity effects in individual fires and over time. Although fire was a regular and important driver of forest dynamics throughout the 1800s, the last widespread fire was in 1879. Without periodic fires, the forest has increased in density and changed in species composition over the past century (Fulé et al. 2003; Vankat 2011). Allowing fires to behave differently on south-facing and north-facing slopes could help promote diversity of tree species and keep the high level of heterogeneity present in these forests.

8.4 Discussion

In each case study, dendroecological analyses were central for understanding past fire frequency, severity and spatial attributes. Although they each feature a mixed-severity fire regime, variations among them demonstrate the importance of ecosystem-specific understanding of historical fires to guide science-based management that aims to maintain forest resilience.

The juxtaposition of traditional management paradigms in the montane forests of the Colorado Rockies and Canadian Cordillera exemplifies this point. In the Colorado Front Range, ponderosa pine is common. Although high-frequency, low-severity fire regimes are often associated with this species and dominate the lower montane forests, mixed- and high-severity fires historically dominated the upper montane forests. For the majority of this landscape, managing for a low-severity fire regime is not consistent with the historical range of variation; moderate- and high-severity fires must be taken into account. In contrast, timber and fire management largely reflect the presumed high-severity fire regime in the Canadian Cordillera. Reconstructions provide strong evidence of historical mixed-severity fire regimes but reduced surface fires during the twentieth century. Despite the high latitude, surface fires were an important component of historical fire regimes and need to be reflected in contemporary management.

Land-cover change and fire exclusion can simplify mixed-severity fire regimes, altering forest composition and structure. For example, the mixed-conifer forests of the North Rim of the Grand Canyon are rich with fire scars indicating surface fires, but high-severity patches burned during periodic widespread fires. In the absence of mixed-severity fires for more than a century, forest density increased and composition shifted toward fire-susceptible species, conditions conducive to high-severity fire. Fire exclusion has had the opposite effect in the lodgepole pine–antelope bitterbrush forests of central Oregon's Pumice Plateau. Historically, its mixed-severity fire regime also included high-severity patches embedded in extensive low-severity fires. The resulting gaps facilitated bitterbrush regeneration, which enabled fire spread, perpetuating a positive feedback. Fire exclusion has disrupted this feedback, driving the decline of bitterbrush and limiting torching fire across this landscape. Although fire exclusion has had contrasting impacts, a mixed-severity fire regime is needed to maintain forest diversity and resilience in both landscapes.

8.5 Future Directions and Research Priorities

Recent large and costly fires in western North America and globally have raised urgent questions about altered fire regimes, including fire causes, effects and feedbacks, as well as post-fire vegetation recovery (Bowman et al. 2009; Stephens et al. 2013). Are contemporary fires more severe or larger than the historical range of variation? To what degree are current high-severity fires driven by climatic conditions *versus* forest structures and fuel availability? Are contemporary forests resilient to fire? Reconstructions of historical fire frequency, severity, spatial variability and extent are essential to gauge changes in contemporary fire regimes. Existing classification methods have proven effective for inferring the severity of individual fires and stand-level fire regimes in a range of forest types. Combining evidence of fire severity with reconstructions of past fire extent provides a powerful framework for quantifying the spatial variability within and among historical fires.

This type of integrated approach could be used to develop empirically-based, quantitative estimates of the relative abundance of different fire severity classes at stand to landscape levels. Such criteria are an important next step toward determining if fire severity is increasing relative to historical conditions.

Knowledge of historical fire regime attributes provides an empirical framework for management strategies to improve forest resilience to the cumulative effects of environmental change (Stephens et al. 2013). Climate change projections for western North America include increased temperatures, earlier snowmelt, longer fire seasons with enhanced drought, potentially driving more frequent, large and severe forest fires (McKenzie et al. 2004; Flannigan et al. 2009). These climatic effects may be exacerbated by land-use change, fire exclusion, and urban expansion (Moritz et al. 2014). In forests where low-severity fires have been reduced, proactive management may include restoration of forest composition, structure and spatial patterns or mitigation of fuels (Stephens et al. 2013). In forests historically dominated by high-severity fires, understanding variation in fire attributes and vegetation patterns across climatic regions helps anticipate changes in forest types and age structures at landscape to regional scales (Stephens et al. 2013). As dendroecological research has shown, forests with mixed-severity fire regimes form a continuum between these extremes. Therefore, ecosystem-specific knowledge of the relative importance of low- *versus* high-severity fires is essential to ensure management enhances rather than compromises forest resilience. Dendroecological reconstructions, combined with multiple lines of corroborating evidence, are key for achieving this goal.

References

Agee JK (1993) Fire ecology of Pacific Northwest forests. Island Press, Washington, DC

Agee JK (1998) The landscape ecology of western forest fire regimes. Northwest Sci 72:24–34

Amoroso MM, Daniels LD (2010) Cambial mortality in declining *Austrocedrus chilensis* forests: implications for stand dynamics studies. Can J For Res 40:885–893. doi:10.1139/X10-042

Amoroso MM, Daniels LD, Bataneih M et al (2011) Evidence of mixed-severity fires in the foothills of the Rocky Mountains of west-central Alberta, Canada. For Ecol Manag 262:2240–2249. doi:10.1016/j.foreco.2011.08.016

Applequist MB (1958) A simple pith locator for use with off-center increment cores. J For 56:141

Baker WL, Veblen TT, Sherriff RL (2007) Fire, fuels and restoration of ponderosa pine–Douglas fir forests in the Rocky Mountains, USA. J Biogeogr 34:251–269. doi:10.1111/j.1365-2699.2006.01592.x

Bowman DM, Balch JK, Artaxo P et al (2009) Fire in the earth system. Science 324:481–484. doi:10.1126/science.1163886

Brown PM, Swetnam TW (1994) A cross-dated fire history from coast redwood near Redwood National Park, California. Can J For Res 24:21–31. doi:10.1139/x94-004

Brown PM, Wu R (2005) Climate and disturbance forcing of episodic tree recruitment in a southwestern ponderosa pine landscape. Ecology 86:3030–3038. doi:10.1890/05-0034

Brown PM, Wienk CL, Symstad AJ (2008) Fire and forest history at Mount Rushmore. Ecol Appl 18:1984–1999. doi:10.1890/07-1337.1

Busse MD, Riegel GM (2009) Response of antelope bitter brush to repeated prescribed burning in Central Oregon ponderosa pine forests. For Ecol Manag 257:904–910. doi:10.1016/j.foreco.2008.10.026

Caprio AC, Swetnam TW (1995) Historic fire regimes along an elevational gradient on the west slope of the Sierra Nevada, California. In: Brown JK, Mutch RW, Spoon CW, Wakimoto RW (tech. coords) Proceedings of the symposium on fire in wilderness and park management: past lessons and future opportunities, Missoula, MT, USA, Mar 30–Apr 1, 1993. USDA Forest Service General Technical Report INT. pp 173–179

Chavardès RD, Daniels LD (2016) Altered mixed-severity fire regime has homogenized montane forests of Jasper National Park. Int J Wildland Fire 25:433–444. doi:10.1071/WF15048

Cherubini P, Fontana G, Rigling D et al (2002) Tree-life history prior to death: two fungal root pathogens affect tree growth differently. J Ecol 90:839–850. doi:10.1046/j.1365-2745.2002.00715.x

Daniels LD, Dobry J, Klinka K et al (1997) Determining year of death of logs and snags of *Thuja plicata* in southwestern coastal British Columbia. Can J For Res 27:1132–1141

Dieterich JH, Swetnam TW (1984) Dendrochronology of a fire scarred ponderosa pine. For Sci 30:238–247

Duncan R (1989) An evaluation of errors in tree-age estimates based on increment cores in kahikatea (*Dacrycarpus dacrydioides*). New Zeal. Nat Sci 16:31–37

Ehle D, Baker WL (2003) Disturbance and stand dynamics in ponderosa pine forests in Rocky Mountain National Park, USA. Ecol Monogr 73:543–566. doi:10.1890/03-4014

Falk DA, Miller C, McKenzie D et al (2007) Cross-scale analysis of fire regimes. Ecosystems 10:809–823. doi:10.1007/s10021-007-9070-7

Falk DA, Heyerdahl EK, Brown PM et al (2011) Multiscale controls of historical fire regimes: new insights from fire scar networks. Front Ecol Environ 9:446–454. doi:10.1890/100052

Farris CA, Baisan CH, Falk DA et al (2010) Spatial and temporal corroboration of a fire-scar-based fire history in a frequently burned ponderosa pine forest. Ecol Appl 20:1598–1614. doi:10.1890/09-1535.1

Farris CA, Baisan CH, Falk DA et al (2013) A comparison of targeted and systematic fire-scar sampling for estimating historical fire frequency in southwestern ponderosa pine forests. Int J Wildland Fire 22:1021–1033. doi:10.1071/WF13026

Finney MA (2006) An overview of FlamMap fire modeling capabilities. In: Andrews PL, Butler BW (eds) Fuels management—how to measure success: conference proceedings, Portland, OR, 28–30 Mar 2006. USDA For. Serv. Proc. RMRS-P-41, Fort Collins, CO, pp 213–220

Flannigan MD, Krawchuk MA, deGroot WJ et al (2009) Implication of changing climate for global wildland fire. Int J Wildland Fire 18:483–507. doi:10.1071/WF08187

Fritts HC, Swetnam TW (1989) Dendroecology: a tool for evaluating variations in past and present forest environments. Adv Ecol Res 19:111–188. doi:10.1016/S0065-2504(08)60158-0

Fulé PZ, Crouse JE, Heinlein TA et al (2003) Mixed-severity fire regime in a high-elevation forest of Grand Canyon, Arizona, USA. Landsc Ecol 18:465–486. doi:10.1023/A:1026012118011

Gartner MH, Veblen TT, Sherriff RL et al (2012) Proximity to grasslands influences fire frequency and sensitivity to climate variability in ponderosa pine forests of the Colorado Front Range. Int J Wildland Fire 21:562–571. doi:10.1071/WF10103

Geist JM, Cochran PH (1991) Influences of volcanic ash and pumice deposition on productivity of western interior forest soils. In: Harvey AE, Neuenschwander LF (eds) Proceedings: management and productivity of western-montane forest soils, Boise, ID, USA April 1990. USDA Forest Service General Technical Report INT-280. pp 82–89

Greene GA, Daniels LD (2017) Spatial interpolation and mean fire interval analyses quantify historical burn metrics in mixed-severity fire regimes. Int J Wildland Fire 26(2):136–147. doi:10.1071/WF16084

Hagmann RK, Franklin JF, Johnson KN (2013) Historical structure and composition of ponderosa pine and mixed-conifer forests in south-central Oregon. For Ecol Manag 304:492–504. doi:10.1016/j.foreco.2013.04.005

Halofsky JE, Donato DC, Hibbs DE et al (2011) Mixed-severity fire regimes: lessons and hypotheses from the Klamath-Siskiyou Ecoregion. Ecosphere 2:1–19. doi:10.1890/ES10-00184.1

Heinselman ML (1973) Fire in the virgin forests of the Boundary Waters Canoe Area, Minnesota. Quat Res 3:329–382. doi:10.1016/0033-5894(73)90003-3

Hessburg PF, Salter RB, James KM (2007) Re-examining fire severity relations in pre-management era mixed conifer forests: inferences from landscape patterns of forest structure. Landsc Ecol 22:5–24. doi:10.1007/s10980-007-9098-2
Hessl AE, Miller J, Kernan JT et al (2007) Mapping paleo-fire boundaries from binary point data: comparing interpolation methods. Prof Geogr 59:87–104. doi:10.1111/j.1467-9272.2007.00593.x
Heyerdahl EK, Brubaker LB, Agee JK (2001) Spatial controls of historical fire regimes: a multiscale example from the interior west, USA. Ecology 82:660–678. doi:10.1890/0012-9658(2001)082[0660:SCOHFR]2.0.CO;2
Heyerdahl EK, Lertzman K, Wong CM (2012) Mixed-severity fire regimes in dry forests of southern interior British Columbia, Canada. Can J For Res 42:88–98. doi:10.1139/x11-160
Heyerdahl EK, Loehman RA, Falk DA (2014) Mixed-severity fire in lodgepole pine dominated forests: are historical regimes sustainable on Oregon's Pumice Plateau, USA? Can J For Res 44:593–603. doi:10.1139/cjfr-2013-0413
Johnson EA, Gutsell SL (1994) Fire frequency models, methods and interpretations. Adv Ecol Res 25:239–287
Johnson EA, Van Wagner CE (1985) The theory and use of two fire history models. Can J For Res 15:214–220. doi:10.1139/x85-039
Jones EL, Daniels LD (2012) Assessment of dendrochronological year-of-death estimates using permanent sample plot data. Tree-Ring Res 68:3–16. doi:10.3959/2010-10.1
Jonsson B, Holm S, Kallur H (1992) A forest inventory method based on density-adapted circular plot size. Scand J Forest Res 7:405–421. doi:10.1080/02827589209382733
Keeley JE, Zedler PH (1998) Evolution of life history in *Pinus*. In: Richardson DM (ed) Ecology and biogeography of Pinus. Cambridge University Press, Cambridge, pp 219–250
Kernan JT, Hessl AE (2010) Spatially heterogeneous estimates of fire frequency in ponderosa pine forests of Washington, USA. Fire Ecol 6:117–135. doi:10.4996/fireecology.0603117
Kipfmueller KF, Baker WL (1998) A comparison of three techniques to date stand-replacing fires in lodgepole pine forests. For Ecol Manag 104:171–177. doi:10.1016/S0378-1127(97)00245-4
Kitzberger T, Brown PM, Heyerdahl EK et al (2007) Contingent Pacific-Atlantic Ocean influence on multi-century wildfire synchrony over western North America. Proc Natl Acad Sci U S A 104:543–548. doi:10.1073/pnas.0606078104
Landres P, Morgan P, Swanson F (1999) Overview of the use of natural variability concepts in managing ecological systems. Ecol Appl 9:1179–1188. doi:10.1890/1051-0761(1999)009[1179:OOTUON]2.0.CO;2
Lertzman K, Fall J, Dorner B (1998) Three kinds of heterogeneity in fire regimes: at the crossroads of fire history and landscape ecology. Northwest Sci 72:4–23
Marcoux HM, Gergel SG, Daniels LD (2013) Mixed-severity fire regimes: How well are they represented by existing fire-regime classification systems? Can J For Res 43:658–668. doi:10.1139/cjfr-2012-0449
Marcoux HM, Daniels LD, Gergel SG et al (2015) Differentiating mixed- and high-severity fire regimes in mixed-conifer forests of the Canadian Cordillera. For Ecol Manag 341:45–58. doi:10.1016/j.foreco.2014.12.027
Margolis EQ, Swetnam TW, Allen CD (2007) A stand-replacing fire history in upper montane forests of the southern Rocky Mountains. Can J For Res 37:2227–2241. doi:10.1139/X07-079
Margolis EQ, Swetnam TW, Allen CD (2011) Historical stand-replacing fire in upper montane forests of the Madrean Sky Islands and Mogollon Plateau, Southwestern USA. Fire Ecol 7:88–107. doi:10.4996/fireecology.0703088
Mast JN, Veblen TT, Linhart YB (1998) Disturbance and climatic influences on age structure of ponderosa pine at the pine/grassland ecotone, Colorado Front Range. J Biogeogr 25:743–755. doi:10.1046/j.1365-2699.1998.2540743.x
McBride JR (1983) Analysis of tree rings and fire scars to establish fire history. Tree-Ring Bull 43:51–67
McKenzie D, Gedalof Z, Peterson DL et al (2004) Climatic change, wildfire, and conservation. Conserv Biol 18:890–902. doi:10.1111/j.1523-1739.2004.00492.x

Moritz MA, Batllori E, Bradstock RA et al (2014) Learning to coexist with wildfire. Nature 515:58–66. doi:10.1038/nature13946

Mowat EL (1960) No serotinous cones on central Oregon lodgepole pine. J Forest 58:118–119

Norton DA, Palmer JG, Ogden J (1987) Dendroecological studies in New Zealand, Part 1. An evaluation of tree age estimates based on increment cores. New Zeal J Bot 25:373–383. doi:10.1080/0028825X.1987.10413355

O'Connor CD, Falk DA, Lynch AM et al (2014) Fire severity, size, and climate associations diverge from historical precedent along an ecological gradient in the Pinaleño Mountains, Arizona, USA. For Ecol Manag 329:264–278. doi:10.1016/j.foreco.2014.06.032

Odion DC, Hanson CT, Arsenault A et al (2014) Examining historical and current mixed-severity fire regimes in ponderosa pine and mixed-conifer forests of western North America. PLoS One 9(2):e87852. doi:10.1371/journal.pone.0087852

Perry DA, Hessburg PF, Skinner CN et al (2011) The ecology of mixed severity fire regimes in Washington, Oregon, and Northern California. For Ecol Manag 262:703–717. doi:10.1016/j.foreco.2011.05.004

Preisler HK, Hicke JA, Ager AA, Hayes JL (2012) Climate and weather influences on spatial temporal patterns of mountain pine beetle populations in Washington and Oregon. Ecology 93:2421–2434. doi:10.1890/11-1412.1

Romme WR (1980) Fire history terminology—report of the ad hoc committee. In: Stokes M, Dietrich JH (eds) Proceedings of the Fire History Workshop; Ft Collins, CO, USA. USDA Forest Service General Technical Report RM-81. pp 135–137

Ruha T, Landsberg J, Martin R (1996) Influence of fire on understory shrub vegetation in ponderosa pine stands. In: Barrow JR, McArthur DE, Sosebee RE, Tausch RJ (eds) Proceedings: Shrubland Ecosystem Dynamics in a Changing Environment, Las Cruces, NM, USA, May 1995. USDA Forest Service General Technical Report INT-GTR-338. pp 108–113

Schoennagel T, Nelson CR (2011) Restoration relevance of recent National Fire Plan treatments in forests of the western United States. Front Ecol Environ 9:271–277. doi:10.1890/090199

Schoennagel T, Veblen TT, Romme W (2004) The interaction of fire, fuels, and climate across Rocky Mountain forests. Bioscience 54:661–676. doi:10.1641/0006-3568(2004)054[0661:TIOFFA]2.0.CO;2

Schoennagel T, Smithwick EAH, Turner MG (2008) Landscape heterogeneity following large fires: insights from Yellowstone National Park, U.S.A. Int J Wildland Fire 17:742–753. doi:10.1071/WF07146

Schoennagel TL, Sherriff RL, Veblen TT (2011) Fire history and tree recruitment in the Colorado Front Range upper montane zone: implications for forest restoration. Ecol Appl 21:2210–2222. doi:10.1890/10-1222.1

Sherriff RL, Veblen TT (2006) Ecological effects of changes in fire regimes in *Pinus ponderosa* ecosystems in the Colorado Front Range. J Veg Sci 17:705–718. doi:10.1111/j.1654-1103.2006.tb02494.x

Sherriff RL, Veblen TT (2007) A spatially-explicit reconstruction of historical fire occurrence in the ponderosa pine zone of the Colorado Front Range. Ecosystems 10:311–323. doi:10.1007/s10021-007-9022-2

Sherriff RL, Platt RV, Veblen TT et al (2014) Historical, observed, and modeled wildfire severity in montane forests of the Colorado Front Range. PLoS One 9:e106791. doi:10.1371/journal.pone.0106971

Sibold JS, Veblen TT, Gonzalez ME (2006) Spatial and temporal variation in historic fire regimes in subalpine forests across the Colorado Front Range in Rocky Mountain National Park, Colorado, USA. J Biogeogr 33:631–647. doi:10.1111/j.1365-2699.2005.01404.x

Simpson M (2007) Forested plant associations of the Oregon East Cascades. USDA Forest Service, Pacific Northwest Region, Technical Paper R6-NR-ECOL-TP-032007

Smith KT, Sutherland EK (2001) Terminology and biology of fire scars in selected central hardwoods. Tree-Ring Bull 57:141–147

Smith KT, Arbellay E, Falk DA et al (2016) Macroanatomy and compartmentalization of recent fire scars in three North American conifers. Can J For Res 46:535–542. doi:10.1139/cjfr-2015-0377

Stephens SL, Fulé PZ (2005) Western pine forests with continuing frequent fire regimes: possible reference sites for management. J For 103:357–362

Stephens SL, Agee JK, Fulé PZ et al (2013) Managing forests and fire in changing climates. Science 342:41–42. doi:10.1126/science.1240294

Stretch V, Gedalof Z, Cockburn J et al (2016) Sensitivity of reconstructed fire histories to detection criteria in mixed-severity landscapes. For Ecol Manag 379:61–69. doi:10.1016/j.foreco.2016.08.009

Swetnam T, Baisan C (1996) Historical fire regime patterns in the southwestern United States since AD 1700. In: Allen CD (ed) Fire effects in Southwestern forests: Proceedings of the 2nd La Mesa Fire Symposium. USDA Forest Service, Rocky Mountain Research Station, General Technical Report RM-GTR-286. pp 11–32

Swetnam T, Allen C, Betancourt J (1999) Applied historical ecology: using the past to manage for the future. Ecol Appl 9:1189–1206. doi:10.1890/1051-0761(1999)009[1189:AHEUTP]2.0.CO;2

Swetnam TL, Falk DA, Hessl AE et al (2011) Reconstructing landscape pattern of historical fires and fire regimes. In: McKenzie D, Miller C, Falk DA (eds) The landscape ecology of fire. Springer Publishing, New York, pp 165–192

Taylor AH, Skinner CN (2003) Spatial patterns and controls on historical fire regimes and forest structure in the Klamath Mountains. Ecol Appl 13:704–719. doi:10.1890/1051-0761(2003)013[0704:SPACOH]2.0.CO;2

Tepley AL, Veblen TT (2015) Spatiotemporal fire dynamics in mixed-conifer and aspen forests in the San Juan Mountains of southwestern Colorado, USA. Ecol Monogr 85:583–603. doi:10.1890/14-1496.1

Turner MG (2010) Disturbance and landscape dynamics in a changing world. Ecology 91:2833–2849. doi:10.1890/10-0097.1

Van Horne ML, Fulé PZ (2006) Comparing methods of reconstructing fire history using fire scars in a southwestern United States ponderosa pine forest. Can J For Res 36:855–867. doi:10.1139/x05-289

Van Wagner CE (1978) Age-class distribution and the forest fire cycle. Can J For Res 8:220–227. doi:10.1139/x78-034

Vankat JL (2011) Post-1935 changes in forest vegetation of Grand Canyon National Park, Arizona, USA: Part 1—ponderosa pine forest. For Ecol Manag 261:309–325. doi:10.1016/j.foreco.2010.05.026

Veblen TT, Lorenz DC (1986) Anthropogenic disturbance and recovery patterns in montane forests, Colorado Front Range. Phys Geogr 7:1–24. doi:10.1080/02723646.1986.10642278

Veblen TT, Lorenz DC (1991) The Colorado Front Range: a century of ecological change. University of Utah Press, Salt Lake City

Veblen TT, Kitzberger T, Donnegan J (2000) Climatic and human influence on fire regimes in ponderosa pine forests in the Colorado Front Range. Ecol Appl 10:1178–1195. doi:10.1890/1051-0761(2000)010[1178:CAHIOF]2.0.CO;2

Villalba R, Veblen TT (1997) Improving estimates of total tree ages based on increment core samples. Ecoscience 4:534–542. doi:10.1080/11956860.1997.11682433

Williams MA, Baker WL (2012) Spatially extensive reconstructions show variable-severity fire and heterogeneous structure in historical western United States dry forests. Glob Ecol Biogeogr 21:1042–1052. doi:10.1111/j.1466-8238.2011.00750.x

Wong CM, Lertzman KP (2001) Error in estimating tree age: implications for studies of stand dynamics. Can J For Res 31:1262–1271. doi:10.1139/x01-060

Yocom Kent LL (2014) An evaluation of fire regime reconstruction methods. ERI Working Paper No. 32. Ecological Restoration Institute and Southwest Fire Science Consortium, Northern Arizona University, Flagstaff, AZ. 15 p

Yocom Kent LL, Fulé PZ, Bunn WA et al (2015) Historical high-severity fire patches in mixed-conifer forests. Can J For Res 45:1587–1596. doi:10.1139/cjfr-2015-0128

Chapter 9
Fire History and Fire Regimes Shifts in Patagonian Temperate Forests

Ignacio A. Mundo, Andrés Holz, Mauro E. González, and Juan Paritsis

Abstract Fire has been a frequent disturbance in Patagonia. The presence of charcoal in sedimentary records covering the last 44,000 years suggests that natural fires played a significant role in shaping the landscape before the arrival of Native Americans ca. 14,500–12,500 years ago. Dendrochronological studies focused on the reconstruction of fire histories have been conducted in the Patagonian forests on both sides of the Andes Cordillera, beginning in the late 1990s. Here, we review the present knowledge of the history of fires in temperate forests in Patagonia, their main drivers, and discuss the evidence and impacts of burns and reburns on post-fire response, as well as possible mechanisms to shift into alternative stable states. Dendrochronology was extremely useful to develop multi-century fire histories in *Araucaria araucana*, *Pilgerodendron uviferum*, *Fitzroya cupressoides* and *Austrocedrus chilensis* and mixed *Austrocedrus-Nothofagus dombeyi* forests in Patagonia. In the case of *Araucaria*, *Austrocedrus* and *Pilgerodendron* forests, dendrochronological reconstructions show diverse and heterogeneous patterns of

I.A. Mundo (✉)
Laboratorio de Dendrocronología e Historia Ambiental, IANIGLA-CONICET, CCT CONICET Mendoza, Av. Ruiz Leal s/n. CC330-M5502IRA Mendoza, Pcia. de Mendoza, Argentina

Facultad de Ciencias Exactas y Naturales, Universidad Nacional de Cuyo, Padre Contreras 1300. M5502JMA Mendoza Pcia. de Mendoza, Argentina
e-mail: iamundo@mendoza-conicet.gob.ar

A. Holz
Department of Geography, Portland State University, 1721 SW Broadway, Cramer Hall Rm. 424, Portland, OR, 97201, USA
e-mail: holzc@pdx.edu

M.E. González
Laboratorio Ecología de Bosques, Instituto de Conservación, Biodiversidad y Territorio, Facultad de Ciencias Forestales y Recursos Naturales, Universidad Austral de Chile, Campus Isla Teja, Casilla 567, Valdivia, Chile

Center for Climate and Resilience Research (CR2), Santiago, Chile
e-mail: maurogonzalez@uach.cl

J. Paritsis
Laboratorio Ecotono, INIBIOMA, CONICET-Universidad Nacional del Comahue, Quintral 1250, R8400FRD, Bariloche, Río Negro, Argentina
e-mail: j.paritsis@gmail.com

M.M. Amoroso et al. (eds.), *Dendroecology*, Ecological Studies 231,
DOI 10.1007/978-3-319-61669-8_9

fire frequency related to changes in human activities and settlement processes over the last centuries. Fire history reconstructions document infrequent events in the *Fitzroya* wet rainforests, with ca. 800-year old in the Costal Range in South-Central Chile and ca. 1000-year old chronologies in the Argentinean Andes. Climate variability has a significant influence on fire occurrence in these Patagonian forests. Fire events have been strongly associated with low moisture availability linked to El Niño – Southern Oscillation (ENSO) and the Southern Annular Mode (SAM), the major climate drivers promoting fire. Future directions and challenges for fire history studies in Patagonian forests are proposed at the end of this chapter.

Keywords Fire history • Dendropyrochronology • Patagonia • *Araucaria araucana* • *Austrocedrus chilensis* • *Fitzroya cupressoides* • *Nothofagus dombeyi* • *Pilgerodendron uviferum*

9.1 Overview of Historical Fire Regimes in Patagonia

9.1.1 Historical Context from Palaeoecological Studies

Fire has been a frequent disturbance driving forest dynamics in Patagonia. The presence of charcoal in sedimentary records covering the last 44,000 years suggests that natural fires have played a significant role in shaping the landscape before the arrival of Native Americans ca. 14,000–12,500 years ago (Heusser 1994; Dillehay et al. 2015). Sedimentary charcoal and archeological records show that fire activity increased significantly after the arrival of humans into this region, with a peak at ca. 11.5 K years ago (Whitlock et al. 2007; Holz et al. 2016). However, during the past 6,000 years fire activity decreased and turned into a more spatially variable disturbance. In this context, the unusually frequent, extensive, and severe wildfires have occurred in Patagonia over recent decades have been mediated by land-use practices (e.g. fire suppression or logging) and climate change (Veblen et al. 2008, 2011).

9.1.2 Recent Unprecedented Fire Activity and Climate Change

Over the past 15 years (2002–2016) Patagonia has experienced several mega-fire events unprecedented in the region compared to the previous 50 years. Large and severe events burned (ca. 36,000 ha) in several national parks and natural reserves in Chile (González 2005; González and Lara 2015), including Tolhuaca NP and Malleco NR (2002 and 2015), China Muerta NR (2015) and Torres del Paine NP (2005 and 2011), and another 47,000 ha more recently in mountainous forests in Neuquén, Río Negro and northern Chubut provinces (2013 and 2015) in Argentina (Mohr-Bell 2015). These megafires and other recent severe fires have been associated with trends of regional warming temperatures and reduced summer precipitation (Holz et al. 2012b). Some of these large events garnered significant public attention, as a few fires burned very old (>800–1,000 years) trees in protected areas. A few of these large events were lighting-ignited, which is an uncommon but steadily increasing weather event in the Patagonian-Andean region (Veblen et al. 2008). As warming trends continue, it is expected that lightning activity will continue to increase the probability of fire ignitions across Patagonia (Garreaud et al. 2014).

There is concern that climate change and change in land-use trends are altering fire regimes in Patagonia and in turn these regimes are transforming ecological function, structure, and composition of forest ecosystems. Ongoing research on post-fire responses and reburn impacts on forests highlights the current concern about future vegetation and associated fire regimes. High severity (i.e. high mortality) events and/or repeated fires are extirpating the seed sources from these forests, which filters out the capability of obligate seeders to recover. Resprouter shrubs and invasive species that are more flammable tend to recover and invade faster than burned tall forest species and, thus, dominate post-fire environments. Microclimatic conditions in these post-fire vegetation types are warmer and dryer than under closed-canopy tall forests, which favor the stabilization of a new alternative stable state, as the likelihood of future events is increased and the tree establishment of obligate tree species is reduced (Cóbar-Carranza et al. 2015; Paritsis et al. 2015; Kitzberger et al. 2016).

9.1.3 Dendrochronology: The Key to Understanding Fire History

Dendrochronology is an effective tool for reconstructing fire history with annual and even seasonal resolution. Dendroecology more specifically, can contribute to determining if the current trends in fire frequency and magnitude and in post-fire establishment are unprecedented over the past few centuries. It can help link modern, spatially-explicit techniques (GIS and imagery analyses) with longer-term, non-spatially explicit approaches (e.g. paleoecological methods). Dendroecology is also useful for understanding the mechanisms behind the effects of broad-scale climate modes (e.g. ENSO) on local climate variability in the past, which continues to affect wildfire activity and post-fire tree recruitment. Here, we review the present knowledge of the history of fires in temperate forests in Patagonia, their main drivers, the evidence of burns and reburns on post-fire response, and possible mechanisms to shift into alternative stable states.

Since the late 1990s, dendrochronological studies focused on the reconstruction of fire histories have been conducted in the Patagonian forests on both sides of the Andes Cordillera (Fig. 9.1). In Patagonia, studies on fire regimes have primarily relied on conifers of the Cupressaceae and Araucariaceae families, which are proven sources for dendroclimatological and dendroecological studies.

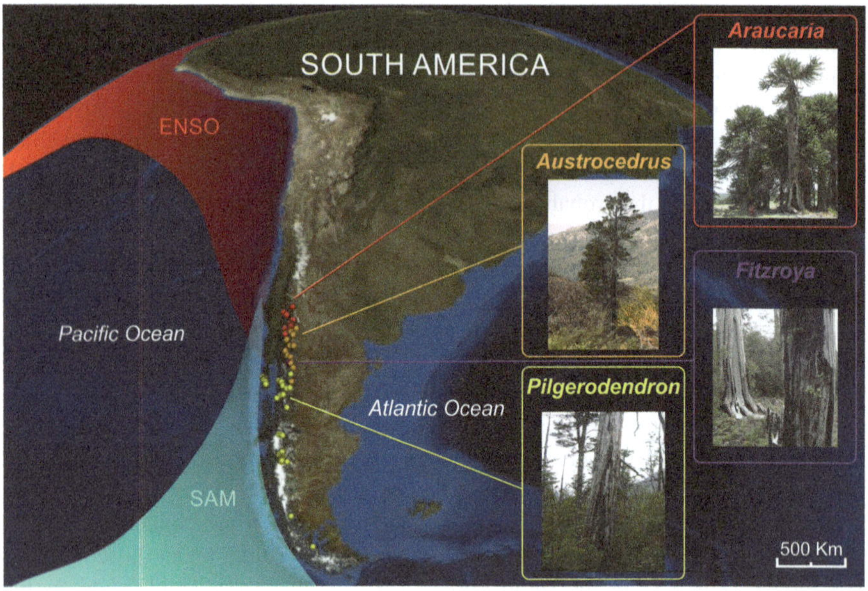

Fig. 9.1 Location of the published fire history reconstructions (by species) and spatial domain of the main climate modes (ENSO and SAM) that affect Patagonia. Latitudinally, the existing reconstructions have been conducted using *Araucaria araucana* (*red*), *Austrocedrus chilensis* (*orange*), *Fitzroya cupressoides* (*purple*) and *Pilgerodendron uviferum* (*yellow*). Photo Credits: I. Mundo (*Araucaria*); J. Landesmann (*Austrocedrus*), M.E. González (*Fitzroya*) and A. Holz (*Pilgerodendron*)

9.2 Reconstructed Fire Histories in the Patagonian Coastal and Andean Forests: Human and Climate Drivers

9.2.1 *Araucaria araucana ((Molina) K. Koch)*

9.2.1.1 Spatio-Temporal Variability and Human Impacts on Past Fire Activity

A. araucana (Araucariaceae, common native name "pehuén") is a coniferous evergreen species that occurs between 900 and 1,600 m altitude in southwestern Argentina and south-central Chile in two disjunct areas: in the coastal range (Nahuelbuta) and the Andean Range (between 37°20′ and 40°20′ S; Veblen 1982). It has evenly spaced nodes of horizontal-spreading branches that are arranged in regular whorls around straight and cylindrical boles. At maturity, *A. araucana* can reach up to 45 m tall and develop trunks of 1.5–2 m dbh (Tortorelli 1942). The tree has a hard and thick bark organized in polyhedral-shaped plates that protect living tissues against fire. *A. araucana* trees are known to have long lifespans of nearly 1,000 years (LaMarche et al. 1979; Mundo et al. 2012; Hadad et al. 2015; Aguilera-Betti et al. 2017). Due to its thick bark, trees withstand fires of low and medium intensity (Alfonso 1941; Tortorelli 1942; Veblen 1982; Burns 1993; González et al. 2010). Wildfire has shaped *Araucaria* forests over long time periods (for the last 3,000 years; Heusser et al. 1988), and is perhaps the only tree species in both Argentina and Chile with unequivocal fire adaptations; e.g. thick fire-resistant bark, protected terminal buds, and a self-pruning growth form that protects sensitive meristems (Heusser et al. 1988; Burns 1993; Veblen et al. 1995). Wildfires can also favor and stabilize the dominance of understory bamboo thickets in *Araucaria-Nothofagus* forests, as resprouting bamboo impede tree regeneration and are highly flammable (Kitzberger et al. 2016).

Dendrochronological reconstructions of fire history conducted at ca. 37°–39° S on the west and east slopes of the Andes show a significant variation in fire activity over the last 500 years (González et al. 2005; Quezada 2008; Mundo et al. 2013). These dendrochronological reconstructions of cohort recruitment and fire scars show that the *Araucaria* forest landscape has been shaped by a mixed-severity fire regime that includes surface and crown fires (González et al. 2010). In both Chile and Argentina fire frequency over the indigenous period (ca. pre-1880) was lower than over the Euro-settling period, when ignition increased due mostly to logging and clearing land for livestock (Fig. 9.2; González et al. 2005). With irregular spatial foci and efficacy, fire suppression efforts have been implemented since the 1930s in Argentina (Mundo et al. 2013) and 1960s in Chile (González et al. 2016).

9.2.1.2 Fire and Climate Relationship

Climate variability has a significant influence on fire occurrence in mesic *Araucaria-Nothofagus* forests on the west side of the Andes (ca. 40° S). Past fire events have been strongly associated with low moisture availability. Years of widespread fire are favored by warmer and drier summers, preceded by 1 or 2 years of dry climatic conditions (González and Veblen 2006). The El Niño – Southern Oscillation (ENSO) is an important factor promoting fire. Years of high fire activity coincide with warm and dry summers following El Niño events. Furthermore, on the east side of the Andes, large fires are also associated with warm and dry years during the La Niña positive phase. Other broad-scale climate drivers such as the Pacific Decadal Oscillation (PDO), as well as the Southern Annular Mode (SAM) are important controls on fire occurrence. High regional fire activity is strongly associated with the negative and positive signature of the PDO and SAM, respectively (Mundo et al. 2013).

9.2.1.3 Post-fire Response and Reburn Impacts on *Araucaria* Forests

Araucaria occurs both in pure stands or mixed with various *Nothofagus* species. In all situations, *Araucaria* forests have been shaped by anthropogenic and natural wildfires. Large *Araucaria* trees are highly resistant to fire, and this species typically survives moderate- to high-severity fires either as dispersed individuals or as small

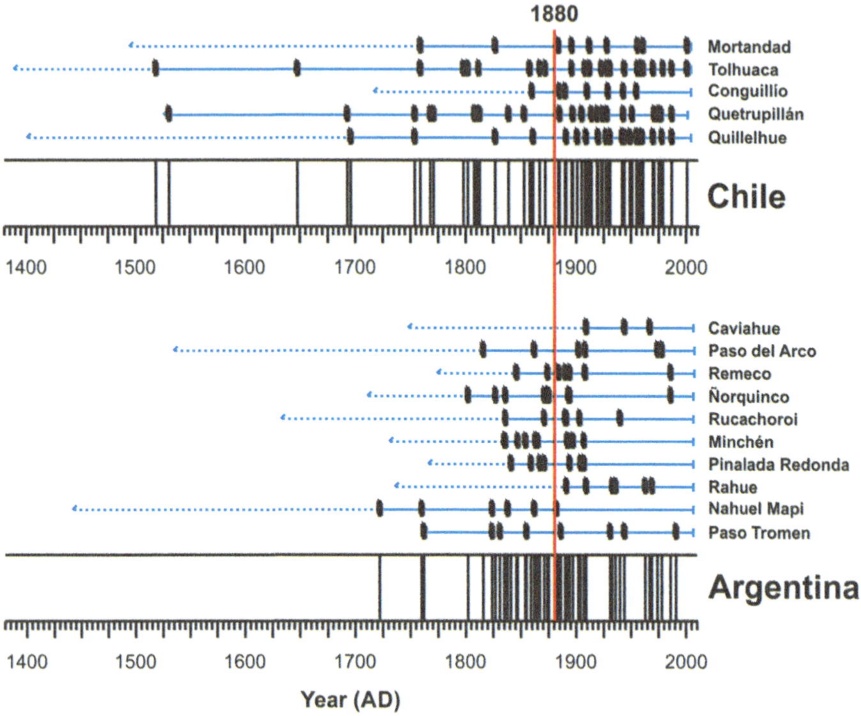

Fig. 9.2 Fire history records from tree-ring fire scar chronologies in the Andean *Araucaria* forests in Chile and Argentina. Each horizontal *light blue line* represents individual sites, where individual fire years (i.e. with min ≥ 2 trees) are indicated by *short vertical lines* in *bolded black*. Individual site chronologies in each country are arranged from north to south. At the *bottom* of charts *long vertical lines* show the composite chronology of all fire records of Chile and Argentina. *Vertical red line* separates the Indigenous from the Euro-settler periods (ca. 1880)

groups of multi-aged trees (Fig. 9.3a, b). Adaptations of *Araucaria* include thick bark, sprouting from epicormic buds and protected terminal buds on branches (Fig. 9.3c, d). Small post-fire cohorts of *Araucaria* may establish, depending on seed availability and the effects of subsequent fires.

In mixed forests, *Araucaria*'s great longevity (often >700 years) and resistance to fire allow some individuals to survive fires that kill and then trigger new *Nothofagus* cohorts. In the case of *Araucaria-N. antarctica* forests, both species are adapted to survive fire. *N. antarctica* resprouts vigorously after fire with multiple shoots. Burns (1993) developed a model of dynamics of this community, particularly in relation to fire. Within the first decade after fire, *Araucaria* (but not *N. antarctica*) seedlings or root suckers began to establish beneath the resprouted *N. antarctica* canopy. However, establishment is spatially clustered around surviving female *Araucaria* or abandoned caches of seed predators. Once established, *Araucaria* grows through the sparse *N. antarctica* canopy, eventually overtopping it. Fire acts as a medium of species co-existence between a vigorously sprouting, shade-intolerant species

Fig. 9.3 *Araucaria* post-fire responses: (**a**) An *Araucaria* stand after 10 months of the 2014 summer fire in Lago Ñorquinco. All trees (adults and saplings) show scorched leaves and surface charcoal over barks. (**b**) Another *Araucaria* stand after 4 years and 9 month of the 2009 fire in Paso Tromen. Trees have almost all the leaves and branches dead. However, some branches were able to restart growth from protected terminal buds (**c**). On other branches, epicormic sprouting was also evident (**d**). After 2002 fire in Tolhuaca National Park, the regeneration strategy was either sexual (**e**) or asexual from root sprouts (**f**). Photo Credits: I. Mundo (**a–d**) and M.E. González (**e–f**)

(*N. antarctica*), and one that partly survives fire above-ground and is more shade-tolerant (*Araucaria*) (Burns 1993). Stand development patterns of subalpine *Araucaria–N. pumilio* forests are largely controlled by moderate- to high-severity fires, and therefore tree regeneration dynamics is strongly dominated by a catastrophic regeneration mode (González and Veblen 2010).

In the austral summer of 2001–2002, extensive fires burned subalpine *Araucaria-Nothofagus* forests at different severities, providing opportunities to study post-fire recruitment of tree seedlings and overall post-fire understory bamboo species responses (González et al. 2016; Kitzberger et al. 2016). Recruitment of obligate seeders *Nothofagus dombeyi* and *N. pumilio* was relatively low in areas that burned with high severity. In contrast, seedlings and saplings of *Araucaria* were able to resprout following both medium- and high-severity conditions and also to establish from seeds under remnant female trees (González and Veblen 2007; González et al. 2010); Fig. 9.3e, f). Following these burns, *Chusquea culeou* bamboo quickly colonized burns of all severities.

In 2015, a human-set severe fire reburned >60% of the same area that burned in the fire season of 2001–2002, almost completely eliminating the post-fire established vegetation (Fig. 9.3f; González et al. 2016). According to our initial assessment this 13-year interval will have long-lasting effects on the overall resilience of *Araucaria-Nothofagus pumilio* forests, altering structure, composition and function, and successional rates and pathways towards more flammable shrubs and resprouting bamboo species. Thus, reburned *Araucaria* forests are expected to switch to new more flammable states, as do most southern hemisphere temperate forests (Kitzberger et al. 2012, 2016; Enright et al. 2015). Changes in climate and land-use trends in Patagonia (Garreaud et al. 2013) might further enhance future reburns (González et al. 2016). Additional pressures to either convert native forests to different land use or to conduct salvage logging can further impact these ecosystems. Plantations can increase fuel amount and fire risk (Cóbar-Carranza et al. 2014) and salvage harvesting can impede or alter the trajectory of recovery of affected areas, and it potentially may decrease forest regeneration and have detrimental impacts on key ecological processes (González and Veblen 2007). Following the 2002 wildfire the Chilean Forest Service (CONAF) has promoted and carried out broad-scale salvage logging operations in *Nothofagus* forests within the Malleco National Reserve, without appropriate pre-/post-logging ecosystem assessment (González and Lara 2016).

9.2.2 Austrocedrus chilensis (D. Don) Florin & Bout

9.2.2.1 Spatial and Temporal Variability of Fire Activity Over Past Centuries

Austrocedrus chilensis (Cupressaceae, common Spanish name "ciprés de la cordillera") is a dioecious conifer that grows in the Argentinean and Chilean Andes from ca. 33° S to ca. 44° S. It typically forms monospecific woodlands towards the eastern xeric portion of its distribution that vary in density depending on precipitation levels. Towards the mesic portion of the gradient it grows in mixed forests with *Nothofagus dombeyi* (Donoso Zegers et al. 2006a). At the dry ecotone of northwestern Patagonia in Argentina, fire regimes of *A. chilensis* forests have experienced marked increase in fire frequency the latter half of the nineteenth century. This increase coincides with the establishment of new Native American groups at the steppe and adjacent open *A. chilensis* woodlands that burned for guanaco hunting. These groups were expelled from their lands in the Chilean Lake and River Districts (ca. 37°–41° S) by European settlers supported by the Chilean government in the late 1890s and early 1900s (Kitzberger et al. 1997; Kitzberger and Veblen 1997; Veblen et al. 1999). At the end of nineteenth century, fire history reconstructions show a noticeable decline in fire frequency following the demise of the Native American population. Fire reconstructions in mesic *A. chilensis-N. dombeyi* forests show a different temporal pattern: in these mixed forests the most noticeable increase in fire frequency occurred in the 1890s coincident with the establishment of European settlers, who extensively burned forested areas to create pastures for livestock (Kitzberger and Veblen 1997). Since the 1920s fire frequency diminished in both mesic and xeric *A. chilensis* forests in association with increasingly effective fire suppression policies.

9.2.2.2 Fire and Climate Relationship

Climate variability plays a crucial role in modulating fire activity in dry *A. chilensis* woodlands. Tree-ring reconstructions show strong influences of climate variability at interannual scales which decline at multidecadal scales (Veblen et al. 1999). Extensive and synchronous fires in dry *A. chilensis* forests tend to occur after a combination of a wet period that increments fine fuel availability followed by a dry period that desiccates coarse fuels (Kitzberger et al. 1997). These climatic sequences are often generated by tele-connections with El Niño Southern Oscillation (ENSO). ENSO events have strong influences on the intensity and latitudinal position of the southeast Pacific subtropical anticyclone that blocks the westerly flow of moist air masses into Patagonia causing temperature and precipitation variations (Daniels and Veblen 2000). Years of widespread fire occurrence in xeric *Austrocedrus* woodlands are associated with both the late stages of La Niña events (i.e. dry winter–springs) and with the warm summers following El Niño events (Kitzberger et al. 1997; Veblen et al. 1999). Years of widespread fire in *Austrocedrus*-dominated sites were also associated with low mean sea level atmospheric pressure at latitudes 50–60° S in the South American–Antarctic sector of the Southern Ocean (Veblen et al. 1999), which implies a positive SAM phase.

9.2.2.3 Post-fire Response and Reburn Impacts

In xeric *A. chilensis* forests and woodlands, post-fire regeneration is strongly influenced by climatic variability. Tree-ring analyses show that tree establishment is favored by periods of cool-wet conditions at decadal scales which are associated with ENSO (Villalba and Veblen 1997). On the contrary, short drought periods of 1 or 2 years cause seedling mortality in the xeric woodlands near the steppe. In most sites, establishment is spatially limited to microsites protected by nurse shrubs and this facilitation is modulated by climate variability. Recruitment of *A. chilensis* during average years requires nurse shrubs. However, during above-average warm and dry seasons, establishment is nominal with or without protection by nurse shrubs. Conversely, in cool and wet years *A. chilensis* establish beneath nurse shrubs as well as in open areas (Kitzberger et al. 2000). Mixed *A. chilensis-N. dombeyi* forests tend to maintain their mixed composition with post-fire recruitment of *A. chilensis* occurring slightly after *N. dombeyi* (Veblen and Lorenz 1988). After ca. 60 years of fire suppression activities, many woodlands of *A. chilensis* have altered their fuel structure. Encroachment of and infilling by *A. chilensis* and exotic pine plantations have increased the potential of these formerly open, dry woodlands to burn at high severity, which is unprecedented in the steppe/woodland ecotone.

9.2.3 *Fitzroya cupressoides* (Molina) Johnston

9.2.3.1 Spatial and Temporal Variability of Fire Activity Over Past Centuries

Fitzroya cupressoides (Cupressaceae, common Spanish name "alerce") is a long-lived conifer endemic to the temperate rain forests of southern Chile and Argentina. *Fitzroya* occurs discontinuously in mixed and pure evergreen forests of the coastal range in south central Chile and on both sides of the Andes at ca. 41–43° S latitudes (Dimitri 1972; Fraver et al. 1999; Donoso Zegers et al. 2006b). It mainly grows on mountain slopes between 300 and 900 m altitude, over shallow, acidic and volcanic soils. It can attain a size of up to 5 m in diameter and 50 m in height, often appearing as an emergent when mixed with other species (Lara 1991). It is one of the longest-lived tree species worldwide, with a lifespan longer than 3600 years (Lara and Villalba 1993). More than two centuries of over-exploitation of *Fitzroya*, harvested for it highly-valued and decay-resistant wood or cleared for agricultural land, has left extensive forests degraded and has significantly reduced and threatened the species persistence (Fraver et al. 1999).

Fire, along with landslides and volcanism, has been an important disturbance that has shaped the structure and dynamics of *Fitzroya* forests (Veblen et al. 1996). Long fire history reconstructions document infrequent events in these wet rainforests, with ca. 800-year old fire chronologies (oldest fire in 1111 AD) in the Costal Range in South-Central Chile (Lara et al. 1999; Urrutia 2002) and ca. 1,000-year old fire chronologies (oldest fire dated in 990 A.D.) in the Argentinean Andes (Veblen et al. 1999).

9.2.3.2 Fire and Climate Relationship

In an Argentine *Fitzroya* forest, Veblen et al. (1999) found that summer temperatures were significantly above average during the fire years; annual precipitation was below average, but not statistically significant. Reduced moisture availability associated mainly with a temperature-induced increase in evapotranspiration during the summer of the fire year appears to promote fire in these rain forests. In terms of regional forcings, years of widespread fire activity in *Fitzroya* forests have co-occurred with high summer temperature favored by late stages of negative La Niña events (Veblen et al. 1999) and/or positive SAM conditions (Holz and Veblen 2011a).

9.2.3.3 Post-fire Response and Reburn Impacts

There is scarce information on the post-fire establishment patterns of *Fitzroya*, but burn size and distance to seed source appears to control its response following fire (Veblen and Ashton 1982; Lara et al. 1999). In many locations where *Fitzroya* forests were burned across broad areas, changes in land-use did not allow for these forests to reestablish. For instance, in the Central Valley of Chile an almost total extirpation of the *Fitzroya* forests occurred due to logging activities and human-set fires dating from the 1850s. These fires that devastated the forests of *Fitzroya* especially after 1880, and led to a matrix of agricultural fields, secondary shrublands, and exotic eucalypt plantations, with a few young stands that represents the only remnants of this forest type outside of the Coastal and Andes cordilleras (Fraver et al. 1999; Lara et al. 2003). Observations of live juveniles growing nearby large fire-scarred individuals at least suggests that non-lethal, fine-scale fires have occurred in the past and that regeneration continues.

9.2.4 *Pilgerodendron uviferum* (D. Don) Florin

9.2.4.1 Spatial and Temporal Variability of Fire Activity over Past Centuries

Pilgerodendron uviferum (Cupressaceae, common Spanish name "ciprés de las Guaitecas"), is a slow-growing, long-lived (ca. 850 years old) conifer endemic to perhumid areas, primarily on the western slope of the Andes in southern Chile and adjacent areas in southwestern Argentina (Donoso 1981; Veblen et al. 1995; Aravena 2007). It is a 1.5–10 m tall, narrow, pyramidal tree with a trunk up to ca. 0.4–1.0 m dbh. *P. uviferum* forms mixed forests with broadleaved tree species, ranging from sea level to 1,200 m above sea level and from 39° 30′ to 55° 30′S, making it the southernmost conifer in the world. Nonetheless, it belongs to some of the least-studied temperate forests in South America (Veblen et al. 1995; Lara et al. 2006). *P. uviferum* was burned by indigenous peoples, who subsequently harvested it for fuel, and later by Euro-settlers who used fire to clear vast areas to facilitate logging access (Lara et al. 2006; Holz and Veblen 2011b). As a result *P. uviferum* is now listed as a protected species under international conservation agreements (Walter and Gillet 1998; Rovere et al. 2002).

Wildfire is an infrequent but important fine-to-broad scale driver of the ecology of *P. uviferum* (Szeicz et al. 2003; Holz et al. 2012b). Our understanding of the fire history of *P. uviferum* forests comes mostly from a fire-scar based chronology network of 27 sites developed primarily in poorly drained peat-forest ecotones, with less work conducted in well drained upland sites in southern South America (ca. 42–48° S; Holz and Veblen 2011b), complimented by charcoal sedimentary records (e.g. Holz et al. 2012a). Fire-history reconstructions from this network indicate that before the arrival of European settlers (i.e. between ca. 1600 and 1850 AD) to western Patagonia, fine-scale fires were relatively frequent (Holz and Veblen 2011a). Upon arrival, European settlers and their decedents quickly increased the frequency of ignitions, favored by more conducive climate conditions (Holz and Veblen 2011b).

9.2.4.2 Fire and Climate Relationship

Fire-scar dates indicate several years of widespread fire activity across ca. 700 km in western Patagonia that was related to climate variability and climate change. Interannual variability in local climate conditions, mediated by ENSO, the Pacific Decadal Oscillation (PDO), and SAM either individually or in combination, strongly affects fire activity in *P. uviferum* forests (Holz and Veblen 2011a, 2012). The fire-enhancing influence of SAM was greatest when PDO was also positive, which indicates warmer conditions throughout the eastern Pacific (Holz and Veblen 2012). Climate-fire analysis using documentary fire records and observed climate data suggests that years of regional drought that coincided with widespread fires co-occur during the positive (above-average) phases of SAM and ENSO (Holz et al. 2012b). A strong increase in wildfire activity in *P. uviferum* forests coincident with warming and drying trends during the twentieth century was found (Holz and Veblen 2012).

9.2.4.3 Post-fire Response and Reburn Impacts

The regeneration dynamics of post-fire *Pilgerodendron* forests both in peat ecotones and upland forests and the understanding of the role that fire has played over its 1700 km-long range (ca. 39–55° S) is a new area of research (Holz 2009; Bannister et al. 2012; Zaret and Holz 2016). Recent and ongoing work indicates that *P. uviferum* forests have a mixed-severity fire regime where surface fires of low intensity are more frequent at the ecotone with peats and wetlands due to the abundance of fine fuels, and moderate- to high-severity fires are more common in upland, closed-canopy forests where conditions are more suitable for fire spread to tree crowns (Holz 2009; Holz and Veblen 2009). In both forest types, tree recruitment of this long-lived, shade-tolerant species depends on source availability, safe sites, both created by fine-scale or patchy fires, and short-term weather conditions Zaret and Holz (2016).

Due to the abrupt increase in burning activity by Euro-settlers, ecological functions, structure, and composition of some *P. uviferum* forests have been fundamentally transformed (Holz 2009; Holz and Veblen 2009; Bannister et al. 2012, 2014; Zaret and Holz 2016). In some places, the frequency of fire in these forests is clearly outside the historical range of variability, as reported from peat-sediment charcoal records recorded over the last 10,000 years (Holz et al. 2012b). In some lowlands peat-forest ecotones, repeated burns have resulted in waterlogging of former *P. uviferum* (Díaz et al. 2007) and the invasion of shade-intolerant *Sphagnum* species. Moreover, *Sphagnum* outcompetes *P. uviferum* seedlings and is more likely to support a higher frequency of fires, since it dries out quickly favoring a positive feedback between fire and vegetation flammability (Holz 2009; Kitzberger et al. 2016; Zaret and Holz 2016).

9.3 Future Directions and Challenges

9.3.1 Fire History in Southern Patagonia

In contrast to the well-documented fire history for Northern Patagonia, there is relatively scarce information about fire frequency and the role of humans and climate on the occurrence of this disturbance in the forests of southern Patagonia (from 48 to 55° S). To address this knowledge gap, new areas and species in under-represented forest types need to be explored throughout Patagonia to better understand fire regimes and post-fire responses, particularly in the southern latitudes. For example, new research is underway in *Nothofagus pumilio* forests to collect fire scars, reconstruct fire regimes and explore the potential influence of climate variability and human occupation history on fire occurrence (Mundo et al. 2016). This work includes sites east and west of the southern Patagonian Andes from Lago Guacho in Chubut province (43°50′S) to Lago Argentino-Brazo Rico (50°28′S) in Santa Cruz province, Argentina. Preliminary results from 16 sites include fire scars back to 1791 and high fire frequency during the twentieth century which is consistent with the pattern of human occupation and colonization in southern Patagonia during the same period (Mundo et al. 2016).

9.3.2 Altered Fire Regimes and Changes in Severity: Impacts of Climate Change, Invasive Plants and Herbivores

The resilience of an ecosystem and its disturbance regime are strongly and reciprocally determined by ecosystem structure, composition and landscape pattern at fine, meso, and broad scales (Scheffer et al. 2001). Changes in any of these factors by gradual perturbation or a large punctuated disturbance can weaken system-level resilience and tip ecosystems across thresholds into new or alternative states that are self-reinforced by new internal feedbacks. In Patagonia, the increase in fire frequency due to land-use change and forest clearing from the 1850s through the 1960s, amplified by post-1950s and 1970s temperature increases has resulted in a regional-scale alteration of fire regimes of many forest types. These regional shifts in fire regimes also are facilitated by the introduction of flammable invasive plants and effects of exotic herbivores on native post-fire vegetation. Thus, changes in vegetation have been observed due to novel fire-vegetation feedbacks (Veblen et al. 2011; Kitzberger et al. 2016). In turn, these changes have affected and will continue to shape future wildfire activity (Veblen et al. 2008). Repeated mid-to-high severity burns have the potential to remove dominant tree vegetation at the landscape-scale, as has been recently documented in mesic/cold subalpine *N. pumilio* (Paritsis et al. 2015), perhumid/cool *P. uviferum* (Zaret and Holz 2016), and dry/cold subalpine *Araucaria* (González and Lara 2015) forests. Ongoing work is aimed at understanding the successional pathway and associated flammability that will increase (decrease) fire frequency/severity through positive (negative) feedbacks.

As in many regions globally, the traditional emphases in dendrochronological studies in Patagonia have been on climatic reconstructions based on the largest and oldest individuals often located in small patches and fire history reconstructed from fire-scar based chronologies collected at areas of low-to-moderate fire severity. Documenting the range of effects that past fires and associated climate conditions have had on tree recruitment would allow a better understanding of the details of forests succession. Knowledge on the relationship between fire frequency and severity, and succession in these forests would allow predictive vegetation models to better simulate the impact of climate change on fire activity, stand structure and species composition. In combination with remote sensing analyses and field manipulative experiments, dendroecology has the capacity to link spatial patterns and ecological mechanisms (e.g. post-fire regeneration dynamics) with long-term sedimentary charcoal records. This powerful multi-method approach can provide invaluable knowledge and datasets to build process-based simulation models to predict the dynamics of Patagonian forests under global change.

9.3.3 Paleo Context for Anticipating the Future: Needs for Long-Term Charcoal/Pollen/Dendroecology Pairing

In the context of climate change impacts and human-altered fire regimes globally, dendropyrochonological proxies have proved to be useful for providing historical reference conditions of the structure and composition prior to European settlement. However, most dendroecologically-based reference condition studies lack information on fire severity, and are unlikely to fully represent scenarios of future climate change (Williams and Jackson 2007), especially when past climate conditions differed from those anticipated in the future. Historical references are, however, critical in helping to identify when and where ecological thresholds are likely to be exceeded, especially as a result of climate-mediated disturbances during decadal- and centennial-scale warm periods in the past. Around the world, studies of historical reference conditions have helped to bind the uncertainty linked to these complex climate-fire-vegetation dynamics (Higuera et al. 2009; Fletcher et al. 2014). For example, in North America, paired pollen and charcoal records indicate that rapid warming during the Medieval Climate Anomaly (ca. 850–1250 AD) resulted in an increased fire severity and frequency with negative feedbacks that caused a shift from mesic to open park-like forests (Calder et al. 2015). This historical warm, dry climate period is comparable to current weather extremes and forecasted climate conditions, providing an example of potential future fire regimes and vegetation changes (Calder et al. 2015). Although comparable work is at an incipient phase in southern South America (e.g. Holz et al. 2012a), there are ongoing efforts to combine evidence from pollen, charcoal, tree rings and other proxies to better understand complex climate-fire-vegetation dynamics.

Acknowledgments We thank the Administración de Parques Nacionales (National Parks Administration of Argentina) and private owners for sampling permissions during the last decades. IAM was supported by the Agencia Nacional de Promoción Científica y Tecnológica de Argentina (grant PICT 2012-1891). AH thanks National Science Foundation (Grant No. 0956552) and National Geographic Society (Grant No.7988-06). MEG thanks the Center for Climate and Resilience Research $(CR)^2$ (CONICYT/FONDAP/15110009) and FONDECYT (Grant No. 1171400). IAM and JP are researchers at CONICET (National Council for Scientific and Technical Research of Argentina).

References

Aguilera-Betti I, Muñoz A, Stahle D, et al (2017) The first millennium-age *Araucaria araucana* in Patagonia. Tree Ring Res 73: 53–56. doi:10.3959/1536-1098-73.1.53

Alfonso JL (1941) El Pehuén, Araucaria o Pino del Neuquén en la Argentina. Ing Agron 3:1–14

Aravena JC (2007) Reconstructing climate variability using tree rings and glacier fluctuations in the southern Chilean Andes. PhD Thesis, University of Western Ontario

Bannister JR, Donoso PJ, Bauhus J (2012) Persistence of the slow growing conifer *Pilgerodendron uviferum* in old-growth and fire-disturbed southern bog forests. Ecosystems 15:1158–1172. doi:10.1007/s10021-012-9574-7

Bannister JR, Wagner S, Donoso PJ, Bauhus J (2014) The importance of seed trees in the dioecious conifer *Pilgerodendron uviferum* for passive restoration of fire disturbed southern bog forests. Austral Ecol 39:204–213. doi:10.1111/aec.12060

Burns BR (1993) Fire-induced dynamics of *Araucaria araucana-Nothofagus antarctica* forest in the Southern Andes. J Biogeogr 20:669–685

Calder WJ, Parker D, Stopka CJ et al (2015) Medieval warming initiated exceptionally large wildfire outbreaks in the Rocky Mountains. Proc Natl Acad Sci 112:13261–13266. doi:10.1073/pnas.1500796112

Cóbar-Carranza AJ, García RA, Pauchard A, Peña E (2015) Efecto de la alta temperatura en la germinación y supervivencia de semillas de la especie invasora *Pinus contorta* y dos especies nativas del sur de Chile. Bosque Valdivia 36:53–60

Cóbar-Carranza AJ, García RA, Pauchard A, Peña E (2014) Effect of *Pinus contorta* invasion on forest fuel properties and its potential implications on the fire regime of *Araucaria araucana* and *Nothofagus antarctica* forests. Biol Invasions 16:2273–2291. doi:10.1007/s10530-014-0663-8

Daniels LD, Veblen TT (2000) ENSO effects on temperature and precipitation of the Patagonian-Andean region: implications for biogeography. Phys Geogr 21:223–243

Díaz MF, Bigelow S, Armesto JJ (2007) Alteration of the hydrologic cycle due to forest clearing and its consequences for rainforest succession. For Ecol Manag 244:32–40. doi:10.1016/j.foreco.2007.03.030

Dillehay TD, Ocampo C, Saavedra J et al (2015) New Archaeological Evidence for an Early Human Presence at Monte Verde, Chile. PLoS One 10:e0141923. doi:10.1371/journal.pone.0141923

Dimitri MJ (1972) La región de los Bosques Andino-Patagónicos. Sinopsis general. INTA, Buenos Aires

Donoso C (1981) Tipos forestales de los bosques nativos de Chile. CONAF-PNUD-FAO, Santiago de Chile

Donoso Zegers C, Escobar B, Pastorino M et al (2006a) *Austrocedrus chilensis* Pic.Ser. et Bizarri; Ciprés de la Cordillera, Len. In: Donoso ZC (ed) Las especies arbóreas de los bosques templados de Chile y Argentina: Autoecología. Marisa Cúneo Ediciones, Valdivia, pp 54–67

Donoso Zegers C, Lara A, Escobar B et al (2006b) *Fitzroya cupressoides* (Molina) I.M. Johnst.; Alerce, Lahuén, Lahuán. In: Donoso ZC (ed) Las especies arbóreas de los bosques templados de Chile y Argentina: Autoecología. Marisa Cúneo Ediciones, Valdivia, pp 68–81

Enright NJ, Fontaine JB, Bowman DMJS et al (2015) Interval squeeze: altered fire regimes and demographic responses interact to threaten woody species persistence as climate changes. Front Ecol Environ 13:265–272. doi:10.1890/140231

Fletcher M-S, Wood SW, Haberle SG (2014) A fire-driven shift from forest to non-forest: evidence for alternative stable states? Ecology 95:2504–2513. doi:10.1890/12-1766.1

Fraver S, González ME, Silla F et al (1999) Composition and structure of remnant *Fitzroya cupressoides* (Alerce) forests of southern Chile's Central Depression. J Torrey Bot Soc 126: 49–57

Garreaud R, Lopez P, Minvielle M, Rojas M (2013) Large-scale control on the Patagonian climate. J Clim 26:215–230. doi:10.1175/JCLI-D-12-00001.1

Garreaud RD, Nicora MG, Bürgesser RE, Ávila EE (2014) Lightning in Western Patagonia. J Geophys Res Atmos 119:4471–4485. doi:10.1002/2013JD021160

González ME (2005) Fire history data as reference information in ecological restoration. Dendrochronologia 22:149–154. doi:10.1016/j.dendro.2005.04.001

González ME, Lara A (2015) Large fires in the Andean *Araucaria* forests: when a natural ecological process becomes a threat. Orix Int J Conserv 49:394. doi:10.1017/S0030605315000575

González ME, Lara A (2016) Restauración post-fuego de los bosques de *Araucaria*: nuevos escenarios, antiguas amenazas. In: 1er Congreso Nacional de Áreas Silvestres Protegidas del Estado, Temuco, Chile

González ME, Muñoz A, Assal T, Sibold JS (2016) Fire regimes and fire effects in Chilean Araucaria forests. Symposium Wildfire regime shifts in Southern temperate forest ecosystems: climate change, anthropogenic influences, and ecological feedbacks. In: Abstract of VIII Southern Connection Congress. Punta Arenas, p 53

González ME, Veblen TT (2006) Climatic influences on fire in *Araucaria araucana-Nothofagus* forests in the Andean cordillera of south-central Chile. Écoscience 13:342–350

González ME, Veblen TT (2007) Incendios en bosques de *Araucaria araucana* y consideraciones ecológicas al madereo de aprovechamiento en áreas recientemente quemadas. Rev Chil Hist Nat 80:243–253

González ME, Veblen TT, Sibold JS (2010) Influence of fire severity on stand development of *Araucaria araucana–Nothofagus pumilio* stands in the Andean cordillera of south-central Chile. Austral Ecol 35:597–615

González ME, Veblen TT, Sibold JS (2005) Fire history of *Araucaria-Nothofagus* forests in Villarrica National Park, Chile. J Biogeogr 32:1187–1202

Hadad MA, Roig Juñent FA, Boninsegna JA, Patón-Domínguez D (2015) Age effects on the climatic signal in *Araucaria araucana* from xeric sites in Patagonia, Argentina. Plant Ecol Divers 8:343–351

Heusser C (1994) Paleoindians and fire during the late Quaternary in southern South America. Rev Chil Hist Nat 67:455–442

Heusser CJ, Rabassa J, Brandani A (1988) Late-Holocene vegetation of the Andean Araucaria region, Province of Neuquén, Argentina. Mt Res Dev 8:53–63

Higuera PE, Brubaker LB, Anderson PM et al (2009) Vegetation mediated the impacts of postglacial climate change on fire regimes in the south-central Brooks Range, Alaska. Ecol Monogr 79:201–219. doi:10.1890/07-2019.1

Holz A, Méndez C, Borrero L, Prieto A, Torrejón F, Maldonado A (2016) Fires: the main human impact on past environments in Patagonia? PAGES Magazine 24:72–73. doi:10.22498/pages.24.2.72

Holz A (2009) Climatic and human influences on fire regime and forest dynamics in temperate rainforests in southern Chile. Department of Geography, University of Colorado, Boulder

Holz A, Haberle S, Veblen TT et al (2012a) Fire history in western Patagonia from paired tree-ring fire-scar and charcoal records. Clim Past 8:451–466. doi:10.5194/cp-8-451-2012

Holz A, Kitzberger T, Paritsis J, Veblen TT (2012b) Ecological and climatic controls of modern wildfire activity patterns across southwestern South America. Ecosphere 3:art103. doi:10.1890/ES12-00234.1

Holz A, Veblen TT (2011a) Variability in the Southern Annular Mode determines wildfire activity in Patagonia. Geophys Res Lett 38:L14710

Holz A, Veblen TT (2011b) The amplifying effects of humans on fire regimes in temperate rainforests in western Patagonia. Palaeogeogr Palaeoclimatol Palaeoecol 311:82–92. doi:10.1016/j.palaeo.2011.08.008

Holz A, Veblen TT (2012) Wildfire activity in rainforests in western Patagonia linked to the Southern Annular Mode. Int J Wildland Fire 21:114–126

Holz A, Veblen TT (2009) *Pilgerodendron uviferum*: The southernmost tree-ring fire recorder species. Écoscience 16:322–329

Kitzberger T, Aráoz E, Gowda JH et al (2012) Decreases in fire spread probability with forest age promotes alternative community states, reduced resilience to climate variability and large fire regime shifts. Ecosystems 15:97–112. doi:10.1007/s10021-011-9494-y

Kitzberger T, Perry GLW, Paritsis J et al (2016) Fire–vegetation feedbacks and alternative states: common mechanisms of temperate forest vulnerability to fire in southern South America and New Zealand. N Z J Bot 54:247–272. doi:10.1080/0028825X.2016.1151903

Kitzberger T, Steinaker DF, Veblen TT (2000) Effects of climatic variability on facilitation of tree establishment in northern Patagonia. Ecology 81:1914–1924

Kitzberger T, Veblen TT (1997) Influences of humans and ENSO on fire history of *Austrocedrus chilensis* woodlands in northern Patagonia, Argentina. Écoscience 4:508–520

Kitzberger T, Veblen TT, Villalba R (1997) Climatic influences on fire regimes along a rain forest-to-xeric woodland gradient in northern Patagonia, Argentina. J Biogeogr 24:35–47

LaMarche VC, Holmes RL, Donwiddie P, Drew L (1979) Tree-ring chronologies of the southern hemisphere: 1. University of Arizona, Tucson

Lara A (1991) The dynamics and disturbance regimes of *Fitzroya cupressoides* forests in the south-central Andes of Chile. PhD Thesis, Department of Geography, University of Colorado

Lara A, Donoso ZC, Escobar B et al (2006) *Pilgerodendron uviferum* (D. Don) Florin. In: Donoso ZC (ed) Las especies arbóreas de los bosques templados de Chile y Argentina. Autoecología. Marisa Cúneo Ediciones, Valdivia, pp 82–91

Lara A, Fraver S, Aravena JC, Wolodarsky-Franke A (1999) Fire and the dynamics of *Fitzroya cupressoides* (alerce) forests of Chile's Cordillera Pelada. Écoscience 6:100–109

Lara A, Villalba R (1993) A 3620-year temperature record from Fitzroya cupressoides tree rings in Southern South America. Science 260:1104–1106

Lara A, Wolodarsky-Franke A, Aravena JC et al (2003) Fire regimes and forest dynamics in the Lake Region of south-central Chile. In: Veblen TT, Baker WL, Montenegro G, Swetnam TW (eds) Fire and climatic change in temperate ecosystems of the Western Americas. Springer, New York, pp 322–342

Mohr-Bell D (2015) Superficies afectadas por incendios en la región Bosque Andino Patagónico durante los veranos de 2013–2014 y 2014–2015. SAyDS-CIEFAP, Nodo Regional Bosque Andino Patagónico

Mundo IA, Kitzberger T, Roig Juñent FA et al (2013) Fire history in the *Araucaria araucana* forests of Argentina: Human and climate influences. Int J Wildland Fire 22:194–206. doi:10.1071/WF11164

Mundo IA, Roig Juñent FA, Villalba R et al (2012) *Araucaria araucana* tree-ring chronologies in Argentina: spatial growth variations and climate influences. Trees Struct Funct 26:443–458. doi:10.1007/s00468-011-0605-3

Mundo IA, Villalba R, Veblen TT, et al (2016) Fire history in Southern Patagonia: human and climate influences on fire activity in *Nothofagus pumilio* forests. In: Meeting Program and Abstracts—AmeriDendro 2016. Mendoza, Argentina, pp 38–39

Paritsis J, Veblen TT, Holz A (2015) Positive fire feedbacks contribute to shifts from *Nothofagus pumilio* forests to fire-prone shrublands in Patagonia. J Veg Sci 26:89–101. doi:10.1111/jvs.12225

Quezada JM (2008) Historia de incendios en bosques de *Araucaria araucana* (Mol.) Koch del Parque Nacional Villarrica, a partir de anillos de crecimiento y registros orales. Tesina de grado, Universidad Austral de Chile, Facultad de Ciencias Forestales

Rovere A, Premoli AC, Newton AC (2002) Estado de conservación de Ciprés de las Guaitecas (*Pilgerodendron uviferum* [D Don] Florin) en Argentina. Bosque 23:11–20

Scheffer M, Carpenter S, Foley JA et al (2001) Catastrophic shifts in ecosystems. Nature 413:591–596. doi:10.1038/35098000

Szeicz JM, Haberle SG, Bennett KD (2003) Dynamics of North Patagonian rainforests from fine-resolution pollen, charcoal and tree-ring analysis, Chonos Archipelago, Southern Chile. Austral Ecol 28:413–422. doi:10.1046/j.1442-9993.2003.01299.x

Tortorelli LA (1942) La explotación racional de los bosques de Araucaria de Neuquén. Su importancia económica. Sep Serv VI:1–74

Urrutia R (2002) Desarrollo de una Cronología de Anchos de Anillos para Alerce (*Fitzroya Cupressoides*) y Reconstrucción de la Historia de Incendios en el Área de Abtao, Parque Nacional Chiloé, X Región. Tesis de Ingeniería Forestal, Pontificia Universidad Católica de Chile. BSc Forest Engineering thesis, Pontificia Universidad Católica de Chile

Veblen TT (1982) Regeneration patterns in *Araucaria araucana* forests in Chile. J Biogeogr 9:11–28

Veblen TT, Ashton DH (1982) The regeneration status of *Fitzroya cupressoides* in the Cordillera Pelada, Chile. Biol Conserv 23:141–161

Veblen TT, Burns BR, Kitzberger T et al (1995) The ecology of the conifers of southern South America. In: Ecology of the Southern Conifers. Melbourne University Press, Parkville, pp 120–155

Veblen TT, Donoso C, Kitzberger T, Rebertus AJ (1996) Ecology of Southern Chilean and Argentinean *Nothofagus* forests. In: Veblen TT, Hill RS, Read J (eds) The Ecology and Biogeography of Nothofagus forests. Yale University Press, New Haven, pp 293–353

Veblen TT, Holz A, Paritsis J et al (2011) Adapting to global environmental change in Patagonia: what role for disturbance ecology? Austral Ecol 36:891–903. doi:10.1111/j.1442-9993.2010.02236.x

Veblen TT, Kitzberger T, Raffaele E et al (2008) The historical range of variability of fires in the Andean–Patagonian *Nothofagus* forest region. Int J Wildland Fire 17:724–741. doi:10.1071/WF07152

Veblen TT, Kitzberger T, Villalba R, Donnegan J (1999) Fire history in Northern Patagonia: the roles of humans and climatic variation. Ecol Monogr 69:47–67

Veblen TT, Lorenz DC (1988) Recent vegetation changes along the forest/steppe ecotone of northern Patagonia. Ann Assoc Am Geogr 78:93–111

Villalba R, Veblen TT (1997) Regional patterns of tree population age structures in northern Patagonia: climate and disturbance influences. J Ecol 85:113–124

Walter KS, Gillet HJ (eds) (1998) 1997 IUCN Red List of Threatened Plants. IUCN, Gland/Cambridge

Whitlock C, Moreno PI, Bartlein P (2007) Climatic controls of Holocene fire patterns in southern South America. Quat Res 68:28–36

Williams JW, Jackson ST (2007) Novel climates, no-analog communities, and ecological surprises. Front Ecol Environ 5:475–482. doi:10.1890/070037

Zaret K, Holz A (2016) Seedling performance of *Pilgerodendron uviferum* (Ciprés de las Guaitecas) in a burned peatland, lower Baker river watershed: implications for restoration. In: Abstract of VIII Southern Connection Congress. Punta Arenas, Chile.

Chapter 10
Creating a Buzz: Insect Outbreaks and Disturbance Interactions

James H. Speer and Dominik Kulakowski

Abstract Dendroentochronology is the study of insect outbreaks as recorded in tree rings. This type of research has been going on in earnest since the 1950s but the number of publications has rapidly increased from the 1980s to the present. Most of the work has been completed in Canada and the United States with some important work in Europe and Asia. Insect outbreak studies have been conducted in Australasia and more recently in South America, but very little work has been done in Africa. From tree-ring research on insect outbreaks we have been able to document the outbreak dynamics of more than 20 species of insect around the world. Insects can be grouped into three classes based on their impacts on trees: defoliators, cambium feeders, and root parasites. Using tree rings, we have been able to document periodic occurrences of insect outbreaks, their effects on the volume of wood production, climatic triggers to outbreaks, and the spread of insect outbreaks across landscapes over decades to more than a millenium. Dendrochronologists are now exploring the interactions of multiple insect outbreak systems and their interactions with fire and weather phenomena. From this work, we can see that the cumulative effects of multiple disturbances can be greater than the individual effects and that one disturbance can hinder or enhance a different type of disturbance. Overall, dendroentochronologists have been able to demonstrate the important role of disturbances in natural systems and provide important insight into the management of these ecosystems.

Keywords Insect outbreaks • Fire • Wind throw • Disturbance • Dendroecology • Dendroentochronology

---

J.H. Speer (✉)
Department of Environmental Systems, Indiana State University, Terre Haute, IN, 47809, USA
e-mail: jim.speer@indstate.edu

D. Kulakowski
Graduate School of Geography, Clark University, Worcester, MA, USA
e-mail: DKulakowski@clarku.edu

M.M. Amoroso et al. (eds.), *Dendroecology*, Ecological Studies 231,
DOI 10.1007/978-3-319-61669-8_10

10.1 Introduction

Disturbance ecology has been and continues to be a major focus for dendrochronologists and much research has focused on reconstructions of fire, insect outbreaks, wind, and drought (Swetnam and Betancourt 1998; Speer 2010). Professional foresters have long noted the effect of some of these disturbances on tree growth, but dendrochronological investigations of disturbance ecology have expanded and flourished since the 1980s. This new research has deepened our understanding of the causes and consequences of important forest dynamics and has helped us understand long-term ecological fluctuations, which in turn has helped to put recent and ongoing environmental change into context.

The specific branch of dendrochronology that deals with reconstructing insect outbreaks (abrupt growths in population) and examining the effect of insects on tree growth is called dendroentochronology (Speer 2010). Insect outbreaks are complex and dynamic, depending on the biology of each insect. Insects that are capable of outbreaks with important effects on forest ecosystems can be classified as defoliators (the most common), cambium feeders (mainly bark beetles), or root parasites (periodical cicadas being the best example of this class) (Table 10.1). There are a wide range of defoliators around the world that are well adapted to individual host species or groups of species. During outbreak events, defoliator populations increase exponentially and the resultant herbivory causes loss of photosynthetic area, sometimes damaging the buds or apical meristem of affected trees (Swetnam et al. 1985). This type of damage usually causes a suppression in tree growth for the duration of the outbreak and during the subsequent recovery time of the tree. This suppression can last from 1 year (e.g. tussock moth) (Swetnam et al. 1995), 10 years for western spruce budworm (Swetnam and Lynch 1989), and as long as 14 years with pandora moth (Speer et al. 2001). In between outbreaks, most populations of defoliators crash to endemic levels, from which populations again erupt. Mortality of trees affected by defoliating insects is usually fairly low, allowing dendrochronologists to record multiple outbreaks through the lifetime of the tree (Speer et al. 2001; Clark et al. 2017).

In contrast to defoliators, wood boring beetles lay their eggs underneath the bark of trees and the larvae consume the cambium as a food source (Safranyik and Wilson 2007; Safranyik et al. 2007). Healthy trees can normally defend themselves from endemic populations of bark beetles by pitching out the insects at the site of entry. However, when populations of bark beetles erupt, individual trees can be swarmed by hundreds to thousands of beetles that preferentially target stressed trees that cannot defend themselves except in the recent mountain pine beetle outbreak when the beetles even attacked healthy trees at the peak of the outbreak. If enough larvae feed in a given tree, that tree will be girdled and killed. Bark beetle outbreaks are harder to reconstruct than outbreaks of defoliators because attacked trees die and eventually fall and decay. It is difficult to confidently identify the cause of mortality and evidence of mortality is lost over time. Researchers have relied on beetle galleries under the bark, strip attack scars, cross-dating of beetle-killed trees,

Table 10.1 Examples of insects that have been studied using dendrochronology (updated from Speer 2010)

Common name	Scientific name	Type	Select publications
Mountain pine beetle	*Dendroctonos ponderosae*	Cambium feeder	Shore et al. (2006), Campbell et al. (2007), Alfaro et al. (2010), Axelson et al. (2010), Hrinkevich and Lewis (2011), Schoennagel et al. (2012), Smith et al. (2012), and Jarvis and Kulakowski (2015)
Spruce beetle	*Dendroctonus rufipennis*	Cambium feeder	Baker and Veblen (1990), Veblen et al. (1991a, b), Eisenhart and Veblen (2000), Berg et al. (2006), and Sherriff et al. (2011)
European spruce beetle	*Ips typographus*	Cambium feeder	Čada et al. (2013)
Douglas-fir bark beetle	*Dendroctonus pseudotsugae*	Cambium feeder	Hadley (1994)
Two-year spruce budworm	*Choristoneura biennis*	Defoliator	Zhang and Alfaro (2002)
Western spruce budworm	*Choristoneura occidentalis*	Defoliator	Blais (1962), Swetnam (1987), Swetnam and Lynch (1993), and Flower et al. (2014)
Eastern spruce budworm	*Choristoneura fumiferana*	Defoliator	Morin et al. (1993) and Boulanger and Arseneault (2004)
Larch casebearer	*Coleophora laricella*	Defoliator	Alfaro et al. (1991)
Pandora moth	*Coloradia pandora*	Defoliator	Speer et al. (2001)
Needleminer	*Epinota subsequana*	Defoliator	Camarero et al. (2003)
Gypsy moth	*Lymantria dispar*	Defoliator	Asshof et al. (1999) and Muzika and Liebhold (1999)
Forest tent caterpillar	*Malacosoma disstria*	Defoliator	Duncan and Hodson (1958), Hildahl and Reeks (1960), Hogg et al. (2002), and Huang et al. (2008)
Western false hemlock looper	*Nepytia freemanii*	Defoliator	Alfaro and Macdonald (1988)
Tussock moth	*Orgyia sp.*	Defoliator	Alfaro and Sheppard (1991), Mason et al. (1997) and Wickman (1963, 1980)
White pine weevil	*Pissodes strobi*	Defoliator	Alfaro (1995)
Larch sawfly	*Pristiphora erichsonii*	Defoliator	Girardin et al. (2001) and Case and MacDonald (2003)
Processionary caterpillars	*Thaumetopoea pityocampa*	Defoliator	Carus (2004, 2009)
Cedar processionary moth	*Tramatocampa ispartaensis*	Defoliator	Avcí and Carus (2005)
Larch budmoth	*Zeiraphera diniana also, Zeiraphera griseana*	Defoliator	Rolland et al. (2001) and Esper et al. (2007)
Periodical cicadas	*Magicicada sp.*	Root parasite	Speer et al. (2010)

and the response of non-host trees that are released from competition due to the death of beetle-killed neighboring canopy trees (Axelson et al. 2010). The latter method works best in mixed-species stands, such as Engelmann spruce-subalpine fir stands of the Rocky Mountains (Veblen et al. 1991a). In these forests, spruce beetle (*Dendroctonus rufipennis*) kill Engelmann spruce, which leads to abrupt and sustained increases in growth of subalpine fir. These growth releases, along with the species-specific mortality, can then be used to definitively reconstruct spruce beetle outbreaks. In contrast, the European spruce beetle (*Ips typographus*) attacks Norway spruce stands, which are often monospecific. The absence of non-host tree species makes distinguishing outbreaks from windstorms in the tree-ring record difficult (e.g., Svoboda et al. 2012).

Dendrochronologists have recently started to study the effect of periodical cicadas on tree growth as a root parasite (Speer et al. 2010). Annual cicadas occur on every continent except for Antarctica, but periodical cicadas are a major disturbance agent in the eastern deciduous forests of the USA that cause some defoliation and make up a major part of the food web in emergence years. This is the first study to examine the effects of a root parasite on tree growth. Applications of dendrochronology are expanding every year and providing more tools that researchers can use to study natural systems.

In addition to studying the history, causes, and consequences of individual disturbances, in the last couple of decades, many researchers have embraced the complexity of ecological systems and have started to grapple with investigating multiple disturbances and their interactions in individual stands. This chapter reviews dendrochronological inquiries into individual insect outbreaks as well as interactions among disturbances and demonstrates how associated research approaches have developed in the literature.

10.2 History of Insect Outbreak Studies

The realization that defoliating insects cause a reduction in tree growth goes back to early researchers such as the German Botanist Ratzbuger (1866 as cited in Studhalter 1995 and Wimmer 2001). An early introductory forestry textbook (Hough 1882) shows a picture of suppressed tree rings due to an insect outbreak, demonstrating that the idea of trees recording insect outbreaks has been known to foresters for more than 100 years. The field of dendroecology dramatically advanced with the work of J.R. Blais in the 1950s through 1980s with his tree-ring documentation of eastern spruce budworm (*Choristoneura fumiferana*) outbreaks in Canada (e.g. Blais 1954, 1965, 1983) (Fig. 10.1). His early work used tree and stand measurements to examine survival and regeneration after budworm outbreaks along with the documentation of suppressions in the tree-ring record (Blais 1954). Duncan and Hodson (1958) were among the first to use tree rings to explore the effect of defoliation from tent caterpillar (*Malacosoma disstria*) in Minnesota. Annual tree growth increment was examined by Hildahl and Reeks (1960) to determine the effect of tent caterpillar on growth of trembling aspen (*Populus tremuloisdes*)

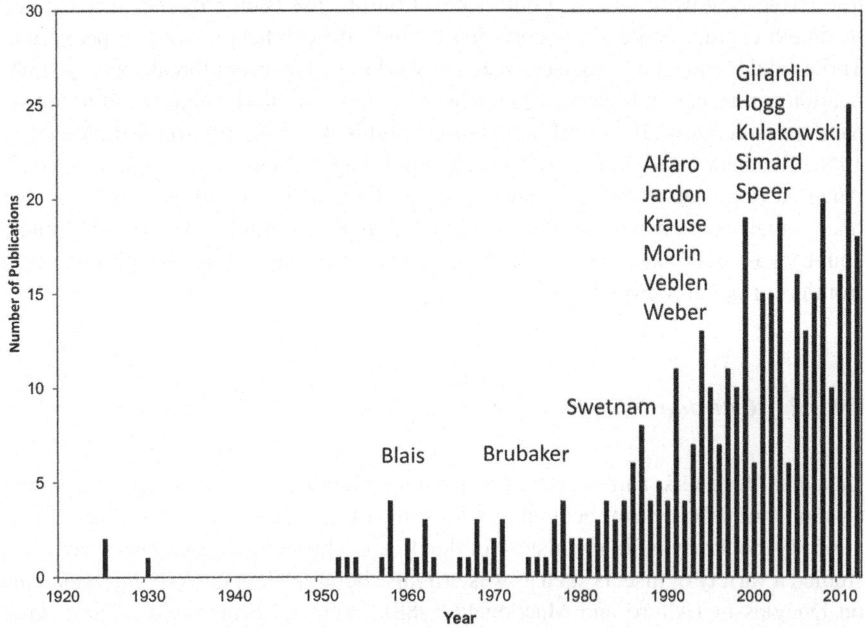

Fig. 10.1 Number of Dendrochronology publications in English on insect outbreaks through time with authors listed that had five or more publications on insect outbreaks based on the data from the Bibliography of Dendrochronology by Henri Grissino-Mayer with the latest data updated through 2012 (Dobbertin and Grissino-Mayer 2004)

in Manitoba and Saskatchewan, Canada. Recorded observations indicated that defoliation from this insect occurred every year from 1923 to 1957 and the tree-ring record showed that 8.4% of the total basal area was decreased due to defoliation over time.

A decade later researchers started to examine the effects of multiple insects in the same tree species by documenting suppressions from Douglas-fir tussock moth (*Orgia pseudotsugata*) and western spruce budworm (*Choristoneura occidentalis*) in grand fir (*Abies grandis*) and Douglas-fir (*Pseudotsuga menziesii*) in Washington state (Brubaker and Greene 1979). Dendrochronological techniques were improved and documented by Swetnam et al. (1985) who demonstrated the use of host and non-host chronologies to differentiate drought from insect outbreaks in western spruce budworm populations.

10.2.1 Regional Consideration of Insect Outbreak Dynamics

Much of the intensive tree-ring work on reconstructing insect outbreaks started in Canada where it has found the greatest application. Work in the United States has examined new insect outbreak systems and multiple disturbances in the same trees. Tree-ring work on insect outbreaks in Europe has been growing since the late twentieth century and early twenty-first century. Researchers in Asia (especially in Turkey and Russia) have been consistently working on insect outbreaks with 16 publications on insect outbreaks in that area as of 2016. Similarly, dendrochronologists have been increasingly publishing on insect outbreaks in Australasia (e.g., Mazanec 1968; Readshaw and Mazanec 1969; Morrow and LaMarche 1978) and in South America (e.g., Delgado 2000; Paritsis et al. 2009; Paritsis and Veblen 2011) but many other locations around the world including the continent of Africa and many countries in Asia have not yet examined insect-tree interactions, which remains a frontier for dendrochronology.

10.2.2 Canada

Canadian foresters and researchers have conducted much research examining eastern spruce budworm, beginning with some of the earliest work by Blais (1954, 1965, 1983). Early on in the history of dendroentochronology researchers have also studied a variety of insects such as western false hemlock looper (*Nepytia freemanii*) on Douglas-fir (Alfaro and Macdonald 1988), larch casebearer (*Coleophora laricella*) found on western larch (*Larix occidentalis*) (Alfaro et al. 1991), larch sawfly (*Pristiphora erichsonsii)* (Jardon et al. 1994; Girardin et al. 2001), and forest tent caterpillar (*Malacosoma disstria*) (Huang et al. 2008). Much of the work completed in Canada focused on eastern spruce budworm (Morin et al. 1993, 2009; Krause and Morin 1995; Boulanger et al. 2012). The majority of the research on insect outbreaks using dendrochronology (based on five or more first authored papers found in the Bibliography of Dendrochronology and other web searches using Google Scholar) has been completed in Canada in the last 30 years with work by Alfaro (Alfaro 1995), Girardin (Girardin et al. 2001), Hogg (Hogg et al. 2002), Jardon (Jardon et al. 1994), Krause (Krause 1997), Morin (Morin 1994), and Simard (Simard et al. 2008) (Fig. 10.1). Canadian researchers have also been on the forefront of using wood anatomical features to identify insect outbreaks (Alfaro 1995; Hogg et al. 2002; Sutton and Tardif 2007; Huang et al. 2008; Axelson et al. 2014).

10.2.3 United States

Researchers from the United States have been integral in defining the field of dendroentochronology (e.g., Swetnam et al. 1985), have worked with new insect systems demonstrating how they could be studied using dendrochronology (e.g., Speer et al. 2001, 2010), and have examined the complexity of natural systems by integrating the study of snow avalanches, fire, and insects in the same landscape (e.g., Veblen et al. 1994). Research has expanded into reconstruction of the history of several key insect species (e.g., Eisenhart and Veblen 2000; Fraver et al. 2007; Flower et al. 2014; Jarvis and Kulakowski 2015) and the examination of disturbances interactions such as fire and bark beetle outbreaks (e.g., Kulakowski et al. 2003, 2012). Brubaker and Greene (1979) documented Douglas-fir tussock moth and western spruce budworm in Douglas-fir and grand fir trees and provided one of the earlier examples of the ability of tree-ring analysis to differentiate two separate insect species in the same chronologies. Lynch (2012) provides an excellent summary of the recent research in dendroentochronology and the insects that researchers have examined.

10.2.4 Europe

Dendrochronological research in Europe has examined outbreaks of various insects, though this body of work is not as large as in other regions, in part due to intense human influence across much of the continent which has obscured evidence of past forest dynamics (Kulakowski et al. 2016a). Despite this, dendrochronological studies in Europe have shed light on several insect systems. Camerero et al. (2003) worked in *Abies alba* in the Aragón Pyrenees in Spain and examined the needle miner, which bore through the needle, causing death of the needles and subsequent defoliation. Baltensweiler (1964) has studied the larch budmoth (*Zeiraphera diniana* also identified as *Zeiraphera griseana*) since 1964 in Europe and has recently started to collaborate on dendrochronological studies of the insect (Rolland et al. 2001; Baltensweiler et al. 2008). Weber (1997) reconstructed 57 outbreaks of the larch budmoth from AD 1503 to 1990 and documented an average periodicity of 8.58 year. One of the longest insect outbreak reconstructions in the world with annual resolution was completed by Esper et al. (2007), which demonstrates the utility of archaeological samples to extend a chronology back in time, as well as the work that can be completed during multiple dendrochronology fieldweeks. In this work, Esper et al. (2007) developed a larch budmoth (*Zeiraphera diniana*) chronology that extended back 1173 years in the European Alps and demonstrated a consistent cycle over this whole period that ended in 1981, suggesting current vulnerability of the ecological system to anthropogenic climate change. Buntgen et al. (2009) examined a network of 70 host chronologies (larch) and 73 non-host chronologies (spruce) across Europe to examine six different reconstruction techniques for larch budmoth, the stability of the outbreak cycle through time, and the potential effect on climate reconstructions. They suggested further collaborations between biologists, ecologists, and climatologists for a better understanding of tree dynamics and climate change.

10.2.5 Asia

Most of the work using dendrochronology in Asia was by Carus, who produced four publications on defoliators in Turkey (Carus 2004, 2009, 2010; Carus and Avci 2005) with much of his work focusing on pine processionary moth (*Thaumetopoea wilkinsoni*). Other important work has been conducted in Russia, Georgia, Mongolia, and India. Sarajishvili (1997) studied new pine knot-horn moth in Georgia near the Black Sea, Sviderskaya and Pal'nikova (2003) studied pine looper (*Bupalus piniarius*) in Russia, and Kucherov (1991) studied gypsy moth (*Lymantria dispar*) in the Ural Mountains. Studies in central Asia include those in Mongolia (Dulamsuren et al. 2010) on gypsy moth and in India (Priya and Bhat 1998) on insect damage in teak (*Tectona grandis*). Whereas Turkey and Russia have been well studied, much more work remains to be done in Asia, especially in the central region of the continent.

10.3 Millennial Length Chronologies

Insect outbreak reconstructions have just broken the millennial boundary with Esper et al. (2007) publishing the first millennial length chronology with a 1173 year-long larch budmoth reconstruction. They reconstructed outbreaks from a network of four sites in the Swiss Alps with the majority of their samples coming from the Lötschental Valley in Switzerland. Much of their data came from archaeological sites which allowed them to extend their chronology so far back in time. They used a decrease in maximum latewood density to reconstruct larch budmoth defoliation events and found an average 9.3 year cycle that lasted throughout their record back to AD 832, but abruptly ended in 1981. They hypothesize that the changing climate has made this natural ecosystem vulnerable and that this change is greater than any effect observed during the medieval warm period.

Clark et al. (2017) developed a 1572 year-long pandora moth (*Coloradia pandora*) reconstruction by collecting samples from some of the oldest documented living ponderosa pine (*Pinus ponderosa*) trees that individually were living for more than 800 years and combining that in Oregon with remnant wood that has existed on the surface of the Mokst Butte lava flow in Oregon for hundreds of years. They were able to reconstruct 13 individual pandora moth outbreaks and the Palmer Drought Severity Index throughout the length of the chronology. They demonstrate that removing the time period of pandora moth suppression during the calibration of the climate response function improved the variance explained in their climate model by 11%. Through this work, they demonstrate the importance of understanding both the ecological effects such as disturbance as well as the climatic effects on tree growth to develop more reliable climate reconstructions.

10.4 New Techniques in Insect Outbreak Detection

Simard et al. (2008) used **stable carbon** isotopes as indicators of past spruce budworm outbreaks in Quebec. They found an enrichment in $\delta^{13}C$ during years of outbreak which they believe is due to needle removal during the defoliation. They found that this indicator could be as good of a measure of spruce budworm outbreaks as traditional ring-width indices. Other researchers used **wood anatomical features** to identify insect outbreaks including white ring formation (Hogg et al. 2002; Sutton and Tardif 2005; Huang et al. 2008) and the number of latewood tracheids (Krause and Morin 1995; Axelson et al. 2014). This is a new area in dendroentochronology during which researchers can use other cell characteristics within the rings as definitive indicators of insect outbreaks back through time. Simard and Payette (2003) used a unique application of **dendroarchitecture** to examine the fine scale effects of spruce budworm defoliation on individual trees. They dated the mortality and initiation of individual limbs on trees in response to bud damage from spruce budworm to track the effects of the outbreak on individual trees. In doing this, they found that they had better resolution in time and space of the first incidence of attack by spruce budworm.

Researchers have now developed enough site-level insect outbreak reconstructions that they are starting to put these together into **networks** which may allow landscape-level analysis of outbreaks. Speer and Jensen (2003) were able to document the initiation and spread of pandora moth outbreaks in Oregon from the first year of suppression across 14 sites in central Oregon. De Graauw (2012) took this further and examined 200 ponderosa pine chronologies from the International Tree-Ring Databank (ITRDB) to explore pandora moth outbreaks across their entire range. From this data set, she could observe the spread of outbreaks across the landscape and document these outbreaks spreading into regions previously thought to have been unaffected by pandora moth outbreaks. This type of research will enable future studies on the driving mechanisms of insect outbreak spread, which in the pandora moth system is thought to be wind dispersal of the adults (Speer 1997; de Graauw 2012). Zhang and Alfaro (2003) developed a network of 2-year cycle spruce budworm outbreaks from eight sites through central British Columbia and documented the size of the outbreaks by the number of sites each outbreak affected with only four events covering all eight sites. Huang et al. (2008) used a latitudinal gradient to document the northern fluctuations of the forest tent caterpillar over time. They used white rings as an indicator of forest tent caterpillar defoliation and documented an outbreak of the insect at higher latitudes (further north) than had previously been documented. Buntgen et al. (2009) examine outbreak amplitude, synchrony, and climate dependence over the past 300 years from 70 larch sites exploring the dynamics of larch budmoth across France, Switzerland, Italy, Austria, and Slovenia. Flower et al. (2014) used 10 sites to examine fire and western spruce budworm outbreaks across Oregon, Idaho, and Montana. They found that the two disturbances could have a subtle synergistic relationship to each other such that they could reduce the severity of each other over time, but they were both driven by climate, and the occurrence of one disturbance did not affect the probability of occurrence of the other. Axelson et al. (2015) developed a network of 11 reconstructions of western spruce budworm in the central interior of British Columbia to determine if recent outbreak history demonstrated a range expansion for the insect because large outbreaks had not been observed in this region since records started in the 1950s. They found that the trees in this area did record previous western spruce budworm outbreaks and that the current dynamics were within the natural range of variability.

10.5 Multiple Disturbance Interactions

Over the past decades ecologists have increasingly turned their attention not only to understanding individual disturbance agents, but also to understanding how multiple disturbances interact (Table 10.2). Dendrochronology has provided an important methodological approach to this line of research. Recent research on disturbance interactions has used remote sensing and field surveys (e.g., reviewed in Kulakowski and Veblen 2015) and has led to important understanding of how disturbances interact. The foundations of this recent research stem from dendroecological investigations of disturbance interactions, which have the advantage of being able to examine disturbance interactions over longer time periods and longer ago than studies based only on remote sensing and field surveys. Most research on disturbance interactions has focused on how the occurrence of one disturbance affects the occurrence of subsequent disturbances, but more recent studies have also looked at how the occurrence of two or more disturbances in short succession cumulatively affect ecosystems and at how the severity of disturbances affects disturbance interactions.

Table 10.2 Disturbance interactions that have been studied using dendrochronology

Disturbance 1	Disturbance 2	Disturbance 3	Publications
Fire	Spruce beetle		Veblen et al. (1994), Kulakowski et al. (2003), Kulakowski and Veblen (2006), Bakaj et al. (2016), and Kulakowski et al. (2016b)
Fire	Wind		Kulakowski and Veblen (2002)
Spruce beetle	Fire		Kulakowski et al. (2003)
Mountain pine beetle	Fire		Kulakowski and Jarvis (2011)
Western spruce budworm	Drought		Flower et al. (2014)
Western spruce budworm and Douglas-fir beetle	Fire		Hadley (1994)
Pandora moth	Fire		Speer (1997)
Climate	Fire	Pandora moth	Pohl et al. (2006)
Climate	Fire	Western spruce budworm	Swetnam and Betancourt (1998)

10.5.1 Effect of Fires on Insect Outbreaks and Other Disturbances

Given the primacy of wildfires in controlling stand structure in many forest ecosystems, much dendrochronological research on disturbance interactions has focused on how fires affect or are affected by other disturbances. Much of this research has been conducted in the Rocky Mountains and in other forests of the Western United States.

Most dendrochronological research on disturbance interactions has combined tree-ring methods to provide a temporal dimension and GIS to provide a spatial dimension. In a seminal study, Veblen et al. (1994) mapped disturbances by fires, insect outbreaks, and avalanches over the last centuries in a contiguous 4000 hectare area using a combination of dendrochronological methods, including reconstructing stand-origin dates, fire-scar dates, mortality of beetle-killed trees, and releases in surviving trees, along with aerial photo interpretation. They examined the spatial patterns of reconstructed disturbance and showed that avalanche tracks can act as breaks for stand-replacing fires and stand-replacing fires can reduce susceptibility to and affect the extent of subsequent spruce beetle (*Dendroctonus rufipennis*) outbreaks. Kulakowski and Veblen (2002) built on this work by using dendrochronological techniques, remote sensing, and GIS to map stand-replacing fires in a contiguous area of subalpine forest in Colorado and showed that the extent and severity of a major wind disturbance in 1997 (as mapped based on aerial photo interpretation) were controlled by fire history over the past centuries. Kulakowski et al. (2003) then showed that stand-replacing fires in the nineteenth century determined the extent of a large outbreak of spruce beetle in the 1940s, suggesting that interactions between stand-replacing fires and bark beetle outbreaks may have been a predictable component of subalpine forests in the Rocky Mountains.

Hadley (1994) explored the interactions of fire, western spruce budworm, Douglas-fir beetle, and topography. He found that fire can be a broad-scale phenomena that removes living trees from the landscape, effectively reducing the age and diameter of the standing tree stock. This in turn can delay western spruce budworm outbreaks, but once budworm outbreaks occur, Douglas-fir beetle will often follow in the weakened trees. He further found that all of this is mediated by topography where the forests would recover more quickly from fire on north facing slopes and more slowly on the drier south facing slopes, although the insect outbreaks seem to be more damaging on north facing slopes due to higher density of host species. All of these factors combined to create a complex mosaic of forest age and structure with disturbances occurring at different spatial scales.

The methodological approach of many dendrochronological studies of disturbance interactions has been based on the premise of patch dynamics that states that landscape heterogeneity results from variations in disturbance history as well as underlying environmental heterogeneity. Methodologically this translates to identifying unique patches on the landscape that are visible on aerial photographs, visiting those patches in the field, and reconstructing the disturbance history of those patches using dendrochronological methods. Various GIS methodologies then can be used to study the spatial relationships among disturbances. These methods work well in wilderness areas and other large, contiguous, forest areas that have been shaped by the disturbance agents of interest.

Studying interactions among natural disturbances in landscapes that have been fragmented by logging or other human land use requires a different methodological approach. Working in the Grand Mesa National Forest, Kulakowski and Veblen (2006) reconstructed the history of stand-replacing fires and a nineteenth century spruce beetle outbreak at dispersed points across the landscape that were located in unlogged forests within a matrix of an otherwise intensely managed landscape. Stands with similar stand-origin dates were then combined and the influence of bark beetle outbreaks in cohorts of stands was compared using simple statistical methods, rather than spatial analyses. As with all dendrochronological methods, research becomes more difficult with increasing time since present. In this regard, the challenge faced by Kulakowski and Veblen (2006) was identifying dates of stand origin and of outbreaks based on fragmentary evidence. Similar methodological approaches have been used to study how stand replacing fires affect susceptibility to mountain pine beetle (Kulakowski et al. 2012).

More recent dendroecological research has looked at how variation in the severity and spatial patterns of fires have affected outbreaks of spruce beetle. Kulakowski et al. (2016b) tackled this issue by reconstructing the occurrence and severity of fires in stands that were burned at high and moderate severity and comparing the occurrence of outbreaks in those stands with unburned stands (Fig. 10.2). Fire severity was determined by pulses of tree establishment, and the presence or absence of remnant trees as documented in the dendrochronological record. The occurrence of beetle outbreak was determined by cross-dated mortality dates of the host spruce and releases in surviving trees. Bakaj et al. (2016) reconstructed the spatial patterns of severe wildfires and spruce beetle outbreaks by mapping all trees in 50 m × 50 m plots and collecting increment core samples from trees within the plots to develop a spatiotemporal picture of disturbance interactions.

10.5.2 Effect of Insect Outbreaks on Fires

As insect outbreaks affect forest fuel complexes, considerable research attention has been devoted to the question of how outbreaks affect fire regimes. This line of inquiry has examined the effects of outbreaks on both low-severity surface fires and high-severity stand replacing fire. The latter question is particularly challenging to address methodologically as severe fires destroy much evidence of pre-fire conditions.

Early work on the question of how outbreaks affect fires looked at the spatial overlap between outbreaks of spruce beetle and low-severity fires (Kulakowski et al. 2003). The spatial extent of outbreaks was reconstructed as described above. The extent of low-severity fires was reconstructed by dating and mapping the location

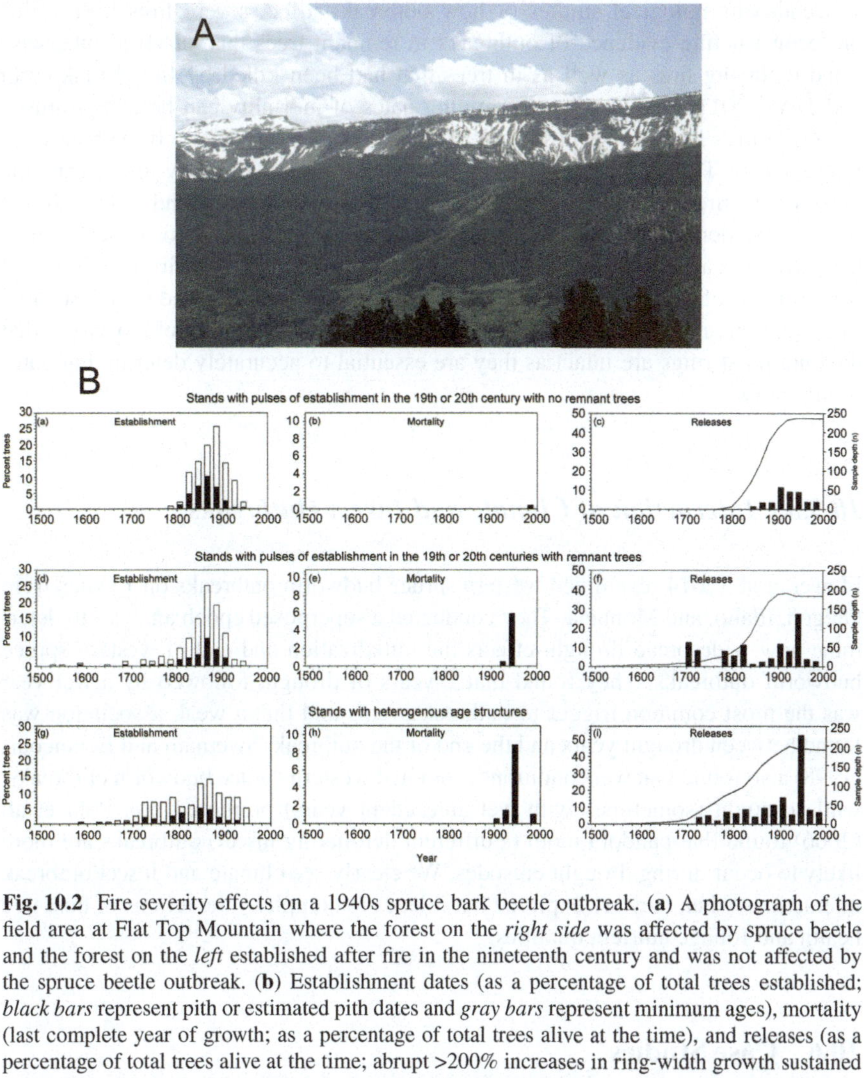

Fig. 10.2 Fire severity effects on a 1940s spruce bark beetle outbreak. (**a**) A photograph of the field area at Flat Top Mountain where the forest on the *right side* was affected by spruce beetle and the forest on the *left* established after fire in the nineteenth century and was not affected by the spruce beetle outbreak. (**b**) Establishment dates (as a percentage of total trees established; *black bars* represent pith or estimated pith dates and *gray bars* represent minimum ages), mortality (last complete year of growth; as a percentage of total trees alive at the time), and releases (as a percentage of total trees alive at the time; abrupt >200% increases in ring-width growth sustained >10 years) of trees in 20 year bins in (**a–c**) stands characterized by a pulse of establishment in the 19th or twentieth century with no remnant trees (data aggregated from 11 stands); (**d–f**) stands characterized by a pulse of establishment in the nineteenth or twentieth centuries with remnant trees (data aggregated from 9 stands); and (**g–i**) stands characterized by heterogeneous age structures (data aggregated from 12 stands) (graphic modified from Kulakowski et al. 2016b)

of fire-scarred trees. At the same time, a photointerpretation key was developed by Bebi et al. (2003) based on the disturbance history mapped dendrochronologically by Kulakowski et al. (2003). Bebi et al. (2003) then analyzed the two-way interactions between outbreaks and fires over a large (hundreds of square km) area using GIS.

Dendrochronological studies of how outbreaks affect severe fires have relied on reconstructing evidence of outbreaks in remnant trees that survived otherwise stand-replacing fires as well as in trees that had been killed by fire (Kulakowski and Jarvis 2011). Accurately determining dates of mortality can be compromised by severe fires as the outermost rings can be consumed, especially if the bark is no longer intact. This problem is not as acute when trees are killed by fire as even the most severe fires normally do not burn through the entire bark and bole of living trees. Thus, dendrochronological reconstructions of outbreaks prior to severe fires becomes increasingly difficult as the time between outbreak and fire increases and bark falls off of beetle-killed trees, thereby leaving the bole exposed to combustion. Therefore, great care is needed in collecting core samples in the field to assure that the outermost rings are intact as they are essential to accurately determining dates of mortality.

10.5.3 Interaction of Climate and Insect Outbreaks

Flower et al. (2014) examined western spruce budworm outbreaks on 13 sites from Oregon, Idaho, and Montana. They conducted a superposed epoch analysis to determine how widespread drought effects the initialization and end of western spruce budworm outbreaks. They found that 2 years of drought followed by a wet year was the most common trigger to outbreak events and that a weak association was found between drought years and the end of the outbreak. Swetnam and Betancourt (1998) also found that wet conditions promoted western spruce budworm outbreaks, while drought (sometimes with wet antecedent years) promoted fire. Pohl et al. (2006) found that pandora moth (a different defoliating insect) outbreaks are more likely to occur during drought episodes. We clearly see climate and insect outbreak interactions when they are explored but it leads to complex interactions of host tree health and foliage nutrient qualities.

10.6 Case Studies

10.6.1 Defoliators

Defoliating insects feed on leaves of host trees, thereby reducing their photosynthetic capacity until the feeding subsides and leaves regrow (Speer et al. 2001). Thus, dendrochronological reconstructions of defoliator outbreaks are typically based on identifying growth suppressions on host trees. A well-studied example of a defoliating insect is pandora moth, which has a 2 year life cycle (Speer 1997; Speer et al. 2001; Speer and Jensen 2003; Speer and Holmes 2004; Pohl et al. 2006; De Graauw 2012). It defoliates the trees as a larva, but spends part of its life as a pupa in the soil. The heaviest defoliation occurs when the larvae are in the latter instar stages (approximately 2 years after hatch) (Speer et al. 2001). As defoliation subsides every other year, the host species (mainly ponderosa pine (*Pinus ponderosa*)) have a chance to produce some needles and live through the outbreak. Only about 2% of trees attacked by Pandora moth die and most trees live to maintain a record of the outbreaks (Schmid and Bennett 1988). As is the case with many insect outbreak systems, the ring-width signature is tied to the biology of the insect. The population of insects slowly grows so that the beginning of the outbreak is hard to detect in the tree rings. After a few generations the populations increase greatly causing heavy defoliation of the trees. This causes the ring in the first year of the tree-ring signature to be approximately half the size of immediately preceding years and the next 2 years are the smallest in the signature (Speer et al. 2001). At this time the active defoliation and the cumulative effect of the past few years of defoliation converge to cause the most damage to the tree. Once the population of moths becomes large, a nucleopolyhedrosis virus spreads through the population through contact amongst the larvae, which eventually causes mortality of the larvae and the end of the outbreak (Schmid and Bennett 1988). The end of the outbreak signature in the tree ring record is a gradual increase of the ring width as the tree replaces its lost photosynthetic area. Additionally, throughout the signature, the tree rings have a thin latewood because the majority of the defoliation occurs during the end of the growing season as the larvae mature. This distinctive signature makes the identification of outbreak events easier in the wood (Speer 1997; Speer et al. 2001).

Because pandora moth does not cause a large amount of mortality in the host tree population, the importance of this defoliating insect can be underestimated, but Speer and Holmes (2004) found that a significant amount of potential biomass does not accrue due to multiple outbreaks over the lifetime of affected trees. The authors conducted a stem analysis on wind-thrown trees to reconstruct the volume lost due to pandora moth outbreaks. Overall, they observed a reduction in volume of 9.912 m^3/ha (1700 board ft/acre) due to suppression from the insects over the lifetime of the trees. This work demonstrated how stem analysis can be used to reconstruct the total effect of insect outbreaks on tree growth.

Some work has been done to examine the interaction of multiple disturbances of pandora moth outbreaks and fire history on the Pringle Fall Experimental Forest (PPF) site in central Oregon. Speer (1997) hypothesizes that pandora moth outbreaks could have been initiated because of stress caused to the trees because of fire (weakening their defenses) or could be ending the outbreaks because pandora moth spend the end of their second year as pupae in the soil. He actually found that neither of these hypotheses were supported by clustering of fire events at the beginning or end of outbreak episodes but that pandora moth was suppressing fire by consuming the needles and reducing the fine fuels in the stand for a 10 year period during outbreaks. Fires were five times less likely during periods of insect outbreaks than during periods without the occurrence of defoliation (only 2 fires in 91 years of suppression vs. 17 fires during 176 years of time without suppression—Fig. 10.3). They further hypothesize that pandora moth outbreaks provided the needed break in

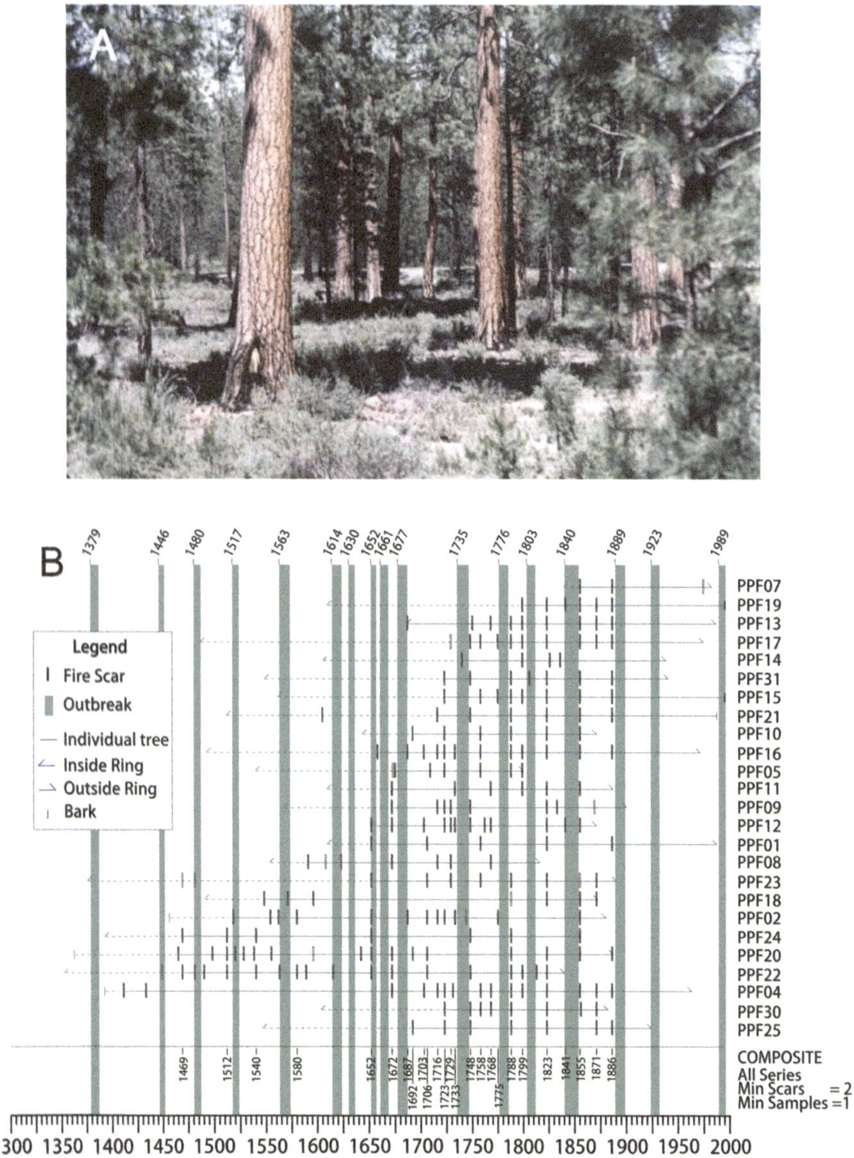

Fig. 10.3 Pringle Falls Experimental Forest, Oregon with fire and pandora moth outbreaks. (**a**) Photograph of the forest showing the older trees that recorded pandora moth outbreaks and the triangular shaped scare at the base of the trees that recorded fire history from the site. (**b**) Fire history chart showing fires as vertical *dashed lines* and the *horizontal lines* of each tree on the site with pandora moth outbreaks shown as *shaded vertical bars* covering all of the trees. Dates at the *top* are of pandora moth outbreaks recorded on the site and you can see very few fire events during outbreak periods compared to frequent fire outside of suppression periods (graphic modified from Speer 1997)

natural fire to allow ponderosa pine to regenerate and grow large enough to survive subsequent fires meaning that pandora moth is a symbiont rather than a parasite of ponderosa pine forests (Speer 1997).

Researchers have started to develop networks of outbreak reconstructions that enable analysis of the spatial dynamics of insect outbreaks (Speer and Jensen 2003) and the population dynamics of insects across their entire range (de Graauw 2012). Speer and Jensen (2003) demonstrate how analysis of pandora moth outbreaks back through time, can show the spread of outbreaks across central Oregon based on a network of 14 sites. De Graauw (2012) built on this work to analyze pandora moth outbreaks across their entire range. She developed a habitat model from soils, host species, and climate to predict where pandora moth could occur and then compared that model to the records from 200 ponderosa pine chronologies in the International Tree-Ring Databank to test the model and determine where pandora moth has occurred over the past 400 years.

10.6.2 Cambium Feeders

Cambium feeders can cause high levels of mortality of host trees over large areas. Among the most important of this group are mountain pine beetle (*Dendroctonus ponderosae*) and spruce beetles (*Dendroctonus rufipennis* and *Ips typographus*). As outbreaks of bark beetles typically cause mortality of host trees, outbreaks are reconstructed based on crossdating mortality dates of host trees and releases of trees that survived the outbreak, typically non-host species and host species that were too small at the time of the outbreak to be attacked (Veblen et al. 1991a, b).

The current mountain pine beetle outbreak is the most severe contiguous insect outbreak recorded historically or reconstructed with dendrochronology (Carroll et al. 2003; Safranyik and Carroll 2006; Kurz et al. 2008; Raffa et al. 2008; Axelson et al. 2010; Lewis et al. 2017). The Canadian Forest Service and National Forestry Database report that over 16 million hectares were affected by mountain pine beetle from 1996 through 2014 with the peak occurring in 2007 at over 10 million hectares affected (NRC 2016). The USDA Forest Service report 2 million hectares affected by mountain pine beetle from 1996 through 2012 in Colorado, South Dakota, and Wyoming with a peak in 2008 at 1.8 million hectares (USDA 2016). This totals 18 million hectares of lodgepole pine forest in which trees have been killed in British Columbia, Colorado, and Wyoming from 1996 through 2014, although the outbreak is now declining because of overconsumption of the food source (NRC 2016; USDA 2016). Although it appears that mountain pine beetle is expanding into a naïve host of jack pine (*Pinus banksiana*) in eastern Alberta and into the Northwest Territories and Saskatchewan as their traditional host species have been decimated (Erbilgin et al. 2014). This massive outbreak was the result of a combination of warmer weather, especially in winter, and the availability of a large number of host trees (Raffa et al. 2008). Axelson et al. (2009, 2010) found that mountain pine beetle had a history of killing mature lodgepole pine trees in continuous and mixed stands, but would leave the younger individuals with less cambial area. After these trees had matured for 20 or 40 more years, subsequent outbreaks would take advantage of this next cohort of trees. An important caveat to the common idea that ongoing outbreaks are ecologically unprecedented is that mountain pine beetle outbreaks are difficult to reconstruct dendrochronologically as lodgepole pine stands are often monospecific, complicating reconstructions based on releases of non-host trees. Thus, past outbreaks may be absent from dendrochronological records not because they did not occur, but because they are difficult to reconstruct. Indeed, recent dendrochronological work in several hundred year old lodgepole pine forests combined with a comprehensive analysis of available dendrochronological reconstructions and historical records indicates that outbreaks of mountain pine beetle were likely synchronous across North America over the past centuries (Alfaro et al. 2003; Jarvis and Kulakowski 2015). Further research is necessary to determine the severity and contiguity of these outbreaks over the past centuries.

Another important example of a cambial feeder is the spruce beetle (*Dendroctonus rufipennis*), which typically causes tree mortality so that identification of the outbreaks is usually reconstructed based on crossdating mortality dates of host species and the release of surrounding trees (Veblen et al. 1991a, b). In nine stands in northwestern Colorado, Eisenhart and Veblen (2000) found evidence of multiple spruce beetle outbreaks in Engelmann spruce (*Picea engelmannii*) over the past centuries, with 1–2 outbreaks per century that lasted 20 years on average. Hart et al. (2015) used tree ring and documentary records to examine spruce beetle in northwestern Colorado and compare outbreak events to climatic parameters and found that spruce beetle outbreaks are more likely to occur under drought conditions of high temperature combined with low precipitation.

As mentioned earlier, other cambium feeders such as the European spruce bark beetle (*Ips typographus*) have similar dynamics to the bark beetles in North American and Canada but have been more difficult to reconstruct as the host species Norway spruce (*Picea abies*) often occurs in monospecific stands, which leads to few survivors that would record releases following outbreaks (Svoboda et al. 2012; Čada et al. 2013).

10.6.3 Root Parasites (or Symbiont)

Root parasites have important effects on forest dynamics and can be studied dendrochronologically. Periodical cicadas (*Magicicada* spp.) are unique to the eastern United States where they emerge en masse every 13 or 17 years depending upon the brood (Johnson and Lyon 1991). The effect on the trees is interesting to contemplate because the majority of the cicadas' life cycle occurs underground as they progress through five nymphal stages each concurrent with rapid growth. Because the preponderance of their life is spent underground, we would expect that their main ecological contributions would be as root parasites, but Speer et al. (2010) found little effect of periodical cicadas on the trees during their underground nymphal stages. The sample trees did decrease growth during the emergence year when the female cicadas lay their eggs in small smooth branches and cause visual flagging or dead leaves on the branch tips. Speer et al. (2010) also documented a consistent pulse of growth 5 years after the emergence whose mechanism is still uncertain. We are not aware of any other work with a root insect parasite in dendrochronology.

10.7 Conclusion

Dendroentochronology is a quickly growing sub-discipline in the field of dendrochronology. We can use the specific biological signature that is left in the wood from the interaction of the insect with the tree to accurately identify past insect outbreaks. This work is more widespread and easier with defoliators, but success also has been found with cambium feeders and root parasites as well, especially in cambium feeding systems. When non-host trees leave records of releases following mortality of host trees in cambium feeding systems. We can accurately date the onset of specific events and examine the return interval and duration of insect outbreaks, which is useful information for forest managers. Using statistical tools like superposed epoch analysis, we can determine climatic events that are more likely to trigger some insect outbreaks and examine broad climatic forcing that have made some conditions more conducive to outbreaks. One example of such a climatic over forcing is the increase in mountain pine beetle with the gradual warming trend over in the past decades. Much tree-ring research has been conducted on the insect outbreak systems in North America and a body of work continues to grow in Europe, Asia, and elsewhere. But intensive research into outbreaking insects needs to be expanded globally so that ecologists and forest managers understand the full dynamics of the systems that they study and manage. Complex interactions between climate, insect outbreaks, and other disturbances often work in conjunction to shape forest systems. Dendrochronological reconstructions of past outbreaks and disturbance interactions provide important insights into on the causes and consequences of forest disturbances, especially in the context of climatic variability and change. These insights can then provide important guidance for management strategies.

Acknowledgements This work was supported by multiple National Science Foundation grants.

References

Alfaro RI (1995) An induced defense reaction in white spruce to attack by the white pine weevil, Pissodes strobi. Can J For Res 25(10):1725–1730. doi:10.1139/x95-186

Alfaro RI, MacDonald RN (1988) Effects of defoliation by the western false hemlock looper on Douglas-fir tree-ring chronologies. Tree Ring Bull 48:3–11

Alfaro RI, Shepherd RF (1991) Tree-ring growth of interior Douglas-fir after one year's defoliation by Douglas-fir tussock moth. For Sci 37(3):959–964
Alfaro RI, Qiwei L, Vallentgoed J (1991) Diameter growth losses in western larch caused by larch casebearer defoliation. Western J Appl For 6(4):105–108
Alfaro RI, Campbell R, Vera P, et al (2003) Dendroecological reconstruction of mountain pine beetle outbreaks in the Chilcotin Plateau of British Columbia. In Shore TL, Brooks JE, Stone JE (eds) Mountain Pine Beetle Symposium: challenges and solutions. Kelowna, BC. pp 245–256
Alfaro RI, Campbell E, Hawkes BC (2010) Historical frequency, intensity and extent of mountain pine beetle disturbance in British Columbia, vol. 2009. Pacific Forestry Centre.
Asshof R, Schweingruber FH, Wermelinger B (1999) Influence of a gypsy moth (*Lymantria dispar* L.) outbreak on radial growth and wood-anatomy of Spanish chestnut (*Castanea sativa* Mill.) in Ticino (Switzerland). Dendrochronologia 16–17:133–145
Avcí M, Carus S (2005) The impact of cedar processionary moth [*Tramatocampa ispartaensis* Dog˘anlar & Avcí (Lepidoptera: Notodontidae)] outbreaks on radial growth of Lebanon cedar (*Cedrus libani* A. Rich.) trees in Turkey. J Pest Sci 78:91–98. doi:10.1007/s10340-004-0073-2
Axelson JN, Alfaro RI, Hawkes BC (2009) Influence of fire and mountain pine beetle on the dynamics of lodgepole pine stands in British Columbia, Canada. Forest Ecol Manage 257(9):1874–1882
Axelson JN, Alfaro RI, Hawkes BC (2010) Changes in stand structure in uneven-aged lodgepole pine stands impacted by mountain pine beetle epidemics and fires in central British Columbia. For Chron 86(1):87–99. doi:10.5558/tfc86087-1
Axelson JN, Bast A, Alfaro R et al (2014) Variation in wood anatomical structure of Douglas-fir defoliated by the western spruce budworm: a case study in the coastal-transitional zone of British Columbia, Canada. Trees 28(6):1837–1846. doi:10.1007/s00468-014-1091-1
Axelson JN, Smith DJ, Daniels LD, Alfaro RI (2015) Multicentury reconstruction of western spruce budworm outbreaks in central British Columbia, Canada. For Ecol Manag 335:235–248. doi:10.1016/j.foreco.2014.10.002
Bakaj F, Mietkiewicz M, Veblen TT, Kulakowski D (2016) The relative importance of tree and stand properties in predicting susceptibility to spruce beetle outbreak in the mid-20th century. Ecosphere 7(10):e01485. doi:10.1002/ecs2.1485
Baker WL, Veblen TT (1990) Spruce beetles and fires in the nineteenth-century subalpine forests of western Colorado, USA. Arct Alp Res 22(1):65–80. doi:10.2307/1551721
Baltensweiler W (1964) Zeiraphera griseana Hübner (Lepidoptera: Tortricidae) in the European Alps. A contribution to the problem of cycles. Can Entomol 96(5):792–800. doi:10.4039/Ent96792-5
Baltensweiler W, Weber UM, Cherubini P (2008) Tracing the influence of larch-bud-moth insect outbreaks and weather conditions on larch tree ring growth in Engadine, Switzerland. Oikos 117:161–172. doi:10.1111/j.2007.0030-1299.16117.x
Bebi P, Kulakowski D, Veblen TT (2003) Interactions between fire and spruce beetles in a subalpine Rocky Mountain forest landscape. Ecology 84(2):362–371
Berg EE, Henry JD, Fastie CL, De Volder AD, Matsuoka SM (2006) Spruce beetle outbreaks on the Kenai Peninsula, Alaska, and Kluane National Park and Reserve, Yukon Territory: relationship to summer temperatures and regional differences in disturbance regimes. For Ecol Manage 227(3):219–232
Blais JR (1954) The recurrence of spruce budworm infestations in the past century in the Lac Seul area of northwestern Ontario. Ecology 35:62–71. doi:10.2307/1931405
Blais JR (1962) Collection and analysis of radial-growth data from trees for evidence of past spruce budworm outbreaks. For Chronicle 38(4):474–484
Blais JR (1965) Spruce budworm outbreaks in the past three centuries in the Laurentide Park, Quebec. For Sci 11:130–138
Blais JR (1983) Trends in the frequency, extent, and severity of spruce budworm outbreaks in eastern Canada. Can J For Res 13:539–547. doi:10.1139/x83-079

Boulanger Y, Arseneault D (2004) Spruce budworm outbreaks in eastern Quebec over the last 450 years. Can J For Res 34(5):1035–1043. doi:10.1139/x03-269
Boulanger Y, Arseneault D, Morin H et al (2012) Dendrochronological reconstruction of spruce budworm (Choristoneura fumiferana) outbreaks in southern Quebec for the last 400 years. Can J For Res 42(7):1264–1276. doi:10.1139/x2012-069
Brubaker LB, Greene SK (1979) Differential effects of Douglas-fir tussock moth and western spruce budworm defoliation on radial growth of grand fir and Douglas-fir. Can J For Res 9(1):95–105. doi:10.1139/x79-016
Buntgen U, Frank D, Liebhold A et al (2009) Three centuries of insect outbreaks across the European Alps. New Phytol 182:929–941. doi:10.1111/j.1469-8137.2009.02825.x
Čada V, Svoboda M, Janda P (2013) Dendrochronological reconstruction of the disturbance history and past development of the mountain Norway spruce in the Bohemian Forest, central Europe. For Ecol Manag 295:59–68. doi:10.1016/j.foreco.2012.12.037
Camarero JJ, Martín E, Gil-Pelegrín E (2003) The impact of a needleminer (Epinotia subsequana) outbreak on radial growth of silver fir (Abies alba) in the Aragón Pyrenees: a dendrochronological assessment. Dendrochronologia 21(1):3–12. doi:10.1078/1125-7865-00035
Campbell EM, Alfaro RI, Hawkes B (2007) Spatial distribution of mountain pine beetle outbreaks in relation to climate and stand characteristics: a dendroecological analysis. J Integr Plant Biol 49(2):168–178. doi:10.1111/j.1744-7909.2007.00423.x
Carroll AL, Taylor SW, Régnière J, Safranyik L (2003) Effect of climate change on range expansion by the mountain pine beetle in British Columbia. In: TL Shore et al (eds) Mountain Pine Beetle Symposium: challenges and solutions, Natural Resources Canada, Information Report BC-X-399, Kelowna, BC, 30–31 Oct 2003, pp 223–232
Carus S (2004) Impact of defoliation by the pine processionary moth (Thaumetopoea pityocampa) on radial, height and volume growth of Calabria pine (Pinus brutia) trees in Turkey. Phytoparasitica 32:459–469. doi:10.1007/BF0298044
Carus S (2009) Effects of defoliation caused by the processionary moth on growth of Crimean pines in western Turkey. Phytoparasitica 37:105–114. doi:10.1007/s12600-008-0018-z
Carus S (2010) Effect of defoliation by the pine processionary moth (PPM) on radial, height and volume growth of Crimean pine (Pinus nigra) trees in Turkey. J Environ Biol 31(4):453–460
Carus S, Avci M (2005) Growth loss of Lebanon cedar (Cedrus libani) stands as related to periodic outbreaks of the cedar shoot moth (Dichelia cedricola). Phytoparasitica 33(1):33–48. doi:10.1007/BF02980923
Case RA, MacDonald GM (2003) Dendrochronological analysis of the response of tamarack (Larix laricina) to climate and larch sawfly (Pristiphora erichsonii) infestations in central Saskatchewan. Ecoscience 10(3):380–388. doi:10.1080/11956860.2003.11682787
Clark P, Speer JH, Winship L (2017) Identifying and separating Pandora moth outbreaks and climate from a 1500-Year long Ponderosa pine chronology from Central Oregon. Tree Ring Res 73(2)
Delgado SC (2000) Aplicaciones estadísticas en estudios dendrocronológicos. In: Roig FA (ed) Dendrocronología en América Latina. EDIUNC, Mendoza, pp 79–102
Dobbertin MK, Grissino-Mayer HD (2004) The online bibliography of dendrochronology. Dendrochronologia 21(2):85–90
Dulamsuren C, Hauck M, Leuschner HH, Leuschner C (2010) Gypsy moth-induced growth decline of Larix sibirica in a forest-steppe ecotone. Dendrochronologia 28(4):207–213. doi:10.1016/j.dendro.2009.05.007
Duncan DP, Hodson AC (1958) Influence of the forest tent caterpillar upon the aspen forests of Minnesota. For Sci 4:71–93
Eisenhart KS, Veblen TT (2000) Dendroecological detection of spruce bark beetle outbreaks in northwestern Colorado. Can J For Res 30(11):1788–1798. doi:10.1139/x00-104

Erbilgin N, Ma C, Whitehouse C et al (2014) Chemical similarity between historical and novel host plants promotes range and host expansion of the mountain pine beetle in a naïve host ecosystem. New Phytol 201(3):940–950. doi:10.1111/nph.12573
Esper J, Büntgen U, Frank DC, Nievergelt D, Liebhold A (2007) 1200 years of regular outbreaks in alpine insects. Proc R Soc Lond B: Biol Sci 274(1610):671–679
Flower A, Gavin DG, Heyerdahl EK et al (2014) Drought-triggered western spruce budworm outbreaks in the interior Pacific Northwest: a multi-century dendrochronological record. For Ecol Manag 324:16–27. doi:10.1016/j.foreco.2014.03.042
Fraver S, Seymour RS, Speer JH, White AS (2007) Dendrochronological reconstruction of spruce budworm outbreaks in northern Maine, USA. Can J For Res 37(3):523–529. doi:10.1139/X06-251
Girardin MP, Tardif J, Bergeron Y (2001) Radial growth analysis of Larix laricina from the Lake Duparquet area, Quebec, in relation to climate and larch sawfly outbreaks. Ecoscience 8(1):127–138. doi:10.1080/11956860.2001.11682638
de Graauw K (2012) Tree-ring analysis of outbreak dynamics across an insect's entire range: the Pandora moth system. Thesis, Indiana State University
Hadley KS (1994) The role of disturbance, topography, and forest structure in the development of a montane forest landscape. Bull Torrey Bot Club 121(1):47–61. doi:10.2307/2996883
Hart SJ, Veblen TT, Mietkiewicz N, Kulakowski D (2015) Negative feedbacks on bark beetle outbreaks: widespread and severe spruce beetle infestation restricts subsequent infestation. PLoS One 10(5):e0127975. doi:10.1371/journal.pone.0127975
Hildahl V, Reeks WA (1960) Outbreaks of the forest tent caterpillar, Malacosoma disstria Hbn., and their effects on stands of trembling aspen in Manitoba and Saskatchewan. Can Entomol 92:199–209. doi:10.4039/Ent92199-3
Hogg EH, Brandt JP, Kochtubajda B (2002) Growth and dieback of aspen forests in northwestern Alberta, Canada, in relation to climate and insects. Can J For Res 32(5):823–832. doi:10.1139/x01-152
Hough FB (1882) The elements of forestry. Robert Clarke and Company, Cincinnati, p 381
Hrinkevich K, Lewis KJ (2011) Northern range limit mountain pine beetle outbreak dynamics in mixed sub-boreal pine forests of British Columbia. Ecosphere 2(10):1–16
Huang JG, Tardif J, Denneler B et al (2008) Tree-ring evidence extends the historic northern range limit of severe defoliation by insects in the aspen stands of western Quebec, Canada. Can J For Res 38(9):2535–2544. doi:10.1139/X08-080
Jardon Y, Filion L, Cloutier C (1994) Tree-ring evidence for endemicity of the larch sawfly in North America. Can J For Res 24(4):742–747. doi:10.1139/x94-098
Johnson WT, Lyon HH (1991) Insects that feed on trees and shrubs, 2nd edn. Cornell University Press Ithaca, New York, pp 490–492
Jarvis DS, Kulakowski D (2015) Long-term history and synchrony of mountain pine beetle outbreaks in lodgepole pine forests. J Biogeogr 42(6):1029–1039
Johnson WT, Lyon HH (1991) Insects that feed on trees and shrubs, 2nd edn. Cornell University Press Ithaca, New York, pp 490–492
Krause C (1997) The use of dendrochronological material from buildings to get information about past spruce budworm outbreaks. Can J For Res 27(1):69–75. doi:10.1139/x96-168
Krause C, Morin H (1995) Impact of spruce budworm defoliation on the number of latewood tracheids in balsam fir and black spruce. Can J For Res 25(12):2029–2034. doi:10.1139/x95-219
Kucherov S (1991) The reconstruction of Lymantria dispar outbreaks by dendrochronological methods in the south Urals. In: Baranchikov YN, Mattson WJ, Hain FP, Payne TL (eds) Forest insect guilds: patterns of interaction with host trees. USDA Forest Service GTR NE-153. pp 205–206
Kulakowski D, Jarvis D (2011) The influence of mountain pine beetle outbreaks on severe wildfires in northwestern Colorado and southern Wyoming: a look at the past century. For Ecol Manag 261(1):1686–1696. doi:10.1016/j.foreco.2011.07.016

Kulakowski D, Veblen TT (2002) Influences of fire history and topography on the pattern of a severe wind blowdown in a Colorado subalpine forest. J Ecol 90:806–819. doi:10.1046/j.1365-2745.2002.00722.x
Kulakowski D, Veblen TT (2006) The effect of fires on susceptibility of subalpine forests to a 19th century spruce beetle outbreak in western Colorado. Can J For Res 36(11):2974–2982. doi:10.1139/x06-182
Kulakowski D, Veblen TT (2015) Bark beetles and high-severity fires in Rocky Mountain subalpine forests. In: Mixed-high severity fires: ecosystem processes and biodiversity. Elsevier, San Francisco, pp 149–174
Kulakowski D, Veblen TT, Bebi P (2003) Effects of fire and spruce beetle outbreak legacies on the disturbance regime of a subalpine forest in Colorado. J Biogeogr 30:1445–1456. doi:10.1046/j.1365-2699.2003.00912.x
Kulakowski D, Jarvis D, Veblen TT, Smith J (2012) Stand-replacing fires reduce susceptibility to mountain pine beetle outbreaks in Colorado. J Biogeogr 39:2052–2060. doi:10.1111/j.1365-2699.2012.02748.x
Kulakowski D, Seidl R, Holeksa J et al (2016a) A walk on the wild side: disturbance dynamics and the conservation and management of European mountain forest ecosystems. For Ecol Manag 388:120–131. doi:10.1016/j.foreco.2016.07.037
Kulakowski D, Veblen TT, Bebi P (2016b) Fire severity controlled susceptibility to a 1940s spruce beetle outbreak in Colorado, USA. PLoS One 11(7):e0158138. doi:10.1371/journal.pone.0158138
Kurz WA, Dymond CC, Stinson G et al (2008) Mountain pine beetle and forest carbon feedback to climate change. Nature 452(7190):987–990. doi:10.1038/nature06777
Lewis KJ, Welsh C, Wong CM, Speer JH (2017) Pathogens, invasive species, and prognosis for the future. In: Amoroso MM, Daniels LD, Baker PJ, Camarero JJ (eds) Dendroecology: tree-ring analyses applied to ecological studies. Springer, Cham
Lynch AM (2012) What tree-ring reconstruction tells us about conifer defoliator outbreaks. In: Barbosa P, Letourneau DK, Agrawal AA (eds) Insect outbreaks revisited. Wiley-Blackwell, Chichester, pp 126–154
Mason RR, Wickman BE, Paul HG (1997) Radial growth response of Douglas-fir and grand fir to larval densities of the Douglas-fir tussock moth and the western spruce budworm. For Sci 43:194–205
Mazanec Z (1968) Influence of defoliation by the phasmatid Didymuria violescens on seasonal diameter growth and the pattern of growth rings in alpine ash. Aust For 32(1):3–14. doi:10.1080/00049158.1968.10675456
Morin H (1994) Dynamics of balsam fir forests in relation to spruce budworm outbreaks in the boreal zone of Quebec. Can J For Res 24(4):730–741. doi:10.1139/x94-097
Morin H, Laprise D, Bergeron Y (1993) Chronology of spruce budworm outbreaks near Lake Duparquet, Abitibi region, Quebec. Can J For Res 23(8):1497–1506. doi:10.1139/x93-189
Morin H, Laprise D, Simard AA et al (2009) Spruce budworm outbreak regimes in eastern North America. Ecosystem management in the boreal forest. In: Gauthier S, Vaillancourt MA, Leduc A, et al (eds) Presse de l'Université du Québec, Québec. pp 155–182.
Morrow PA, LaMarche VC Jr (1978) Tree ring evidence for chronic Insect suppression of productivity in subalpine Eucalyptus. Science 201:1244–1246. doi:10.1126/science.201.4362.1244
Muzika RM, Liebhold AM (1999) Changes in radial increment of host and nonhost tree species with gypsy moth defoliation. Can J For Res 29(9):1365–1373. doi:10.1139/x99-098
NRC (2016) Natural Resources Canada and National Forestry Database. http://nfdp.ccfm.org/data/graphs/graph_41_b_e.php. Accessed 9 Aug 2016
Paritsis J, Veblen TT (2011) Dendroecological analysis of defoliator outbreaks on Nothofagus pumilio and their relation to climate variability in the Patagonian Andes. Glob Chang Biol 17(1):239–253. doi:10.1111/j.1365-2486.2010.02255.x
Paritsis J, Veblen TT, Kitzberger T (2009) Assessing dendroecological methods to reconstruct defoliator outbreaks on Nothofagus pumilio in northwestern Patagonia, Argentina. Can J For Res 39(9):1617–1629. doi:10.1139/X09-085

Pohl KA, Hadley KS, Arabas KB (2006) Decoupling tree-ring signatures of climate variation, fire, and insect outbreaks in Central Oregon. Tree Ring Res 62(2):37–50. doi:10.3959/1536-1098-62.2.37

Priya PB, Bhat KM (1998) False ring formation in teak (Tectona grandis Lf) and the influence of environmental factors. For Ecol Manag 108(3):215–222. doi:10.1016/S0378-1127(98)00227-8

Raffa KF, Aukema BH, Bentz BJ et al (2008) Cross-scale drivers of natural disturbances prone to anthropogenic amplification: the dynamics of bark beetle eruptions. Bioscience 58(6):501–517. doi:10.1641/B580607

Readshaw JL, Mazanec Z (1969) Use of growth rings to determine past phasmatid defoliations of alpine ash forests. Aust For 33(1):29–36. doi:10.1080/00049158.1969.10675483

Rolland C, Baltensweiler W, Petitcolas V (2001) The potential for using Larix decidua ring widths in reconstructions of larch budmoth (Zeiraphera diniana) outbreak history: dendrochronological estimates compared with insect surveys. Trees 15(7):414–424. doi:10.1007/s004680100116

Safranyik L, Carroll AL (2006) The biology and epidemiology of the mountain pine beetle in lodgepole pine forests. In: Safranyik L, Wilson WR (eds) The mountain pine beetle: a synthesis of biology, management, and impacts on lodgepole pine. Natural Resources Canada, Canadian Forest Service, Pacific Forestry Centre, Victoria, British Columbia, 304 p

Safranyik L, Wilson B (2007) The mountain pine beetle: a synthesis of biology, management and impacts on lodgepole pine. Canadian Forest Service, Victoria. ISBN:0662426231

Safranyik L, Carroll AL, Wilson B (2007) The biology and epidemiology of the mountain pine beetle in lodgepole pine forests. In: The mountain pine beetle: a synthesis of biology, management and impacts on lodgepole pine. Natural Resources Canada, Canadian Forest Service, Pacific Forestry Centre, Victoria, pp 3–66

Sarajishvili KG (1997) Dendrochronological indication of entomoresistance in Pinus eldarica against Dioryctria slyvestrella Ratz (Lepidoptera, Pyralidae) in Georgia. Acta Phytopathol Entomol Hung 32(1–2):245–250

Schmid JM, Bennett DD (1988) The North Kaibab pandora moth outbreak, 1978–1984. USDA Forest Service Rocky Mountain Forest and Range Experimental Station Research Note RM-153

Schoennagel T, Veblen TT, Negron JF, Smith JM (2012) Effects of mountain pine beetle on fuels and expected fire behavior in lodgepole pine forests, Colorado, USA. PLoS One 7(1):e30002

Sherriff RL, Berg EE, Miller AE (2011) Climate variability and spruce beetle (Dendroctonus rufipennis) outbreaks in south-central and southwest Alaska. Ecology 92(7):1459–1470

Shore TL, Safranyik L, Hawkes BC, Taylor SW (2006) Effects of the mountain pine beetle on lodgepole pine stand structure and dynamics. In: Safranyik L, Wilson B (eds) The mountain pine beetle: a synthesis of biology, management, and impacts on lodgepole pine. Canadian Forest Service, Victoria, pp 95–114

Simard M, Payette S (2003) Accurate dating of spruce budworm infestation using tree growth anomalies. Ecoscience 10(2):204–216. doi:10.1080/11956860.2003.11682768

Simard S, Elhani S, Morin H et al (2008) Carbon and oxygen stable isotopes from tree-rings to identify spruce budworm outbreaks in the boreal forest of Québec. Chem Geol 252(1):80–87. doi:10.1016/j.chemgeo.2008.01.018

Smith AL, Hewitt N, Klenk N, Bazely DR, Yan N, Wood S, Henriques I, MacLellan JI, Lipsig-Mummé C (2012) Effects of climate change on the distribution of invasive alien species in Canada: a knowledge synthesis of range change projections in a warming world. Environ Rev 20(1):1–16

Speer JH (1997) A dendrochronological record of Pandora Moth (Coloradia pandora, Blake) outbreaks in Central Oregon. Thesis, The University of Arizona

Speer JH (2010) Fundamentals of tree-ring research. University of Arizona Press, Tucson

Speer JH, Holmes RL (2004) Effects of pandora moth outbreaks on ponderosa pine wood volume. Tree Ring Res 60(2):69–76. doi:10.3959/1536-1098-60.2.69

Speer JH, Jensen RR (2003) A hazards approach towards modelling pandora moth risk. J Biogeogr 30(12):1899–1906. doi:10.1111/j.1365-2699.2003.00951.x

Speer JH, Swetnam TW, Wickman BE et al (2001) Changes in pandora moth outbreak dynamics during the past 622 years. Ecology 82:679–697. doi:10.1890/0012-9658(2001)082[0679:CIPMOD]2.0.CO;2

Speer JH, Clay K, Bishop G et al (2010) The effect of periodical cicadas on growth of five tree species in midwestern deciduous forests. Am Midl Nat 164(2):173–186. doi:10.1674/0003-0031-164.2.173

Studhalter RA (1955) Tree growth: some historical chapters. Bot Rev 21(1–3):1–72

Sutton A, Tardif J (2005) Distribution and anatomical characteristics of white rings in Populus tremuloides. IAWA J 26(2):221–238

Sutton A, Tardif JC (2007) Dendrochronological reconstruction of forest tent caterpillar outbreaks in time and space, western Manitoba, Canada. Can J For Res 37(9):1643–1657

Sviderskaya IV, Pal'nikova EN (2003) Radial growth of Pinus sylvestris trees related to defoliation by Bupalus piniarius. Lesovedenie 5:44–53. doi:10.1139/X07-021

Svoboda M, Janda P, Nagel TA et al (2012) Disturbance history of an old-growth subalpine Picea abies stand in the Bohemian Forest, Czech Republic. J Veg Sci 23(1):86–97. doi:10.1111/j.1654-1103.2011.01329.x

Swetnam TW (1987) A dendrochronological assessment of western spruce budworm (*Choristoneura occidentalis* Freeman) in the Southern Rocky Mountains (Colorado, New Mexico). Dissertation, The University of Arizona

Swetnam TW, Betancourt JL (1998) Mesoscale disturbance and ecological response to decadal climatic variability in the American Southwest. J Clim 11:3128–3147. doi:10.1175/1520-0442(1998)011<3128:MDAERT>2.0.CO;2

Swetnam TW, Lynch AM (1989) A Tree-Ring reconstruction of western spruce budworm outbreaks in the Southern Rocky Mountains. For Sci 35(4):962–986

Swetnam TW, Lynch AM (1993) Multi-century, regional-scale patterns of western spruce budworm history. Ecol Monogr 63(4):399–424. doi:10.2307/2937153

Swetnam TW, Thompson MA, Sutherland EK (1985) Using dendrochronology to measure radial growth of defoliated trees, USDA Forest Service Agriculture Handbook, vol 639. USDA Forest Service, Washington, DC

Swetnam TW, Wickman BE, Paul HG, Baisan CH (1995) Historical patterns of western spruce budworm and Douglas-fir tussock moth outbreaks in the northern Blue Mountains, Oregon, since AD 1700 Pacific Northwest Research Station PNW-RP-484

USDA (2016) USDA Forest Service Aerial Detection Survey. http://www.fs.usda.gov/detail/r2/forest-grasslandhealth/?cid=stelprdb5408531. Accessed 9 Aug 2016

Veblen TT, Hadley KS, Reid MS, Rebertus AJ (1991a) Methods of detecting past spruce beetle outbreaks in Rocky Mountain subalpine forests. Can J For Res 21(2):242–254

Veblen TT, Hadley KS, Reid MS, Rebertus AJ (1991b) The response of subalpine forests to spruce beetle outbreak in Colorado. Ecology 72(1):213–231

Veblen TT, Hadley KS, Nel EM et al (1994) Disturbance regime and disturbance interactions in a Rocky Mountain Subalpine Forest. J Ecol 82(1):125–135. doi:10.2307/2261392

Weber UM (1997) Dendroecological reconstruction and interpretation of larch budmoth (Zeiraphera diniana) outbreaks in two central alpine valleys of Switzerland from 1470–1990. Trees 11(5):277–290. doi:10.1007/PL00009674

Wickman BE (1963) Mortality and growth reduction of white fir following defoliation by the Douglas-fir tussock moth USDA Forest Service Research Paper PSW-7

Wickman BE (1980) Increased growth of white fire after a Douglas-fir tussock moth outbreak. J For 78:31–33

Wimmer R (2001) Arthur Freiherr von Sechendorff-Gudent and the early history of tree-ring crossdating. Dendrochronologia 19(1):153–158

Zhang QB, Alfaro RI (2002) Periodicity of two-year cycle spruce budworm outbreaks in central British Columbia: a dendro-ecological analysis. For Sci 48(4):722–731

Zhang QB, Alfaro RI (2003) Spatial synchrony of the two-year cycle budworm outbreaks in central British Columbia, Canada. Oikos 102(1):146–154

Chapter 11
Pathogens, Invasive Species, and Prognosis for the Future

Kathy J. Lewis, Cedar Welsh, Carmen M. Wong, and James H. Speer

Abstract Organisms such as fungi and insects, can attack trees and cause reductions in growth due to reallocation of resources to defense responses, or reductions in photosynthetic capacity. These organisms can also kill trees, which can enhance the growth of neighbor trees due to release from competition. Changes in growth brought about by such attacks can be detected and studied using tree-ring analysis. Three case studies are presented in this chapter. The first is on *Dothistroma* needle blight in British Columbia, Canada where it is a native pathogen that causes growth reduction of pine. Dendrochronology is used to reconstruct past outbreaks of the pathogen, to identify weather variables that enhance disease development and spread, and to link changes in outbreak dynamics with climate change. The second case study is on white pine blister rust in North America, where the fungus was introduced into high elevation forests of whitebark pine. Dendrochronology enabled approximation of the date of introduction, disease intensification and spread dynamics, interactions with other insects and climate, as well as projections of future stand composition and structure. This is an important application of tree-ring analysis given the importance of whitebark pine for ecosystem function. The third case study is on emerald ash borer, a wood boring insect recently introduced to North America. In this case, tree-ring analysis is used to determine the efficacy of

K.J. Lewis (✉)
Ecosystem Science and Management Program, University of Northern British Columbia 3333 University Way, Prince George, BC, Canada V2N 4Z9
e-mail: kathy.lewis@unbc.ca

C. Welsh
Department of Geography, University of Victoria, B203 David Turpin Building 3800 Finnerty Road, Victoria, BC, Canada V8P5C2
e-mail: welsh@uvic.ca

C.M. Wong
Yukon Field Unit, Parks Canada 205-300 Main Street, Whitehorse, Yukon Territory, Canada Y1A 2B5
e-mail: carmen.wong@pc.gc.ca

J.H. Speer
Department of Environmental Systems, Indiana State University 600 Chestnut Street, Terre Haute, IN, USA 47809
e-mail: jim.speer@indstate.edu

M.M. Amoroso et al. (eds.), *Dendroecology*, Ecological Studies 231,
DOI 10.1007/978-3-319-61669-8_11

applications of insecticide in reducing growth loss and mortality rates. Collectively these case studies demonstrate that dendrochronology enables integrated analysis of host, climate and pathogen or insect effects on tree growth and survival, and ultimately on ecosystem function and resilience.

Keywords Pathogen • Insect • Invasive species • Outbreak • Dothistroma • Blister rust • Emerald ash borer • Emerging disease

11.1 Introduction and Background

Pathogens are organisms that cause disease, and disease in trees is often defined as a persistent disruption of normal physiological function (Agrios 2005; Desprez-Loustau et al. 2016). Pathogens that cause disease in trees include biotic agents such as fungi, bacteria, nematodes and parasitic plants. Tree disease can also be caused by environmental stress factors (abiotic agents) including competition, lack or excess of nutrients, and air or water pollution. This chapter will focus on biotic agents of tree disease. These agents can affect many different physiological functions in trees. For example, there are fungi, parasitic plants, and bacteria that interfere with a tree's ability to translocate carbon and nutrients or transpire, and fungi that affect reproductive or photosynthetic structures. Initially, infection of the tree by the agent results in host cells being damaged or altered such that their ability to carry out normal processes, including radial growth, is impaired. As the agent spreads to more cells, the tree will exhibit symptoms of disease, which can include reduced or abnormal growth, distress cone crops, discoloration and/or loss of foliage, among others. In some cases, trees can ultimately be killed by the pathogen, and in others the tree is only damaged (Dietze and Matthes 2014).

Pathogens have important ecological roles in forest ecosystems. Those that are host specific, for example some root pathogens, can serve as agents of diversity by killing some species and enabling growth of others. Other pathogens, such as rust or canker fungi serve as natural thinning agents. Decay fungi provide habitat for cavity nesting species, and are important for nutrient cycling.

Trees and forests have evolved in the presence of pathogens and other disturbance agents, and have adapted to infection and attack at the tree level with defensive responses (Franceschi et al. 2005), and at the forest level with successional processes that enable persistence of healthy, functioning forest ecosystems. Invasive species are pathogens or other disturbance agents such as insects, that have been transported or have spread naturally into new environments, and exposed to trees and forests that have not co-evolved with the agent. Successful establishment of invasive organisms requires susceptible hosts, suitable environments for the agent, and the ability to reproduce and spread (Liebhold et al. 1995). There are notable examples of invasive organisms that have successfully established in new environments and have caused considerable economic and environmental damage. One of these, white pine blister rust, caused by the fungus *Cronartium ribicola*, is included as a case study in this chapter. Chestnut blight, caused by the fungus *Cryphonectria parasitica*, was introduced to North America in 1904 on nursery stock from Asia. Known as a "super pathogen", it found a highly susceptible host in American chestnut (*Castanea dentata*) and many others on which it can survive (Schumann 1991). It has two means of spread—one windborne and the other insect or rain-splash dispersed. Within a few decades it essentially eliminated mature chestnut from eastern North American forests, with trees surviving only as stump sprouts until they too are attacked and killed (Hepting 1974). Pine wilt nematode (*Bursaphelenchus xylophilus*) was introduced to Japan most likely from North America, and has caused extensive mortality of Japanese red and black pines (*Pinus densiflora* and *P. thunbergii* respectively, Mamiya 1983). The nematode attacks the parenchyma cells surrounding the resin canals of infested trees, disrupting water transport. Upon death, trees become breeding habitat for *Monochamus* beetles, and emerging young beetles carry nematodes to new hosts. This invasive species has resulted in trade barriers for North American wood products into Europe and Scandinavia (Bergdahl 1988). A final example comes from an interesting group of organisms that are in their own Kingdom (*Chromista*) due to fungal, plant, and animal-like characteristics. *Phytophthora* species have had dramatic impact on global communities since the Irish potato famine in the mid-1800s where *P. infestans* attacked introduced potato varieties causing millions of deaths and emigrations. Introductions of *P. lateralis* and *P. ramorum* to the west coast of the US has decimated local populations of Lawson cypress (*Chamaecyparis lawsoniana*), oaks (*Quercus* sp.) and other species, in addition to causing significant impacts on trade in nursery species. In addition to the known risk of epidemics resulting from introductions of exotic pathogens, is the associated, but less understood risk of pathogen evolution due to genetic exchange among introduced species. A new *Phytophthora* pathogen of alder has arisen from interspecific hybridization (Brasier et al. 1999), and the chance of other emerging hybrids increases with world trade and time.

Native or introduced, pathogens must be virulent, have a susceptible host, and a suitable environment for disease to occur. This is conceptualized in the "disease triangle" which can be expanded to other disturbance agents including insects (Agrios 2005). Disease management is focused on reducing or eliminating one of the factors through activities that negatively impact the pathogen, raise the resistance of the host, or alter the environment to be less conducive to disease development. The disease triangle conceptual model is also useful in understanding impacts, real or potential, of climate change on disease epidemics. For example, as plant diseases are influenced by environmental conditions, we know that climate change will have effects on the pathogen, through changes in growth rates, reproduction, dispersal and other functions, and that climate change will also affect the host through changes in vigour and host distribution (Anderegg et al. 2015).

Dendrochronology provides a useful tool to study historic relationships between trees and agents of disturbance, and how these past relationships are changing as a result of climate change and other human impacts. The application of tree-ring analysis to the study of pathogens and invasive species depends on how the agent affects the tree. Those that can cause mortality or pronounced but temporary growth reductions can be studied using mortality dates and growth suppressions and releases, and those that cause localized cambial necrosis (e.g. canker fungi) can be studied by determining the first year of infection of the cambium. However, disturbance agents that cause subtle changes in growth (e.g. slow acting root diseases), do not lend themselves as well to tree-ring analysis.

The remainder of this chapter presents three case studies. One is on *Dothistroma septosporum,* a foliar pathogen of pine worldwide. It causes growth reductions during outbreak periods and has a very strong relationship with climate. The second case study is on white pine blister rust, caused by an introduced pathogen from Asia that can kill native five-needle pines in only a few years due to lack of resistance in the host. The final case study is on emerald ash borer (*Agrilus planipennis*), an insect introduced to eastern North America in the 1990s. Native to Asia, the insect has caused serious damage to native ash trees.

11.2 *Dothistroma* Needle Blight Case Study

Dothistroma needle blight (DNB) is regarded as the most serious needle disease of pine (*Pinus* spp.) in the world. Two species of fungi cause the disease: *Dothistroma septosporum* (telemorph: *Mycosphaerella pini*) and *D. pini* (teleomorph unknown) (Barnes et al. 2004). The fungus infects the needles, causing necrotic lesions and premature needle drop (Bradshaw et al. 1997). DNB was first studied as an introduced pathogen in the Southern Hemisphere, responsible for widespread defoliation to native and exotic pines (Harrington and Wingfield 1998; Barnes et al. 2004). In recent years, the prevalence of the disease has increased, and damage by the disease is reported from new geographic locations in the Northern Hemisphere (Woods 2003; Bradshaw 2004). Severe damage has been reported in parts of western North America, where the disease has long been established and historically considered a benign pine needle pathogen (Barnes et al. 2004, 2008; Woods et al. 2005; Welsh et al. 2009). Moreover, extensive mortality has also been reported in plantations and mature pine forests in these areas, a previously undocumented condition of the disease. DNB has now been recorded in over 63 countries, infecting over 82 different species of pine (Barnes et al. 2014). Human-mediated movement of *Dothistroma*-infected trees is considered the primary means of spread to new host populations (Bradshaw 2004; Woods et al. 2016), but the recent range expansion and increase in severity in the Northern Hemisphere appear to be linked to environmental change (Woods et al. 2005; Welsh et al. 2014).

Dothistroma spreads by means of splash-dispersed sexual (ascospores) and asexual spores (conidiospores), with conidia reported as the dominant state of the pathogen (Gibson 1972). Studies have shown that conidia may be released and germinate any time temperatures are above 5 °C and under continuous moisture regimes (Sinclair et al. 1987). In northwest British Columbia (BC), Canada where severe damage from *Dothistroma* has been recorded, dispersal of conidia begins in early- to mid-June, and ceases in mid-September (Boateng and Lewis 2015).

Ascospores have also been observed in BC and there is strong evidence of sexual reproduction in the pathogen population (Dale et al. 2011), but ascospores are much less common. Infection can occur over a wide range of temperatures, but the severity of infection remains low except at warm temperatures (15–20 °C) during prolonged wet periods (Peterson 1967; Gadgil 1974). The incidence of *Dothistroma* infection is thus highly sensitive to yearly differences in weather (Peterson 1973) and rapid development of outbreaks can occur during periods of prolonged wet weather during the growing season (Harrington and Wingfield 1998). The life cycle of *Dothistroma* generally requires 1–2 years to complete, but can vary depending on the presence of optimal temperature and precipitation conditions, host species and provenance (Gibson 1972). Given our understanding of the close relationship between DNB and weather, it seems likely that the recent emergence of the disease could be due to changes in climate.

Ongoing climate changes are likely to increase the risk of forest diseases (Coakley 1995; Chakraborty et al. 1998; Coakley et al. 1999; Sturrock et al. 2011). Ecosystems most vulnerable to damage may be those where pathogens are at low disease levels because of previously unfavourable climate conditions (Coakley et al. 1999; Woods et al. 2005). A recent epidemic of DNB in northwest BC presented an opportunity to study an emerging disease and its relationship to climate. Woods et al. (2016) emphasized a possible link between the timing of the strong El Niño event of 1997/1998 (Cai et al. 2014) in the form of increased summer precipitation and the recent epidemic in the area. While singular events attract attention, consideration of temporal and spatial scales are critical in any attempt to link environmental factors to disease development. For instance, a longer temporal and spatial perspective is necessary to identify past environmental drivers and associated trends in disease dynamics, and to provide a historical context to compare changes in severity and extent.

11.2.1 Applications of Tree-Ring Analysis

Dendrochronology enables us to explore outbreak dynamics at multiple temporal, and broad spatial scales. A dendrochronological approach adapted from studies of outbreaks of defoliating insects (Swetnam et al. 1985; Swetnam and Lynch 1989) was used to gain this long-term view of past outbreak dynamics of DNB in northwest BC.

Lodgepole pine (*Pinus contorta*) stands identified in historical survey records as showing signs of past *Dothistroma* infection were used to develop a distinct dendrochronological signature associated with defoliation patterns (Welsh et al. 2009). Increment cores were collected from both host lodgepole pine and non-host Roche spruce (*Picea glauca* x *sitchensis*) or hybrid white spruce (*Picea glauca* x *engelmannii*). A host-non-host correction procedure was performed to separate the observed reduced growth in the host due to disease from climate variations (Swetnam et al. 1985). The reliability of this "correction" procedure is based on two major assumptions: (1) the non-host species are unaffected by the agent causing the outbreak, and (2) the host and non-host species respond similarly to environmental influences (Swetnam et al. 1985). Climate-tree growth response function analyses, chronology statistics and correlation analyses demonstrated that the host and the non-host series were not measurably different in response to climate (Welsh et al. 2009). The verified DNB outbreaks consistently appeared as periods with declining trends in the pine chronologies and relatively unchanging growth trends in the non-host spruce chronologies, suggesting that the observed periods of growth reduction in the host trees were not caused by a climate event.

The resulting growth reductions associated with historical survey records of outbreaks in the 1960s and 1980s were quantified and used to develop outbreak criteria for the software program OUTBREAK (Holmes and Swetnam 1996). The criteria were applied simultaneously to identify probable pre-record outbreak periods at 16 individual sites. A 40% threshold was used to distinguish between periods of reduced growth attributable to the defoliation events and smaller events expressed in the corrected host chronology. The resulting regional time series was developed by summing the number of trees recording an outbreak as defined by the statistical criteria and the 40% threshold for each year across all sites. The regional time series was then standardized by computing the percentage of trees recording an outbreak each year.

The regional reconstructions demonstrated that DNB outbreaks have occurred periodically over the last 174 years in northwest BC, with an increase in outbreak incidence and extent since the 1940s (Fig. 11.1). This subsequent change in disease severity since the 1940s has also been documented in *Dothistroma* literature worldwide and suggests a climate trigger is at play (Woods et al. 2016). Outbreaks identified after the low sample depth period (Fig. 11.1a) started in mid-1910, mid-1930s, mid-1940s, mid-1960s, late 1970s, late 1980s, and 2000. Peak percentages in outbreak incidence (i.e., defined by the outbreak criteria) were 32%, 9%, 17%, 24%, 29%, 29%, and 62%, respectively (Fig. 11.1b). These peaks in disease occurrence also correspond to documented records of DNB incidence in the area during the 1960s and 1980s. The most distinct change observed in the outbreak history was the greater severity and synchrony among sites affected during the most recent outbreak. The regional outbreak reconstruction shows that *Dothistroma* has had a long history in the study area, and in the past has persisted at low levels in the host population. These findings supported survey reports of an increase in disease spread and severity in the study area since the mid-1990s (Woods et al. 2005).

Temporal and spatial scales are a critical consideration in any attempt to link climatic variability with biological effects (Scherm and Yang 1995), therefore a number of biologically relevant climate variables were modeled with the outbreak reconstruction to determine their role in relation to disease occurrence (Welsh et al. 2014). The results indicated that past outbreaks were associated with prolonged periods of wet and warm conditions, suggesting that climate trends drive *Dothistroma* behavior at the regional scale. Graphical comparisons showed that the

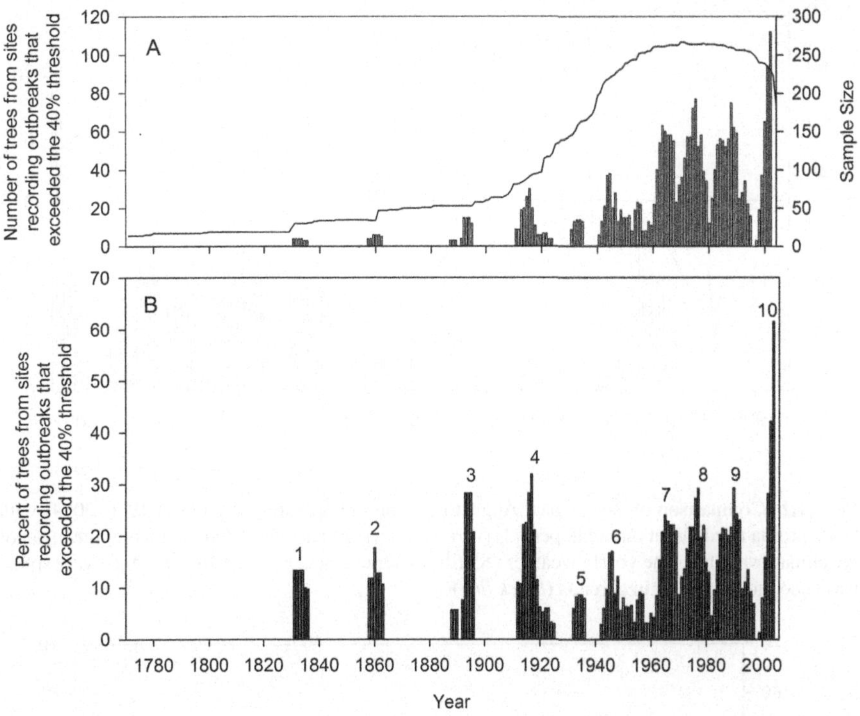

Fig. 11.1 (**a**) Time series of the number of trees from sites recording outbreaks identified by the outbreak criteria (i.e., statistical criteria and 40% threshold) and the sample size curve (or the number of trees included in the regional outbreak reconstruction for each year). Maximum sample depth represented by 267 host trees. (**b**) Percent trees from sites recording an outbreak identified by the outbreak criteria. Numbers indicate 10 outbreak periods

periodic outbreaks were in phase with spring precipitation and summer minimum temperatures. The trend towards increased minimum temperatures appears to be the most important climate factor contributing to outbreaks of the disease; peaks in the outbreaks directly coincide with peaks in August minimum temperature (Fig. 11.2). The greatest increases in temperature associated with climate change are expected to occur in overnight minimums, and as a result dipping less often to record lows (Meehl et al. 2009). It also appears that peaks in these warm periods correspond to increases in spring precipitation levels, and both show an increasing trend over the last 50 years (Fig. 11.3). Even more pronounced was a spike in spring precipitation to previously unrecorded levels and its association with the recent outbreak. The climate in northwest BC has become increasingly warmer and wetter, and the correspondence of outbreaks with periods of wet and warm conditions in northwest BC suggests that regional climate is becoming more favourable for the disease resulting in widespread, synchronous outbreaks.

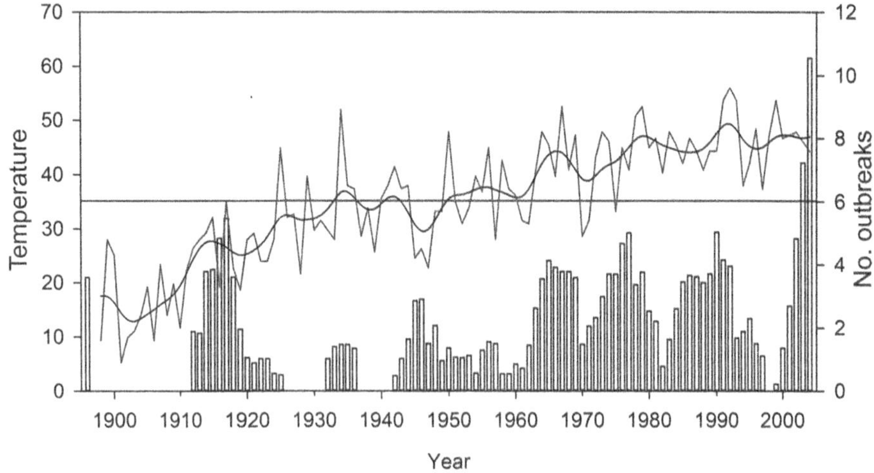

Fig. 11.2 Comparison of 2-year past August minimum temperature (°C) (AD 1986–2004) with Dothistroma need blight outbreak periods (*vertical bars*). The *horizontal line* indicates the mean of the climate variable. The yearly weather records are represented by the *gray line*. A 10-year spline was added to help visualize trends (*black line*)

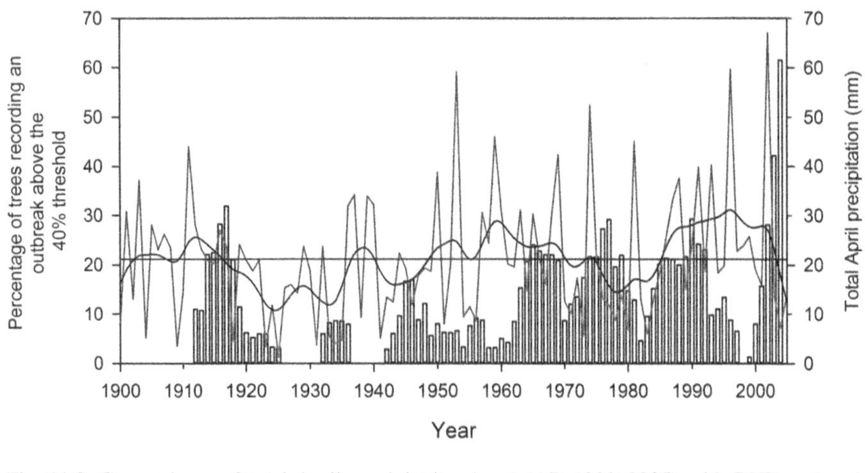

Fig. 11.3 Comparisons of total April precipitation (mm) (AD 1900–2005) with DNB outbreak periods (*vertical bars*). The *horizontal line* indicates the mean of the climate variable. The yearly weather records are represented by the *gray line*. A 10-year spline was added to help visualize trends (*black line*)

11.2.2 Prognosis and Research Opportunities Using Tree Rings

The extent and severity of the recent outbreak raises the possibility that the key factors, which have limited outbreaks in the past, have changed. This has serious implications for the spread of disease into new areas as climate change creates conditions that are more favourable for disease development (Woods et al. 2016). Lodgepole pine forests in BC could be severely impacted on a wider geographical scale and increase the risk of spreading to new areas. The results of this work highlight the unpredictable nature of climate change on disease development and expansion.

The validity of reconstruction of pathogen outbreaks is challenged by the potential to confuse pathogen outbreaks with past insect outbreaks. Careful analysis of the tree-ring structure and wood anatomy from documented outbreaks will help ensure reliability of the reconstruction (Welsh et al. 2009). For instance, the observed reduction in total annual growth during *Dothistroma* events is unique in comparison to many insect defoliators, such as larch sawfly (*Pristiphora erichsonii*), gypsy moth (*Lymantria dispar*), and western spruce budworm (*Choristoneura occidentalis*). The effects of insect defoliation in the first year of reduced growth manifests as only reduced latewood production (Jardon et al. 1994; Muzika and Liebhold 1999). Numerous insect defoliators utilize lodgepole pine trees as a host in BC but all are regarded as uncommon colonial defoliators in the study area. In addition, numerous foliar pathogens defoliate lodgepole pine in BC, but only three are considered common in the northwest: *Lophodermella concolor*, *L. montivaga*, and *Dothistroma*. Outbreaks of *L. concolor* and *L. montivaga* can be locally severe, where cool, moist environmental conditions persist. However, *Dothistroma* has greater potential to cause serious and more widespread damage. The consistent recurrence of the *Dothistroma*-specific signature in association with the historically documented records and the utilization of a minimum percent threshold eliminate the possibility of defoliation events caused by weaker fungal pathogens (Welsh et al. 2009). Tree-ring chronologies usually encompass single species systems (sometimes two insect species simultaneously), and no research has combined the interactions of a pathogen and an insect defoliator in a tree-ring study. More research on the long-term interactions of multiple disturbance agents will be important to understand how forests and their pests respond to environmental change and determine the extent to which current conditions have departed from past trajectories (Speer and Kulakowski 2017). It is possible that dendrochronology can provide a means to study these interacting effects.

There are a growing number of reports of DNB in new countries in the northern latitudes suggesting an increasingly favourable climate for the pathogen (Woods et al. 2016). Without long-term records of past outbreak occurrences, it is difficult to draw conclusions about disease dynamics and relationships to climate patterns, and predict change. Dendrochronological techniques give us the opportunity to identify historical foliar pathogen outbreaks and establish long-term records of disease.

11.3 White Pine Blister Rust Case Study

The invasive fungal pathogen, *Cronartium ribicola*, causal agent of white pine blister rust, is specific to five-needled pines of the subgenus *Strobus*, and is capable of infecting all nine species in North America including three common commercial species, western white pine *(Pinus monticola)*, sugar pine (*P. lambertiana*) and eastern white pine (*P. strobus*), and two high elevation species, limber pine (*P. flexilis*) and whitebark pine (*P. abicaulis*) (Geils et al. 2010). *C. ribicola* is originally from Asia and is believed to have been introduced to eastern North America around 1904 from imported currants (*Ribes* spp.) (Geils et al. 2010) and to western North America around 1910 from imported pine seedlings from Europe (Hunt 2009). It rapidly infected native five-needled pines with reports of white pine blister rust (hereafter called "blister rust") on western white pine in Washington by 1913, Oregon by 1917 and British Columbia by 1921. By 1930, blister rust was reported on multiple pine species in the interior states of Idaho, Montana and Wyoming, and the province of Alberta. Now, a century since its introduction to North America, there are few places with five-needled pines where blister rust is absent (Geils et al. 2010; Tomback and Achuff 2010).

The life cycle of *C. ribicola* is comprised of a five-needled white pine as a primary host and an alternate host from the *Ribes*, *Castilleja* (paintbrush), and *Pedicularis* (lousewort) genera (Colley 1918; McDonald et al. 2006). Infected alternate hosts produce basidiospores in the fall that infect white pines by entering through their needles. Infected pines develop swelling and cankers around infection sites and the infection moves towards the stem. The first aeciospores on pines are typically released in the spring after a latency period of 3–5 years and annually thereafter (Lachmund 1933). Flagging (dead branches) is symptomatic of cankers girdling branches and top kill (mortality of the majority of the crown) occurs when cankers girdle the stem. Additional symptoms of blister rust include evidence of squirrel feeding on the phloem and spermagonial exudate from the canker (Hoff 1992). Once infected, trees cannot recover and will eventually die from progressive girdling of the stem (Geils et al. 2010).

The incidence and impact of blister rust varies by species and region (Schwandt et al. 2010). Whitebark pine is the most susceptible species (Hoff et al. 1980) and blister rust has caused its near range-wide decline (Tomback et al. 2011). Whitebark pine is a slow-growing, long-lived tree found at tree line and subalpine elevations in Washington, Oregon, Nevada, California, Idaho, Montana, and Wyoming, and in British Columbia and Alberta, Canada. Whitebark pine is considered a keystone or foundation species in these high elevation forests for various ecosystem services and fostering biodiversity (Ellison et al. 2005; Tomback et al. 2011).

Blister rust threatens the resilience of whitebark pine populations by: (1) impacting recruitment from direct mortality of saplings and decreased cone production, and (2) inciting further mortality by other disturbance agents. Blister rust affects most ages of whitebark pine trees from saplings to mature trees (Smith et al. 2008; Field et al. 2012). For mature trees, often the top half or more of the tree is killed, leaving the lower stem and branches alive. Unfortunately for whitebark pine, cones are produced only on the top few branches, meaning that top-killed trees, although alive, are reproductively dead. Reduced cone production also affects those mammal species that rely on the seeds as a major seasonal source of protein—Clark's nutcracker (*Nucifraga columbiana*), grizzly bears (*Ursus arctos*) and squirrels (Mattson et al. 1992; McKinney et al. 2009; Barringer et al. 2012). Indirectly, blister rust can be an inciting factor to forest decline, reducing tree resistance to bark beetles, primarily mountain pine beetle (*Dendroctonus ponderosae*) or other stresses such as drought (Bockino and Tinker 2012; Wong and Daniels 2016). High levels of mortality of mature whitebark pine across its range have been caused by various outbreaks of mountain pine beetle, likely in conjunction with blister rust (Campbell and Antos 2000; Larson 2011). Concern about whitebark pine resilience has led to it being listed as an endangered species in Canada (Government of Canada 2012) and monitored for a potential similar listing in the United States (United States Fish and Wildlife Service 2015).

11.3.1 Applications of Tree-Ring Analyses

Blister rust can cause hydraulic failure in five-needled pines by occluding the xylem and phloem (Hudgins et al. 2005). It also can cause carbon starvation by girdling the stem, which reduces the production and flow of photosynthates from the foliage, and by inducing the production of defense metabolites (Hudgins et al. 2005) which diverts carbon away from growth (Herms and Mattson 1992; Bonello et al. 2006). Because blister rust requires living host cells for its nutrients and reproduction, trees can be infected but remain alive for decades (Wong and Daniels 2016). Thus, infected trees likely grow slower and are less resilient to drought or variation in precipitation than healthy trees (McDowell et al. 2011; Anderegg and Callaway 2012).

Reduced growth over several decades since the 1950s has been observed in whitebark pine ring-width chronologies in areas infected with blister rust (Kipfmueller and Salzer 2010; Youngblut and Luckman 2013). This growth pattern was confirmed as a dendrochronological signature of blister rust by Wong and Daniels (2016), who compared the radial growth of live, healthy whitebark pines to those with blister rust. Since drought can also cause prolonged reduced growth in whitebark pine (Millar et al. 2012), comparing growth between healthy and infected trees is key for identifying the blister rust signature in tree rings.

The dendrochronological signature of blister rust allows accurate dating of the introduction of blister rust to various areas. Wong and Daniels (2016) were able to date the introduction of blister rust to whitebark pines in their study area in southern Alberta, Canada, using segmented regression on ring-width chronologies. The breakpoint indicated the approximate year when a significant decline in growth began. The time between a decline in radial growth and the date of the first infection of a needle by the fungus requires further research but is likely a few years. In southern Alberta, the dendrochronological determination of the date of introduction was a decade earlier than in written records. While this method remains to be tested with chronologies from other locations, it offers promise for determining accurate dates of introduction and time-to-death, particularly because blister rust can go undetected for years (Kearns et al. 2009) and evidence of blister rust does not persist long after tree death (Wong 2012).

Within a tree, the year of cambial necrosis can be used to date the age of individual cankers. This has yet to be done on whitebark pine, but has been examined on limber pine by Kearns et al. (2009). Rates of canker expansion were determined as well as the rate of spread throughout a tree. Kearns et al. (2009) suggest that a sample of canker ages can be used to determine how long a stand has been infested with blister rust and whether its spread was constant versus episodic and correlated to certain climatic variables.

Analyses of tree rings in non-host species and surviving whitebark pines can project future stand composition after mortality from blister rust. The loss of mature whitebark pine trees will decrease the competition pressure on smaller subalpine fir (*Abies lasiocarpa*) trees and cause a subsequent growth release evident in tree rings (Wong 2012). A similar response has been found in non-host species to the mortality of sugar pine from blister rust (Waring and O'Hara 2009). While Wong (2012) found that the growth of some subalpine fir trees was facilitated by whitebark pine trees and thus, responded to the death of their whitebark pine neighbor with a decline in radial growth, this decline was not prolonged. The predominant response of growth release will accelerate the dominance of subalpine fir in these high-elevation forests and threaten whitebark pine existence (Wong 2012).

Perhaps the most important application of tree-ring analyses to blister rust infection of whitebark pine is to examine multi-way interactions between blister rust, insects and climate over various time scales. Millar et al. (2012) examined the climatic drivers behind whitebark pine mortality from mountain pine beetle from 2004 to 2010 in eastern California where no blister rust was present. They found that trees prior to their death from mountain pine beetle grew significantly slower than non-attacked trees over the past 115 years. Further, non-attacked trees were more responsive to fluctuations in water deficit and temperatures during the growing season than trees ultimately attacked and killed by mountain pine beetle. They concluded that whitebark pines more stressed by drought had lower resistance to mountain pine beetle.

In their study area in southern Alberta, Wong and Daniels (2016) found dead trees formed 72% of whitebark pine stand density: 31% died from mountain pine beetle, 25% from *Ips* spp. bark beetles, and 41% attributed to blister rust. Tree mortality peaked in the 1970s; however, the annual basal area increment of these trees began to decline significantly in the late 1940s and persisted for up to 30 years until the trees died. Wong and Daniels (2016) proposed multiple factors interacted over several decades to cause unprecedented rates of whitebark pine mortality. Correlations between climatic variables and radial growth indicated that the growth of healthy whitebark pines was limited by temperature during the warm phases of the Pacific Decadal Oscillation (PDO) and by spring precipitation during the cool phase of the PDO (1946–1976). This suggests that whitebark pines were drought stressed during the cool phase of the PDO. During this phase, the climate-growth relations of the disturbed trees began to diverge from living, healthy trees and were less responsive to spring precipitation. Trees infected by blister rust but alive at the time of sampling diverged least, while trees killed by mountain pine beetle diverged most. Wong and Daniels (2016) concluded that climatic variation during the cool PDO phase caused drought stress that likely further stressed trees infected by blister rust. Ultimately, bark beetles and blister rust contributed to tree death.

11.3.2 Prognosis and Research Opportunities Using Tree Rings

To understand the future impacts of blister rust on whitebark pine, it is critical to improve existing epidemiological models predicting the spread of blister rust (e.g., Leung and Kot 2015)—in particular, how fast blister rust spreads within and between trees and how this may change with climate change. Preliminary modelling indicates that assumptions behind the latency period and how long an infected individual tree contributes to blister rust spread significantly impacts the invasion speed (Leung and Kot 2015). Models predict that if disease-induced mortality exceeds the rate of spread, invasion by blister rust will stop. Further use of dendrochronology to determine accurate dates of introduction of blister rust and how this may be affected by climate stress will help parameterize these models and test these hypotheses.

Blister rust is not acting independently in whitebark pine forests. Most studies have examined two-way interactions between blister rust and bark beetles or climate at one point in time. Further dendrochronological studies can help examine multi-way disease-insect-climate interactions over longer time scales in different high-elevation forests.

11.4 Emerald Ash Borer Case Study

The emerald ash borer (EAB; *Agrilus planipennis*), a phloem-feeding beetle native to Asia, has become the most destructive and economically costly forest insect to invade North America (Aukema et al. 2011). Since it was identified in 2002 in Detroit, Michigan, and Windsor, Ontario, EAB has killed tens of millions of ash (*Fraxinus* spp.) trees across much of the upper US Midwest, southern Ontario and Quebec, Canada (Cappaert et al. 2005; Poland and McCullough 2006; EAB 2016). In addition to the natural expansion of existing EAB infestations, inadvertent transport of infested ash firewood, logs, and nursery trees has resulted in the establishment of dozens of localized "outlier" populations. To date, EAB populations have been found in a total of 28 states and two Canadian provinces.

Adult EAB are active throughout the summer. Beetles feed on ash foliage for roughly 5–7 days before mating and females feed for another 10–14 days before oviposition begins. Broad-scale field studies have shown most beetles lay eggs less than 150 m from their emergence point, but a small proportion of mature females disperse considerably farther (Siegert et al. 2010; Mercader et al. 2012). Larvae begin feeding in midsummer on phloem and cambium tissue in tunnels that also score the outer xylem. Over time, larval density builds, which eventually disrupts transport of nutrients and water, leading to dieback and eventual mortality. In its native range in China, EAB attacks only severely stressed trees, functioning as a secondary pest. In North America, EAB will preferentially colonize stressed trees (McCullough et al. 2009a; b), but will also readily colonize and kill healthy ash trees. By 2019, at least 38 million ash trees in urban landscapes will need to be replaced or treated with insecticide, at a projected cost of over $10.7 billion (Kovacs et al. 2010).

On the campus of Indiana State University, located in Terre Haute (USA), EAB was first observed in 2012 and suspected (based on observations at other locations) to have been present for 1–2 years by the time it was sighted. The campus had approximately 400 ash trees, of which 181 were treated with injected pesticide (emamectin benzoate) in 2013.

11.4.1 Applications of Tree-Ring Analysis

Dendrochronology was used in a pilot study to examine tree health in ash trees treated with insecticide, untreated ash trees, and oak trees as a climatic control. Five treated trees, five untreated trees, and five oak trees were sampled with two cores at breast height on opposite sides of the tree. The samples were mounted and sanded with progressively finer sandpaper, and were dated using skeleton plots (Stokes and Smiley 1968; Speer 2010). Cores were measured using a Velmex measuring machine and visual dating of the cores was checked with the computer program COFECHA (Holmes 1983). The series were standardized using the computer program ARSTAN (Cook and Holmes 1986) and individual tree-level chronologies and the master chronologies were examined for a reduction in ring width associated with EAB attacks.

We found that coexisting oak trees were good climate controls for the ash trees, tracking their growth closely with the extreme 2012 drought year showing up in all three chronologies (Fig. 11.4). Once we subtracted the oak chronology from the treated and untreated ash chronologies, we compared the resultant corrected growth indices between the two treatments. We found that the 2014 year of growth was the second largest decrease in growth in untreated trees for the entire record (with only 2010 having a slightly larger decrease in growth) with no similar reduction in growth of treated trees. Despite obvious canopy damage to the untreated trees, the growth decline was not as pronounced as expected. It is possible that trees damaged by EAB compensate by producing extra growth on the stem as a wound wood response, but then die quickly once damage reaches a particular threshold.

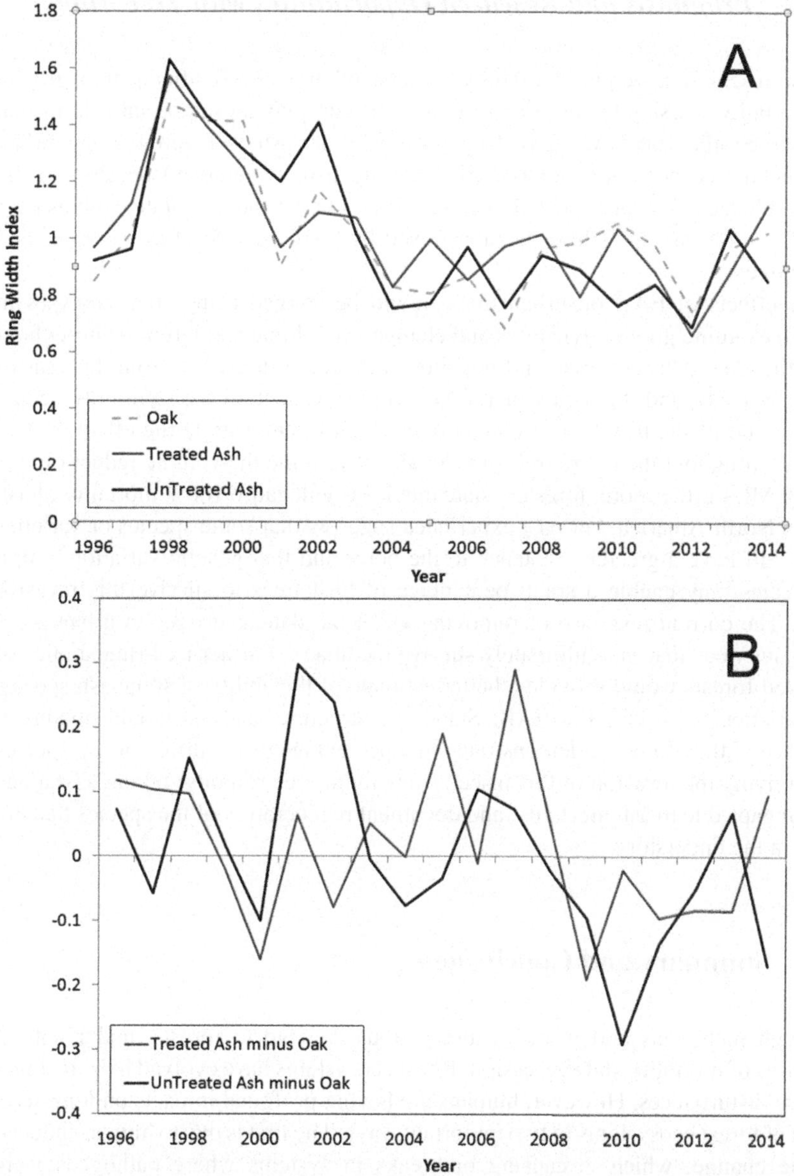

Fig. 11.4 Oak and ash tree growth at the Indiana State University Campus for the past 18 years. (**a**) Ring width indices for oak, treated ash, and untreated. (**b**) Ring width difference chronologies where each oak chronology was subtracted from the ash chronology to control for climate

11.4.2 Prognosis and Research Opportunities with Tree Rings

Further research is required to fully understand how EAB affects tree growth and mortality thresholds, and tree-ring analysis can play an important role in that research. Studies on EAB may face challenges as ash is favored as an urban ornamental tree but trees attacked by EAB are usually removed because of the potential hazard they may pose in these settings. On the Indiana State campus over 200 of the 400 ash trees have been removed by 2016 because they were deemed hazard trees.

The effects of EAB on urban trees should be studied using stem analyses in order to examine growth dynamics and changes in volume and biomass throughout the entire tree. Attacked trees did not show a decrease in growth from the year of arrival by EAB, and it appears that a reduction in growth in the stem only occurs near the end of the tree's life. A stem analysis will demonstrate the effect on tree growth throughout the entire tree and can also determine the volume reduction due to the EAB. Furthermore, foresters state that EAB will cause 100% mortality of ash trees in North America, but past experience suggests that some species or varieties of ash will have a greater resistance to the borer and that genetic variation within the species may enable a small percentage of host trees to survive this invasive insect. The current response of removing all of the damaged trees in urban areas eliminates trees that may ultimately survive the attack. Further tree-ring studies of untreated forests would provide a better estimate of the ability of some ash species and varieties to survive the EAB. Stand-age structure analysis would document mortality of the ash trees, demonstrate what percentage of the different ash species are surviving the invasion of this insect, show living tree response taking advantage of light gaps due to ash mortality, and document regeneration of the species that are living in the understory.

11.5 Summary and Conclusions

Although pathogens and insects cause serious damage to forests, under natural processes of mortality and succession, forest ecosystems have evolved to be resilient to these disturbances. However, humans are having profound impacts on long-term health of forest ecosystems in two important ways. The first is due to human-induced climate change, which is causing outbreaks in systems where pathogens were historically limited by unfavourable climatic conditions such as low temperatures. Examples of this are becoming common and often have serious impacts; witness the recent mountain pine beetle epidemic in North America, and the *Dothistroma* needle blight outbreak described above. The second human influence is through increasing global trade and with that, greater chances of pathogens interacting with native hosts. Examples abound, and while many introduced species fail to establish, those that do have had devastating impacts on forest ecosystems. Past examples include Dutch elm disease, chestnut blight, and sudden oak death, with a more recent but very serious outbreak of emerald ash borer. These two human-mediated influences do not function in isolation, and the whitebark pine blister rust case study in this chapter has highlighted the combined influence of an introduced pathogen, and changing environmental conditions that have enabled upward movement of a native beetle into regions previously too cold to sustain outbreaks.

Tree-ring analysis provides a critical tool that will help with understanding the existing and potential future impacts of climate change and global trade on forest ecosystems. Evaluation of historic patterns of tree mortality or changes in growth can be determined using tree-ring analysis. As well, environmental drivers of these changes can be deduced through a combination of historic climate records and tree-ring analysis. Further, carefully designed sampling techniques and access to preserved wood (e.g. wood in use, wood remnants in glaciers) enables examination of environmental drivers and pest outbreaks at various spatial and temporal scales, which is critical for addressing future challenges of emerging pests due to climate change, introduced species, and interactions among different biotic stressors.

References

Agrios GN (2005) Plant pathology, 5th edn. Elsevier Academic Press, New York

Anderegg WL, Callaway ES (2012) Infestation and hydraulic consequences of induced carbon starvation. Plant Physiol 159:1866–1874

Anderegg WL, Hicke JA, Fisher RA, Allen CD, Aukema J, Bentz B, Hood S, Lichstein JW, Macalady AK, McDowell N, Pan Y, Raffa K, Sala A, Sha JD, Stephenson NL, Tague C, Zeppel M (2015) Tree mortality from drought, insects, and their interactions in a changing climate. New Phytol 208:674–683

Aukema JE, Leung B, Kovacs K, Chivers C, Britton KO, Englin J, Frankel SJ, Haight RG, Holmes TP, Liebhold A, McCullough DG, Von Holle B (2011) Economic impacts of non-native forest insects in the continental United States. PLoS One 6:e24587. doi:10.1371/journal.pone.0024587

Barnes I, Crous PW, Wingfield BD, Wingfield M (2004) Multigene phylogenies reveal that red band needle blight of *Pinus* is caused by two distinct species of *Dothistroma, D. septosporum* and *D. pini*. Stud Mycol 50:551–565

Barnes I, Kirisits T, Akulov A, Chhetri DB, Wingfield BD, Bulgakov TS, Wingfield MJ (2008) New host and country records of the Dothistroma needle blight pathogens from Europe and Asia. For Pathol 38:178–195

Barnes I, Wingfield MJ, Carbone I, Kiristis T, Wingfield BD (2014) Population structure and diversity of an invasive pine needle pathogen reflects anthropogenic activity. Ecol Evol 4:3642–3661

Barringer LE, Tomback DF, Wunder MB, McKinney ST (2012) Whitebark pine stand condition, tree abundance, and cone production as predictors of visitation by Clark's Nutcracker. PLoS One 7:e37663

Bergdahl DR (1988) Impact of pinewood nematode in North America: present and future. J Nematol 20:260–265

Boateng K, Lewis KJ (2015) Spore Dispersal by *Dothistroma septosporum* in northwest British Columbia. Phytopathology 105:69–79

Bockino NK, Tinker DB (2012) Interactions of white pine blister rust and mountain pine beetle in whitebark pine ecosystems in the southern Greater Yellowstone Area. Nat Areas J 32:31–40

Bonello PTR, Gordon DA, Herms DL, Wood EN (2006) Nature and ecological implications of pathogen-induced systemic resistance in conifers: a novel hypothesis. Physiol Mol Plant Pathol 68:95–104

Bradshaw RE (2004) Dothistroma (red-band) needle blight of pines and the dothistromin toxin: a review. For Pathol 34:163–185

Bradshaw RE, Bidlake A, Forester N, Scott DB (1997) Transformation of the fungal forest pathogen *Dothistroma pini* to hygromycin resistence. Mycol Res 101:1247–1250

Brasier CM, Cooke DEL, Duncan JM (1999) Origin of a new *Phytophthora* pathogen through interspecific hybridization. Proc Natl Acad Sci U S A 96:5878–5883

Cai W, Borlace S, Lengaigne M, van Rensch P, Collins M, Vecchi G, Timmerman A, Santoso A, McPhaden MJ, Wu L, England MH, Wang G, Guilyardi E, Jin FF (2014) Increasing frequency of extreme El Niño events due to greenhouse warming. Nat Clim Chang 4:111–116

Campbell EM, Antos JA (2000) Distribution and severity of white pine blister rust and mountain pine beetle on whitebark pine in British Columbia. Can J For Res 30:1051–1059

Cappaert D, McCullough DG, Poland TM, Siegert NW (2005) Emerald ash borer in North America: a research and regulatory challenge. Am Entomol 51:152–165

Chakraborty S, Murray GM, Magarey PA, Yonow T, O'Brien RG, Croft BJ, Barbetti MJ, Sivasithamparam K, Old KM, Dudzinski MJ, Sutherst RW, Penrose LJ, Archer C, Emmett RW (1998) Potential impact of climate change on plant diseases of economic significance to Australia. Australas Plant Pathol 27:15–35

Coakley SM (1995) Biospheric change: will it matter in plant pathology? Can J Plant Pathol 17:147–153

Coakley SM, Scherm H, Chakraborty S (1999) Climate change and plant disease management. Annu Rev Phytopathol 37:399–426

Colley RH (1918) Parasitism, morphology, and cytology of *Cronartium ribicola*. J Agric Res 15:619–659

Cook ER, Holmes RL (1986) Users manual for program ARSTAN. Laboratory of Tree-Ring Research, University of Arizona, Tucson

Dale AL, Lewis KJ, Murray BW (2011) Sexual reproduction and gene flow in the pine pathogen *Dothistroma septosporum* in British Columbia. Phytopathology 101:68–76

Desprez-Loustau ML, Aguayo J, Dutech C, Hayden KJ, Husson C, Jakushkin B, Marçais B, Piou D, Robin C, Vacher C (2016) An evolutionary ecology perspective to address forest pathology challenges of today and tomorrow. Ann For Sci 73:45–67

Dietze MC, Matthes JH (2014) A general ecophysiological framework for modelling the impact of pests and pathogens on forest ecosystems. Ecol Lett 17:1418–1426

Ellison AM, Bank MS, Clinton BD, Colburn EA, Elliott K, Ford CR, Foster DR, Kloeppel BD, Knoepp JD, Lovett GM, Mohan J, Orwig DA, Rodenhouse NL, Sobczak WV, Stinson KA, Stone JK, Swan CM, Thompson J, Holle BV, Webster JR (2005) Loss of foundation species: consequences for the structure and dynamics of forested ecosystems. Front Ecol Environ 3:479–486

Emerald Ash Borer Information Network (2016) http://www.emeraldashborer.info/. Accessed 30 Sept 2016

Field SG, Schoettle AW, Klutsch JG, Tavener SJ, Antolin MF (2012) Demographic projection of high-elevation white pines infected with white pine blister rust: a nonlinear disease model. Ecol Appl 22:166–183

Franceschi VR, Krokene P, Christiansen E, Krekling T (2005) Anatomical and chemical defenses of conifer bark against bark beetles and other pests. New Phytol 167:353–375

Gadgil PD (1974) Effect of temperature and leaf wetness period on infection of *Pinus radiata* by *Dothistroma pini*. N Z J For Sci 4:495–501

Geils BW, Hummer KE, Hunt RS (2010) White pines, *Ribes*, and blister rust: a review and synthesis. For Pathol 40:147–185

Gibson IAS (1972) Dothistroma blight of *Pinus radiata*. Annu Rev Phytopathol 10:51–72

Government of Canada (2012) Order amending Schedule 1 to the Species at Risk Act. Canada Gazette. Part II, vol 146, No. 14, SOR/2012-113

Harrington TC, Wingfield MJ (1998) Disease and the ecology of indigenous and exotic pines. In: Richardson DM (ed) Ecology and biogeography of pinus. Cambridge University Press, Cambridge, pp 381–406

Hepting GH (1974) Death of the American chestnut. J For Hist 18:60–67

Herms DA, Mattson WJ (1992) The dilemma of plants: to grow or defend. Q Rev Biol 67:282–335

Hoff RJ (1992) How to recognize blister rust on whitebark pine. United States Department of Agriculture, Forest Service, Intermountain Research Station, Ogden

Hoff RJ, Bingham RT, McDonald GI (1980) Relative blister rust resistance of white pines. Eur J For Pathol 10:307–316

Holmes RL (1983) Computer-assisted quality control in tree-ring dating and measurement. Tree-Ring Bull 43:69–78

Holmes RL, Swetnam TW (1996) Program OUTBREAK user's manual. Unpublished paper, on file at the Laboratory of Tree-ring Research, University of Arizona, Tucson, Arizona

Hudgins JW, McDonald GI, Zambino PJ, Klopfenstein NB, Franceschi VR (2005) Anatomical and cellular responses of *Pinus monticola* stem tissues to invasion by *Cronartium ribicola*. For Pathol 3:423–443

Hunt RS (2009) History of western white pine and blister rust. For Chron 85:516–520

Jardon Y, Filion L, Cloutier C (1994) Long-term impact of insect defoliation on growth and mortality of eastern larch in boreal Quebec. Ecoscience 1:231–238

Kearns H, Jacobi W, Geils B (2009) A method for estimating white pine blister rust canker age on limber pine in the central Rocky Mountain. For Pathol 39:177–191

Kipfmueller KF, Salzer MW (2010) Linear trend and climate response of five-needle pines in the western United States related to treeline proximity. Can J For Res 40:134–142

Kovacs KF, Haight RG, McCullough DG, Mercader RJ, Siegert NW, Liebhold AM (2010) Cost of potential emerald ash borer damage in US communities, 2009–2019. Ecol Econ 69:569–578

Lachmund HG (1933) Mode of entrance and periods in the life cycle of *Cronartium ribicola* on *Pinus monticola*. J Agric Res 47:791–805

Larson ER (2011) Influences of the biophysical environment on blister rust and mountain pine beetle, and their interactions, in whitebark pine forests. J Biogeogr 38:453–470

Leung MR, Kot M (2015) Models for the spread of white pine blister rust. J Theor Biol 382:328–336

Liebhold AM, MacDonald WL, Bergdahl D, Mastro VC (1995) Invasion by exotic forest pests: a threat to forest ecosystems. For Sci Monogr 30:1–49

Mamiya Y (1983) Pathology of the pine wilt disease caused by Bursaphelenchus xylophilus. Annu Rev Phytopathol 21:201–220

Mattson DJ, Blanchard BM, Knight RR (1992) Yellowstone grizzly bear mortality, human habituation, and whitebark pine seed crops. J Wildl Manag 56:432–442

McCullough DG, Poland TM, Anulewicz TC, Cappaert D (2009a) Emerald ash borer (Coleoptera: Buprestidae) attraction to stressed or baited ash trees. Environ Entomol 38:1668–1679

McCullough DG, Poland TM, Cappaert D, Anulewicz AC (2009b) Emerald ash borer (*Agrilus planipennis*) attraction to ash trees stressed by girdling, herbicide and wounding. Can J For Res 39:1331–1345

McDonald GI, Richardson BA, Zambino PJ, Klopfenstein NB, Kim MS (2006) *Pedicularis* and *Castilleja* are natural hosts of *Cronartium ribicola* in North America: a first report. For Pathol 36:73–82

McDowell NG, Beerling DJ, Breshears DD, Fisher RA, Raffa KF, Sitt M (2011) The interdependence of mechanisms underlying climate-driven vegetation mortality. Trends Ecol Evol 26:523–532

McKinney ST, Fielder CE, Tomback DF (2009) Invasive pathogen threatens bird–pine mutualism: implications for sustaining a high-elevation ecosystem. Ecol Appl 19:597–607

Meehl GA, Tebaldi C, Walton G, Easterling D, McDaniel L (2009) The relative increase of record high maximum temperatures compared to record low minimum temperatures in the US. Geophys Res Lett 36:L23701

Mercader RJ, Siegert NW, McCullough DG (2012) Estimating the influence of population density and dispersal behavior on the ability to detect and monitor *Agrilus planipennis* (Coleoptera: Buprestidae) populations. J Econ Entomol 105:272–281

Millar CI, Westfall RD, Delany DL, Bokach MJ, Flint AL, Flint LE (2012) Forest mortality in high-elevation whitebark pine (*Pinus albicaulis*) forests of eastern California, USA; influence of environmental context, bark beetles, climatic water deficit, and warming. Can J For Res 42:749–765

Muzika RM, Liebhold AM (1999) Changes in radial increment of host and nonhost tree species with gypsy moth defoliation. Can J For Res 29:1365–1373

Peterson GW (1967) Dothistroma needle blight of Austrian and ponderosa pines: epidemiology and control. Phytopathology 57:437–441

Peterson GW (1973) Infection of Austrian and Ponderosa pine by *Dothistroma pini* in eastern Nebraska. Phytopathology 62:1060–1063

Poland TM, McCullough DG (2006) Emerald ash borer: invasion of the urban forest and the threat to North America's ash resource. J For 104:118–124

Scherm H, Yang XB (1995) Interannual variations in wheat rust development in China and the United States in relation to the El Niño/Southern Oscillation. Phytopathology 85:970–976

Schumann GL (1991) Plant diseases: their biology and social impact. APS Press, St. Paul, pp 267–272

Schwandt JW, Lockman IB, Kliejunas JT, Muir JA (2010) Current health issues and management strategies for white pines in the western United States and Canada. For Pathol 40:226–250

Siegert NW, McCullough DG, Williams DW, Fraser I, Poland TM (2010) Dispersal of *Agrilus planipennis* (Coleoptera: Buprestidae) from discrete epicenters in two outlier sites. Environ Entomol 39:253–265

Sinclair WA, Lyon HH, Johnson WT (1987) Diseases of trees and shrubs. Cornell University Press, Ithaca

Smith C, Wilson B, Rasheed S, Walker RC, Carolin T, Shepherd B (2008) Whitebark pine and white pine blister rust in the Rocky Mountains of Canada and northern Montana. Can J For Res 38:982–995

Speer JH (2010) Fundamentals of tree-ring research. University of Arizona Press, Tucson

Speer JH, Kulakowski D (2017) Creating a buzz: insect outbreaks and disturbance interactions. In: Amoroso MM, Daniels LD, Baker PJ, Camarero JJ (eds) Dendroecology: tree-ring analyses applied to ecological studies. Springer, Cham

Stokes MA, Smiley TL (1968) An introduction to tree-ring dating. University of Chicago Press, Chicago

Sturrock RN, Fankel SJ, Brown AV, Hennon PE, Kliejunas JT, Lewis KJ, Worrall JJ, Woods AJ (2011) Climate change and forest diseases. Plant Pathol 60:133–149

Swetnam TW, Lynch AM (1989) A tree-ring reconstruction of western spruce budworm history in the southern Rocky Mountains. For Sci 35:962–986

Swetnam TW, Thompson MA, Sutherland EK (1985) Using dendrochronology to measure radial growth of defoliated trees, Agricultural Handbook, vol 639. US Department of Agriculture, Washington, DC

Tomback DF, Achuff P (2010) Blister rust and western forest biodiversity: ecology, values and outlook for white pines. For Pathol 40:186–225

Tomback DF, Achuff PL, Schoettle AW, Schwandt JW, Mastrogiuseppe RJ (2011) The magnificent high-elevation five-needle white pines: ecological roles and future outlook. In: Keane RE, Tomback DF, Murray MP, Smith CM (eds) The future of high-elevation, five-needle white pines in Western North America: Proceedings of the High Five Symposium. US Department of Agriculture, Forest Service, Rocky Mountain Research Station, Missoula, MT, pp 154–164

United States Fish and Wildlife Service (2015) Review of native species that are candidates for listing as endangered or threatened. Notice of Review, 80 Federal Register, 80583, pp 80583–80614

Waring KM, O'Hara KL (2009) Stand development and tree growth response to sugar pine mortality in Sierran mixed-conifer forests. Northwest Sci 83:89–100

Welsh C, Lewis KJ, Woods AJ (2009) The outbreak history of Dothistroma needle blight: an emerging forest disease in northwestern British Columbia, Canada. Can J For Res 39:2505–2519

Welsh C, Lewis KJ, Woods AJ (2014) Regional outbreak dynamics of Dothistroma needle blight linked to weather patterns in British Columbia, Canada. Can J For Res 44:212–219

Wong CM (2012) Understanding disturbance, facilitation, and competition for conservation of whitebark pine in the Canadian Rockies. Dissertation, University of British Columbia

Wong CM, Daniels L (2016) Novel forest decline triggered by multiple interactions among climate, an introduced pathogen and bark beetles. Glob Chang Biol 23(5):1926–1941. doi:10.1111/gcb.13554

Woods AJ (2003) Species diversity and forest health in northwest British Columbia. For Chron 79:892–897

Woods AJ, Coates KD, Hamann A (2005) Is an unprecedented *Dothistroma* needle blight epidemic related to climate change? Bioscience 55:761–769

Woods AJ, Martín-García J, Bulman L, Vasconcelos MW, Boberg J, La Porta N, Peredo H, Vergara G, Ahumada R, Brown A, Diez JJ (2016) *Dothistroma* needle blight, weather and possible climatic triggers for the disease's recent emergence. For Pathol 46:443–452. doi:10.1111/efp.12248

Youngblut DK, Luckman BH (2013) Evaluating the temperature sensitivity of radial growth patterns from whitebark pine in the western Canadian Cordillera. Dendrochronologia 31:1–74

Chapter 12
Deciphering Dendroecological Fingerprints of Geomorphic Process Activity

Markus Stoffel, Juan A. Ballesteros-Cánovas, Christophe Corona, and Karel Šilhán

Abstract The initial employment of tree rings in geomorphic studies was simply as a dating tool and rarely exploited other environmental information and records of damage induced by earth surface processes within the tree. However, these unique, annually resolved, tree-ring records preserve valuable archives of past process activity on timescales of decades to centuries. As many of these processes also represent significant natural hazards, understanding their distribution, timing and controls provides valuable information that can assist in the prediction, mitigation and defence against these hazards and their effects on society. This chapter provides an introduction to the topics, and illustrates it with three case-study examples, demonstrating the application of tree-ring records in studying earth-surface processes and illustrating the breadth and diverse applications of contemporary dendrogeomorphology. It also underlines the growing potential to expand dendrogeomorphic research, possibly leading to the establishment of a range of techniques and approaches that may become standard practice in the analysis and understanding of earth-surface processes and related natural hazards in the future.

Keywords Natural hazards • Dendrogeomorphology • Earth-surface processes • Floods • Erosion • Landslides

M. Stoffel (✉) • J.A. Ballesteros-Cánovas
Climatic Change and Climate Impacts, Institute for Environmental Sciences, University of Geneva, 13 Rue des Maraîchers, Geneva, CH-1205, Switzerland

Dendrolab.ch, Department of Earth Sciences, University of Geneva, 66 Boulevard Carl-Vogt, Geneva, CH-1205, Switzerland
e-mail: markus.stoffel@dendrolab.ch; juan.ballesteros@dendrolab.ch

C. Corona
Geolab, UMR 6042 CNRS, Université Blaise Pascal 4 Rue Ledru, F-63057, Clermont-Ferrand, France
e-mail: Christophe.CORONA@univ-bpclermont.fr

K. Šilhán
Department of Physical Geography and Geoecology, University of Ostrava, Ostrava, Czech Republic
e-mail: Karel.Silhan@osu.cz

M.M. Amoroso et al. (eds.), *Dendroecology*, Ecological Studies 231,
DOI 10.1007/978-3-319-61669-8_12

12.1 Introduction

A major key to the understanding of natural hazards and risks is the documentation of past geomorphic process activity (Stoffel and Huggel 2012). As a result of the absence of documentary records, this information must often be developed from natural archives or "silent witnesses" (Aulitzky 1992) that remain visible in the landscape after an event. The significant contribution of tree rings to these endeavors lies in their capacity to preserve evidence of past geomorphic activity, as well as the interactions and linkages of hydrological processes with landforms or the interaction of geomorphic processes with water. In many climates, tree-ring records may thus represent one of the most valuable and precise natural archives for the reconstruction and understanding of past and ongoing processes during the past several hundred years (Stoffel and Bollschweiler 2008; Stoffel et al. 2010; Stoffel and Corona 2014).

The initial employment of tree rings in geomorphic studies was simply as a dating tool (Douglass 1941; Stokes and Smiley 1968)—but only rarely exploited environmental information that could be derived from studies of ring-width variations and records of anatomical anomalies contained within the tree. However, these unique, annually resolved, tree-ring records preserve potentially valuable archives of past geomorphic processes. As many geomorphic processes are also significant natural hazards, understanding their distribution, timing and controls provides valuable information that can assist in the development of mitigation and defense against these hazards and their effects on society (Osterkamp et al. 2012; Stoffel and Wilford 2012).

Apart from the site-specific information common to many trees at any site, individual trees also record the effects of mechanical disturbance caused by external processes. This contribution aims at presenting (1) how dendroecological techniques can help in unraveling and dating past geomorphic disturbance, (2) which growth anomalies can be identified and attributed to past geomorphic process activity, and at providing (3) a suite of examples with selected recent applications and issues in the field of dendrogeomorphology.

12.2 Dendroecological Consequences of Geomorphic Disturbances

In his seminal work, Alestalo (1971) illustrated that the occurrence of earth-surface processes will typically injure trees, tear off their crown or branches, tilt their stems, partially bury them or expose their roots. Evidence of these events is recorded in growth-ring records of affected trees (Shroder 1978) and can be used for the dating of past geomorphic process activity, provided that tree-ring series are cross-dated properly. Based on the principles presented in these seminal papers (Butler and

Stoffel 2013), a set of characteristic growth disturbances has been typically used in dendrogeomorphic studies (Stoffel et al. 2013a), with a clear focus on injuries, reaction wood and growth suppression (Stoffel and Corona 2014). In the following, we provide an overview of how geomorphic process activity usually affects trees and how trees will react to these disturbances.

12.2.1 *Injuries and Callus Tissue*

Partial bark removal and wood-penetrating injuries are a common feature in trees affected by geomorphic processes (Trappmann and Stoffel 2013, 2015). Wounds can occur on the tree's stem, its branches or on roots. If impacts locally destroy the cambium, incremental cell formation will become disrupted and new cell formation will cease in the injured segment of the tree. To minimize rot and the negative effects of insect attacks after damage, the injured tree will compartmentalize the wound (Shigo 1984; Stoffel and Klinkmüller 2013) and start the production of chaotic callus tissue at the edges of the injury so as to continuously close the wound. The extent of wound healing will, however, greatly depend on the annual increment rate, tree age and on scar size (Bollschweiler et al. 2008; Schneuwly et al. 2009a). The presence of injuries and chaotic callus tissue is commonly regarded as a valuable and reliable indicator of past geomorphic process activity. Whereas scars will remain largely visible on the stem surface of tree species with smooth bark (*e.g. Abies*, *Alnus*, *Betula*, *Fagus*; Trappmann and Stoffel 2013), they may become fully obscured in species with thicker bark structures such as *Larix*, *Picea*, *Pinus* or *Quercus* (Stoffel and Perret 2006; Trappmann et al. 2013).

12.2.2 *Tangential Rows of Traumatic Resin Ducts*

Following cambium disturbance, tangential rows of traumatic resin ducts (TRDs) are produced in the developing secondary xylem of certain conifer species such as *Larix*, *Picea*, *Pseudotsuga* or *Abies* (Bannan 1936; Jacoby 1997; Stoffel 2008), where they extend both tangentially and axially from the injury (Bollschweiler et al. 2008; Schneuwly et al. 2009a, b). When wounding occurs during the vegetation period of the tree, resin production will start within a few days after the impact and ducts will emerge within 3 weeks after the disturbance). Therefore, when analyzing cross-sections, the intra-seasonal position of the first series of TRDs can be used to reconstruct previous events with monthly precision (Stoffel et al. 2005b, 2008; Stoffel and Beniston 2006; Schneuwly-Bollschweiler and Stoffel 2012), provided that disturbance occurred during the vegetation period. With increasing axial and tangential distance from the impact, however, TRDs tend to migrate to later portions of the tree ring (Bollschweiler et al. 2008; Schneuwly et al. 2009a). The intra-seasonal dating with monthly precision thus has to be based on cross-sections or a large number of increment cores at the same elevation on the stem. This technique cannot be used in *Pinus* because TRDs do not normally occur in this genus, which produces copious amounts of resin and resin ducts unrelated to mechanical wounding (Blodgett et al. 2007; Ballesteros et al. 2010a).

12.2.3 Tracheid and Vessel Anomalies

Anomalies in tracheids and vessels have only rarely been used in the past to extract signals of geomorphic activity. Most work has been realized on riparian trees affected by floods and/or debris flows (*e.g.* St. George et al. 2002; Ballesteros et al. 2010b; Arbellay et al. 2010a, b; Wertz et al. 2013). Other processes, such as snow avalanches, have not been studied frequently in the past (Arbellay et al. 2013). Tree microscopic response to wounding was primarily studied between rings formed in the year of disturbance and subsequent years as well as in uninjured control rings. Injured rings were characterized by much smaller (but more) vessels as compared with uninjured rings, yet fiber and parenchyma cells did not differ significantly in numbers and size between injured and uninjured rings. These results highlight the existence of anatomical tree-ring signatures related to past geomorphic process activity and address an innovative methodological approach to date injuries inflicted on broadleaved trees with minimally destructive techniques. Arbellay et al. (2012a, b) have also expanded their approach to analyze thickness-to-span ratios of vessels, xylem relative conductivity and xylem vulnerability to cavitation, and state that the degree to which the wound-induced anatomical changes in wood structure express the functional need of trees to improve xylem hydraulic safety and mechanical strength at the expense of water transport. They conclude that xylem hydraulic efficiency was restored in 1 year, while xylem mechanical reinforcement and resistance to cavitation and decay lasted over several years.

Research on tracheid changes in conifers has been basically limited to root exposure, with the exception of Stoffel and Hitz (2008) who identified changes in lumina of earlywood tracheids after wounding by snow avalanches and rockfalls or Arbellay et al. (2014a, b) who analyzed ecophysiological traits next to scars in North American conifer trees affected by fire. Exposed roots will continue to grow and fulfill their functions as long as their outer tips remain in the ground. In the exposed portion of the root, anatomical changes will occur (Stoffel et al. 2013b; Ballesteros-Cánovas et al. 2017a) and individual growth rings similar to those in the stem or branches will be formed. The localization of such changes in the tree-ring record allows determination of the moment of exposure (Corona et al. 2010, 2011; Lopez Saez et al. 2011, Stoffel et al. 2012).

12.2.4 Reaction Wood

Inclination of the stem may result from the sudden pressure induced by hydrogeomorphic processes directly, by the associated deposition of material (*e.g.* avalanche snow, debris-flow material), or by the slow but ongoing destabilization of a tree through landslide activity or erosion. Tilted trees are common in most areas affected by geomorphic processes and have therefore been used in many publications focusing on the dating of event histories (Braam et al. 1987; Fantucci and Sorriso-Valvo 1999; Lopez Saez et al. 2012a, b, 2013; Sorg et al. 2015).

Subsequent growth in the trunk of a tilted tree will attempt to restore its vertical position, and the reaction will be most clearly visible in that segment of the tree to which the center of gravity has been moved through the inclination of the stem axis (Mattheck 1993). In the tree-ring record, eccentric growth will be visible after a tilting event and thus will allow accurate dating of the disturbance. In conifers, compression wood (also referred to as reaction wood) will be produced on the underside of the trunk. Individual rings will be considerably larger and slightly darker in appearance as compared to the upslope side. The difference in color results from much thicker and rounded cell walls of earlywood and latewood tracheids (Du and Yamamoto 2007). Compression wood also tends to have a higher proportion of latewood, higher lignin content and higher density (Timell 1986). Multiple tilting events in the same stem may be recognized by changes in the amount, color or orientation of reaction wood series in the tree-ring record. By contrast, stem tilting in broadleaved trees leads to the formation of tension wood (Westing 1965) on the upper side facing the tilting agent. Broadleaved trees react upon tilting with ultra-structural modifications (*e.g.* fewer vessels of smaller diameter, higher cellulose content and a gelatinous layer oriented nearly parallel to the fiber axis) that are typically only visible when studied on micro-sections (Pilate et al. 2004).

12.2.5 Growth Reduction

Debris flows, floods, or landslides may bury trees by depositing material around their stem base. Growth suppression after burial with debris is caused, on the one hand, by a reduced activity of the roots, and on the other hand, by mechanical effects caused by the enormous weight of debris. The pressure on the cambium exerted by bark and phloem impedes the cell division and leads to a reduced number of cells with narrower lumen (Kny 1877). The supply of water and nutrients will be temporarily disrupted or at least limited (LaMarche 1966; Hupp et al. 1987; Friedman et al. 2005), and the yearly increment will be diminished (Kogelnig-Mayer et al. 2013). Reductions in annual ring widths in tilted trees are thought to be related to the partial destruction of root mass in the case of unstable slopes (Mayer et al. 2010). If stem burial exceeds a certain threshold, trees will die from a shortage of water and nutrient supply.

Bouncing rocks and boulders, debris flows and lahars or the windblast of snow avalanches may cause decapitation of trees or the removal of branches. The loss of the crown or branches is more common in bigger trees, when stems have lost their flexibility. Apex loss has also been observed as a result of rockfall impacts close to the ground level. In such cases, the sinusoidal propagation of shockwaves in the stem results in the break-off of the crown (whiplash; Lundström et al. 2009). Trees react upon decapitation or branch loss with distinct radial growth suppression following the impact. One or several lateral branches will form a "leader" that replaces the broken crown, resulting in the tree morphology called "candelabra" growth (Butler and Malanson 1985; Stoffel et al. 2005a). Leaders may also be formed from prostrated trunks knocked over by geomorphic events.

12.2.6 Process Dating with Dendrogeomorphic Evidence

The expression of disturbance in the tree-ring record may vary in intensity as well as in the spatial and temporal extent between processes, species and age/size classes of trees. As a consequence, and based on the process and tree species analyzed, different classification systems have been used in the past. In Table 12.1, we present a synthesis of indicators commonly used in dendrogeomorphic research and make a proposal on how to analyze and interpret growth anomalies (Stoffel and Corona 2014). Distinction is made between weak, moderate and strong growth anomalies, and thresholds are typically used to distinguish geomorphic signals from noise (e.g., anthropogenic disturbances, insect attacks). An overview of thresholds and guidelines on how to reconstruct time series of geomorphic events can be found in previous publications (Corona et al. 2012, 2014; Schneuwly-Bollschweiler et al. 2013; Stoffel et al. 2013a; Stoffel and Corona 2014; Trappmann et al. 2014; Chiroiu et al. 2015; Morel et al. 2015); we do not provide further details in this paper but rather present three case studies exemplifying recent progress and remaining questions in the research field.

12.3 Tree Reactions to Floods and Erosion: Two Case Studies

In the following, we present two cases where the application of dendroecological disturbances in trees can be used to infer past mass movement activity. These examples represent the state-of-the-art in the respective research fields, but also point to remaining challenges—thus calling for new research.

Table 12.1 Intensities of growth anomalies (GDs) based on their appearance and/or persistence in the tree-ring series

GD	Parameter	Weak GD	Moderate GD	Strong GD
Injury, callus tissue		N/A	N/A	Clear indicator of an event
TRDs		Tangentially aligned with clear gaps between ducts	Compact, but not continuous row	Extremely compact and continuous row
Vessel anomaly	Decreased lumen area (%)	N/A	N/A	≥30%
	Decreased vessel number (%)	N/A	≥30%	≥50%
Kill date		N/A	N/A	Clear indicator of an event
Reaction wood	≥50% of ring width consists of compression wood cells			
	Duration	≥3 year	3–8 year	≥8 year
Growth suppression	Change in ring width (%)	<60%	≥100%	≥200%
	Duration	≥4 year	<8 year but ≥4 year	≥8 year
Growth release	Change in ring width (%)	<50%	≥100%	≥150%
	Duration	≥4 year	<8 year but ≥4 year	≥8 year

Note that tangential rows of traumatic resin ducts (TRDs) are formed in some conifer species (e.g. *Abies*, *Larix*, *Picea*, *Pseudotsuga*), but not in most *Pinus* or in broadleaved species. Vessel anomalies are related to injuries and typical for broadleaved tree species (modified from Stoffel and Corona 2014)

12.3.1 River Floods and Disturbance in Trees

The interaction between trees and fluvial processes allow the dating of past flood activity, but also contribute to an improved understanding of ecological feedbacks in floodplains. The exploratory analysis of tree-ring responses to hydrology initially started in the first half of the twentieth century (Hardman and Reil 1936). Early work primarily focused on streamflow reconstructions by using the relation between tree-ring growth and season and/or annual flow discharge record. This approach has since become a research direction of its own with applications for water resources management (Woodhouse and Lukas 2006). The application of tree-ring records in flood reconstructions was initiated by Robert S. Sigafoos, who described, in his seminal publications in the early 1960s (Sigafoos 1961, 1964), linkages between riparian vegetation and flood frequency in the Potomac River

(Washington, USA). He concluded that botanical evidence of past floods events has a clear economic value as it allows improved flood frequency analysis, and consequently risk estimation. Sigafoos (1964) also contributed to the understanding of the dissimilar adaptability of tree species to flood variability, as well as to its influence on seed establishment in floodplains, which has since been considered as a benchmark in subsequent research on riparian ecology (Hupp and Osterkamp 1996; Naiman et al. 2010; Stoffel and Wilford 2012). Tree-ring records have since been used as a valuable source of information in flood hazard assessments, especially in poorly or ungauged mountain catchments (see Ballesteros-Cánovas et al. 2015a, b, c, d, e for a recent review).

Floods can leave a large set of anomalous growth patterns in trees. In floodplains, the distribution, nature and age of riparian species can yield comprehensive information on flow dynamics and competence (Hupp and Osterkamp 1985, 1996), including analyses of changes in fluvial geomorphology (Marston et al. 1995; Garófano-Gómez et al. 2012). At the level of individual trees, disturbances can very frequently be observed in steep, high-gradient streams. The most commonly observed set of botanical evidence in these environments includes abrasion scars caused by sediment, ice or wood transported during floods (Zielonka et al. 2008; Ruiz-Villanueva et al. 2010; Ballesteros-Cánovas et al. 2011a, b; Lagadac et al. 2015), sprouts (new leaders) growing from downed stems (Sigafoos 1964; Gottesfeld and Gottesfeld 1990), trunks tilted by unidirectional flow pressure (Ballesteros-Cánovas et al. 2015d) roots exposed by the erosion of the root-plate system (Malik 2006; Casteller et al. 2015). In low gradient streams, trees growing in the endorheic zone can document past floods as well, even in the absence of other physical evidence, as the transport and production of growth hormones in root systems will suffer from the prolonged influence of water, thus leading to growth abnormalities (St. George and Nielsen 2003; St. George et al. 2002; Wertz et al. 2013; Copini et al. 2016).

The identification and use of the growth abnormalities mentioned above allows reconstructions of multi-centennial flood occurrence records with (sub-)annual resolution, and thereby contributes to an improved understanding of flood variability and flood-climate linkages. Flood reconstructions in low-gradient streams are often based on the detection of anatomical structures linked to anoxic conditions during persistent flood (Yanosky 1983, 1984; Astrade and Bégin 1997; St. George and Nielsen 2003; Ballesteros et al. 2010a, b; Therrell and Bialecki 2015).

These reconstructions have enabled the creation of flood chronologies extending back for several centuries. Recent work in low-gradient river systems, for instance, included the assessment of spring floods in the lower Mississippi River going back to the late eighteenth century (Therrell and Bialecki 2015). In mountain streams, Ballesteros-Cánovas et al. (2015b) used forest management to minimize noise and to maximize flash-flood signals in a set of small catchments in Central Spain. Stoffel et al. (2012) and Casteller et al. (2015), in turn, combined growth anomalies from stems and roots to distinguish erosional from sedimentation processes during flash floods in the Patagonian Andes. Comparable approaches have also been used for

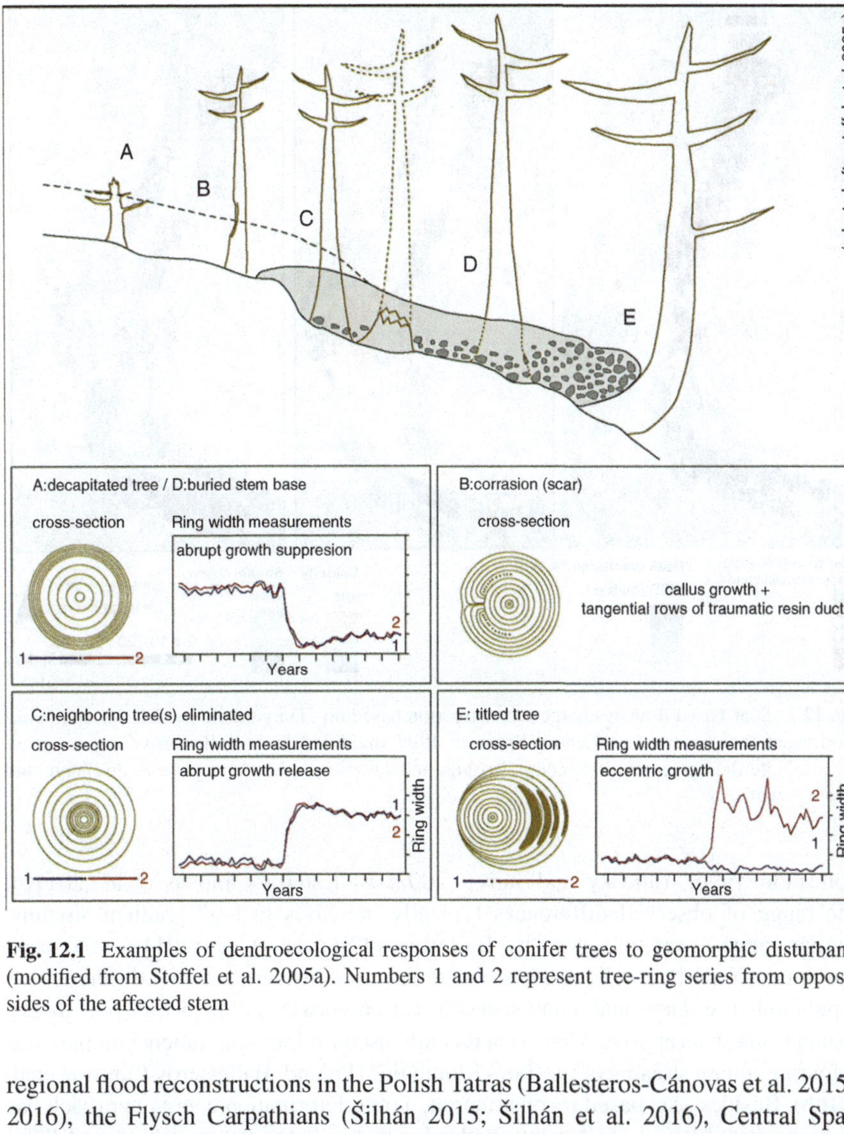

Fig. 12.1 Examples of dendroecological responses of conifer trees to geomorphic disturbance (modified from Stoffel et al. 2005a). Numbers 1 and 2 represent tree-ring series from opposite sides of the affected stem

regional flood reconstructions in the Polish Tatras (Ballesteros-Cánovas et al. 2015c, 2016), the Flysch Carpathians (Šilhán 2015; Šilhán et al. 2016), Central Spain (Rodriguez-Morata et al. 2016), and in the Indian Himalayas (Ballesteros-Cánovas et al. 2017b) (Fig. 12.1).

In combination with hydraulic models, the height of scars in trees—also referred to as paleostage indicators or PSI—are often used to reconstruct flood discharge and hence flood magnitudes (Fig. 12.2), as they indicate the minimum water level during any past event (Ballesteros-Cánovas et al. 2011a, b, 2015a). Several studies have highlighted good agreement between fresh high water marks (HWM) and scar height; discrepancies have been attributed mostly due to uncertainties in the timing of scar generation within the hydrograph (Smith and Reynolds 1983;

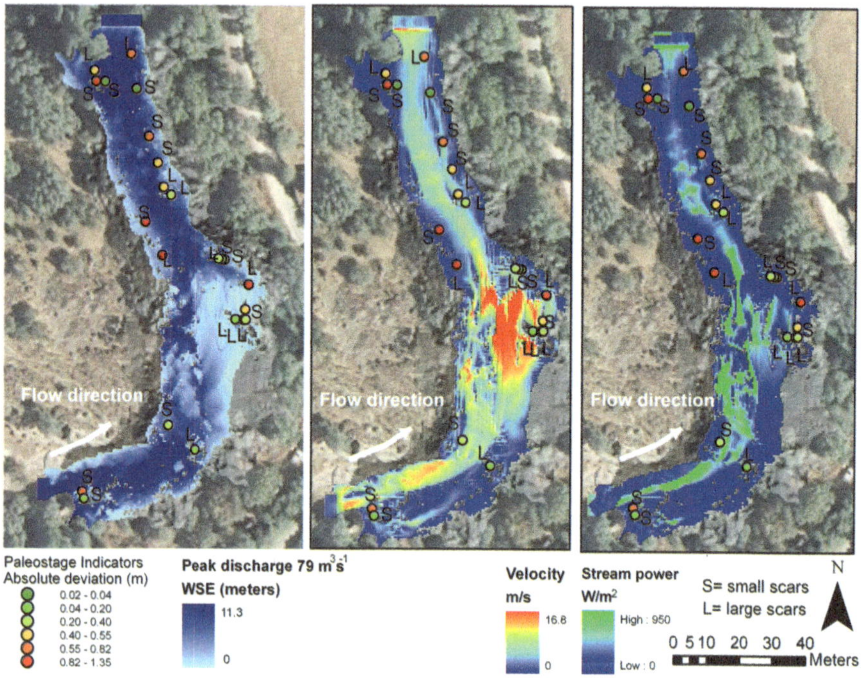

Fig. 12.2 Scar-based flow discharge reconstruction based on 2D hydraulic models for an intense flood event that occurred in Central Spain in 1997 (modified from Ballesteros-Cánovas et al. 2011a). Note the agreement between paleostage indicators (scar height) and peak discharge and stream power

Gottesfeld 1996; Yanosky and Jarrett 2002; Ballesteros-Cánovas et al. 2011b). The range of observed differences typically increases in high-gradient streams. Similar ranges were observed by Ballesteros-Cánovas et al. (2011a) who used two-dimensional hydraulic models and took into consideration two different scar populations (i.e. large and small scars) based on classic sediment transport theory assumptions. This approach has been recently used for intra-catchment comparisons in four mountain streams of the Tatra Mountains (Poland, Ballesteros-Cánovas et al. 2016b). Besides scar-based reconstruction, a tree-deformation energy approach has been suggested based on the expected relationship between tree tilting and flood magnitude (Ballesteros-Cánovas et al. 2015c).

12.3.2 *Denudation Processes and Erosion Signals in Roots*

Erosion is a major threat to soil resources and may impair their ability to deliver a range of ecosystem goods and services in various environments worldwide (Verheijen et al. 2009). It not only leads to a loss of soil fertility, but also causes off-site effects in the form of downstream sedimentation (de Vente and Poesen 2005), reduced hydraulic capacity of rivers and drainage ditches, increased flood risks (Sinnakaudan et al. 2003), the blocking of irrigation channels, as well as a reduction of design life of reservoirs (Romero-Díaz et al. 2012). Soil erosion also leads to the transport of chemicals and thereby alters biogeochemical cycling, which in turn may cause eutrophication of water bodies (Quinton et al. 2010). Rates of soil loss have been measured, modelled or inferred for most types of soil erosion in a variety of landscapes. Yet, traditional field-based methods of monitoring erosion (sediment traps, erosion pins, and bridges) are labor intensive and therefore are generally limited in both time (Cantón et al. 2011) and/or spatial extent (Gillan et al. 2016), thus limiting our knowledge on how erosion processes evolve over time and on what their impacts are on changes in environmental conditions (Poesen et al. 2003; Nadal-Romero et al. 2011). The development of methodologies capable of quantifying erosion rates retrospectively and at different spatio-temporal scales therefore remains a major scientific challenge (Poesen et al. 2003). In this regard, bio-indicators—such as exposed roots—represent an alternative method to determine erosion rates at short to medium term timescales (Stoffel et al. 2013a, b), especially in vegetated, and ungauged regions.

The potential of roots as an indicator of degradation was recognized since the mid-twentieth century. Schulman (1945) was the first to successfully synchronize ring widths from large roots of *Pseudotsuga menziesii* (Mirb.), thereby obtaining a series of soil moisture changes and a detailed record of year-to-year fluctuations in runoff for the upper Colorado River. In the White Mountains of California, LaMarche (1963) realized that the vertical buttress form observed in exposed roots was the result of (1) bark and cambium stripping after abrasion and weathering from the upper root surface and (2) continued secondary growth on the lower root surface. Based on these observations of shape changes and discontinuity of growth rings, he produced the first estimate of root ages at the time of initial cambium reduction. In the following years, LaMarche (1968) and Carrara and Carroll (1979) used the ratio between the minimum depth of erosion—obtained from the reconstructed root diameter at the moment of denudation—and the time (i.e., number of growth rings) passed since root exposure, to establish total vertical erosional losses from hillslopes over lifetimes of individual, long-lived trees. Since these seminal papers, the dendrogeomorphic method has been widely accepted as a powerful tool in reconstructing rates of soil erosion from analysis of exposed root (e.g., Krause and Eckstein 1993; McAuliffe et al. 2006; Chartier et al. 2009). A similar approach—combined with the sprouting of adventitious roots, scars or growth-ring shape changes from concentric to eccentric following exposure—was also used to quantify gully (e.g., Pérez-Rodríguez et al. 2007), sheet (Bodoque et al. 2005), meander bank (Malik 2006), or shore (Fantucci 2007) erosion. All these studies focused on the variation in tree-ring structure (width, concentricity, scars) at a macroscopic level.

More recently, microscopic approaches and equations have been introduced, determining the year of exposure based on changes in (1) the anatomical structure of tracheids in conifer roots (Gärtner 2007; Rubiales et al. 2008; Corona et al. 2011; Lopez Saez et al. 2011) and in (2) the formation of vessels in roots of broadleaved trees (Hitz et al. 2008a, b). Two criteria are usually retained to assess the first year of exposure, namely the (1) abrupt reduction in tracheid size (Corona et al. 2011); and the (2) abrupt increase in the cell wall thickness (Stoffel and Corona 2014). These changes are interpreted as a response of root cells in order to limit their vulnerability to freezing and water stress-induced embolism caused by higher temperature and moisture variations following exposure (Pitterman and Sperry 2003; Mayr et al. 2007). In the case of sudden erosion processes, these typical signatures coincide with the onset of exposure (Gärtner 2007). Erosion rates are calculated using the rings formed since exposure (NRex) and the thickness of the eroded soil layer (Er), whereby Er is obtained via the height of the exposed part of the root measured with a depth gauge (Ex). These factors, along with root growth (Gr1, Gr2) since exposure and bark thickness on the upper and lower sides of the root (B1, B2) can be related as follows:

$$Er = \left(Ex - (Gr1 + Gr2) + \frac{(B1 + B2)}{2} \right) / NRex$$

In the case of continuous denudation, anatomical changes in roots start to emerge as soon as denudation had reduced soil cover to less than a few centimeters (Corona et al. 2011; Lopez Saez et al. 2011). As a consequence, a bias (ε) related to root response under thin soil layers should be estimated systematically. In addition, high-resolution microtopographic information around roots—derived from e.g. Terrestrial Laser Scanning—are essential in order to reduce uncertainties related to the quantification of eroded soil layers (Ballesteros-Cánovas et al. 2013a, b; Stoffel et al. 2013a, b), which can be as high as 50% of the eroded soil layer in silica badlands (Bodoque et al. 2015). Finally, Corona et al. (2011) points out that reconstructed soil thickness values also depend on the stability of the root axis. If the root axis remains stable over time, a relative uplift of the root center will occur because root increment in its lower part will be balanced by positive adjustment of root curvature (Coutts 1989; Polacek et al. 2006). In this case, an overestimation of Ex will result from the subsequent growth of the upper part of the root and Er has to be calculated as follows:

$$Er = \left(Ex - (Gr1) + \frac{(B1 + B2)}{2} + \varepsilon \right) / NRex$$

Based on these anatomical considerations, the reconstruction of erosion rates focused on (1) precise quantification of continuous erosion processes in badlands and on (2) the validation of root-ring measurements by comparison with other records. Corona et al. (2011) thus used exposed roots from *Pinus sylvestris* to quantify sheet erosion processes on interfluve and gully slopes in marly badlands of the Southern French Alps and obtained bias-corrected (ε) erosion rates of 5.9 ± 2.6 mm year^{-1}. Comparison of reconstructed erosion rates with a series of systematic measurements performed across a network of marking stakes not only shows almost identical rates (5.7 ± 2.3 mm year^{-1}), but also points to the fact that values would have been underestimated on the interfluves and gully slopes without the bias-adjustment. Similarly, in experimental sandy badlands located in Central Spain, Bodoque et al. (2011) and Ballesteros-Cánovas et al. (2015a, b, c, d, e) demonstrated that root-ring records (6.6–8.8 mm year^{-1}) differed by 36% from those obtained through direct observation (60 erosion pins and 12 pedestals; 11.9 mm year^{-1} for a period of observation of three hydrological years (Lucía et al. 2011), thereby pointing to the strong influence of a high-intensity rainfall event on monitoring data and the smoothing of extremes in medium-term rates of erosion.

In gully and torrential environments, abrupt changes in cell lumina following sudden root exposure have mainly been used to quantify channel incision, headcut retreat and sidewall processes (see e.g. Malik 2008; Lucía et al. 2011; Silhan et al. 2016). In Central Spain, dendrogeomorphic analyses from 120 exposed roots thus enabled the 3D reconstruction of channel incision, widening and retreatment during the last 23 years and with annual precision of a gully developed in silica sand (Ballesteros Canovas et al. n.d.). Annual average gully incision rates were estimated of 0.09 ± 0.1 m year^{-1} while lateral widening as well as soil erosion rate due to concentrated runoff reached 0.2 m year^{-1} and 6 cm year^{-1}, respectively. In a same manner, roots from torrent and river systems have been for instance used in Poland, Czech Republic, Switzerland to date the occurrence of bank erosion processes (e.g., Hitz et al. 2008a, b; Malik 2008; Stoffel et al. 2012; Silhan et al. 2016). Based on the evidence conserved in the root-ring record of various species, the authors were also successful in linking the information contained in the roots with the occurrence of extraordinary floods in the wider study region.

With respect to shore erosion, despite the demonstration of the obvious potential of the approach by Bégin et al. (1991) along the St. Lawrence stream (Canada), very limited attention has been paid to the dendrogeomorphic reconstruction of processes in the past. Fantucci (2007) established a map of coastline erosion for the Bolsena Caldera Lake in central Italy based on roots of broadleaved trees and demonstrated that erosion rates were obviously related to the intensity and frequency of winds. In an attempt to quantify shore erosion in the Mediterranean Sea, Rovéra et al. (2013) analyzed anatomical evidence of exposure in *Pinus halepensis* Mill., roots hanging from decametric, detritic cliffs of Porquerolles Island (Var, France). Reconstructed rates of erosion range from 13 to 34.5 mm year^{-1} (Fig. 12.3) and are in the same order of magnitude as erosion rates derived from photogrammetric analysis on sandstone cliffs in the Algarve region (Marques 1997), or the Larache region (Morocco, Marques 2003).

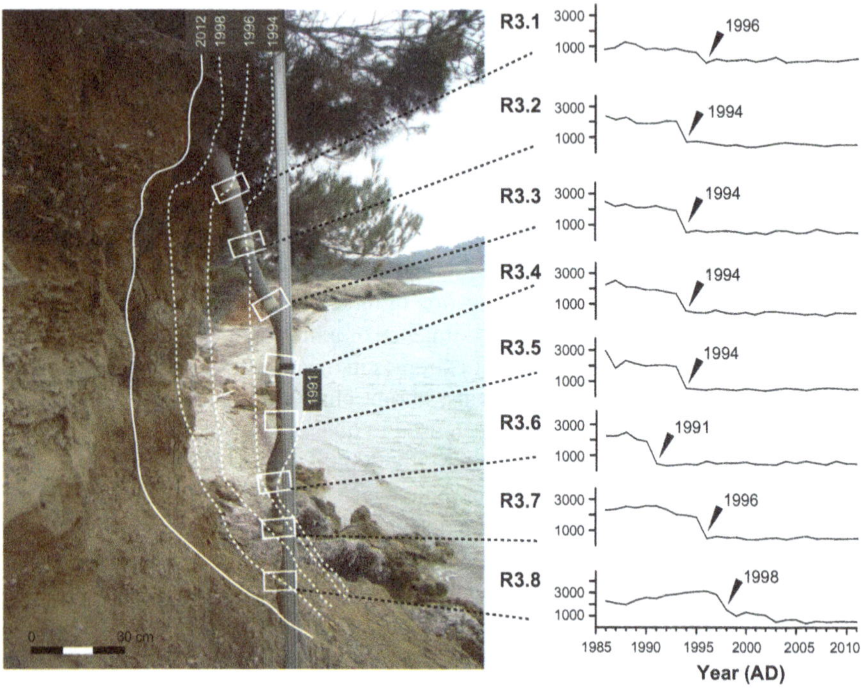

Fig. 12.3 Reconstructed rates of cliff erosion derived from exposed *Pinus halepensis* roots hanging from decametric, detritic cliffs of Porquerolles Island (Var, France). The analysis of oblique roots also allows analysis of temporal changes in cliff profiles, and thus could become a valuable tool for the assessment of the temporal frequency or changes in storminess (Rovéra et al. 2013)

12.4 Recent Progress in Process Dating: Tree Age, Tree Sensitivity and Geomorphic Disturbances

Dendroecological reconstructions of geomorphic processes are based on the identification and dating of growth anomalies to external disturbances. The variability of different types of growth responses is closely related to the nature and extent of the external disturbance (Stoffel and Bollschweiler 2008), with responses occurring with different intensities (Stoffel and Corona 2014).

Several authors have suggested that the nature and intensity of responses will change with tree age (Bollschweiler and Stoffel 2010), as it was described for climate fluctuations (Carrer and Urbinati 2004; Esper et al. 2008; Wu et al. 2013). Quantitative data about possible age-dependent changes in tree sensitivity to geomorphic disturbances is not, however, readily available. The main assumed drivers influencing tree sensitivity are changes in bark thickness, loss of stem elasticity, or changes in stem diameter, with the latter increasing target size. The assumed changes in sensitivity would thus have a direct impact on the ability of a tree to record geomorphic process activity in their ring sequences. As a consequence, the age-dependent tree sensitivity to geomorphic disturbances could significantly affect tree-ring-based chronologies of geomorphic processes.

Some of the most frequently studied geomorphic processes in Alpine environments are debris flows. The dynamic movement of solid blocks, fine materials, water and air significantly influences tree growth and can kill individual trees. Using trees affected by this process, Šilhán et al. (2015) focused on differences in the type and nature of growth disturbances recorded by "young" and "old" trees on the southern slopes of the Crimean Mountains. In total, 1122 tree-ring series from 566 black pines (*Pinus nigra* ssp. *pallasiana*) were sampled in 8 debris flow catchments: 361 "young" trees with a post-1930 innermost ring and 205 "old" trees with pre-1930 germination. The year 1930 was selected due to the strong earthquake in 1927 (magnitude $M = 6.8$), when massive rockfalls reduced the number of trees in the study area, leading to the colonization of the newly created bare surfaces. In total, 1271 growth anomalies could be attributed to debris flow. More than 65% of all growth disturbances were identified between 1939 and 2010, when primarily "young" trees helped the construction of the debris-flow chronology. The dominant growth response of "young" trees was in the form of compression wood (71%). By contrast, the same anomaly represented only 13% of all growth disturbances in the "old" trees during the same time period. "Young" trees did not react by abrupt growth releases to events, whereas this growth response represented 27% of all growth responses in the "old" trees. Considerable differences were also found for growth suppression with 13% in "young" and 53% in "old" trees. Scars and callus tissue represented 16% of all responses in "young" and 6% in "old" trees.

The significant differences in the formation of compression wood can be related to the more flexible stems of "young" trees. Moreover, "old" trees with larger root plates are better anchored on the slopes and can better resist tilting due to the pressure of debris flow material. The next significant difference in the proportion of scars and callus tissue can be attributed to the divergence of the bark thickness of *P. nigra*, because "young" trees with thinner bark can be easily injured. In the results, less flexible, well-anchored stems with the thick bark of "old" trees react preferentially with abrupt growth changes.

Another common process in mountain environments is rockfall (defined here as a free fall, bouncing or rolling of individual or a few rocks and boulders). Šilhán et al. (2013) analyzed 140 heavily damaged individuals of *P. nigra* on the Taraktash talus slope, Crimean Mountains, which allowed identification of 977 growth disturbances and the reconstruction of 702 rockfall events. Here, abrupt changes in growth dominated with 69% (abrupt growth suppression, 51%; abrupt growth release, 18%), whereas compression wood (16%) and callus tissue (15%) were much scarcer. Changes in tree sensitivity to rockfall were studied by calculating the number of rockfall events for each decade a tree had at the time of impact (Fig. 12.4a). The first decade of a tree's life was not taken into account as trees are, in general, very sensitive to non-geomorphic factors (e.g., snow creeping) during their juvenile growth. Results reveal a quite unambiguous pattern of changes in the

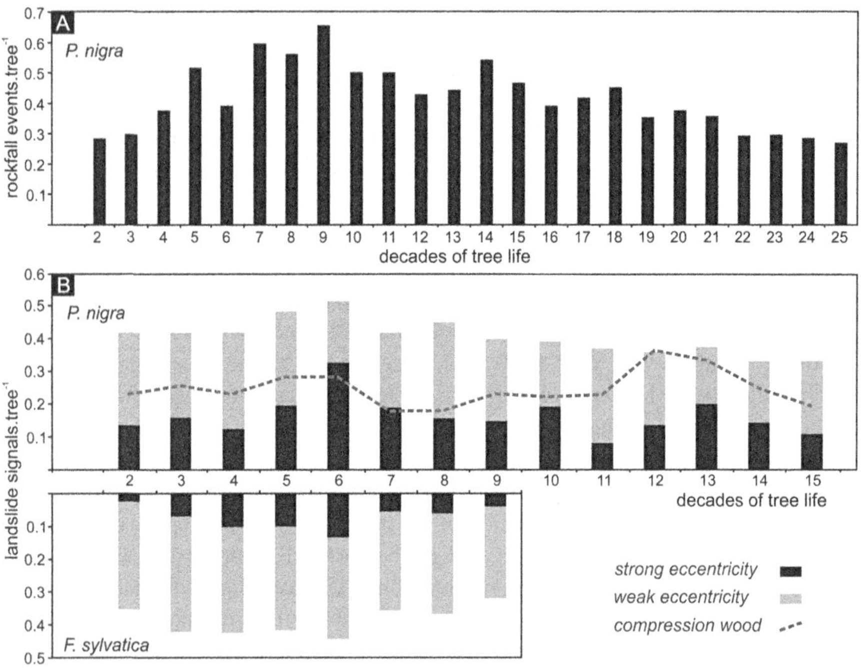

Fig. 12.4 Changes in the number of recorded events of geomorphic processes with tree age: (**a**) number of rockfall events recorded in individual decades of *P. nigra* life; (**b**) number of landslide events recorded in individual decades of *P. nigra* and *F. sylvatica* life

number of recorded rockfall events per individual tree and for individual decades of its life. In general, all trees revealed an increasing number of recorded rockfalls up to an age of 90 years (an average of 0.43 rockfall events per tree and decade for this time period), with the ninth decade being the one with the largest number of events. Thereafter, the relative number of recorded rockfalls decreases continuously. Although the thinner bark in the youngest trees can be expected to be more sensitive to geomorphic impacts, the sensitivity is not the highest in the early decades of tree life. As rockfall is a discrete process affecting trees with individual falling clasts, tree diameter seems to be the key factor influencing rockfall frequency as increasingly larger trees also represent an increasingly larger target for falling rocks. The stem diameter probably dominates the ability of trees to record impacts up to its ninth decade. As annual increment rates generally decrease with increasing age, the increase in target size will be reduced as well, whereas bark of increasing thickness will gradually provide more protection and thus reduce tree sensitivity to rockfall, at least in the case of *P. nigra*.

Landslides represent yet another geomorphic process with a long history of study. Šilhán and Stoffel (2015) evaluated the sensitivity of *P. nigra* and *Fagus sylvatica* (L.) to landslides by analyzing one case study each in the Outer Western Carpathians and Crimean Mountains. In total, 39 *P. nigra* and 119 *F. sylvatica* were analyzed, with a focus on eccentric growth and compression wood (*P. nigra*) and eccentric growth (*F. sylvatica*). Sensitivity was assessed by calculating the number of recorded landslide events per tree and decade of its life (Fig. 12.4b). In both species, the number of landslide signals increased in *F. sylvatica* up to the sixth decade, when the number of recorded events per tree was highest. Moreover, the intensity of landslide signals (the abruptness of eccentricity changes) also culminated around six decades. The analysis of compression wood, by contrast, yielded somewhat different results. Here, we observed a second peak in the number of recorded landslide events after twelve to thirteen decades. The decrease of tree sensitivity can be attributed to changes in stem elasticity and the decrease in annual increment. Resources are likely allocated to the growing trunk and branches, which in turn limits eccentric growth. This assumed effect would not evidently influences the ability of trees to form compression wood as a reaction to stem tilting. It is also assumed that young *P. nigra* trees influenced by landslide events would create compression wood to regain the original, vertical growth position, whereas older trees would produce compression wood rather for stabilization than for the straightening of what has become a much larger stem.

12.5 Concluding Remarks

The original employment of tree rings in natural hazard studies was simply as a dating tool and rarely exploited other environmental and/or ecological information that could be derived from studies of ring-width variations and records of damage contained within the tree. However, these unique, annually resolved, tree-ring records preserve potentially valuable archives of past geomorphic events on timescales of decades to centuries. As many of these processes are significant natural hazards, understanding their distribution, timing and controls provides valuable information that can assist in the prediction, mitigation and defence against these hazards and their effects on society. This chapter has provided some recent examples of dendroecological dating geomorphic process activity, but also pointed to some limitations and knowledge gaps in the field. The age-dependent sensitivity of trees to geomorphic disturbances yet needs more attention, and protocols yet need to be defined to increase confidence in time-series of past events.

As shown with the example on flood reconstructions, the great advantage of tree-based approaches is that they yield reliable evidence of past flood frequency and magnitude that took place prior to the instrumental period. Consequently, this information can be incorporated into flood hazard and risk assessments. By analyzing trees affected by past floods, one can quantify fluvial processes and anticipate potential geomorphic changes in the river channel. This understanding can be incorporated to hazard zonation (Brooks and St. George 2015), and it allows identification of non-stationary catchment behavior. In parallel, the inclusion of reconstructed data in flood-frequency analyses has other considerable advantages (Benito and Thorndycraft 2004; Greenbaum et al. 2014). Similar to historical or sedimentary records, peak discharge reconstructions based on tree-ring records have produced distinct percentile estimations (Ballesteros-Cánovas et al. 2013a, b, 2016b). The advantage of tree-ring records over other paleoflood sources (Baker 2008) is that they provide annually resolved data which can easily be referenced to systematic records. The ubiquity of trees, especially in mountain regions, as well as their high temporal and spatial resolution makes tree-based flood reconstructions highly suitable for hazard assessments. Ballesteros-Cánovas et al. (2016a) recently demonstrated that tree-ring approaches, combined with regional flood-frequency assessments, maximize available information and thus can reduce uncertainties related to flood quartile estimation.

The examples presented on root erosion merely represent snapshots of recent developments in root-based reconstructions of various types of erosion processes in different geographic settings around the globe. They illustrate the substantial progress achieved in the field of dendrogeomorphic erosion analysis. Through the systematic comparison of exposure signals in roots and mean erosion rates with data obtained from monitoring devices, the precision of dendrogeomorphic approaches has been improved substantially. These methods have clear advantages over the shorter time series obtained with repeat monitoring or over longer, but more coarsely resolved records. The resolution of dendrogeomorphic data and the time windows typically covered by roots also facilitate the comparison of averaged erosion rates with meteorological records, and the analyses also exhibit a much better cost–benefit ratio than most other techniques used to infer erosion. Despite limitations primarily related to the presence of trees and shrubs in the study area, to climate seasonality and to the age of roots available for analysis, we believe that root-based erosion assessments therefore constitute a valuable alternative to empirical models, especially in regions where data for calibration and validation are completely missing.

Last but not least, this chapter also clearly calls for more linkages between disciplines and an improved understanding of tree reactions and the formation of growth anomalies with relation to tree age or tree size. A first set of quite preliminary answers has been provided in one of the sections of this chapter, but much more research is still needed in this field.

Acknowledgements The authors are very grateful to the editors of this book for their invitation to write a chapter, and to the referees for constructive comments on an earlier version of this manuscript.

References

Alestalo J (1971) Dendrochronological interpretation of geomorphic processes. Fennia 105:1–140

Arbellay E, Stoffel M, Bollschweiler M (2010a) Wood anatomical analysis of *Alnus incana* and *Betula pendula* injured by a debris-flow event. Tree Physiol 30:1290–1298

Arbellay E, Stoffel M, Bollschweiler M (2010b) Dendrogeomorphic reconstruction of past debris-flow activity using injured broad-leaved trees. Earth Surf Process Landf 35:399–406

Arbellay E, Fonti P, Stoffel M (2012a) Duration and extension of anatomical changes in wood structure after cambial injury. J Exp Bot 63:3271–3277

Arbellay E, Corona C, Stoffel M, Fonti P, Decaulne A (2012b) Defining an adequate sample of earlywood vessels for retrospective injury detection in diffuse-porous species. PLoS One 7(6):e38824. doi:10.1371/journal.pone.0038824

Arbellay E, Stoffel M, Decaulne A (2013) Dating of snow avalanches by means of wound-induced vessel anomalies in subarctic *Betula pubescens*. Boreas 42:568–574

Arbellay E, Stoffel M, Sutherland EK, Smith KT, Falk DA (2014a) Resin duct size and density as ecophysiological traits in fire scars of *Pseudotsuga menziesii* and *Larix occidentalis*. Ann Bot 114:973–980

Arbellay E, Stoffel M, Sutherland EK, Smith KT, Falk DA (2014b) Changes in tracheid and ray traits in fire scars of North American conifers and their ecophysiological implications. Ann Bot 114:223–232

Astrade L, Bégin Y (1997) Tree-ring response of Populus tremula L. and Quercus robur L. to recent spring floods of the Saône River, France. Ecoscience 4:232–239

Aulitzky H (1992) Die Sprache der "stummen Zeugen". In International Symposium Interpraevent "Schutz des Lebensraumes vor Hochwasser Muren und Lawinen" Bern, June 29–July 3, 1992, Klagenfurt, Forschungsgesellschaft für vorbeugende Hochwasserbekämpfung 6, pp 139–174

Baker VR (2008) Paleoflood hydrology: origin, progress, prospects. Geomorphology 101:1–13

Ballesteros JA, Stoffel M, Bodoque JM, Bollschweiler M, Hitz OM, Diez A (2010a) Changes in wood anatomy in tree rings of *Pinus pinaster* Ait. following wounding by flash floods. Tree Ring Res 66:93–103

Ballesteros JA, Stoffel M, Bollschweiler M, Bodoque JM, Díez A (2010b) Flash-flood impacts cause changes in wood anatomy of *Alnus glutinosa*, *Fraxinus angustifolia* and *Quercus pyrenaica*. Tree Physiol 30:773–781

Ballesteros-Cánovas JA, Eguibar M, Bodoque JM, Díez-Herrero A, Stoffel M, Gutiérrez-Pérez I (2011a) Estimating flash flood discharge in an ungauged mountain catchment with 2D hydraulic models and dendrogeomorphic paleostage indicators. Hydrol Process 25:970–979

Ballesteros-Cánovas JA, Bodoque JM, Díez-Herrero A, Sanchez-Silva M, Stoffel M (2011b) Calibration of floodplain roughness and estimation of palaeoflood discharge based on tree-ring evidence and hydraulic modelling. J Hydrol 403:103–115

Ballesteros-Cánovas JA, Sanchez-Silva M, Bodoque JM, Díez-Herrero A (2013a) An example of integrated approach to flood risk management: the case of Navaluenga (Central Spain). Water. Res Manag 27(8):3051–3069

Ballesteros-Cánovas JA, Bodoque JM, Lucía A, Martín-Duque JF, Díez-Herrero A, Ruiz-Villanueva V, Rubiales JM, Génova M (2013b) Dendrogeomorphology in badlands: methods, case studies and prospects. Catena 106:113–122

Ballesteros-Cánovas JA, Stoffel M, St. George S, Hirschboeck K (2015a) A review of flood records from tree rings. Prog Phys Geogr 39(6):794–816

Ballesteros-Cánovas JA, Rodríguez-Morata C, Garófano-Gómez V, Rubiales JM, Sánchez-Salguero R, Stoffel M (2015b) Unravelling past flash flood activity in a forested mountain catchment of the Spanish Central System. J Hydrol 529:468–479

Ballesteros-Cánovas JA, Czajka B, Janecka K, Lempa M, Kaczka RJ, Stoffel M (2015c) Flash floods in the Tatra Mountain streams: frequency and triggers. Sci Total Environ 511:639–648

Ballesteros-Cánovas JA, Márquez-Peñaranda JF, Sánchez-Silva M, Díez-Herrero A, Ruiz-Villanueva V, Bodoque JM, Stoffel M (2015d) Can tree tilting be used for palaeoflood discharge estimation? J Hydrol 529:480–489

Ballesteros-Cánovas JA, Corona C, Stoffel M, Vela AL, Bodoque JM, Martin Duque JF (2015e) Combining terrestrial laser scanning and root exposure to estimate erosion rates. Plant Soil 394:127–137

Ballesteros-Cánovas JA, Stoffel M, Spyt B, Janecka K, Kaczka RJ, Lempa M (2016) Paleoflood discharge reconstruction in Tatra Mountain streams. Geomorphology 272:92–101

Ballesteros-Cánovas JA, Stoffel M, Martín-Duque JF, Corona C, Lucía A, Bodoque JM, Montgomery DR (2017a) Gully evolution and geomorphic adjustments of badlands to reforestation. Nat Sci Rep **7**:45027

Ballesteros Cánovas JA, Trappmann D, Shekhar M, Bhattacharyya A, Stoffel M (2017b) Regional flood-frequency reconstruction for Kullu district, Western Indian Himalayas. J Hydrol 546:140–149

Bannan MW (1936) Vertical resin ducts in the secondary wood of the Abietineae. New Phytol 35:11–46

Bégin Y, Langlais D, Cournoyer L (1991) Tree-ring dating of shore erosion events (Upper St Lawrence stream, eastern Canada). Geogr Ann 73A:53–59

Benito G, Thorndycraft VR (eds) (2004) Systematic, palaeoflood and historical data for the improvement of flood risk estimation, methodological guidelines. CSIC, Madrid, pp 1–115

Blodgett JT, Eyles A, Bonello P (2007) Organ-dependent induction of systemic resistance and systemic susceptibility in *Pinus nigra* inoculated with *Sphaeropsis sapinea* and *Diplodia scrobiculata*. Tree Physiol 27:511–517

Bodoque JM, Díez-Herrero A, Martín-Duque JF, Rubiales JM, Godfrey A, Pedraza J, Carrasco RM, Sanz MA (2005) Sheet erosion rates determined by using dendrogeomorphological analysis of exposed tree roots: two examples from Central Spain. Catena 64:81–102

Bodoque JM, Lucía A, Ballesteros JA, Martín-Duque JF, Rubiales JM, Genova M (2011) Measuring medium-term sheet erosion in gullies from trees: a case study using dendrogeomorphological analysis of exposed pine roots in central Iberia. Geomorphology 134:417–425

Bodoque JM, Ballesteros JA, Lucía A, Díez-Herrero A, Martín-Duque JF (2015) Source of error and uncertainty in sheet erosion rates estimated from dendrogeomorphology. Earth Surf Process Landf 40:1146–1157

Bollschweiler M, Stoffel M (2010) Tree rings and debris flows: recent developments, future directions. Prog Phys Geogr 34:625–645

Bollschweiler M, Stoffel M, Schneuwly DM, Bourqui K (2008) Traumatic resin ducts in *Larix decidua* stems impacted by debris flows. Tree Physiol 28:255–263

Braam RR, Weiss EEJ, Burrough A (1987) Dendrogeomorphological analysis of mass movement: a technical note on the research method. Catena 14:585–589

Brooks GR, St. George S (2015) Flooding, structural flood control measures, and recent geomorphic research along the Red River, Manitoba, Canada. In: Hudson P, Middelkoop H (eds) Geomorphic approaches to integrated floodplain management of lowland fluvial systems in North America and Europe. Springer, New York, pp 87–117

Butler DR, Malanson GP (1985) A history of 643 high-magnitude snow avalanches, southern Glacier National Park, Montana, U.S.A. Mt Res Dev 5:175–182

Butler DR, Stoffel M (2013) Classics in physical geography revisited: John F. Shroder, Jr.'s 1978 and 1980 papers on dendrogeomorphology. Prog Phys Geogr 4:161–188

Cantón Y, Solé-Benet A, de Vente J, Boix-Fayos C, Calvo-Cases A, Asensio C, Puigdefábregas J (2011) A review of runoff generation and soil erosion across scales in semiarid south-eastern Spain. J Arid Environ 75:1254–1261

Carrara PE, Carroll TR (1979) The determination of erosion rates from exposed tree roots in the Piceance Basin, Colorado. Earth Surf Process 4:307–317

Carrer M, Urbinati C (2004) Age-dependent tree ring growth responses to climate in *Larix decidua* and *Pinus cembra*. Ecology 85:730–740

Casteller A, Stoffel M, Crespo S, Villalba R, Corona C, Bianchi E (2015) Flash floods in the Patagonian Andes: a dendrogeomorphological approach. Geomorphology 228:116–123

Chartier MP, Rostagno CM, Roig FA (2009) Soil erosion rates in rangelands of northeastern Patagonia: a dendrogeomorphological analysis using exposed shrub roots. Geomorphology 106:344–351

Chiroiu P, Stoffel M, Onaca A, Urdea P (2015) Testing dendrogeomorphic approaches and thresholds to reconstruct snow avalanche activity in the Fagara? Mountains (Romanian Carpathians). Quat Geochronol 27:1–10

Copini P, den Ouden J, Robert EM, Tardif J, Loesberg WA, Goudzwaard L, Sass-Klaassen U (2016) Flood-ring formation and root development in response to experimental flooding of young *Quercus robur* trees. Front Plant Sci 7:775

Corona C, Lopez J, Rovéra G, Astrade L, Stoffel M, Berger F (2010) Quantification des vitesses d'érosion au moyen de racines déchaussées: validation de la méthode dans les badlands marneux des bassins versants expérimentaux de Draix (Alpes de Haute-Provence). Géomorphologie Relief Processus Environ 11:83–94

Corona C, Lopez J, Rovéra G, Stoffel M, Astrade L, Berger F (2011) High resolution, quantitative reconstruction of erosion rates based on anatomical changes in exposed roots (Draix, Alpes de Haute-Provence)—critical review of existing approaches and independent quality control of results. Geomorphology 125:433–444

Corona C, Lopez Saez J, Stoffel M, Bonnefoy M, Richard D, Astrade L, Berger F (2012) How much of the real avalanche activity can be captured with tree rings? An evaluation of classic dendrogeomorphic approaches and comparison with historical archives. Cold Reg Sci Technol 74–75:31–42

Corona C, Lopez Saez J, Stoffel M (2014) Defining optimal sample size, sampling design and thresholds for dendrogeomorphic landslide sampling. Quat Geochronol 22:72–84

Coutts MP (1989) Factors affecting the direction of growth of tree roots. Ann Sci For 46:277–287

de Vente J, Poesen J (2005) Predicting soil erosion and sediment yield at the basin scale: scale issues and semi-quantitative models. Earth-Sci Rev 71:95–125

Douglass AE (1941) Notes on the technique of tree-ring analysis. Tree Ring Bull 7:28–34

Du S, Yamamoto F (2007) An overview of the biology of reaction wood formation. J Integr Plant Biol 49:131–143

Esper J, Niederer R, Bebi P, Frank D (2008) Climate signal age effects—evidence from young and old trees in the Swiss Engadin. For Ecol Manag 255:3783–3789

Fantucci R (2007) Dendrogeomorphological analysis of shore erosion along Bolsena lake (Central Italy). Dendrochronologia 24:69–78

Fantucci R, Sorriso-Valvo M (1999) Dendrogeomorphological analysis of a slope near Lago, Calabria (Italy). Geomorphology 30:165–174

Friedman JM, Vincent KR, Shafroth PB (2005) Dating floodplain sediments using tree-ring response to burial. Earth Surf Process Landf 30:1077–1091

Garófano-Gómez V, Martínez-Capel F, Bertoldi W, Gurnell A, Estornell J, Segura-Beltrán F (2012) Six decades of changes in the riparian corridor of a Mediterranean river: a synthetic analysis based on historical data sources. Ecohydrology 6:536–553

Gärtner HW (2007) Tree roots—methodological review and new development in dating and quantifying erosive processes. Geomorphology 86:243–251

Gillan JK, Karl JW, Barger NN, Elaksher A, Duniway MC (2016) Spatially explicit rangeland erosion monitoring using high-resolution digital qerial imagery. Rangel Ecol Manag 69:95–107

Gottesfeld AS (1996) British Columbia flood scars: maximum flood-stage indicator. Geomorphology 14:319–325

Gottesfeld AS, Gottesfeld LMJ (1990) Floodplain dynamics of a wandering river, dendrochronology of the Morice River, British Columbia, Canada. Geomorphology 3:159–179

Greenbaum N, Harden TM, Baker VR et al (2014) A 2000 year natural record of magnitudes and frequencies for the largest Upper Colorado River floods near Moab, Utah. Water Resour Res 50:5249–5269

Hardman G, Reil OE (1936) The relationship between tree growth and stream runoff in the Truckee River Basin. Nevada Agricultural Experiment Station. California-Nevada Bulletin 141

Hitz OM, Gärtner HW, Heinrich I, Monbaron M (2008a) Application of ash (*Fraxinus excelsior* L.) roots to determine erosion rates in mountain torrents. Catena 72:248–258

Hitz OM, Gärtner HW, Heinrich I, Monbaron M (2008b) Wood anatomical changes in roots of European ash (*Fraxinus excelsior* L.) after exposure. Dendrochronologia 25:145–152

Hupp CR, Osterkamp WR (1985) Bottomland vegetation distribution along Passage Creek, Virginia, in relation to fluvial landforms. Ecology 66:670–681

Hupp CR, Osterkamp WR (1996) Riparian vegetation and fluvial geomorphic processes. Geomorphology 14:277–295
Hupp CR, Osterkamp PB, Thornton JL (1987) Dendrogeomorphic evidence and dating of recent debris flows on Mount Shasta, Northern California. US Geological Survey Professional Paper 1396-B, 39 pp
Jacoby GC (1997) Application of tree ring analysis to paleoseismology. Rev Geophys 35:109–124
Kny L (1877) Das Dickenwachsthum des Holzkörpers an beblätterten Sprossen und Wurzeln und seine Abhängigkeit von äusseren Einflüssen, insbesondere von Schwerkraft und Druck. Bot Z 35:415–423
Kogelnig-Mayer B, Stoffel M, Bollschweiler M, Hübl J, Rudolf-Miklau F (2011) Possibilities and limitations of dendrogeomorphic time-series reconstructions on sites influenced by debris flows and frequent snow avalanche activity. Arctic Antarc Alp Res 43:619–658
Kogelnig-Mayer B, Stoffel M, Schneuwly-Bollschweiler M (2013) Four-dimensional growth response of mature *Larix decidua* to stem burial under natural conditions. Trees 27:1217–1223
Krause C, Eckstein D (1993) Dendrochronology of roots. Dendrochronologia 11:9–23
Lagadec A, Boucher E, Germain D (2015) Tree ring analysis of hydro-climatic thresholds that trigger ice jams on the Mistassini River, Quebec. Hydrol Process 29:4880–4890
LaMarche VC (1963) Origin and geologic significance of buttress roots of bristlecone pines, White Mountains, California. US Geological Survey Professional Paper 475–C, pp 149–150
LaMarche VC (1966) An 800-year history of stream erosion as indicated by botanical evidence. US Geological Survey Professional Paper 550D, pp 83–86
LaMarche VC (1968) Rates of slope degradation as determined from botanical evidence, White Mountains, California. US Geological Survey Professional Paper 352–I
Lopez Saez J, Corona C, Stoffel M, Rovéra G, Astrade L, Berger F (2011) Mapping of erosion rates in marly badlands based on a coupling of anatomical changes in exposed roots with slope maps derived from LiDAR data. Earth Surf Process Landf 36:1162–1171
Lopez Saez J, Corona C, Stoffel M, Schoeneich P, Berger F (2012a) Probability maps of landslide reactivation derived from tree-ring records: Pra Bellon landslide, southern French Alps. Geomorphology 138:189–202
Lopez Saez J, Corona C, Stoffel M, Astrade L, Berger F, Malet JP (2012b) Dendrogeomorphic reconstruction of past landslide reactivation with seasonal precision: the Bois Noir landslide, southeast French Alps. Landslides 9:189–203
Lopez Saez J, Corona C, Stoffel M (2013) Climate change increases the frequency of snowmelt-induced landslides in the French Alps. Geology 41:619–622
Lucía A, Laronne JB, Martín-Duque JF (2011) Geodynamic processes on sandy slope gullies in central Spain—field observations, methods and measurements in a singular system. Geodin Acta 24:61–79
Lundström T, Jonsson MJ, Volkwein A, Stoffel M (2009) Reactions and energy absorption of trees subject to rockfall: a detailed assessment using a new experimental method. Tree Physiol 29:345–359
Malik I (2006) Contribution to understanding the historical evolution of meandering rivers using dendrochronological methods: example of the Mała Panew River in southern Poland. Earth Surf Process Landf 31:1227–1245
Malik I (2008) Dating of small gully formation and establishing erosion rates in old gullies under forest by means of anatomical changes in exposed tree roots (southern Poland). Geomorphology 93:421–436
Marques F (1997) As arribas do litoral do Algarve. Dinamica, processos e mecanismos. Ph.D. thesis, University of Lisbon
Marques FMSF (2003) Landslide activity in upper palaeozoic shale sea cliffs: a case study along the western coast of the Algarve (Portugal). Bull Eng Geol Environ 62:299–313
Marston RA, Girel J, Pautou G, Piegay H, Bravard AC (1995) Channel metamorphosis, floodplain disturbance, and vegetation development: Ain River, France. Geomorphology 13:121–131
Mattheck C (1993) Design in der Natur. Rombach, Freiburg, p 242

Mayer B, Stoffel M, Bollschweiler M, Hübl J, Rudolf-Miklau F (2010) Frequency and spread of debris floods on fans: a dendrogeomorphic case study from a dolomite catchment in the Austrian Alps. Geomorphology 118:199–206

Mayr S, Cochard H, Améglio T, Kikuta SB (2007) Embolism formation during freezing in the wood of *Picea abies*. Plant Physiol 143:60–67

McAuliffe JR, Scuderi LA, McFadden LD (2006) Tree-ring record of hillslope erosion and valley floor dynamics: landscape responses to climate variation during the last 400 yr in the Colorado Plateau, northeastern Arizona. Glob Planet Chang 50:184–201

Morel P, Trappmann D, Corona C, Stoffel M (2015) Defining sample size and sampling strategy for dendrogeomorphic rockfall reconstructions. Geomorphology 236:79–89

Nadal-Romero E, Martínez-Murillo JF, Venmaaercke M, Poesen J (2011) Scale dependency of sediment yield from badland areas in Mediterranean environments. Prog Phys Geogr 35:297–332

Naiman RJ, Decamps H, McClain ME (2010) Riparian ecology, conservation, and management of streamside communities. Academic, Burlington. ISBN: 0-12-663315-0

Osterkamp WR, Hupp CR, Stoffel M (2012) The interactions between vegetation and erosion: new directions for research at the interface of ecology and geomorphology. Earth Surf Process Landf 37:23–36

Pérez-Rodríguez R, Marques MJ, Bienes R (2007) Use of dendrochronological method in *Pinus halepensis* to estimate the soil erosion in the South East of Madrid (Spain). Sci Total Environ 378:156–160

Pilate G, Chabbert B, Cathala B, Yoshinaga A, Leple JC, Laurans F, Lapierre C, Ruel K (2004) Lignification and tension wood. C R Biol 327:889–901

Pitterman J, Sperry J (2003) Tracheid diameter is the key trait determining the extent of freezing-induced embolism in conifers. Tree Physiol 23:907–914

Poesen J, Nachtergaele J, Verstraeten G, Valentin C (2003) Gully erosion and environmental change: importance and research needs. Catena 50:91–113

Polacek D, Kofler W, Oberhuber W (2006) Radial growth of *Pinus sylvestris* growing on alluvial terraces is sensitive to water-level fluctuations. New Phytol 169:299–308

Quinton JN, Govers G, Van Oost K, Bardgett RD (2010) The impact of agricultural soil erosion on biogeochemical cycling. Nat Geosci 3:311–314

Rodriguez-Morata C, Ballesteros-Cánovas JA, Trappmann D, Beniston M, Stoffel M (2016) Regional reconstruction of flash flood history in the Guadarrama range (Central System, Spain). Sci Total Environ 550:406–417

Romero-Díaz A, Marín-Sanleandro P, Ortiz-Silla R (2012) Loss of soil fertility estimated from sediment trapped in check dams. South-eastern Spain. Catena 99:42–53

Rovéra G, Lopez Saez J, Corona C, Stoffel M (2013) Preliminary quantification of the erosion of sandy-gravelly cliffs on the island of Porquerolles (Provence, France) through dendrogeomorphology, using exposed roots of Aleppo pine (*Pinus halepensis* Mill.) Geogr Fis Din Quat 36:181–187

Rubiales JM, Bodoque JM, Ballesteros JA, Díez A (2008) Response of *Pinus sylvestris* roots to sheet-erosion exposure: an anatomical approach. Nat Hazards Earth Syst Sci 8:223–231

Ruiz-Villanueva V, Díez-Herrero A, Stoffel M, Bollschweiler M, Bodoque JM, Ballesteros JA (2010) Dendrogeomorphic analysis of flash floods in a small ungauged mountain catchment (Central Spain). Geomorphology 118:383–392

Schneuwly DM, Stoffel M, Bollschweiler M (2009a) Formation and spread of callus tissue and tangential rows of resin ducts in *Larix decidua* and *Picea abies* following rockfall impacts. Tree Physiol 29:281–289

Schneuwly DM, Stoffel M, Dorren LKA, Berger F (2009b) Three-dimensional analysis of the anatomical growth response of European conifers to mechanical disturbance. Tree Physiol 29:1247–1257

Schneuwly-Bollschweiler M, Stoffel M (2012) Hydrometeorological triggers of periglacial debris flows in the Zermatt Valley (Switzerland) since 1864. J Geophys Res Earth Surf 117:F02033. doi:10.1029/2011JF002262

Schneuwly-Bollschweiler M, Corona C, Stoffel M (2013) How to improve dating quality and reduce noise in tree-ring based debris-flow reconstructions. Quat Geochronol 18:110–118
Schulman E (1945) Root growth-rings and chronology. Tree Ring Bull 12:2–5
Shigo AL (1984) Compartmentalization—A conceptual framework for understanding how trees grow and defend themselves. Annu Rev Phytopathol 22:189–214
Shroder J (1978) Dendrogeomorphological analysis of mass movement on Table Cliffs Plateau, Utah. Quat Res 9:168–185
Sigafoos RS (1961) Vegetation in relation to flood frequency near Washington, D.C. US Geological Survey Professional Paper 424-C, pp 248–249
Sigafoos RS (1964) Botanical evidence of floods and flood-plain deposition. US Geological Survey Professional Paper 485A, 35 pp
Šilhán K (2015) Frequency, predisposition, and triggers of floods in flysch Carpathians: regional study using dendrogeomorphic methods. Geomorphology 234:243–253
Šilhán K, Stoffel M (2015) Impacts of age-dependent tree sensitivity and dating approaches on dendrogeomorphic time series of landslides. Geomorphology 236:34–43
Šilhán K, Pánek T, Hradecký J (2013) Implications of spatial distribution of rockfall reconstructed by dendrogeomorphological methods. Nat Hazards Earth Syst Sci 13:1817–1826
Šilhán K, Pánek T, Hradecký J, Stoffel M (2015) Tree-age control on reconstructed debris-flow frequencies: examples from a regional dendrogeomorphic reconstruction in the Crimean Mountains. Earth Surf Process Landf 40:243–251
Šilhán K, Galia T, Škarpich V (2016) Detailed spatio-temporal sediment supply reconstruction using tree roots data. Hydrol Process 30:4139–4153
Silhan K, Ruzek I, Burian L (2016) Dynamics of gully side erosion: a case study using tree roots exposure data. Open Geosci 9:108–116
Sinnakaudan SK, Ab Ghani A, Ahmad MSS, Zacaria NA (2003) Flood risk mapping for Pari River incorporating sediment transport. Environ Model Softw 18:119–130
Smith DG, Reynolds DM (1983) Tree scars to determine the frequency and stage of high magnitude river ice drivers and jams, Red Deer, Alberta. Can Water Res J 8:77–94
Sorg A, Kääb A, Roesch A, Bigler C, Stoffel M (2015) Contrasting responses of Central Asian rock glaciers to global warming. Sci Rep 5:8228. doi:10.1038/srep08228
St. George S, Nielsen E (2003) Palaeoflood records for the Red River, Manitoba, Canada, derived from anatomical tree-ring signatures. The Holocene 13:547–555
St. George S, Nielsen E, Conciatori F, Tardif J (2002) Trends in *Quercus macrocarpa* vessel areas and their implications for tree-ring paleoflood studies. Tree Ring Res 58:3–10
Stoffel M (2008) Dating past geomorphic processes with tangential rows of traumatic resin ducts. Dendrochronologia 26:53–60
Stoffel M, Beniston M (2006) On the incidence of debris flows from the early Little Ice Age to a future greenhouse climate: a case study from the Swiss Alps. Geophys Res Lett 33:L16404. doi:10.1029/2006GL026805
Stoffel M, Bollschweiler M (2008) Tree-ring analysis in natural hazards research—an overview. Nat Hazards Earth Syst Sci 8:187–202
Stoffel M, Corona C (2014) Dendroecological dating of geomorphic disturbance in trees. Tree Ring Res 70:3–20
Stoffel M, Hitz OM (2008) Snow avalanche and rockfall impacts leave different anatomical signatures in tree rings of *Larix decidua*. Tree Physiol 28:1713–1720
Stoffel M, Huggel C (2012) Effects of climate change on mass movements in mountain environments. Prog Phys Geogr 36:421–439
Stoffel M, Klinkmüller M (2013) 3D analysis of anatomical reactions in conifers after mechanical wounding: first qualitative insights from X-ray computed tomography. Trees 27:1805–1811
Stoffel M, Perret S (2006) Reconstructing past rockfall activity with tree rings: some methodological considerations. Dendrochronologia 24:1–15
Stoffel M, Wilford DJ (2012) Hydrogeomorphic processes and vegetation: disturbance, process histories, dependencies and interactions. Earth Surf Process Landf 37:9–22

Stoffel M, Lievre I, Monbaron M, Perret S (2005a) Seasonal timing of rockfall activity on a forested slope at Täschgufer (Swiss Alps)—a dendrochronological approach. Z Geomorphol 49:89–106
Stoffel M, Schneuwly D, Bollschweiler M, Lièvre I, Delaloye R, Myint M, Monbaron M (2005b) Analyzing rockfall activity (1600–2002) in a protection forest—a case study using dendrogeomorphology. Geomorphology 68:224–241
Stoffel M, Conus D, Grichting MA, Lièvre I, Maître G (2008) Unraveling the patterns of late Holocene debris-flow activity on a cone in the Swiss Alps: chronology, environment and implications for the future. Glob Planet Chang 60:222–234
Stoffel M, Bollschweiler M, Butler DR, Luckman BH (2010) Tree rings and natural hazards—a state-of-the-art. Springer, New York
Stoffel M, Casteller A, Luckman BH, Villalba R (2012) Spatiotemporal analysis of channel wall erosion in ephemeral torrents using tree roots—an example from the Patagonian Andes. Geology 40:247–250
Stoffel M, Corona C, Ballesteros-Cánovas JA, Bodoque JM (2013a) Dating and quantification of erosion processes based on exposed roots. Earth-Sci Rev 123:18–34
Stoffel M, Butler DR, Corona C (2013b) Mass movements and tree rings: a guide to dendrogeomorphic field sampling and dating. Geomorphology 200:106–120
Stokes MA, Smiley TL (1968) An introduction to tree-ring dating. University of Chicago Press, Chicago
Therrell MD, Bialecki MB (2015) A multi-century tree-ring record of spring flooding on the Mississippi River. J Hydrol 529:490–498
Timell TE (1986) Compression wood in gymnosperms. Springer, Berlin
Trappmann D, Stoffel M (2013) Counting scars on tree stems to assess rockfall hazards: a low effort approach, but how reliable? Geomorphology 180–181:180–186
Trappmann D, Stoffel M (2015) Visual dating of rockfall scars in *Larix decidua* (Mill.) trees. Geomorphology 245:62–72
Trappmann D, Corona C, Stoffel M (2013) Rolling stones and tree rings: a state of research on dendrogeomorphic reconstructions of rockfall. Prog Phys Geogr 37:701–716
Trappmann D, Stoffel M, Corona C (2014) Achieving a more realistic assessment of rockfall hazards by coupling three-dimensional, process based models and field-based tree-ring data. Earth Surf Process Landf 39:1866–1875
Verheijen FGA, Jones RJA, Rickson RJ, Smith CJ (2009) Tolerable versus actual soil erosion rates in Europe. Earth-Sci Rev 94:23–38
Wertz EL, St. George S, Zeleznik JD (2013) Vessel anomalies in *Quercus macrocarpa* tree rings associated with recent floods along the Red River of the North, United States. Water Resour Res 49:630–634. doi:10.1029/2012WR012900
Westing AH (1965) Formation and function of compression wood in gymnosperms II. Bot Rev 34:51–78
Woodhouse CA, Lukas JJ (2006) Multi-century tree-ring reconstructions of Colorado streamflow for water resource planning. Clim Chang 78:293–315
Wu G, Xu G, Chen T, Liu X, Zhang Y, An W, Wang W, Fang Z, Yu S (2013) Age-dependent tree-ring growth responses of Schrenk spruce (*Picea schrenkiana*) to climate—a case study in the Tianshan Mountain, China. Dendrochronologia 31:318–326
Yanosky TM (1983) Evidence of flood on Potomac River from anatomical abnormalities in the wood of flood plain trees. US Geological Survey Professional Paper 1296, pp 1–42
Yanosky TM (1984) Documentation of high summer flows on the Potomac River from the wood anatomy of ash trees. Water Resour Bull 20:241–250
Yanosky TM, Jarrett RD (2002) Dendrochronologic evidence for the frequency and magnitude of paleofloods. Water Sci Appl 5:77–89
Zielonka T, Holeksa J, Ciapala S (2008) A reconstruction of flood events using scarred tree in the Tatra Mountains, Poland. Dendrochronologia 26:173–183

Part III
Forest Decline

Credit: M. M. Amoroso

Chapter 13
The Multiple Causes of Forest Decline in Spain: Drought, Historical Logging, Competition and Biotic Stressors

J. Julio Camarero, Juan Carlos Linares, Gabriel Sangüesa-Barreda, Raúl Sánchez-Salguero, Antonio Gazol, Rafael M. Navarro-Cerrillo, and José Antonio Carreira

Abstract Climate warming is predicted to intensify drought stress in forests by amplifying the severity and frequency of droughts. This drying trend will potentially trigger forest dieback, characterized by tree growth decline and mortality. In drought-prone Spain, forest decline is mainly attributed to severe water shortage. This inciting factor causes a loss in tree vigor and interacts with predisposing or contributing factors that usually act at long time scales. Combined, these factors drive complex patterns of forest decline. Here, we illustrate four case studies on forest decline in Spain, evaluated using dendroecology. These declines were triggered by drought, interacting with other causal factors including past logging in Pyrenean silver fir, increased competition induced by land-use changes in Spanish

---

J.J. Camarero (✉) • G. Sangüesa-Barreda • A. Gazol
Pyrenean Institute of Ecology (IPE-CSIC), Avenida Montañana 1005, Zaragoza, 50059 Spain
e-mail: jjcamarero@ipe.csic.es; gsanguesa@ipe.csic.es; agazolbu@gmail.com

J.C. Linares
Department of Physical, Chemical and Natural Systems, Universidad Pablo de Olavide, Ctra. Utrera km. 1, Seville 41013, Spain
e-mail: jclincal@upo.es

R. Sánchez-Salguero
Pyrenean Institute of Ecology (IPE-CSIC), Avenida Montañana 1005, Zaragoza, 50059 Spain

Department of Physical, Chemical and Natural Systems, Universidad Pablo de Olavide, Ctra. Utrera km. 1, Seville 41013, Spain
e-mail: rsanchez@upo.es

R.M. Navarro-Cerrillo
Department of Forest Engineering, ERSAF, University of Córdoba, Campus de Rabanales, Ctra. IV, km 396, Córdoba 14071, Spain
e-mail: rmnavarro@uco.es

J.A. Carreira
Centro de Estudios Avanzados de la Tierra (CEACTierra), Universidad de Jaén, 23071 Jaén, Spain
e-mail: jafuente@ujaen.es

M.M. Amoroso et al. (eds.), *Dendroecology*, Ecological Studies 231,
DOI 10.1007/978-3-319-61669-8_13

fir, variation in drought vulnerability of planted pine species, and mistletoe infection in Scots pine. These case studies demonstrate that multiple causes drive forest decline in drought-prone regions. Differentiating the causal factors and inferring their interactions was feasible through tree-ring analyses, which provide robust reconstructions of changes in tree vitality through time.

Keywords Drought • Forest dieback • Stress • Tree death • Tree growth

13.1 Introduction

Climatic change is among the major drivers of forest decline at present (Allen et al. 2015) and in the past (Foster et al. 2006). Forest decline or dieback is a phenomenon affecting trees worldwide, causing a premature loss of vitality, growth decline, canopy dieback, and rising mortality rates (Muller-Dombois 1992). In water-limited regions, climate warming may intensify drought stress and accelerating forest decline processes, preferentially affecting the most vulnerable individuals, populations and species (McDowell et al. 2008; Anderegg et al. 2016). Climate warming without increases in precipitation is accompanied by aridification and extreme drought events can lead to abrupt and rapid changes in forest dynamics including severe forest decline (Allen et al. 2015).

Drought-triggered forest decline can be also predisposed by other factors, which act synergistically by reducing tree vitality (*sensu* Manion 1991). Competition is a major contributor to tree mortality in temperate (Das et al. 2011; Amoroso and Larson 2010) and dry Mediterranean forests (Bravo-Oviedo et al. 2006). In regions with intensively managed forests, such as Europe and north-eastern America (Reams and Huso 1990), past thinning, harvesting, and reforestation shape the current forest structure and composition, often resulting in forests more susceptible to drought. For instance, reforestation with high initial densities or using genotypes with a low tolerance to drought can contribute to drought-induced decline (Sánchez-Salguero et al. 2012). In stands exploited in the past then abandoned, the establishment of conservation measures may lead to dense forests that are more prone to drought-induced decline and death (Franklin et al. 1987; Peet and Christensen 1987; Waring 1987). In addition to stand history, climate warming may increase the incidence of biotic stressors as mistletoes, insects and pathogenic fungi, making forests more prone to drought-associated decline (Ayres and Lombardero 2000).

Most Spanish forests are subjected to periodic drought stress, which makes them a good system to study forest decline (Martínez-Vilalta and Piñol 2002; Camarero et al. 2015a). The application of dendroecology to study forest decline corroborates observations that trees with pronounced leaf shedding or exhibiting crown dieback also exhibit low rates of growth, which often portend tree death (LeBlanc and Raynal 1990; Ogle et al. 2000; Bigler and Bugmann 2003; Dobbertin 2005;

Amoroso and Larson 2010). Here, we illustrate four case studies on forest decline in Spain, evaluated using dendroecology. We show how drought stress triggers decline and interacts with predisposing or contributing factors (*sensu* Manion 1991) in four case studies: past logging in the case of Pyrenean silver fir (*Abies alba* Mill.), increased competition induced by land-use changes in the case of Spanish fir (*Abies pinsapo* Boiss.), variation in drought vulnerability of several planted pine species, and mistletoe infection in the case of natural Scots pine stands. We use these case studies to: (i) show the complex etiology of forest decline since tree dieback and death are attributed to drought interacting with other stressors; and (ii) to demonstrate how tree-ring analyses applied to this ecological issue pinpoints the timing and temporal patterns of forest decline.

13.2 Silver-fir Decline: Predisposed by Historical Logging and Triggered by Drought

Since the 1980s, silver fir decline has been documented in the central and western Spanish Pyrenees. Loss in tree vitality is characterized by needle yellowing and shedding, abnormal crown architecture, growth decline and elevated mortality, which has been mainly attributed to severe late-summer water deficit accentuated by warmer temperatures (Camarero et al. 2011). Interestingly, growth rates of silver fir exhibit a post-1980s decline only in the Spanish Pyrenees, whereas in mesic sites of the species distribution in central and eastern Europe and Italy, silver fir shows enhanced growth during the twentieth century (Büntgen et al. 2014; Gazol et al. 2015).

Many of the most affected Pyrenean stands are characterized by summer drought causing sub-optimal climatic conditions, low productivity, and low growth rates (Camarero et al. 2011). However, in these declining stands, many dying or recently dead trees showed growth releases in the 1950s during a period of intense logging in the area (Cabrera 2001) (Fig. 13.1a). The post-logging releases lead us to postulate that past logging predisposed the declining stands to drought-triggered dieback (Camarero et al. 2011). Stand thinning by high-grading the large-diameter trees would have favored the persistence of small-diameter and slow-growing trees, which may be the most vulnerable to drought stress. In heavily-thinned stands, opening of large gaps also alters forest microclimatic conditions, increasing water demand at tree level which negatively affects silver-fir vigor (Aussenac 2002). At some sites, thinning can favor the spread of root-rot fungi (Oliva and Colinas 2007), but their abundance is not higher in declining than non-declining forests (Sangüesa-Barreda et al. 2015).

This case study demonstrates the need to account for predisposing historical factors, which act as legacies and condition the responses of trees to drought. Growth responses to predisposing factors can be quantified and used to evaluate

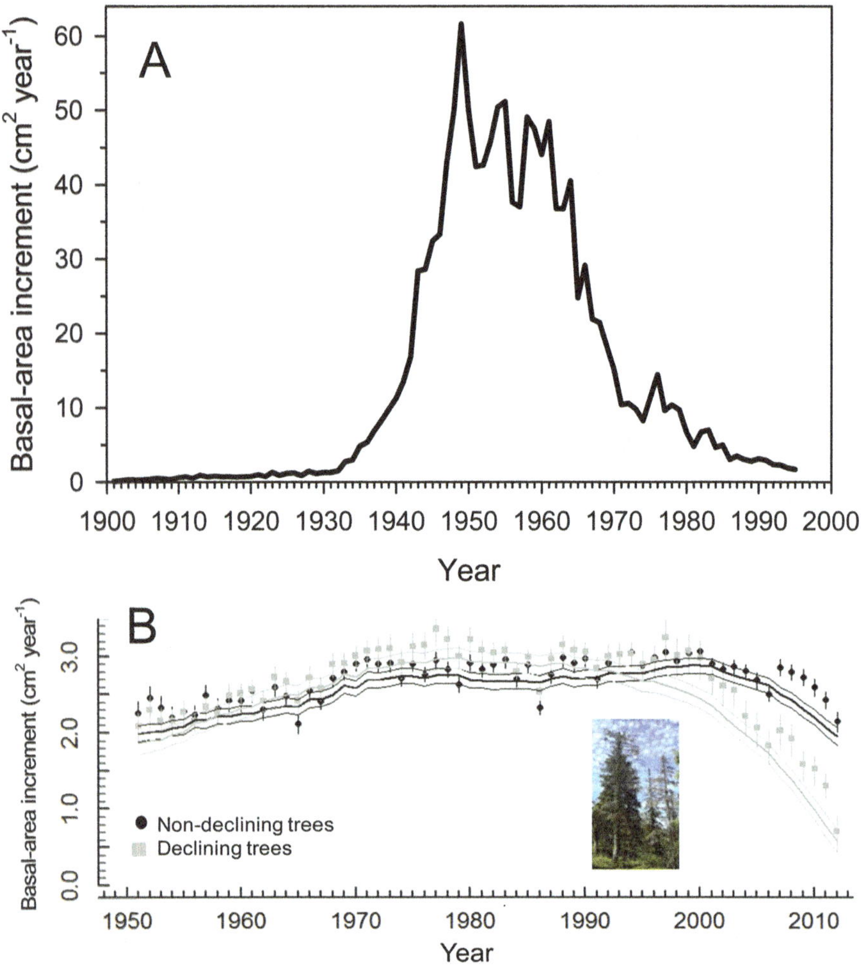

Fig. 13.1 Tree-ring signatures characterizing silver-fir (*Abies alba*) dieback in the Spanish Pyrenees. The *upper plot* (**a**) shows a typical growth curve of a dead silver fir which showed a growth release in the 1950s related to past thinning. The tree was sampled in a declining silver fir forest (Paco Ezpela, Huesca, northern Spain; 42° 45′N, 0° 52′W, 1230 m a.s.l.) where mortality reached 50% of trees after the 2012 severe drought. The *lower* plot (**b**) shows how growth (ln-transformed basal area increment) diverges between coexisting declining (inset image) and non-declining trees at this forest after the 1990s. Points show growth data of each vigor class and lines correspond to statistical models (General Additive Mixed Models) fitted to these data (*lines* show mean fitted values with standard errors) to highlight long-term growth trends. Figures modified from Camarero et al. (2011, 2015a)

tree-ring series from non-declining trees or stands to search for the early-warning signals of decline that portending tree death (Pedersen 1998; Bigler et al. 2004; Cailleret et al. 2016). In this case study, silver fir behaves as a long-term memory species (Fig. 13.1b) because its tree-ring growth patterns are useful predictors of tree death and can be used as early-warning signals of decline and impending mortality (Camarero et al. 2015a).

13.3 Spanish fir Decline: Predisposed by Intense Competition and Triggered by Drought

The recent decline of Spanish fir, a drought-sensitive conifer that is endemic to southern Spain and northern Morocco, shows how drought stress and increased stand density, promoted by changes in land use and stand development, can contribute to a loss of tree vigor and tree death (Linares et al. 2009). In dense stands with trees showing closed canopies, competition between trees drives self-thinning and density-dependent (e.g., crowding-dependent) mortality occurs (Oliver and Larson 1996). Trees with different diameters, basal areas and height classes compete differently for light, water and other resources (Peet and Christensen 1987; Orwig and Abrams 1997; Martin-Benito et al. 2008). In general, larger and taller trees are better competitors and more likely to survive.

Linares et al. (2009, 2010) showed the importance of competition in stand dynamics after land-use changes, as a contributing factor in the decline as a contributing factor in the decline of Spanish fir. The long-term changes in basal area increment of Spanish fir in relict stands were reconstructed from tree rings to investigate how density-dependent factors modulate the responses of radial growth to climatic stress (Linares et al. 2009). The likelihood of mortality of Spanish fir increased with decreasing elevation, indicating the influence of drought stress on mortality. Mortality was not greater in smaller tree size classes; rather, high mortality was linked to stands with closed overstories and homogeneous structures (Linares and Carreira 2009). Conversely, Spanish fir stands with lower mortality rates corresponded to more open and structurally heterogeneous stands (Linares et al. 2009). This example illustrates that both drought and stand structure have the potential to profoundly alter tree growth, vigor and mortality.

Several studies have shown that tree mortality rates are inversely related to radial-growth rates (Waring 1987; Wyckoff and Clark 2000; van Mantgem et al. 2003; van Mantgem and Stephenson 2007; Cailleret et al. 2016). Similarly, the dendroecological assessment of the growth trends in Spanish fir was useful to identify trees with higher probabilities of decline and death (Linares et al. 2009). In this study, the low mean growth rate of Spanish fir was mainly determined by high levels of competition (Fig. 13.2), whereas growth trends indicating decline preceding death were similar among dominant, codominant and suppressed trees and related to climatic variation. Trends of radial growth decline were strongly related to long-term warming trends, while inter-annual growth variability of trees was related to precipitation. These two climate-growth relations decreased in significance as

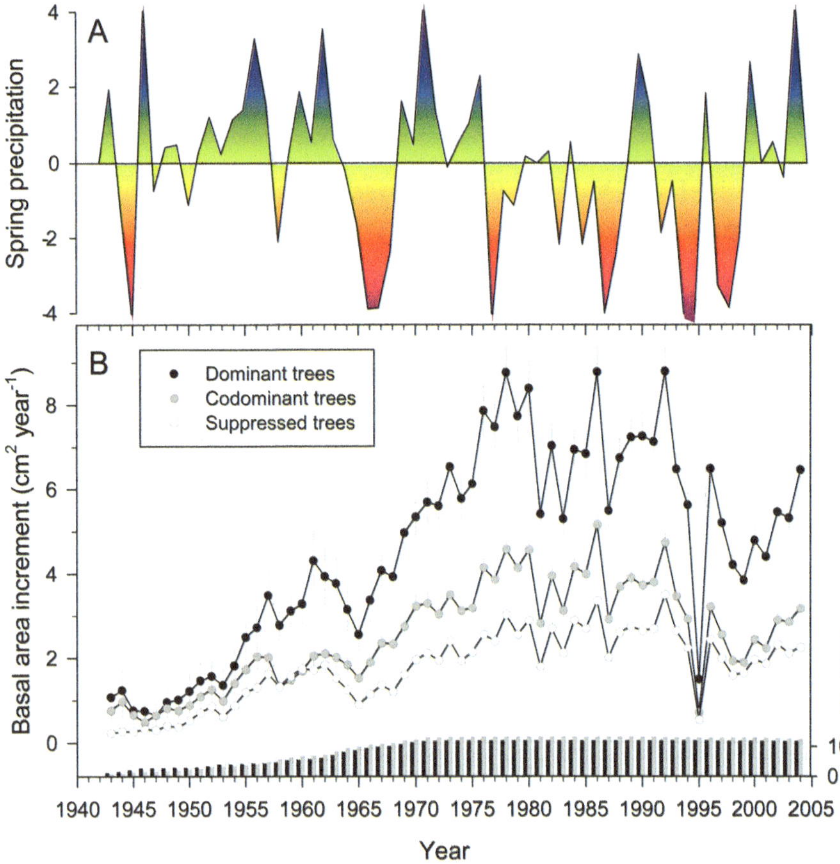

Fig. 13.2 Spring (March to May) precipitation (**a**) expressed as standardized values) and basal-area increment (**b**) measured in dominant, co-dominant, and suppressed Spanish fir (*Abies pinsapo*) trees located near the lowermost elevation limit (36° 43′N, 4° 57′W, 1200 m a.s.l.) of the species' altitudinal range in southern Spain. Competition is estimated using a spatially-explicit competition index values (Linares et al. 2010). *Bottom bars* indicate the number of tree-rings series measured. *Error bars* represent the standard error. Figures modified from Linares et al. (2010)

competition (indicated by tree density) increased. To address potential confounding effects of changing phenology on climate-growth relationships, Linares et al. (2012) showed that climate warming during the second half of the twentieth century did not lead to extended growing seasons in the mid- to low-elevation Spanish fir populations where mortality was greatest. Specifically, non-linear responses of growth cessation to water availability offset the linear effects of increasing temperature on earlier budburst date. These results support the contention that observed growth decline and death of Spanish fir result from interactions between competition over long time scales and drought over short time scales (Linares et al. 2010).

The response to only the most recent drought raises questions about cumulative predisposing stress factors interacting with climatic stress. Extreme climate events, such as successive and severe droughts, may cause abrupt growth declines and pulses of elevated tree mortality. These events are expected to more severely affect tree populations growing near their climatic tolerance limit. Very low annual precipitation was detected in 1994–1995, with similar low rainfall during 1945–1946, 1954–1955, 1965–1967, 1977 and 1987 (Fig. 13.2). However, only the 1994–1995 droughts caused decline in Spanish fir populations (Linares and Carreira 2009).

In summary, increased competition due to changes in land use predisposed Spanish fir to growth decline triggered by drought stress (Linares et al. 2010). Furthermore, various stressors lead to increases in severity of drought impacts on the forest, without corresponding increases in drought intensity. For example, if tree-to-tree competition reduced growth in a significant proportion of a population of susceptible trees, any drought could severely damage or kill more trees than it would in a more heterogeneous stand (Suarez et al. 2004; Linares et al. 2010). Thus, stressors may interact with competition predisposing trees to being killed by drought, as trees subjected to high competition are more likely surpassing their physiological thresholds of drought tolerance (McDowell et al. 2008, 2011).

13.4 Species-Specific Decline in Pine Plantations Triggered by Drought

In many countries in the Mediterranean Basin, intensive land uses such as logging and overgrazing by rural populations caused deforestation that motivated governments to support reforestation. In Spain, extensive plantations of Scots pine (*Pinus sylvestris* L.), Aleppo pine (*Pinus halepensis* Mill.), black pine (*Pinus nigra* Arn.) and maritime pine (*Pinus pinaster* Ait.) were established during the 1950s (FAO 2006). In the semi-arid southeastern Spain, many pine plantations were planted beyond the species natural climatic limits. During cool and wet climate of the 1970s, growth rates of planted pines were high but they were not sustained when climate became warmer and droughts occurred in 1994–1995, 1999, and 2005 (Sánchez-Salguero et al. 2012). During the warm and dry climate of the 1990s and 2000s several of the planted pine stands showed pronounced needle shedding or defoliation, growth decline and extensive mortality (Navarro-Cerrillo et al. 2007).

The four planted pine species differ in their vulnerability to drought, indicated by their natural distribution, ecophysiology and functional traits, which affected their success in semi-arid, southeast Spain. Scots pine showed the most intense growth decline and highest mortality rates (Fig. 13.3) as it is naturally distributed in Eurosiberia and is most vulnerable to drought-induced xylem embolism (Martínez-Vilalta and Piñol 2002). Drought-tolerant Aleppo pine, naturally distributed in the Mediterranean, was least affected (Camarero et al. 2015a). The other two Mediterranean species, black pine and maritime pine, occupied intermediate positions

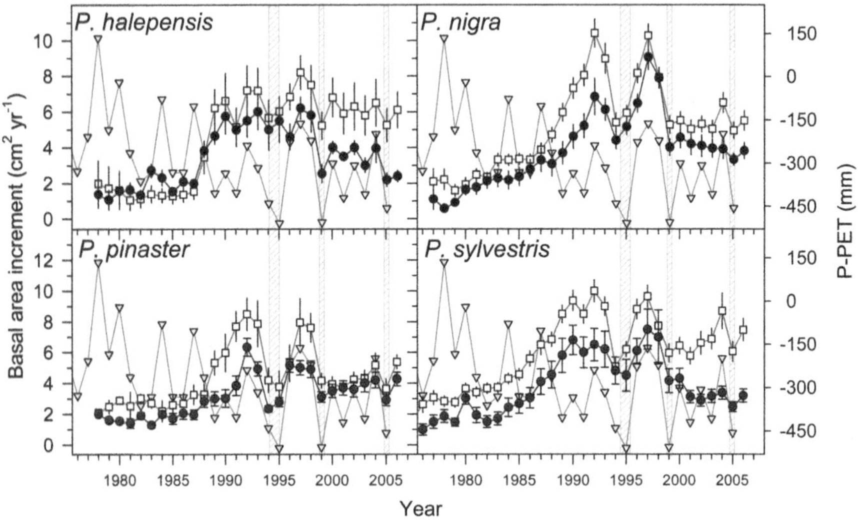

Fig. 13.3 Recent growth trends of basal area increment of four pine species planted at south-eastern Spain (37° 22′ N, 02° 50′ W) according to their recent crown defoliation (*empty squares*, defoliation <50%; *filled circles*, defoliation ≥50%). Years characterized by droughts (1994–1995, 1999 and 2005) correspond to growth decline, particularly in the most defoliated trees and the species less tolerant to drought as Scots pine (*Pinus sylvestris*). The right *y*-axis shows the annual water balance (*lines with triangles*, P − PET, difference between precipitation and potential evapotranspiration; note the reverse scale). The *vertical error bars* show standard errors. Figures modified from Sánchez-Salguero et al. (2012)

within this vulnerability ranking and did not present such marked growth loss. However, the decline of trees due to warmer temperatures and droughts causing water shortage during the growing season was not homogeneous within or among stands. Some individuals showed more defoliation and growth loss than conspecifics at the same site and apparently subjected to similar climatic, topographic, and edaphic conditions (Sánchez-Salguero et al. 2012).

Dendroecology allowed investigation of the decline processes of different pine species and tree vigor classes in response to successive droughts. A divergence in radial growth of trees with defoliation levels higher or lower than the 50% crown transparency was most evident in Scots pine. In contrast, growth differences were lacking between maritime pine with high and low defoliation levels (Fig. 13.3). This difference suggests that maritime pine can be considered a drought-tolerant species, which is consistent with the observed low growth sensitivity of natural maritime pine forests in response to temporal variability in drought (Camarero et al. 2015b).

Local site conditions may have predisposed some planted trees to drought more than others. Topographic variation would explain why most defoliated trees were growing on steeper sites with shallow soils, where soil water availability is lowest (Navarro-Cerrillo et al. 2007). Such predisposing factors may cause persistent differences in vigor among trees of the same species and explain the positive relationships between wet site conditions, tree size and growth rates reported by Sánchez-Salguero et al. (2012). These differences in vigor are particularly relevant to forecast the probability of recovery of defoliated trees, especially for species less resistant to drought-induced xylem embolism such as Scots pine (Martínez-Vilalta and Piñol 2002). The decline of Scots pine plantations described in south-eastern Spain is characterized by a rapid growth reduction and widespread defoliation making the future existence of these plantations unlikely under the warmer and drier conditions predicted for most semi-arid regions of the Mediterranean Basin (Giorgi and Lionello 2008).

13.5 Scots Pine Decline driven by Interactions Between Mistletoe Infestation and Drought

An increasing number of forest declines involving several causal factors including drought and biotic stressors such as mistletoes, insects, and root-rot fungi are being reported worldwide (Ayres and Lombardero 2000; Carnicer et al. 2011). This research contrasts recent reviews on forest decline that conclude there is a research gap on the roles of biotic stressors in forest decline (Allen et al. 2010, 2015). However, studies on interactions between drought and biotic stressors as drivers of forest decline are notably scarce. These types of interactions may be complex and difficult to unravel because the drivers influence tree growth and vigor at different spatial and temporal scales and biotic stressors affect trees in different ways. For example, insect outbreaks cause defoliation, root-rot fungi affect roots, and mistletoes alter crown architecture and branch growth.

Hemiparasitic plants, such as mistletoes, obtain water and nutrients from host trees. Consequently, mistletoe infection may make host trees more sensitive to the negative effects of drought by enhancing water loss and reducing carbon gain of infected branches (Zweifel et al. 2012). Drought and mistletoe infection have been mainly studied separately; how drought and mistletoe interact and induce forest decline is not fully understood (but see Sangüesa-Barreda et al. 2013). Tree-rings allow reconstruction of the growth of infected trees to infer when mistletoe infection started and its effect on growth, and how it predisposes trees to drought-induced decline.

European pine mistletoe (*Viscum album* L. ssp. *austriacum*) is limited by low temperatures and rising temperatures may increase its abundance and promote an upslope or altitudinal expansion of its range (Dobbertin et al. 2005). Increased incidence and spread could accelerate forest decline in the most infected pine stands at dry locations (Dobbertin and Rigling 2006). Similar impacts have been described for other mistletoe affecting conifer species elsewhere (Alfaro et al. 1985; Vallauri 1998; Noetzli et al. 2004; Tsopelas et al. 2004).

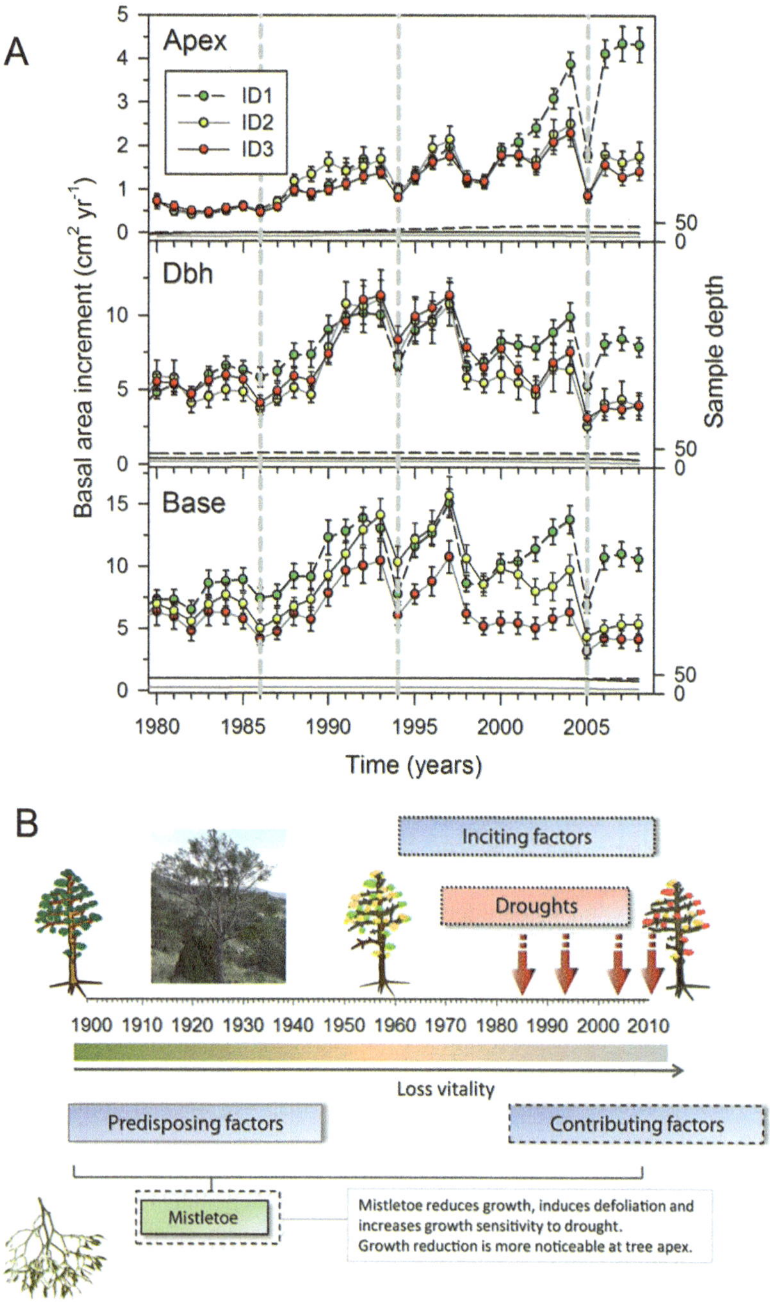

Fig. 13.4 Drought and mistletoe infestation contribute to growth decline in Scots pine. Growth patterns of mistletoe-infested trees (**a**) and scheme showing the Scots pine dieback process of

Dendroecology was applied to determine how mistletoe infection and drought stress modified the long-term growth patterns in Scots pine by exploring changes in growth at different heights, since mistletoe is mainly present in upper exposed tree crowns (Sangüesa-Barreda et al. 2012). Long-term changes in radial growth were assessed at the tree apex, diameter at breast height (dbh), and at the tree base. Basal area increment was used to quantify and compare growth trends of trees with different degrees of mistletoe infestation: trees without mistletoe mats (class ID1), moderately infected trees with mistletoe present in one or two thirds of the crown (class ID2), and severely infected trees with mistletoe present throughout the whole crown (class ID3). As well, drought sensitivity and drought recovery were calculated as the relative changes in growth for the three most severe droughts affecting the study area over the last three decades. In 1986, 1994–1995 and 2005 growing-season precipitation was $\leq$48% of the mean for the 1950–2015 period.

Mistletoe infection negatively affected growth from the apex to the stem base (Fig. 13.4a), but the effect was more noticeable near the tree apex where most mistletoes occur and induce more needle shedding than at other sampled heights. Mistletoe infection of Scots pine caused an average loss of 64% in basal area increment at the tree apex, whereas the loss of basal area increment was c. 51% at the other two heights. Notably, mistletoe most commonly affected the tallest trees with biggest crowns and diameters, which are most exposed to mistletoe seed deposition because birds prefer to perch in the top of tall trees (Sangüesa-Barreda et al. 2012, 2013). These canopy-dominant trees also had the highest growth rates, supporting the hypothesis that large and fast-growing trees are more prone to mistletoe infection and growth decline.

The basal area increment of severely and moderately infected trees diverged from that of non-infected trees since 2001 and the growth loss of infected trees was magnified by drought in 2005. These results indicated a strong interaction between mistletoe infection and drought stress that lead to a short-term reduction of tree growth, and preceded the death of many of the most severely infected trees. Mortality rates were 7–10% in the most affected stands (Sangüesa-Barreda et al. 2012). Interestingly, the growth divergence between infected and non-infected trees was relatively recent (c.10 years). Further research will test if the timing of divergence is different at wetter sites, but preliminary data suggest that divergence was much longer (c. 30 years) as drought stress diminishes. Tree-rings provide information on radial growth and wood formation and represent a proxy of water-use

Fig. 13.4 (continued) infested trees in drought-prone sites following the Manion (1991) model. (**b**) Radial-growth data was quantified at the tree apex, at 1.3 m where the diameter at breast height (dbh) is measured, and at the base for trees with different mistletoe infestation degree (**a**) and responses to drought (*dashed vertical lines*). In the *upper* plot the *right y* axis shows the sampled depth (number of measured radii) and error bars are standard errors. We classified the infested-trees in three classes: ID1, trees without mistletoe; ID2, moderately infested trees; and ID3, severely infested trees. In the *lower plot* (**b**) the image shows a tree heavily infested by mistletoe (class ID3). Figures modified from Sangüesa-Barreda et al. (2012)

efficiency and long-term changes in tree functioning. The discrimination between the carbon of atmospheric CO_2 and the carbon fixed by trees as woody tissues can be quantified by analyzing in $^{13}C/^{12}C$ isotope ratios in tree-ring wood or cellulose (McCarroll and Loader 2004; Saurer et al. 2004). This approach was followed to assess how the combined effect of mistletoe infection and drought affected radial growth and water-use efficiency in Scots pine. The water-use efficiency of the most severely infected trees decreased during the last 5 years prior to sampling, although their radial growth rates were low for more than 10 years (Sangüesa-Barreda et al. 2013). Growth of severely infected trees was more sensitive to drought stress than non-infected trees. Although infected and non-infected trees displayed similar increasing trends of water-use efficiency, the synergistic effect of drought stress and mistletoe in severely infested trees caused a marked growth loss and reversed any potential increase of water-use efficiency due to the rise in atmospheric CO_2 concentrations. Consequently, rising atmospheric CO_2 concentrations could not compensate for the negative impacts of drought and mistletoe on tree growth and water-use efficiency in Scots pine forest under xeric conditions. Finally, these findings suggest mistletoe infection requires more time to cause severe growth loss at mesic than at dry sites, i.e., the growth divergence between infected and non-infected trees should last longer and be less harmful to trees if drought stress is alleviated.

This is not the first study showing that potential fertilization effect of increasing atmospheric CO_2 concentrations do not override the negative effects of stress factors, in this case drought and mistletoe infection on growth and water-use efficiency of affected trees. In declining silver fir trees growth decreased and water-use efficiency stopped rising as a consequence of severe droughts in the 1980s and 1990s, despite rapidly increasing atmospheric CO_2 concentration during the late twentieth century (Linares and Camarero 2012). Both the mistletoe-pine and the silver fir forest-decline examples show that drought and other stress factors override the increased availability of atmospheric carbon, thus causing a persistent reduction in growth rates and often leading to tree death. If warmer and drier conditions persist, Scots pine and silver fir stands currently showing severe forest decline will likely experience changes in stand structure and composition of dominant trees.

13.6 Summary and Further Remarks

Several recent studies have documented forest decline and increased rates of tree mortality worldwide, often associated with drought in recent decades (Allen et al. 2010, 2015; van Mantgem et al. 2009). There is concern that climate warming will exacerbate drought stress by rising evapotranspiration rates, causing more widespread and intense forest decline (Williams et al. 2013). In addition, some authors suggest that warmer conditions could favour the spread and incidence of biotic stressors (e.g., mistletoe, herbivore insects and invasive pathogens) making them more aggressive to trees (Ayres and Lombardero 2000). Dendroecology provides a long-term perspective to ascertain if and how forest decline is connected to recent climate warming and drying trends, in particular in arid areas of the northern hemisphere (Dai 2013).

Here, four case studies of forest decline in Spain illustrated various growth patterns and responses of declining trees to different combinations of stressors. In all four cases, drought stress was an inciting factor (*sensu* Manion 1991), which is common in declining Mediterranean (Sarris et al. 2007) and other drought-prone forests (Suarez et al. 2004; Williams et al. 2013). Severe drought-triggered abrupt growth reductions were observed in mesic Pyrenean silver-fir forests, where water deficit was amplified by warmer summer temperatures and increased evapotranspiration rates (Vicente-Serrano et al. 2015). Evidently, rare but severe droughts could be relevant disturbances shaping the dynamics of humid temperate forests, particularly if climate continues warming (Pederson et al. 2014). The case studies also demonstrate that factors other than drought cause forest decline, such as previous logging (silver fir), competition (Spanish fir), species-specific tolerance to drought (pine plantations), and mistletoe infestation (Scots pine). Such predisposing or contributing factors (*sensu* Manion 1991) often act at longer time scales than punctuated droughts, and their attribution is complicated because forest decline and tree death can occur years or decades after those factors started affecting trees. In this sense, a major application of dendroecology is to forecast tree vigor or predict the probability of tree death based on measured tree-ring variables (Cailleret et al. 2016), including width, anatomy (Pellizzari et al. 2016), or isotopes (Levanič et al. 2011). In addition to statistical characteristics of the tree-ring series, forecasts of vigor may be enhanced by data on climatic stress (e.g., drought severity), (Camarero et al. 2015a), and variables reflecting long-term predisposing and contributing factors (e.g., management and use, competition, some biotic stressors). Including the long-term stressors would require novel approaches and tools to consider legacy effects and decipher how the ecological memory is imprinted on tree-ring records (Ogle et al. 2015).

The use of dendroecology to pinpoint the start of drought-triggered growth decline has several important outcomes and applications. First, identifying the timing of growth decline is helpful to attribute stressors (e.g., drought or logging) and to quantify tree recovery from their impact stressors. Second, if the tree-ring variables exceed a critical threshold, they can be used as early-warning indicators of portending tree death (Camarero et al. 2015a). Consequently, growth trends or growth responses to climate, including post-drought growth resilience, could be used as valuable traits to predict the probability of drought-induced damage in forests (Anderegg et al. 2015; Gazol et al. 2016). Third, since radial growth is considered a long-term proxy of tree vigor (Dobbertin 2005), future directions on forest-dieback research could bridge post-drought changes in radial growth with primary growth. Primary growth (e.g., shoot and root elongation, bud and needle production, and reproduction) is a major component of tree performance and determines canopy architecture (Jalkanen 1995; Vennetier et al. 2013). Fourth, dendroecological studies can address forest decline from a modern pathological perspective by quantifying critical thresholds of disease and tree death required to maintain dynamic and resilient stands able to withstand climate change and new threats from biotic stressors (Manion 2003).

Finally, tree-ring data can be used to parameterize empirical models of tree growth to project forest responses to CO_2-concentration and climate change scenarios and produce probabilistic predictions of dieback events and maps of forest vulnerability (Charney et al. 2016). These outcomes would contribute to science-based adaptive management strategies under future climate and management scenarios.

Acknowledgments We thank several Spanish (CGL2015-69186-C2, FUNDIVER; CGL2013-48843-C2-1-R, CoMo-ReAdapt) and European (COST Action STReESS—Studying Tree Responses to extreme Events: a SynthesiS, FP1106) projects for funding and promoting the development of this research. A. Gazol is supported by a Postdoctoral grant (MINECO-FPDI 2013-16600, FEDER funds). We sincerely thank L. Daniels for her insightful comments on a previous draft of this chapter.

References

Alfaro RI, Bloomberg WJ, Smith RB, Thomson AJ (1985) Epidemiology of dwarf mistletoe in western hemlock stands in south coastal British Columbia. Can J For Res 15:909–913

Allen CD, Macalady AK, Chenchouni H, Bachelet D, McDowell N et al (2010) A global overview of drought and heat-induced tree mortality reveals emerging climate change risks for forests. For Ecol Manag 259:660–684

Allen CD, Breshears DD, McDowell NG (2015) On underestimation of global vulnerability to tree mortality and forest die-off from hotter drought in the Anthropocene. Ecosphere 6:129

Amoroso MM, Larson BC (2010) Stand development patterns as a consequence of the mortality in *Austrocedrus chilensis* forests. For Ecol Manag 259:1981–1992

Anderegg WRL, Schwalm C, Biondi F, Camarero JJ, Koch G, Litvak M, Ogle K, Shaw JD, Shevliakova E, Williams AP, Wolf A, Ziaco E, Pacala S (2015) Pervasive drought legacies in forest ecosystems and their implications for carbon cycle models. Science 349:528–532

Anderegg WRL, Klein T, Bartlett M, Sack L, Pellegrini AFA, Choat B, Jansen S (2016) Meta-analysis reveals that hydraulic traits explain cross-species patterns of drought-induced tree mortality across the globe. PNAS 113:5024–5029

Aussenac G (2002) Ecology and ecophysiology of circum-Mediterranean firs in the context of climate change. Ann For Sci 59:823–832

Ayres MP, Lombardero MJ (2000) Assessing the consequences of climate change for forest herbivore and pathogens. Sci Total Environ 262:263–286

Bigler CJ, Bugmann H (2003) Growth-dependent tree mortality models based on tree rings. Can J For Res 33:210–221

Bigler CJ, Gričar J, Bugmann H, Čufar K (2004) Growth patterns as indicators of impending tree death in silver fir. For Ecol Manag 199:183–190

Bravo-Oviedo A, Sterba H, del Rio M, Bravo F (2006) Competition-induced mortality for Mediterranean *Pinus pinaster* Ait. and *P. sylvestris* L. For Ecol Manag 222:88–98

Büntgen U, Tegel W, Kaplan JO, Schaub M, Hagedorn F, Bürgi M, Brázdil R, Helle G, Carrer M, Heussner K-U, Hofmann J, Kontic R, Kyncl T, Kyncl J, Camarero JJ, Tinner W, Esper J, Liebhold A (2014) Placing unprecedented recent fir growth in a European-wide and Holocene-long context. Front Ecol Environ 12:100–106

Cabrera M (2001) Evolución de abetares del Pirineo aragonés. Cuad Soc Esp Ciencias Forestales 11:43–52

Cailleret M, Bigler C, Bugmann H, Camarero JJ, Cufar K, Davi H, Mészáros I, Minunno F, Peltoniemi M, Robert EMR, Suarez ML, Tognetti R, Martínez-Vilalta J (2016) Towards a common methodology for developing logistic tree mortality models based on ring-width data. Ecol Appl 26:1827–1841

Camarero JJ, Bigler C, Linares JC, Gil-Pelegrín E (2011) Synergistic effects of past historical logging and drought on the decline of Pyrenean silver fir forests. For Ecol Manag 262:759–769

Camarero JJ, Gazol A, Sangüesa-Barreda G, Oliva J, Vicente-Serrano SM (2015a) To die or not to die: early-warning signals of dieback in response to a severe drought. J Ecol 103:44–57

Camarero JJ, Gazol A, Tardif JC, Conciatori F (2015b) Attributing forest responses to global-change drivers: limited evidence of a CO_2-fertilization effect in Iberian pine growth. J Biogeogr 42:2220–2233

Carnicer J, Coll M, Ninyerola M, Pons X, Sánchez G, Peñuelas J (2011) Widespread crown condition decline, food web disruption, and amplified tree mortality with increased climate change-type drought. PNAS 108:1474–1478

Charney ND, Babst F, Poulter B, Record S, Trouet VM, Frank D, Enquist BJ, Evans MEK (2016) Observed forest sensitivity to climate implies large changes in 21st century North American forest growth. Ecol Lett 19:1119–1128

Dai A (2013) Increasing drought under global warming in observations and models. Nat Clim Chang 3:52–58

Das A, Battles J, Stephenson NL, van Mantgem PJ (2011) The contribution of competition to tree mortality in old-growth coniferous forests. For Ecol Manag 261:1203–1213

Dobbertin M (2005) Tree growth as indicator of tree vitality and of tree reaction to environmental stress: a review. Eur J For Res 124:319–333

Dobbertin M, Hilker N, Rebetez M, Zimmermann NE, Wohlgemuth T, Rigling A (2005) The upward shift in altitude of pine mistletoe (*Viscum album* ssp *austriacum*) in Switzerland—the result of climate warming? Int J Biometeorol 50:40–47

Dobbertin M, Rigling A (2006) Pine mistletoe (*Viscum album* ssp. *austriacum*) contributes to Scots pine (*Pinus sylvestris*) mortality in the Rhone valley of Switzerland. For Pathol 36:309–322

FAO (2006) Global planted forests thematic study: results and analysis, Food and Agriculture Organization of the United Nations. In: Del Lungo A, Ball J, Carle J (eds) Planted forests and trees. Working Paper 38, Rome, Italy

Foster DR, Oswald WW, Faison EK, Doughty ED, Hansen BCS (2006) A climatic driver for abrupt mid-holocene vegetation dynamics and the hemlock decline in New England. Ecology 87:2959–2966

Franklin JF, Shugart HH, Harmon ME (1987) Tree death as an ecological process. Bioscience 37:550–556

Gazol A, Camarero JJ, Gutiérrez E, Popa I, Andreu-Hayles L, Motta R, Nola P, Ribas M, Sangüesa-Barreda G, Urbinati C, Carrer M (2015) Distinct effects of climate warming on populations of silver fir (*Abies alba*) across Europe. J Biogeogr 42:1150–1162

Gazol A, Camarero JJ, Anderegg WRL, Vicente-Serrano SM (2016) Impacts of droughts on the growth resilience of Northern Hemisphere forests. Glob Ecol Biogeogr 26(2):166–176.

Giorgi F, Lionello P (2008) Climate change projections for the Mediterranean region. Glob Planet Chang 63:90–104

Jalkanen R (1995) Needle Trace Method (NTM) for retrospective needle retention studies on Scots pine (Pinus sylvestris L.). Acta Universitatis Ouluensis, Oulu, p 84

LeBlanc DC, Raynal DJ (1990) Red spruce decline on Whiteface Mountain, New York. II. Relationships between apical and radial growth decline. Can J For Res 20:1415–1421

Levanič T, Cater M, McDowell NG (2011) Associations between growth, wood anatomy, carbon isotope discrimination and mortality in a *Quercus robur* forest. Tree Physiol 31:298–308

Linares JC, Camarero JJ, Carreira JA (2009) Interacting effects of climate and forest-cover changes on mortality and growth of the southernmost European fir forests. Glob Ecol Biogeogr 18:485–497

Linares JC, Camarero JJ, Carreira JA (2010) Competition modulates the adaptation capacity of forests to climatic stress: insights from recent growth decline and death in relict stands of the Mediterranean fir *Abies pinsapo*. J Ecol 98:592–603

Linares JC, Camarero JJ (2012) From pattern to process: linking intrinsic water-use efficiency to drought-induced forest decline. Glob Chang Biol 18:1000–1015
Linares JC, Carreira JA (2009) Temperate-like stand dynamics in relict Mediterranean-fir (*Abies pinsapo*, Boiss.) forests from Southern Spain. Ann For Sci 66:610–619
Linares JC, Covelo F, Carreira JA, Merino JA (2012) Phenological and water-use patterns underlying maximum growing season length at the highest elevations: implications under climate change. Tree Physiol 32:161–170
Manion PD (1991) Tree disease concepts. Prentice Hall, Englewood Cliffs
Manion PD (2003) Evolution of concepts in forest pathology. Phytopathology 93:1052–1055
Martin-Benito D, Cherubini P, del Río M, Cañellas I (2008) Growth response to climate and drought in *Pinus nigra* Arn. trees of different crown classes. Trees Struct Funct 22:363–373
Martínez-Vilalta J, Piñol J (2002) Drought-induced mortality and hydraulic architecture in pine populations of the NE Iberian Peninsula. For Ecol Manag 161:247–256
McCarroll D, Loader NJ (2004) Stable isotopes in tree rings. Quat Sci Rev 23:771–801
McDowell N, Pockman WT, Allen CD, Breshears DD, Cobb N, Kolb T, Plaut J, Sperry J, West A, Williams DG, Yepez EA (2008) Mechanisms of plant survival and mortality during drought: why do some plants survive while others succumb to drought? New Phytol 178:719–739
McDowell NG, Beerling DJ, Breshears DD, Fisher RA, Raffa KF, Stitt M (2011) The interdependence of mechanisms underlying climate-driven vegetation mortality. Trends Ecol Evol 26:523–532
Muller-Dombois D (1992) A natural dieback theory, cohort senescence as an alternative to the decline disease theory. In: Manion PD, Lachance D (eds) Forest decline concepts. The American Phytopathological Society, St. Paul, pp 26–37
Navarro-Cerrillo RM, Varo MA, Lanjeri S, Hernández Clemente R (2007) Cartografía de defoliación en los pinares de pino silvestre (*Pinus sylvestris* L.) y pino salgareño (*Pinus nigra* Arn.) en la Sierra de los Filabres. Ecosistemas 16:163–171
Noetzli KP, Müller B, Sieber TN (2004) Impact of population dynamics of white mistletoe (*Viscum album* ssp. *abietis*) on European silver fir (*Abies alba*). Ann For Sci 60:773–779
Ogle K, Whitham TC, Cobb NS (2000) Tree-ring variation in pinyon predicts likelihood of death following severe drought. Ecology 81:3237–3243
Ogle K, Barber JJ, Barron-Gafford GA, Bentley LP, Young JM, Huxman TE, Loik ME, Tissue DT (2015) Quantifying ecological memory in plant and ecosystem processes. Ecol Lett 18:221–235
Oliva J, Colinas C (2007) Decline of silver fir (*Abies alba* Mill.) stands in the Spanish Pyrenees: role of management, historic dynamics and pathogens. For Ecol Manag 252:84–97
Oliver CD, Larson BC (1996) Forest stand dynamics. Wiley, New York
Orwig DA, Abrams MD (1997) Variation in radial growth responses to drought among species, site, and canopy strata. Trees Struct Funct 11:474–484
Pedersen BS (1998) Modeling tree mortality in response to short and long-term environmental stresses. Ecol Model 105:347–351
Pederson N, Dyer JM, McEwan RW, Hessl AE, Mock CJ, Orwig DA, Rieder HE, Cook BI (2014) The legacy of episodic climatic events in shaping temperate, broadleaf forests. Ecol Monogr 84:599–620
Peet RK, Christensen NL (1987) Competition and tree death. Bioscience 37:586–594
Pellizzari E, Camarero JJ, Gazol A, Sangüesa-Barreda G, Carrer M (2016) Wood anatomy and carbon-isotope discrimination support long-term hydraulic deterioration as a major cause of drought-induced dieback. Glob Chang Biol 22:2125–2137
Reams GA, Huso MMP (1990) Stand history: an alternative explanation of red spruce radial growth reduction. Can J For Res 20:250–253
Sánchez-Salguero R, Navarro-Cerillo RM, Camarero JJ, Fernández-Cancio A (2012) Selective drought-induced decline of pine species in southeastern Spain. Clim Chang 113:767–785
Sánchez-Salguero R, Camarero JJ, Gutiérrez E, González Rouco F, Gazol A, Sangüesa-Barreda G, Andreu-Hayles L, Linares JC, Seftigen K (2017) Assessing forest vulnerability to climate warming using a process-based model of tree growth: bad prospects for rear-edges. Glob Chang Biol 23:2705–2719. doi:10.1111/gcb.13541

Sangüesa-Barreda G, Linares JC, Camarero JJ (2012) Mistletoe effects on Scots pine decline following drought events: insights from within-tree spatial patterns, growth and carbohydrates. Tree Physiol 32:585–598
Sangüesa-Barreda G, Linares JC, Camarero J (2013) Drought and mistletoe reduce growth and water-use efficiency of Scots pine. For Ecol Manag 296:64–73
Sangüesa-Barreda G, Camarero JJ, Oliva J, Montes F, Gazol A (2015) Past logging, drought and pathogens interact and contribute to forest dieback. Agric For Meteorol 208:85–94
Sarris D, Christodoulakis D, Körner C (2007) Recent decline in precipitation and tree growth in the eastern Mediterranean. Glob Chang Biol 13:1187–1200
Saurer M, Siegwolf R, Schweingruber F (2004) Carbon isotope discrimination indicates improving water-use efficiency of trees in northern Eurasia over the last 100 years. Glob Chang Biol 10:2109–2120
Suarez ML, Ghermandi L, Kitzberger T (2004) Factors predisposing episodic drought-induced tree mortality in *Nothofagus*—site, climatic sensitivity and growth trends. J Ecol 92:954–966
Tsopelas P, Angelopoulos A, Economou A, Soulioti N (2004) Mistletoe (*Viscum album*) in the fir forest of Mount Parnis, Greece. For Ecol Manag 202:59–65
Vallauri D (1998) Dynamique parasitaire de *Viscum album* L. sur pin noir dans le bassin du Saignon (Préalpes françaises du sud). Ann For Sci 55:823–835
van Mantgem PJ, Stephenson NL, Mutch LS, Johnson VG, Esperanza AM, Parsons DJ (2003) Growth rate predicts mortality of *Abies concolor* in both burned and unburned stands. Can J For Res 33:1029–1038
van Mantgem PJ, Stephenson NL (2007) Apparent climatically induced increase of tree mortality rates in a temperate forest. Ecol Lett 10:909–916
van Mantgem PJ, Stephenson NL, Byrne JC et al (2009) Widespread increase of tree mortality rates in the western United States. Science 323:521–524
Vennetier M, Thabeet A, Didier C, Girard F, Cailleret M, Taugourdeau O, Ouarmim S, Sabatier S-A, Caraglio Y, Sabatier S-A, Cailleret M, Ouarmim S, Thabeet A (2013) Climate change impact on tree architectural development and leaf area. In: Singh BR (ed) Climate change—realities, impacts over ice cap, sea level and risks. http://www.intechopen.com/books/export/citation/BibTex/climate-change-realities-impacts-over-ice-cap-sea-level-and-risks/climate-change-impact-on-tree-architectural-development-and-leaf-area. Accessed 14 Nov 2016
Vicente-Serrano SM, Camarero JJ, Zabalza J, Sangüesa-Barreda G, López-Moreno JI, Tague CL (2015) Evapotranspiration deficit controls net primary production and growth of silver fir: implications for Circum-Mediterranean forests under forecasted warmer and drier conditions. Agric For Meteorol 206:45–54
Waring RH (1987) Characteristics of trees predisposed to die. Bioscience 37:569–574
Williams AP, Allen CD, Macalady AK, Griffin D et al (2013) Temperature as a potent driver of regional forest drought stress and tree mortality. Nat Clim Chang 3:8–13
Wyckoff PH, Clark JS (2000) Predicting tree mortality from diameter growth: a comparison of maximum likelihood and Bayesian approaches. Can J For Res 30:156–167
Zweifel R, Bangerter S, Rigling A, Sterck FJ (2012) Pine and mistletoes: how to live with a leak in the water flow and storage system? J Exp Bot 63:2565–2578

Chapter 14
Forest Decline in Northern Patagonia: The Role of Climatic Variability

Mariano M. Amoroso, Milagros Rodríguez-Catón, Ricardo Villalba, and Lori D. Daniels

Abstract Forest decline is driven by several factors interacting in complex ways, and it is often exacerbated by climate, adding complexity to the process and making it more difficult to identify the causing factors. The long-term perspective provided by tree rings has proven to be successful for disentangling the causes of forest decline worldwide. We present recent dendroecological studies developed to determine the influences of climatic variation on the radial growth patterns and death of trees in declining *Austrocedrus chilensis* and *Nothofagus pumilio* forests in northern Patagonia, Argentina. These results were used to distinguish between possible causes and interactions of abiotic and biotic stress factors versus stand dynamics processes in the development of the forest declines. For our study forests, we found complex interactions between abiotic and biotic factors acting at different spatial and temporal scales that predispose trees, then incite and contribute to the development of the symptoms leading to their subsequent death. We summarize our findings in a conceptual model presenting forest decline as a spiral of processes driven by interactions between abiotic and biotic stress factors that predispose, incite and contribute to stand-level decline in these forests. Our framework formalizes an alternate hypothesis to single causal agents of decline in Patagonian forests by emphasizing the role and importance of climatic variability as a driver of decline.

M.M. Amoroso (✉)
Instituto de Investigaciones en Recursos, Naturales, Agroecología y Desarrollo Rural, Universidad Nacional de Río Negro, El Bolsón, Río Negro, Argentina

CCT CONICET Patagonia Norte, Consejo Nacional de Investigaciones, Científicas y Técnicas, San Carlos de Bariloche, Rio Nego, Argentina
e-mail: mariano.amoroso@gmail.com

M. Rodríguez-Catón • R. Villalba
Instituto Argentino de Nivología, Glaciología y Ciencias Ambientales (IANIGLA), CONICET, CCT-Mendoza, C.C. 330, 5500 Mendoza, Argentina
e-mail: mrodriguez@mendoza-conicet.gob.ar; ricardo@mendoza-conicet.gob.ar

L.D. Daniels
Department of Forest and Conservation Sciences, University of British Columbia, Vancouver, BC, Canada
e-mail: lori.daniels@ubc.ca

M.M. Amoroso et al. (eds.), *Dendroecology*, Ecological Studies 231,
DOI 10.1007/978-3-319-61669-8_14

Keywords Dendroecology • Forest decline • Climatic variability • Northern Patagonia

14.1 Introduction

Tree death normally involves multiple processes that can be difficult to understand and reconstruct. Death occurs as a direct result of discrete events or disturbances (White and Pickett 1985; Oliver and Larson 1996), in response to environmental stress (Condit et al. 1995; Villalba and Veblen 1998; Williamson et al. 2000; Suarez et al. 2004), due to direct competition with other trees (Waring 1987; Kobe et al. 1995; Lutz and Halpern 2006), and as a result of complex interactions of different factors (Mueller-Dombois et al. 1983; Hennon et al. 1990; Minorsky 2003). In the latter case, mortality can involve environmental stress factors interacting with different biotic agents in forests of various ages and stages of development, resulting in broad-scale forest decline (Houston 1981; Mueller-Dombois 1983; Manion 1991). Unlike discrete disturbances such as fire, landslides or avalanches, forest declines around the globe are driven by many factors that interact in complex ways making it difficult to identify or isolate specific causes (e.g. Pedersen 1998; Cherubini et al. 2002; Hartmann and Messier 2008; Camarero et al. 2015). Several conceptual models have been developed to describe the process of mortality in trees and forests based on different biotic agents and/or environmental stress (Houston 1981; Mueller-Dombois 1983; Manion 1991). It is now generally accepted that mortality usually involves complex processes and multiple interactions among disturbance agents and stress factors acting at different time scales (Franklin et al. 1987).

Dendrochronology provides a novel method to estimate the timing of tree death, to assess the patterns of tree growth prior to death, and to evaluate the role of climatic events as a driving force contributing to tree death and forest decline (Jump et al. 2006; Sánchez-Salguero et al. 2012). In declining forests, individual trees exhibiting partial crown mortality often have persistent low rates of growth prior to death (LeBlanc and Raynal 1990; Ryan et al. 1994; Bigler and Bugmann 2003; Amoroso and Larson 2010). Temporal changes in tree growth can be used to determine the onset of the decline in symptomatic trees and quantify differences in the vitality of healthy (asymptomatic) versus affected (symptomatic) trees. The substantial reduction in radial growth often exhibited by declining trees indicates reduced water transport through the stem creating a physiological imbalance between the water-demanding foliage and the water-providing roots (Manion 2003). This imbalance in the water-transport system alters the carbon balance and biochemical composition, increasing tree susceptibility to pest, diseases and other stressors (Mattson and Haack 1987; McDowell et al. 2008; Wang et al. 2012). This physiological change initiates a cascade of interconnected events contributing to the process of tree mortality (Manion 1991, 2003). In this context, the dendroecological assessment of the variations in tree growth over time provides useful information on timing and duration of decline processes and can be used to identify signature radial growth patterns indicating trees prone to decline.

In recent decades, several species in the Andean-Patagonian forests of Argentina have exhibited widespread forest decline and tree death. In this chapter, we present recent dendroecological research investigating temporal aspects of decline in *Austrocedrus chilensis* and *Nothofagus pumilio* forests, emphasizing the role of climatic variability as a major triggering factor.

14.2 The Decline of *Austrocedrus chilensis* Forests

14.2.1 Background

Austrocedrus chilensis (D.Don) Pic. Sern.et Bizarri forests exhibit stand-level decline and mortality of overstory trees on mesic sites throughout the species' distributional range in Argentina. This disturbance, locally referred as 'mal del ciprés' (cypress sickness), was first documented about 60 years ago on Victoria Island and near the town of Epuyén in northern Patagonia (Varsavsky et al. 1975; Hranilovic 1988) and has continued to expand up to the present. Because of its magnitude and the ecological and economic consequences, the decline in *A. chilensis* forests represents one of the most important disturbances in Andean-Patagonian forests.

The occurrence of *A. chilensis* decline is generally associated with particular site conditions, specifically stands on low elevation sites with poor soil water drainage (Filip and Rosso 1999, La Manna and Rajchenberg 2004, La Manna et al. 2008a, b, 2012), as well as those at moderate elevations with greater precipitation (Baccalá et al. 1998). In these forests, decline either appears as aggregations of declining and dead trees forming patches of varying sizes immersed in a matrix of asymptomatic trees (Rosso et al. 1994; La Manna and Rajchenberg 2004) or as dispersed declining and dead trees forming a disaggregated pattern within stands (La Manna and Rajchenberg 2004; La Manna et al. 2008a). The decline of the trees often begins in isolated parts of a stand, and then expands forming patches or aggregations of trees in various stages of decline (Havrylenko et al. 1989; Rosso et al. 1994). In declining forests, all individuals are prone to symptoms and eventual death, regardless of their age and size (Hranilovic 1988; Havrylenko et al. 1989; Rajchenberg and Cwielong 1993; Loguercio et al. 1998).

Trees in declining *A. chilensis* forests present diverse symptoms before dying. Tree-level symptoms are manifested below and above ground. Below-ground symptoms include brown rot that can affect the main roots or the entire root system and can spread to the root collar and base of the tree stem (Rajchenberg and Cwielong 1993; Barrotaveña and Rajchenberg 1996). The first visible symptom is the loss of tree vigor manifested as progressive discoloration and wilting of leaves, followed by gradual defoliation (Havrylenko et al. 1989). It is believed below-ground symptoms precede the loss of tree vigor and crown symptoms (Varsavsky et al. 1975; Havrylenko et al. 1989; Hennon and Rajchenberg 2000). In general, the loss of vigor of the crown leads to slow death of trees; in some cases, trees die quickly after a sudden change in foliage color (Filip and Rosso 1999). Another symptom associated with loss of vigor and subsequent death of *A. chilensis* trees is the reduction in radial growth of affected trees (Calí 1996; Mundo et al. 2010; Amoroso et al. 2012a, 2015). Recent dendroecological studies have shown the radial growth decline, an internal symptom, actually precedes external crown symptoms (Amoroso et al. 2012a, 2015).

14.2.2 Dendroecological Studies

Tree-ring research over the past decade has enhanced our understanding of the timing, causes and effects of *A. chilensis* forest decline. Investigations have addressed tree-level radial-growth decline, mortality rates, gap formation and stand dynamics, and the role of climatic variability on these processes. A reduction in radial growth is associated with vigor loss and subsequent death of *A. chilensis* trees in declining forests (Calí 1996; Mundo et al. 2010; Amoroso et al. 2012a, 2015). Mundo and collaborators (2010) reported significant differences in the radial growth patterns of trees with and without crown symptoms in declining *A. chilensis* forests. These authors found that the growth of 50 symptomatic living trees was consistently lower than paired asymptomatic trees, and the growth reductions were prolonged over four to six decades. Studying 12 declining stands in the same area, Amoroso and collaborators (2012a) reported over half of the *c.*1000 *A. chilensis* trees that they sampled presented a decline in radial growth relative to asymptomatic trees. They reconstructed the diameters of 632 *A. chilensis* to estimate relative tree size at the onset of decline, showing trees of all diameters were susceptible and radial growth decline was independent of the canopy position of trees. More than half were overstory trees at the onset of decline, 43% of which were living but had crown symptoms and 39% were dead at the time of sampling. The remaining 18% of overstory trees with radial growth decline were classified as having healthy, asymptomatic crowns in the field. This noteworthy finding was common among study sites; 22–40% of trees per stand showed a declining growth pattern but did not exhibit crown symptoms. The sequential order of these two symptoms supports the hypothesis that the reduction in radial growth precedes the loss of crown vigor of affected trees.

The temporal patterns of radial growth decline in overstory trees varied greatly among trees and stands (Amoroso et al. 2012a). At the tree level, two common patterns in the rate of decline of radial growth were found. The majority (78%) of trees exhibited gradual and constant decline in radial growth over time. The other 22% of trees exhibited an abrupt decline over 1–3 years, followed by relatively constant narrow rings over time. The length of the radial growth decline (e.g. the number of years since onset of decline) varied among trees and sites. It averaged 27 years for all trees, with a maximum of 80 years. Scaling up to the stand level, the onset of decline started as early as the 1920s in some stands and as late as the

1960s in others. Within stands, temporal patterns of the number of trees exhibiting radial growth decline also varied. In some stands, 30–50 years elapsed with a steady increase in the percentage of trees exhibiting decline, eventually exceeding 50% of trees. Alternately, the percentage increased abruptly to more than 50% of trees in less than 20 years in other stands.

Several studies have described and attempted to quantify mortality rates in declining *A. chilensis* forests (Havrylenko et al. 1989; Rosso et al. 1994; Baccalá et al. 1998; Loguercio et al. 1998; Rajchenberg et al. 1998), but only a few have estimated years of death of the trees using information obtained from tree-rings (Relva et al. 2009; Amoroso and Larson 2010). Amoroso and Larson (2010) crossdated ring width series of individual trees to determine years of death, then quantified and described long-term patterns of overstory mortality over several decades. These authors found the death of adult trees was erratic, unpredictable and highly variable over time, even in stands exhibiting forest decline. As a result, mortality patterns and growth rates of surviving trees were highly variable within and among stands.

The mortality of overstory *A. chilensis* trees created gaps in the canopy, driving stand dynamics. New trees successfully established in the understory in all stands; however, the abundance of regeneration and response of residual overstory trees was variable (Amoroso and Larson 2010). Understory establishment varied among stands, depending on the density of the overstory. In some stands, new establishment enhanced an ongoing increase in the abundance of *Nothofagus dombeyi* ((Mirb.) Oerst.) relative to *A. chilensis* regeneration, representing an important shift in composition and increase in complexity of stand structure (Amoroso et al. 2012b). Similarly, the radial growth response of the residual overstory trees varied among stands. While trees in almost all stands released in the gaps created by the mortality of overstory trees, no distinctive patterns in timing, number and magnitude were discernable. In some stands, growth releases occurred after single or multiple tree deaths suggesting a coupled relationship between these processes, while in other stands this was not the case (Amoroso and Larson 2010).

Recent dendroecological research has helped understanding another symptom associated with *A. chilensis* forest decline: the partial death of the cambium of living trees (Rajchenberg and Cwielong 1993; Calí 1996). Amoroso and Daniels (2010) reported that partial death of the cambium affected about a third of *c*. 800 living *A. chilensis* trees in declining stands, with greatest affects in subcanopy, symptomatic trees. Living trees can persist for years to decades despite partial death of their cambium (Amoroso and Daniels 2010); Calí (1996) showed partial cambial mortality can precede tree death by 75 years.

14.2.3 Climatic Influence on Radial Growth Decline and Mortality of Trees

Tree-ring studies have shed light on the role of climatic variation as a causal agent of decline and death of A. *chilensis* trees. Mundo et al. (2010) found that the onset of a persistent reduction in radial growth in trees with external symptoms of decline (e.g., chlorotic leaves or defoliated crowns) was influenced by adverse climatic conditions during the previous growing season. Amoroso and collaborators (2015) expanded on this preliminary result and examined the role of inter-annual climatic variation as a factor contributing to the radial growth decline and death of trees. More precisely, these authors examined the effects of drought (a) on the onset of radial growth decline on a large number of dead and living trees stratified by canopy position at the onset of decline, and (b) on the time of death of the trees stratified by radial growth patterns and canopy position at death. The onset of growth decline in a large proportion of trees was significantly associated with moisture deficit. Late spring and summer droughts triggered immediate radial growth decline in overstory trees (65% of the declining or dead overstory trees). Based on the frequency of trees exhibiting an onset of decline in each year, 17 high-magnitude onset-of-decline events were shown to be concurrent with major and moderate droughts in the twentieth century (Fig. 14.1). Similarly, high-magnitude overstory-tree mortality events were concurrent with significant droughts (Amoroso et al. 2015).

Dendroecological studies have shown climatic variability contributes to stand-level forest decline in *A. chilensis* forests (Mundo et al. 2010; Amoroso et al. 2015). In an alternate interpretation, *A. chilensis* tree mortality results from complex interactions among biotic and abiotic factors, including climatic variation (La Manna and Rajchenberg 2004; Greslebin et al. 2007; El Mujtar 2009; Greslebin and Hansen 2009, 2010; Amoroso and Larson 2010; Mundo et al. 2010; Amoroso et al. 2012a, b). Amoroso and collaborators (2015) present a framework for interpreting tree decline and mortality as a forest decline process driven by multiple factors (sensu Manion 1991; Manion and Lachance 1992). Several factors may simultaneously predispose trees to decline, including site condition, genetic variation and the sex of trees, as *A. chilensisis* is a dioecious species. The occurrence of decline and mortality is not random among stands, but associated with sites at low and moderate elevations with poor soil water drainage and those with abundant precipitation (Havrylenko et al. 1989; Baccalá et al. 1998; Filip and Rosso 1999; La Manna and Rajchenberg 2004; La Manna et al. 2008a, b, 2012). Within stands, genetic variation and the sex of trees appear to be important as predisposing factors. El Mujtar (2009) found genetic differences between individuals with and without symptoms and El Mujtar et al. (2011) reported that female trees were more prone to develop symptoms of decline than male trees. Inciting or triggering factors can be abiotic and biotic agents. Dendroecological studies have demonstrated that drought is associated with the onset of decline in radial growth, potentially rendering individuals more susceptible to subsequent factors, including biotic agents (El Mujtar 2009; Mundo et al. 2010; El Mujtar et al. 2011; Amoroso et al. 2015). Other research has implicated the fungus *Phytophthora austrocedrae* (Grel. & E.M. Hansen) as an important agent of forest decline as it causes necrotic lesions in the root system and conduction tissues at the base of the trunk of *A. chilensis* trees (Greslebin et al. 2007; Greslebin and Hansen 2009, 2010; Velez et al. 2012). Both climatic variation and *P. austrocedrae* can incite or ultimately contribute to tree death, and have potential to can act synergistically. Drought can incite stress making trees more susceptible to

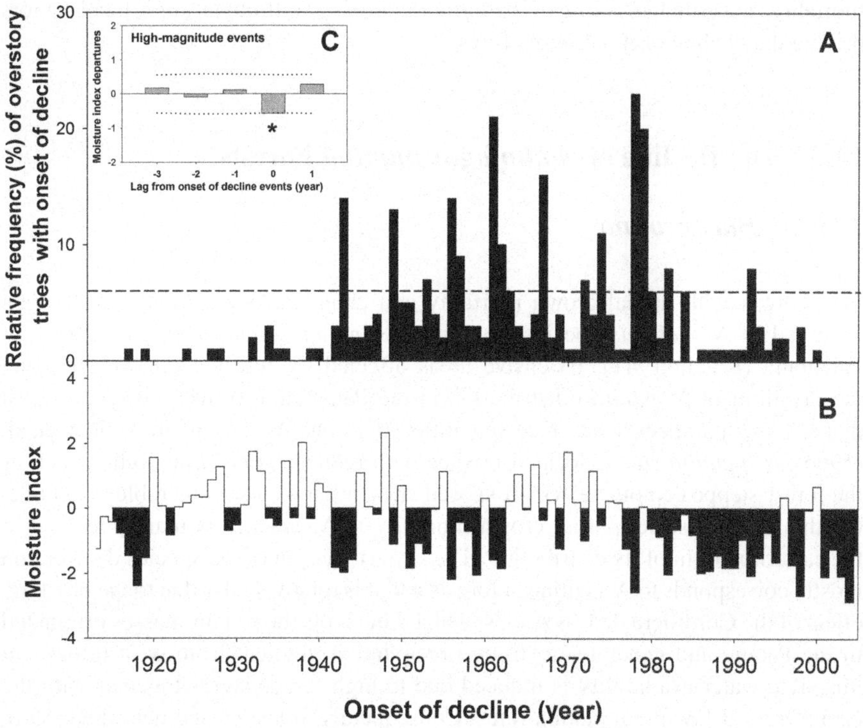

Fig. 14.1 Onset of decline of 301 overstory *Austrocedrus chilensis* trees (**a**) in comparison with the December–February moisture index (**b**) from 1914 to 2005. In the *top panel* (**a**), relative frequency is the percentage of overstory trees exhibiting onset of decline in each year relative to the number of living overstory trees that did not exhibit decline but had the potential to exhibit decline. The *horizontal dashed line* indicates the threshold value above which relative frequencies indicate high-magnitude onset of decline events. In the *bottom panel* (**b**), negative departures (*black bars*) indicate low moisture indices or drought years resulting from low precipitation and/or warm temperature; positive departures (*white bars*) indicate high moisture indices resulting from abundant precipitation and/or cool temperature. In the *upper left* inset (**c**), the association between the onset of decline of high-magnitude events and the December–February moisture is investigated. *Dashed lines* represent the 95% confidence limits based on 1000 Monte Carlo simulations. The *asterisk* indicates a statistically significant departure ($P < 0.05$) from the mean during periods of high-magnitude onset of decline

P. austrocedrae, which generates necrotic lesions that ultimately contribute to tree death (El Mujtar 2009). Alternately, drought and/or infection by *P. austrocedrae* can incite stress, causing trees to exhibit symptoms of decline and making them more susceptible to extreme droughts that ultimately contribute to their death. Amoroso et al. (2015) conclusively showed climatic variation both incites and contributes to the death of *A. chilensis* trees. Similarly, extreme droughts have resulted in widespread death of *A. chilensis* in other parts of northern Patagonia (Villalba and Veblen 1998). It remains possible that other factors predispose, incite and contribute to tree mortality; nevertheless, climatic variation and extreme drought are critical factors driving the decline of *A. chilensis* forests.

14.3 The Decline of *Nothofagus pumilio* Forests

14.3.1 Background

Large areas with partial crown mortality and abundant tree mortality have been observed in *Nothofagus* species since the beginning of the twentieth century in Patagonia (Kalela 1941). Extensive areas affected by forest decline exist along the dry limit of *N. pumilio* distribution in the Patagonian Andes; however, forest dieback in this species has received little attention. According to Veblen et al. (1996), *N. pumilio* stands located in sites with reduced water availability, such as the forest-steppe ecotone or wind-exposed sites, are the most vulnerable to decline. On the regional scale, partial crown mortality in *N. pumilio* is mainly located in the eastern margin of its distribution. The eastern margin of the species distribution mostly corresponds to Argentinean forests and it is relatively dry due to the blocking effect of the Cordillera de Los Andes, which intercepts the wet air masses originated in the Pacific and generates a strong precipitation gradient from west to east. In this area, water availability is reduced due to high temperatures together with the occurrence of low precipitation rates. Consequently, trees located near the eastern boundary of *N. pumilio* distribution are more frequently affected by water stress than those located to the west, which is translated in lower growth rates in these forests (Srur et al. 2013).

At regional to global scales, most studies have focused on the effect of climate variations on tree mortality events (Suarez et al. 2004; van Mantgem et al. 2009; McDowell et al. 2010; Michaelian et al. 2011; Anderegg et al. 2013 and references therein). Comparatively, the influence of climate as a trigger of forest decline related to partial tree mortality and growth decline has been poorly addressed (Leaphart and Stage 1971; Di Filippo et al. 2010; Liu et al. 2013). Extreme climatic events, such as temporary saturation of soils during very wet years (Ogden et al. 1993) or severe droughts (Grant 1984) have been proposed as triggers of forest decline in *Nothofagus* forests (Ogden et al. 1993). Using a combination of dendrochronological methods and instrumental data, Rodríguez-Catón and collaborators (2015, 2016) linked extreme climatic events with past changes in *N. pumilio* tree growth associated with forest decline. They showed that the onset of long-sustained periods of *N. pumilio* growth reduction was induced by adverse climate events such as droughts.

14.3.2 Integrating Stand-Level to Regional-Level Forest Decline

In *N. pumilio* stands affected by forest decline, trees with advanced crown mortality and those with reduced foliage coexist with healthy trees. As a consequence, declining *Nothofagus* forests encompass a whole range of external traits among trees or groups of trees within the same stand; therefore, the particular growth patterns of each tree may contribute differentially when characterizing decline at the stand-level (Bossel 1986). Rodríguez-Catón and collaborators (2015) described in detail the radial growth patterns of all *N. pumilio* individuals at 11 decline-affected stands in northern Patagonia. The study sites were distributed across northern Patagonia, Argentina, along a 500-km transect on the eastern boundary of *N. pumilio* distribution, between 38° and 43° S. Most selected sites were located at pure *N. pumilio* stands between 1000 and 1500 m elevation. Using principal component analysis, Rodríguez-Catón and collaborators (2015) identified the dominant growth patterns in each plot, which surprisingly diverge from the stand mean growth pattern typically calculated to represent tree growth at the stand level. At most sites, three dominant patterns of growth, expressed either as ring widths (RWs) or basal area increments (BAIs), were identified: (1) trees with a sustained reduction in radial growth starting in early 1940s; (2) trees with a gradual reduction from the late 1970s to present; and (3) trees with increasing growth rates since the 1960s. The onset of negative trends in radial growth associated with forest decline (patterns 1 and 2) occurs simultaneously at most stands and negative growth trends were present in all sampled stands. Interestingly, those trees with the highest rates of growth during the early to mid-twentieth century persistently reduced their radial growth to show the lowest rates in recent years. On the other hand, those trees with the lowest growth rates during the early to mid-twentieth century increased their radial growth in the 1960s to show the highest rates of growth at the stand-level today.

The *local growth patterns* exhibiting similar trends in radial growth were subsequently grouped to obtain *regional dominant patterns* of tree growth (Fig. 14.2a). Based on ring widths and basal area increments, two and three *regional dominant patterns* of *N. pumilio* tree growth, respectively, were identified by Rodríguez-Catón and collaborators (2015). Two BAI regional patterns show high rates of growth from early to mid-twentieth century, followed by sustained negative trends over the last 3–6 decades, respectively, whereas the third pattern is characterized by a positive trend since the 1960s (R1–R3 in Fig. 14.2a).

Since tree size or age may be related to long-term trends in radial growth, Rodríguez-Catón and collaborators (2015) calculated mean and standard deviations in the diameter at breast height (DBH) and tested the statistical differences between trees in different pattern. At most stands, trees showing negative trends in growth had significantly larger DBHs. Consistently, larger trees were more frequently associated with declining patterns at the regional scale.

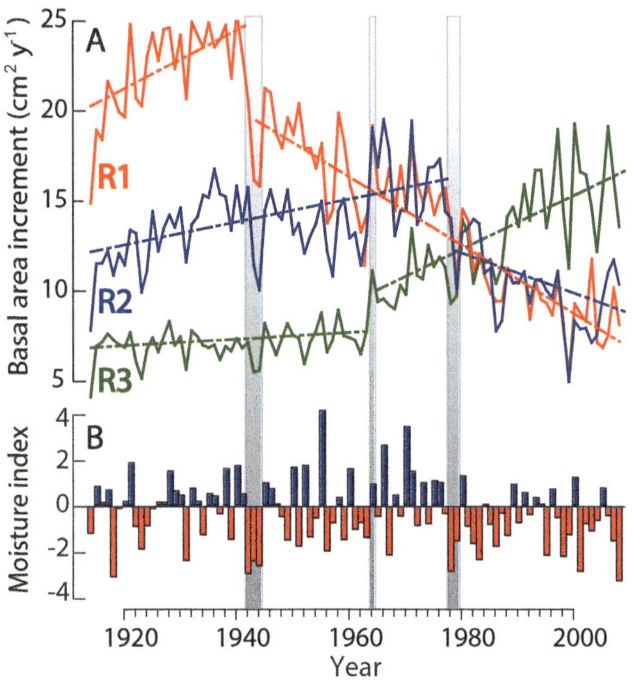

Fig. 14.2 Breakpoints recorded in regional patterns of tree growth in *Nothofagus pumilio* forests with external decline in northern Patagonia (Argentina) during the period 1914–2008. (**a**) Regional basal-area increment patterns R1, R2 and R3. Breakpoints in each growth pattern, were determined using piecewise regression models. (**b**) Barplots represent moisture index departures for December–January. Climatic conditions associated with breakpoints are highlighted with vertical bars. Adapted from Rodríguez-Catón et al. (2016)

These results illustrate several interesting features of tree response to forest decline. First, the assessment of growth variations over time can provide useful information on the temporal evolution of dieback processes and the identification of radial growth patterns in trees more prone to decline. The analysis based on the complete range of health conditions can identify distinctive growth patterns within declining stands and provide a more comprehensive view of the temporal variations in growth. Additionally, dominant growth patterns within a stand should be identified at the individual level to correctly assess dynamic processes such as forest decline.

Another interesting feature documented in Rodríguez-Catón and collaborators (Rodríguez-Catón et al. 2015) addresses the effect of the pervasive trend of decreasing ring width with increasing DBH. The decrease in ring width with age limits the ability to properly discriminate growth trends associated with forest decline versus those related to tree age. Due to this limitation, these authors concluded that patterns of tree growth related to forest decline are better discriminated using basal-area increments instead of tree-ring widths. As the estimation of BAIs takes into account the diameter of each tree, the age-related noise is eliminated, and the trends in growth associated with forest-decline are better captured.

14.3.3 Climate as a Regional Synchronizing Factor of Growth Decline

In a recent contribution, Rodríguez-Catón and collabortors (2016) compared the tree-growth patterns related to forest decline with climatic variations across the northern Patagonia Andes. The most significant changes in BAI trends over the past 100 years, determined using piecewise linear regression, occurred in 1942 and 1978 (Fig. 14.2a). For pattern R1, the positive trend in radial growth from the early twentieth century until 1942 contrasts sharply with the persistent negative trend from 1942 to the present. Similarly, a change in trend was observed for pattern R2 in 1978. In contrast to the declining R1 and R2 patterns during the past decades, the growth pattern R3 shows a positive trend that starts around 1964 and continues to the present with greater variability in recent years.

The breakpoint in growth trend of the first pattern, R1, was associated with three consecutive extreme dry-warm climate events in 1942–1943, 1943–1944 and 1944–1945 (Fig. 14.2b). Similarly, the abrupt change in growth trend shown in pattern R2 was concurrent with an extreme summer drought during 1978 (Fig. 14.2b). Both events corresponded to severe spring-summer droughts that were preceded by up to 10 years of wet periods (Rodríguez-Catón et al. 2016). In response to the humid environmental conditions from mid-late 1930s to early 1940s, canopy trees recorded the highest rates of growth but were the most affected by the dry-warm extreme event 1942–1944 (Fig. 14.2a, b). After the dry conditions prevailing during 1942–1944, trees in the pattern-R2 group also showed a reduction in the positive trend in BAI. On the other hand, R2 trees maintained a relatively stable rate of growth during the following two decades to reach the highest growth rates during the moist-cool environmental conditions during late 1960s to mid 1970s. These trees were the most affected by the dry conditions during spring-summer 1978, the driest year of the previous 36 years (Fig. 14.2a, b). In contrast, the change in trend to a sustained growth increase in pattern R3 around 1964 was not clearly related to an extreme weather event. A relatively wet and cold period from 1963 to 1977 had possibly favored the steady increase in growth of trees in R3. However, the positive growth trend has continued during the last decades, concurrent with dry and warm conditions in northern Patagonia. In this context, R3 trees may have been favored by the gradual release of resources, such as light and water, due to the progressive dieback of individuals with declining R1 and R2 patterns.

The documented changes in tree growth recorded by Rodriguez-Catón and collaborators (2016) show differences in growth responses of *N. pumilio* trees coexisting at the same stand to unfavorable weather events. Larger trees showing higher growth rates are the most susceptible to show decline in response to abrupt unfavorable climate events as dry spells. During severe droughts, water provided by roots fail to meet the water requirements of large crowns, resulting in the death of some branches and sectors of the cambium. In addition, large trees under warmer conditions increase their respiration rates and their vulnerabilities to adverse climatic events (Gillooly et al. 2001, McDowell et al. 2008, but see Reich et al. 2006). Consistent with our observations, other studies suggest that large, dominant trees are most at risk of suffering drought-driven growth decline (Piovesan et al. 2008; McDowell and Allen 2015). Therefore, the size and current rate of growth of *N. pumilio* trees are related to tree decline by extreme climatic events. In addition, our observations are consistent with the concept of "decline disease stabilizing selection" (Manion 2003). Manion (2003) stated that dominant trees are most likely to be affected by mortality to genetically stabilize forests. According to Manion (2003), short-term environmental disturbances, such as droughts, affecting highly competitive dominant trees, facilitate stress-tolerant trees to attain the upper canopy and, in consequence, breeding positions in the population.

Rodríguez-Catón and collaborators (2016) also noted a reduction in radial growth sensitivity to climate variations following the onset of decline. Statistically significant correlations between spring-summer moisture index and radial growth were observed before the dry event in both R1 and R2 patterns, whereas a loss of growth sensitivity to climatic variations takes place following the dry events. Johnson et al. (1988) postulated that the weaker responses of tree growth to climate after extreme climate events are another manifestation of forest decline. In consequence, stands with forest decline should be used cautiously for tree-ring based climatic reconstructions (Wilmking et al. 2004; Andreu et al. 2007; Büntgen et al. 2010).

Since similar trends in tree growth were recorded between 11 stands of *N. pumilio* along 500 km in northern Patagonia, Rodríguez-Catón and collaborators (2015) concluded that meso- to macro-scale environmental forcings modulate regional forest decline at regional scales.

14.3.4 Predisposing and Triggering Factors According to Manion's Spiral of Forest Decline

We have shown how severe droughts occurring after wet periods trigger the decline of large, dominant *N. pumilio* trees with high rates of growth. Following our results, we propose that tree size and age are factors related to the vulnerability to forest decline. Other predisposing factors in these forests are periods of high growth rates favored by wet periods followed by severe droughts (Rodríguez-Catón et al. 2016). Extreme droughts affecting *N. pumilio* forests along dry environments could be considered the triggering factors of forest decline. Following severe droughts, and in the absence of any observed or reported pathogenic activity, contributing factors in *N. pumilio* may be the repeated occurrence of dry events or other environmental changes.

Rodríguez-Catón and collaborators (2016) work provides valuable insights to assess the susceptibility of trees to climatic impacts. For instance, if the positive trend in radial growth recorded in R3 trees continues, resulting in large trees with fast rates of growth in the near future, these trees will be the most vulnerable to future drought-induced decline, following the pattern recorded in the declining R1 and R2 trees. Thus, early-warning indicators of forest decline in *N. pumilio*, such as big size and high growth rates, should be taken into account to predict future events of drought induced forest decline or mortality (Lenton 2011; Camarero et al. 2015).

14.4 Summary and Conceptual Models of Forest Decline

Forest declines worldwide are driven by many factors interacting in complex ways. Forest decline is often exacerbated by climate, adding complexity to the process and making more difficult to identify the causing factors (e.g. Pedersen 1998; Cherubini et al. 2002; Hartmann and Messier 2008; Camarero et al. 2015). In this context, the long-term perspective provided by tree rings have proved to be successful for disentangling the causes of forest decline (e.g. McClenahen 1995; Beier et al. 2008; Sánchez-Salguero et al. 2012). In this manuscript we present recent dendroecological studies that were used to determine the influences of climatic variation on the radial growth patterns and death of trees in declining *A. chilensis* and *N. pumilio* forests in northern Patagonia. Furthermore, these results were used to distinguish between possible causes and interactions of abiotic and biotic stress factors and stand development processes in the development of the forest declines. For the studied forests we found complex interactions between abiotic and biotic factors acting at different spatial and temporal scales that predispose trees, and incite and contribute to the development of the symptoms leading to the subsequent death of trees.

We summarize our findings within the framework of a the conceptual model (spiral) of forest decline (sensu Manion 1991; Manion and Lachance 1992) presenting processes driven by interactions between abiotic and biotic stress factors that predispose, incite and contribute to stand-level decline in these forests (Fig. 14.3). Site conditions, genetic variation and sex of trees are predisposing factors in the development of the *A. chilensis* decline while site location, wet periods and tree size predispose the occurrence of decline in *N. pumilio* forests. In both forest declines, extreme droughts incite or trigger immediate changes in the radial-growth patterns of trees resulting in consistent declining growth rates over time. For *A. chilensis*, particularly, drought-induced stress could become susceptible to the pathogen *P. austrocedrae*, which would act as a secondary agent of decline but could also act independently affecting healthy trees. Ultimately, adverse conditions due to

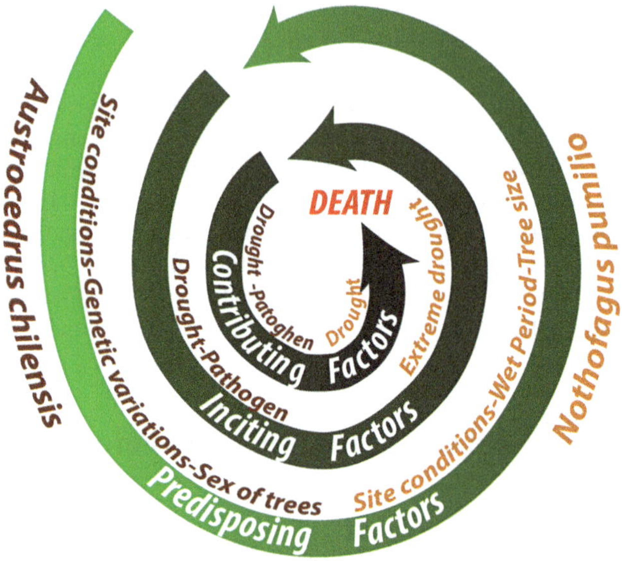

Fig. 14.3 The Manion decline Spiral depicting abiotic and biotic factors that predispose, incite and contribute to stand-level decline and tree death in *A. chilensis* and *N. pumilio* forests in northern Patagonia

climatic variability will contribute leading to tree death in both forests; alternately, the lesions caused by *P. austrocedrae* in *A. chilensis* would eventually kill the trees. Our frameworks formalize the alternate hypothesis to single causal agents of decline in these forests by emphasizing the role and importance of climatic variability as a driver of decline.

There is increasing concern about the effects of climatic variability on forest ecosystems given recently documented climatic change and forecast global warming in the twenty-first century (Dale et al. 2001; Adams et al. 2009; Allen 2009). Numerous studies have documented increased rates of tree mortality and forest decline in diverse forests worldwide seemingly due to the occurrence of droughts in recent decades (Williamson et al. 2000; Guarin and Taylor 2005; Allen et al. 2010; van Mantgem et al. 2009). Furthermore, the effects of global warming are usually exacerbated by extreme events which are difficult to predict but are expected to become more common in the future (IPCC 2013). Climatic predictions suggest an increase in the frequency and intensity of drought events not only in areas where *A. chilensis* and *N. pumilio* forests experience stand-level decline but also in unaffected areas. Based on these considerations, we predict the stand-level processes of decline above described for *A. chilensis* and *N. pumilio* Patagonian forests will exacerbate in the future.

References

Allen CD (2009) Climate-induced forest dieback: an escalating global phenomenon? Unasylva 60:43–49

Adams HD, Guardiola-Claramonte M, Barron-Gafford GA (2009) Temperature sensitivity of drought-induced tree mortality: implications for regional die-off under globalchange-type drought. Proc Natl Acad Sci U S A 106:7063–7066

Allen CD, Macalady AK, Chenchouni H et al (2010) A global overview of drought and heat-induced tree mortality reveals emerging climate change risks for forests. For Ecol Manag 259:660–684

Amoroso MM, Daniels LD (2010) Cambial mortality in declining *Austrocedrus chilensis* forests: implications for stand dynamics studies. Can J For Res 40:885–893

Amoroso MM, Larson BC (2010) Stand development patterns as a consequence of the mortality in *Austrocedrus chilensis* forests. For Ecol Manag 259:1981–1992

Amoroso MM, Daniels LD, Larson BC (2012a) Temporal patterns of radial growth in declining *Austrocedrus chilensis* forests in Northern Patagonia: the use of tree-rings as an indicator of forest decline. For Ecol Manag 265:62–70

Amoroso MM, Suarez ML, Daniels LD (2012b) *Nothofagus dombeyi* regeneration in declining *Austrocedrus chilensis* forests: effects of overstory mortality and climatic events. Dendrochronologia 30:105–112

Amoroso MM, Daniels LD, Villalba R et al (2015) Does drought incite tree decline and death in *Austrocedrus chilensis* forests? J Veg Sci 26:1171–1183

Anderegg WRL, Kane JM, Anderegg LDL (2013) Consequences of widespread tree mortality triggered by drought and temperature stress. Nat Clim Chang 3:30–36

Andreu L, Gutiérrez E, Macias M et al (2007) Climate increases regional tree-growth variability in Iberian pine forests. Glob Chang Biol 13:804–815

Baccalá N, Rosso P, Havrylenko M (1998) *Austrocedrus chilensis* mortality in the Nahuel Huapi National Park (Argentina). For Ecol Manag 109:261–269

Barrotaveña C, Rajchenberg M (1996) Hongos *Aphyllophorales* (Basidiomycetes) que causan pudriciones en *Austrocedrus chilensis* en pie. Bol Soc Argent Bot 31:201–216

Beier CM, Sink SE, Hennon PE et al (2008) Twentieth-century warming and the dendroclimatology of declining yellow-cedar forests in southeastern Alaska. Can J For Res 38:1319–1334

Bigler C, Bugmann H (2003) Growth-dependent tree mortality models based on tree rings. Can J For Res 33:210–221

Bossel H (1986) Dynamics of forest dieback: systems analysis and simulation. Ecol Model 34:259–288

Büntgen U, Frank D, Trouet V et al (2010) Diverse climate sensitivity of Mediterranean tree-ring width and density. Trees 24:261–273

Calí SG (1996) *Austrocedrus chilensis*: estudio de los anillos de crecimiento y su relación con la dinámica del "Mal del Ciprés" en el P.N. Nahuel Huapi, Argentina. Undegraduate Thesis, Universidad Nacional del Comahue

Camarero JJ, Gazol A, Sangüesa-Barreda G et al (2015) To die or not to die: early warnings of tree dieback in response to a severe drought. J Ecol 103:44–57

Cherubini P, Fontana G, Rigling D et al (2002) Treelife history prior to death: two fungal root pathogens affect tree-ring growth differently. J Ecol 90:839–850

Condit R, Hubbell SP, Foster RB (1995) Mortality rates of 205 neotropical tree and shrub species and the impact of severe drought. Ecol Monogr 65:419–439

Dale VH, Joyce LA, McNulty S et al (2001) Climate change and forest disturbances. BioSci 51:723–734

Di Filippo A, Alessandrini A, Biondi F et al (2010) Climate change and oak growth decline: Dendroecology and stand productivity of a Turkey oak (*Quercus cerris* L.) old stored coppice in Central Italy. Ann For Sci 67:706–706

El Mujtar V (2009) Análisis integrado de factores genéticos, bióticos y abióticos para la formulación de una nueva hipótesis sobre la etiología del "mal del ciprés". Dissertation, Universidad Nacional de la Plata

El Mujtar V, Andenmatten E, Perdomo M et al (2011) Temporal progression trends of cypress mortality at permanent plots in a National forest reserve of *Austrocedrus chilensis* (Patagonia, Argentina). For Syst 20:209–217

Filip GM, Rosso PH (1999) Cypress mortality (mal del ciprés) in the Patagonian Andes: comparisons with similar forest diseases and declines in North America. Eur J For Path 29:89–96

Franklin JF, Shugart HH, Harmon ME (1987) Tree death as an ecological process. Bioscience 27:550–556

Gillooly JF, Brow JH, West GB et al (2001) Effects of size and temperature on metabolic rate. Science 293:2248–2251

Grant PJ (1984) Drought effect on high-altitude forests, Ruahine Range, North Island, New Zealand. N Z J Bot 22:15–27

Greslebin AG, Hansen EM (2009) The decline of *Austrocedrus* forests in Patagonia (Mal del Ciprés): another *Phytophthora*-caused forest disease. In: Goheen EM, Frankel SJ (eds) Proceedings of the fourth meeting of the International Union of Forest Research Organizations (IUFRO) Working Party S07.02.09, *Phytophthoras* in forests and natural ecosystems, Albany, 2009

Greslebin AG, Hansen EM (2010) Pathogenicity of *Phytophthora austrocedrae* on *Austrocedrus chilensis* and its relation with "Mal del Ciprés" in Patagonia. Plant Pathol 59:604–612

Greslebin A, Hansen E, Sutton W (2007) *Phytophthora austrocedrae* sp. nov., a new species associated with *Austrocedrus chilensis* mortality in Patagonia (Argentina). Mycol Res 11:308–316

Guarin A, Taylor AH (2005) Drought triggered tree mortality in mixed conifer forests in Yosemite National Park, California, USA. For Ecol Manag 218:229–244

Hartmann H, Messier C (2008) The role of forest tent caterpillar defoliations and partial harvest in the decline and death of sugar maple. Ann Bot 10:377–387

Havrylenko M, Rosso PH, Fontenla SB (1989) *Austrocedrus chilensis*: contribución al estudio de su mortalidad en Argentina. Bosque 1:29–36

Hennon P, Rajchenberg M (2000) El "mal del ciprés". Algunas observaciones, comparaciones e ideas. CIEFAP-Patagonia Forestal 2:4–6

Hennon PE, Hansen EM, Shaw CGIII (1990) Dynamics of decline and mortality of Chamaecyparis nootkatensis in southeast Alaska. Can J Bot 68:651–662

Houston DR (1981) Stress triggered tree diseasesthe diebacks and declines. USDA Forest Service, Broomall

Hranilovic S (1988) Informe histórico sobre el mal del ciprés de la cordillera (*Austrocedrus chilensis*). Rev Asoc For Argent 42:58–62

IPCC (2013) Climate change 2013: the physical science basis. Contribution of Working Group I to the Fifth Assessment Report of the Intergovernmental Panel on Climate Change. Cambridge University Press, Cambridge

Johnson AH, Cook ER, Siccama TG (1988) Climate and red spruce growth and decline in the northern Appalachians. Proc Natl Acad Sci U S A 85:5369–5373

Jump AS, Hunt JH, Peñuelas J (2006) Rapid climate change-related growth decline at the southern range edge of Fagus sylvatica. Glob Chang Biol 12:2163–2174

Kalela E (1941) Über die holzarten und die durch die klimatischen Verhältnisse verursachten Holzartenwechsel in den Wäldern Ostpatagoniens. Helsinki, Annales Academiae Scientiarum Fennicae

Kobe RK, Pacala SW, Silander JA et al (1995) Juvenile tree survivorship as a component of shade tolerance. Ecol Appl 5:517–532

La Manna L, Rajchenberg M (2004) The decline of *Austrocedrus chilensis* forests in Patagonia, Argentina: soil features as predisposing factors. For Ecol Manag 190:345–357

La Manna L, Matteucci SD, Kitzberger T (2008a) Abiotic factors related to the incidence of the *Austrocedrus chilensis* disease syndrome at a landscape scale. For Ecol Manag 256:1087–1095

La Manna L, Collantes M, Bava J et al (2008b) Natural regeneration in declining *Austrocedrus chilensis* forests of Patagonia. Ecol Austral 18:27–41
La Manna L, Matteucci SD, Kitzberger T (2012) Modelling *Phytophthora* disease risk in *Austrocedrus chilensis* forests of Patagonia. Eur J For Res 131:323–337
Leaphart CD, Stage AR (1971) A factor in the origin of the pole blight disease of *Pinus monticola* Dougl. Ecology 52:229–239
LeBlanc DC, Raynal DJ (1990) Red spruce decline on Whiteface Mountain, New York. II. Relationships between apical and radial growth decline. Can J For Res 20:1415–1421
Lenton TM (2011) Early warning of climate tipping points. Nat Clim Chang 1:201–209
Liu H, Williams AP, Allen CD et al (2013) Rapid warming accelerates tree growth decline in semi-arid forests of Inner Asia. Glob Chang Biol 19:2500–2510
Loguercio GA, Urretavizcaya MF, Rey M et al (1998) El "Mal del ciprés" como condicionante de la silvicultura del ciprés de la cordillera *Austrocedrus chilensis* (D. Don) Florin et Boutelje en el norte de la Patagonia Argentina. Paper presented at the I Congreso Latinoamericano IUFRO, Valdivia. Chile 1988.
Lutz JA, Halpern CB (2006) Tree mortality during early forest development: a long-term study of rates, causes, and consequences. Ecol Monogr 76:257–275
Manion PD (1991) Tree disease concepts. Prentice-Hall, Englewood Cliffs
Manion PD (2003) Evolution of Concepts in Forest Pathology. Phytopathology 93:1052–1055
Manion PD, Lachance D (1992) Forest decline concepts: an overview. In: Manion PD, Lachance D (eds) Forest decline concepts. American Phytopathological Society, St. Paul, pp 181–190
Mattson WJ, Haack RA (1987) The role of drought in outbreaks of plant-eating insects. Bioscience 37:110–118
McClenahen JR (1995) Potential dendroecological approaches to understanding forest decline. In: Lewis TE (ed) Tree rings as indicators of ecosystem health. CRC Press, Boca Raton, pp 59–79
McDowell NG, Allen CD (2015) Darcy's law predicts widespread forest mortality under climate warming. Nat Clim Chang 20:12–20
McDowell NG, Pockman WT, Allen CD et al (2008) Mechanisms of plant survival and mortality during drought: why do some plants survive while others succumb to drought? New Phytol 178:719–739
McDowell NG, Allen CD, Marshall L (2010) Growth, carbon-isotope discrimination, and drought-associated mortality across a *Pinus ponderosa* elevational transect. Glob Chang Biol 16:399–415
Michaelian M, Hogg EH, Hall RJ et al (2011) Massive mortality of aspen following severe drought along the southern edge of the Canadian boreal forest. Glob Chang Biol 17:2084–2094
Minorsky PV (2003) The decline of sugar maples (*Acer saccharum*). Plant Physiol 133:441–442
Mueller-Dombois D (1983) Canopy dieback and successional processes in pacific forests. Pac Sci 37:317–325
Mueller-Dombois D, Canfield JE, Holt RA et al (1983) Tree-group death in North American and Hawaiian forests: a pathological problem or a new problem for vegetative ecology. Phytocoenologia 11:117–137
Mundo IA, El Mujtar VA, Perdomo MH et al (2010) *Austrocedrus chilensis* growth decline in relation to drought events in northern Patagonia, Argentina. Trees 24:561–570
Ogden J, Lusk CH, Steel MG (1993) Episodic mortality, forest decline and diversity in a dynamics landscape: Tongariro National Park, New Zealand. In: Huettl RF, Mueller-Dombois D (eds) Forest decline in the Atlantic and Pacific regions. Springer, Berlin, pp 261–274
Oliver CD, Larson BC (1996) Forest stand dynamics. Wiley, New York
Pedersen BS (1998) Modeling tree mortality in response to short- and long-term environmental stresses. Ecol Model 105:347–351
Piovesan G, Biondi F, Di Filippo A et al (2008) Drought-driven growth reduction in old beech (*Fagus sylvatica* L.) forests of the central Apennines, Italy. Glob Chang Biol 14:1–17
Rajchenberg M, Cwielong P (1993) El mal del ciprés (*Austrocedrus chilensis*): su relación con las pudriciones radiculares y el sitio. In: Abstracts of the Congreso Forestal Argentino y Latinoamericano, Paraná, Entre Ríos, Argentina, 1993

Rajchenberg M, Barroetaveña C, Cwielong PP et al (1998) Fungal species associated with the decline of *Austrocedrus chilensis* in Patagonia, Argentina: Preliminary results. In: Abstracts of the 9th International Conference on Root and & butt Rots. Carcans, Francia, 1998.

Reich PB, Tjoelker MG, Machado J-L, Oleksyn J (2006) Universal scaling of respiratory metabolism, size and nitrogen in plants. Nature 439:457–461

Relva MA, Westerholm CL, Kitzberger T (2009) Effects of introduced ungulates on forest understory communities in northern Patagonia are modified by timing and severity of stand mortality. Plant Ecol 201:11–22

Rodríguez-Catón M, Villalba R, Srur AM et al (2015) Long-term trends in radial growth associated with *Nothofagus pumilio* forest decline in Patagonia: integrating local- into regional-scale patterns. For Ecol Manag 339:44–56

Rodríguez-Catón M, Villalba R, Morales MS et al (2016) Influence of droughts on *Nothofagus pumilio* forest decline across northern Patagonia, Argentina. Ecosphere 7:e01390

Rosso PH, Baccalá NB, Havrylenko M et al (1994) Spatial pattern of *Austrocedrus chilensis* wilting and the scope of autocorrelation analysis in natural forests. For Ecol Manag 67:273–279

Ryan DAJ, Allen OB, McLaughlin DL et al (1994) Interpretation of sugar maple (Acer saccharum) ring chronologies from central and southern Ontario using a mixed linear model. Can J For Res 24:568–575

Sánchez-Salguero R, Navarro-Cerrillo RM, Camarero JJ et al (2012) Selective drought-induced decline of pine species in southeastern Spain. Clim Chang 113:767–785

Srur A, Villalba R, Rodríguez-Catón, M et al (2013) Differences in Nothofagus pumilio forest structure and tree-ring growth along altitudinal gradients at different aspect in Southern Patagonia. Paper presented at the II AmeriDendro Conference, University of Arizona, Tucson, 2013

Suarez ML, Ghermandi L, Kitzberger T (2004) Factors predisposing episodic drought-induced tree mortality in *Nothofagus*—site, climatic sensitivity and growth trends. J Ecol 92:954–966

van Mantgem PJ, Stephenson NL, Byrne JC et al (2009) Widespread increase of tree mortality rates in the western United States. Science 323:521–524

Varsavsky E, Bettucci L, Rodríguez García D, Gómez C (1975) Observaciones preliminares sobre la mortalidad del ciprés (Austrocedrus chilensis) en los bosques patagónicos. Fundación Patagonia, San Carlos de Bariloche

Veblen TT, Kitzberger T, Burns BR et al (1996) Perturbaciones y regeneración en bosques andinos del sur de Chile y Argentina. In: Armesto JJ, Arroyo MK, Villagrán C (eds) Ecología del Bosque Nativo de Chile. Universidad de Chile Press, Santiago, pp 169–198

Velez ML, Silva PV, Troncoso OA et al (2012) Alteration of physiological parameters of *Austrocedrus chilensis* by the pathogen *Phytophthora austrocedrae*. Plant Pathol 61:877–888

Villalba R, Veblen TT (1998) Influences of large-scale climatic variability on episodic tree mortality at the forest-steppe ecotone in northern Patagonia. Ecology 79:2624–2640

Wang W, Peng C, Kneeshaw DD et al (2012) Drought-induced tree mortality: ecological consequences, causes, and modeling. Environ Rev 20:109–121

Waring RH (1987) Characteristics of trees predisposed to die. Bioscience 37:569–574

White PS, Pickett STA (1985) Natural disturbance and patch dynamics: an introduction. In: Pickett STA, White PS (eds) The ecology of natural disturbance and patch dynamics. Academic Press, Inc., Orlando, pp 3–13

Williamson GB, Laurance WF, Oliveira AA et al (2000) Amazonian tree mortality during the 1997 El Niño drought. Conserv Biol 14:1538–1542

Wilmking M, Juday G, Barber V et al (2004) Recent climate warming forces contrasting growth responses of white spruce at treeline in Alaska through temperature thresholds. Glob Chang Biol 10:1–13

Part IV
Human-Environment Interactions

Credit: A. Hessl

Chapter 15
Hydraulic Cities, Colonial Catastrophes, and Nomadic Empires: Human-Environment Interactions in Asia

Amy E. Hessl, Caroline Leland, Thomas Saladyga, and Oyunsanaa Byambasuren

Abstract Recent and projected changes in climate highlight the need to understand and predict human-environment interactions. However the diversity of cultures, livelihoods, and political formations today and in the past indicate that relationships between climate, ecosystems, and societies are likely non-linear, complex, and variable over time. A growing network of multi-millennial, absolutely dated annual tree-ring records from Asia provide climatic context for several important historic events which emphasize the diversity of human-environment interactions. Tree rings are an ideal proxy for understanding these relationships due to their extensive spatial coverage, temporal resolution relevant to human systems, and their ability to integrate climate in ways that relate directly to human livelihoods. Herein, we focus on five examples of human-environment interactions in two climatically and culturally distinct regions of Asia: Monsoon and Arid Central Asia. Over the last three millennia, societies have adjusted to climate variability in diverse and (mal)adaptive ways. In Monsoon Asia, drastic swings in moisture availability, notably megadroughts associated with monsoon failure, interacted with socio-political and technical institutions to spur the disintegration of the fourteenth century Khmer kingdom at Angkor and engender continental scale famine in nineteenth

A.E. Hessl (✉)
Department of Geology and Geography, West Virginia University, 98 Beechurst Ave., Morgantown, WV, USA 26501
e-mail: amy.hessl@mail.wvu.edu

C. Leland
Lamont-Doherty Earth Observatory, Columbia University 61 Route 9W, Palisades, NY 10964-1000, USA
e-mail: cleland@ldeo.columbia.edu

T. Saladyga
Department of Geography, Concord University, 1000 Vermillion Street, Athens, WV 24712, USA
e-mail: saladygat@concord.edu

O. Byambasuren
National University of Mongolia, University Str. 3, Ulaanbaatar 14201, Mongolia
e-mail: oyunsanaa@seas.num.edu.mn

M.M. Amoroso et al. (eds.), *Dendroecology*, Ecological Studies 231,
DOI 10.1007/978-3-319-61669-8_15

century colonial Asia. In Arid Central Asia, elevated temperatures and moisture were both a boon and a limitation for historic and modern nomadic pastoralists, depending on the historical context in which climatic events occurred. Future efforts to statistically model and predict human-environment relationships over the arc of human history in Asia will need to account for the diversity of economic, political, and cultural features that filter, dampen, and amplify the effects of climate change on society.

Keywords human-environment interaction • drought • Monsoon Asia • Arid Central Asia • historical climatology

15.1 Introduction

The study of human-environment relationships has been central to tree-ring science for nearly a century (Douglass 1921, 1929; Wissler 1921). During this time, dendrochronological investigations of interactions between society, climate, and ecosystems have primarily centered on three broad topics: (1) tree-ring dating of archeological sites and structures; (2) complex societies and climatic variability; and (3) ecosystem interactions with human activities. In recent decades, concerns about climate change and its diverse effects on societies have renewed interest in how past peoples responded to and interacted with environmental variability. Despite major differences in technology, energy sources, and modes of production, studies of past civilizations and climate change can help us understand resilience and vulnerability of society to future climate (deMenocal 2001). At the same time, current responses and reactions to climate change may help us understand the actions and decisions of human actors in the past (Orlove 2005; Adger et al. 2013). Similar inferences can be made by studying how and to what extent humans modified past ecosystems through fire, agriculture, and other means (Swetnam et al. 1999). These inferences help us contextualize the state of current ecosystems and their potential future trajectories, suggesting management strategies that might be applied today.

Human-environment interactions are multi-directional and often complex relationships between human actors, human institutions, climate, and the environment (Endfield 2012; Adger et al. 2013). Human systems depend on and respond to changing environmental conditions as well as adapt to and modify those conditions. This creates the potential for feedbacks to both the human system and the environment. The diversity of human cultures and actors present today and in the past suggests that human responses, adaptations, and modifications of the environment are not consistent over time. Like the climate system and its forcing mechanisms (Lorenz 1968), the same climatic or environmental driver can have differential effects on societies, particularly as cultures evolve and change over time. These effects and interactions vary across different segments of societies. While powerful elites may benefit from environmental changes, the masses of agricultural peasants may suffer (Davis 2002; Orlove 2005). Societies respond to a variety of forces and stressors, including climate, making it difficult to disentangle climate from other drivers of change. These interactions and complexities make it challenging, though not impossible, to work in a traditional hypothesis testing mode. Individual events or societies cannot be treated as sample units and there are no established metrics of climate response that can be applied or derived for different societies across culture, space, and time. Thus, detailed case studies with local and appropriate paleoenvironmental proxies have been the primary means of understanding relationships between humans and their environment in the pre-instrumental period.

Tree-ring records offer several distinct advantages over many other paleoecological data sources for studying past human-environment interactions. Because they are annually dated, tree rings and their many proxies can be directly related to calendar-dated historical and archeological materials (e.g. inscriptions, journals, births and deaths, tax and trade records). Further, the temporal signals stored in tree-ring records (typically seasonal to centennial), allow us to identify climatic excursions that occur on time frames meaningful for people and ecosystems. Similar arguments can be made for the overlapping spatial scales represented by tree-ring records of past climate, ecosystem processes, and human activities. The abundance of trees with annual rings, particularly in temperate and high latitudes, allows for spatially gridded reconstructions of past temperature or soil moisture capable of resolving climatic features and phenomena at regional to continental spatial scales. Perhaps the most relevant contribution of tree rings to the study of society and climate is that because trees are terrestrial plants, tree growth integrates many of the same environmental conditions important for the production of crops, domesticated animals, and other organisms on which people depend (e.g. Stahle et al. 1988; Pederson et al. 2013, 2014; Bocinsky and Kohler 2014).

Like other paleoclimatic and paleoecological approaches, tree-ring studies of human-environment relationships are vulnerable to some important pitfalls, including neo-environmental determinism (Coombes and Barber 2005), the logical fallacy of correlation meaning causation, and other 'reductionist' analyses and interpretations that simplify the complex relationships between people and their environments (Erickson 1999). These critiques are valid as many studies have ignored social and cultural complexity, human agency, historical legacies, and myriad other interesting relationships between people and their environments, all worthy of investigation. These concerns are not just theoretical as they may strongly affect the way we as a society formulate and respond to future climate (Hulme 2011). Provided these limitations are addressed and explored, the strengths of dendrochronology, including precise dating, high fidelity to climate, and environmental integration, allow for detailed case studies capable of responding to these critiques by adding a new source of historical and archeological data, directly tied to human time scales and ecologies. Historical events taking place over months to decades can be situated within the climate—and even the weather—under which they occurred, sometimes revealing unexpected responses and relationships.

With its long human history, diverse ecosystems, and recent economic growth, Asia has been the subject of a growing body of paleoecological research and data development over the last few decades. These efforts have culminated in two large-scale, gridded climate reconstructions as well as several multi-millennial records of past climate. Of particular note is the Monsoon Asia Drought Atlas (MADA) which has been used to explore the hydroclimatic context of several historical events (Cook et al. 2010). Similarly, a network of 422 tree-ring chronologies (Asia 2k) has been used to reconstruct summer temperatures over the past 1200 years, showing persistent periods of cooling (e.g. Little Ice Age; 1350–1880 CE) and warming (e.g. Medieval Climate Anomaly; 850–1050 CE and twentieth century warming) over the continent with important social and economic consequences (Cook et al. 2013). Where site replication is ample, these spatially explicit reconstructions can provide a climatic backdrop for diverse cultures and livelihoods across Asia. Further, Asia hosts several ultra-long chronologies of past moisture extending more than 1000 years (Buckley et al. 2010; Myglan et al. 2012a; Pederson et al. 2014), suitable for addressing questions about climate variability and society over the Common Era.

Here we focus on five case studies of human-environment interactions during the last 3000 years bound by two major geographic regions in Asia: Monsoon Asia and Arid Central Asia (Fig. 15.1). These regions are climatologically distinct

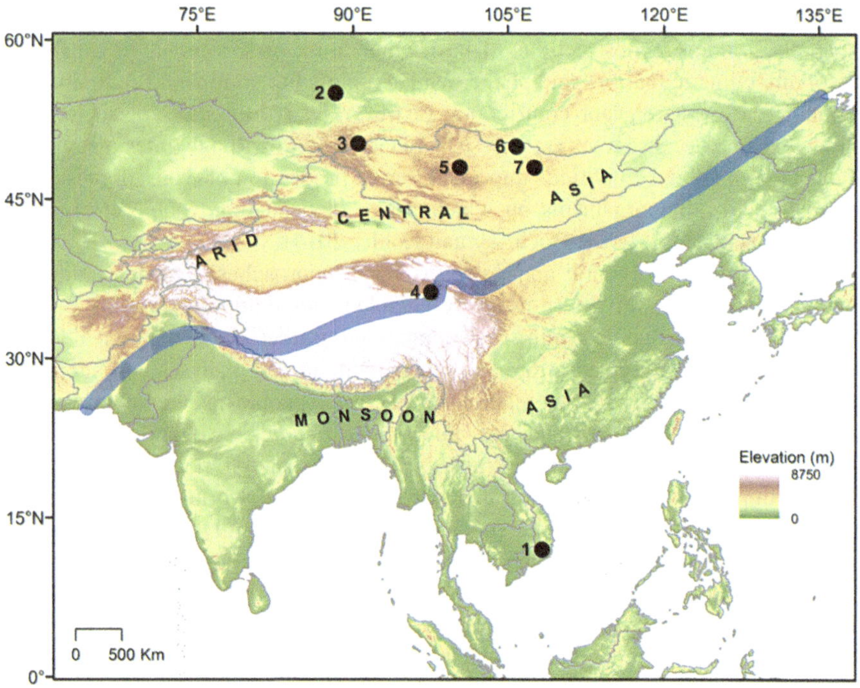

Fig. 15.1 Monsoon and Arid Central Asia regions based on Chen et al. 2008 (*blue* band). Location of tree-ring study sites (numbered circles) including: (*1*) *Fokienia hodginsii* sites (Angkor); (*2 & 3*) *Larix sibirica* (Siberian Scythians); (*4*) *Sabina przewalskii* (Xiatatu Tombs); (*5*) *Pinus sibirica* (Mongol Empire); (*6*) *Pinus sylvestris* and *L. sibirica* (Mongolian Breadbasket); (*7*) *P. sylvestris and L. sibirica* (fire history)

and allow the exploration of two cultural counterpoints: the largely sedentary agricultural societies of Monsoon Asia and the predominantly nomadic pastoral societies of Arid Central Asia. Monsoon Asia is influenced by the strength of the Asian Monsoon systems, with strong seasonal rainfall regimes capable of supporting extensive and complex agricultural societies. For Monsoon Asia, we review two case studies that highlight adaptations and vulnerabilities of ancient and historical agricultural societies: (1) hydroclimatic variability and the decline of Angkor and (2) Asian monsoon failure and the late Victorian droughts. In Arid Central Asia, we explore the continental and arid to semi-arid region of northern Asia where several complex and expansive empires arose from nomadic pastoralists during the last 3000 years. We also explore the interactions between recent socio-economic and environmental change. For Arid Central Asia, we review: (1) dendroarcheology and the development of ultra-long chronologies; (2) climate and environment of the Mongol Empire; and (3) socio-economic and environmental synergies in modern Mongolia. All of these case studies provide opportunities to investigate the diverse ways climate and environment have interacted with culture and society during the last 3000 years and point to the diverse and changing relationships between climate, environment, and society.

15.2 Monsoon Asia

Moisture variability in Monsoon Asia is influenced by the relative strength of the dominant monsoon systems, which are linked to land-sea temperature imbalances, resulting in strong, seasonal moisture delivery across much of the continent. Tree-ring data have been used to infer monsoon activity across this region over centuries to millennia. In particular, tree-ring reconstructions identified periods of monsoon failure and their associated megadroughts (Cook et al. 2010). Beyond ring width, other tree-ring parameters, such as stable isotopes and latewood density have been used to infer changes in monsoon intensity over time (e.g. Bräuning and Mantwill 2004; Grießinger et al. 2011), but have yet to be applied directly to human history. It is important to recognize that transitional regions between Monsoon Asia and Arid Central Asia, such as the Tibetan Plateau (TP), can be influenced by multiple climate systems, and these systems can vary in influence over time and space (e.g. Yang et al. 2011; An et al. 2012). As shown in the following case studies, variations in monsoonal moisture have played an important role in the development and disintegration of past civilizations across southern Asia.

15.2.1 Hydroclimatic Variability and the Decline of Angkor

The decline of the "hydraulic city" of Angkor, the capitol of the Khmer kingdom in Cambodia during the ninth to fourteenth centuries, has been the subject of great debate for over a century. In particular, there has been controversy over the causes, rate, and timing of Angkor's disintegration. The Khmer kingdom ruled over a large swath of continental Southeast Asia deriving much of its success by capitalizing on Summer Asian Monsoon rainfall to support complex agriculture, including massive irrigation complexes for which the ancient city of Angkor is renown. Given their elaborate irrigation systems, the Khmer would have been vulnerable to the vicissitudes of the monsoon; however until recently, few (if any) local records of past climate extended to the time of the Khmer, therefore studies of their demise have emphasized political, economic, and technological factors. Addressing this lack of paleoclimatic data, Buckley and others developed two tree-ring width reconstructions of past hydroclimate from the rare Fujian cypress, *Fokienia hodginsii*, that extended 979 years and covered the period of the Khmer kingdom (Buckley et al. 2007, 2010; Sano et al. 2008). Their reconstruction revealed that decades-long periods of weak Asian summer monsoons and related drought have occurred over the past millennium. One of these periods of weakened monsoon and drought in the mid- to late fourteenth century and another more severe drought in the early fifteenth century correspond to the time of Angkor's demise (Fig. 15.2). The fourteenth century droughts were referenced in historical chronicles of the time (Thera 1962) suggesting a direct impact on the Khmer, but the importance

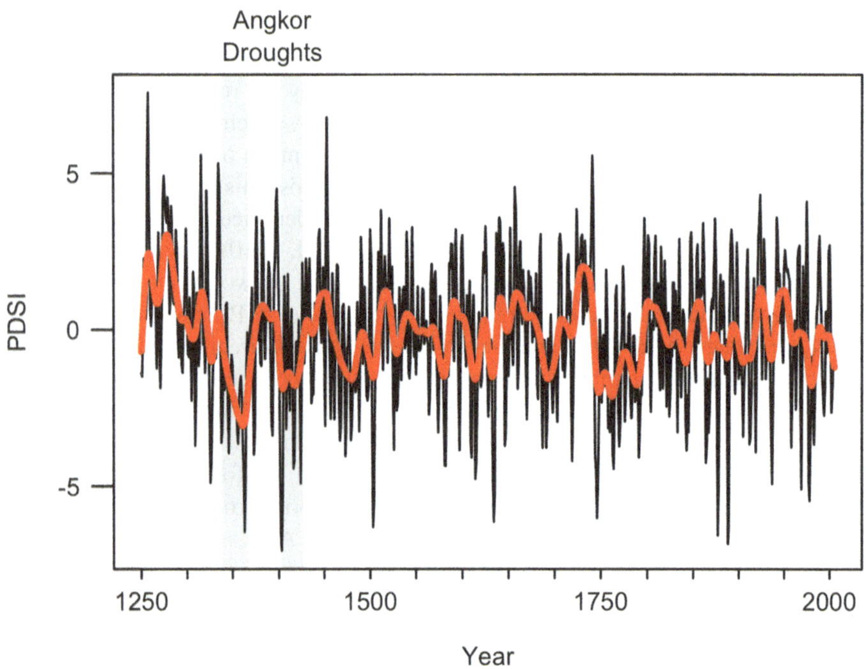

Fig. 15.2 Tree-ring reconstructed PDSI developed from *Fokienia hodginsii* trees in Vietnam (Buckley et al. 2010). The gray bars highlight the severe Angkor droughts of the mid fourteenth and early fifteenth centuries (adapted from Buckley et al. (2010))

of these droughts and intervening floods was not recognized until the *Fokienia* reconstructions were available.

Buckley et al. (2010) inferred that the complex nature of Angkor's canals and reservoirs, though highly adaptive, left the city inflexible and vulnerable to the droughts of the fourteenth and fifteenth centuries. Evidence of occasional extreme floods and depositional events in the canals during these droughts pointed to a suite of climatic conditions that pushed the hydraulic city beyond its adaptive capacity. Though previous archeological research on the canal building at Angkor suggested that the elaborate engineering had led to the decline of the city (Evans et al. 2007), it was the combination of annual tree-ring records, historical documents, and radiocarbon dated depositional material that allowed Buckley et al. (2010) to describe the precise timing of episodic floods and low flow that led to the failure of the irrigation systems. Following fourteenth and fifteenth century droughts and floods, the regional economic focus of the Khmer shifted from intensive agriculture to maritime trade, in part due to the climatic extremes, but also as a result of changing geopolitical forces and economic pressures. Rather than a single, deterministic "collapse" caused by drought, the Khmer appeared to have exhausted their adaptive capacity and realigned the empire in response to a combination of environmental variability and socio-political factors. The hydraulic innovations of the early centuries of the empire became maladaptive under monsoon failure, flood, and socio-economic change. The limits of innovation under multiple environmental and social stressors is a critical inference highly relevant to today's social and environmental challenges.

15.2.2 Spatial and Temporal Context of the Late Victorian Drought

The monsoon failures and associated droughts that were critical for the development and the disintegration of the Khmer, affected Monsoon Asia repeatedly in recorded history with massive consequences for millions of people dependent on seasonal rainfall for agriculture. During the late Victorian drought of 1876–1878, up to 30 million people are thought to have died across Monsoon Asia when the monsoons failed repeatedly. To what extent was this famine caused by climate versus major socio-econmic changes after centuries of colonialism? By the late 1800s, colonial-era imperialism had disrupted local food production throughout the region leaving communities highly vulnerable to local environmental stress and the fluctuations of distant markets (Davis 2002). The late Victorian drought occurred during one of the most severe El Niño events of the last 150 years (Davis 2002) and strong El Niño events have been linked to monsoon failure (Walker and Bliss 1932; Maraun and Kurths 2005) suggesting a possible climatic driver. However, scarce and patchy instrumental records made it difficult to place this historic drought and associated famine in context given the relatively short instrumental record. Longer, spatially explicit records of the monsoon and its drivers were needed to assess the relative contributions and interactions between monsoon failure and colonial-era disruptions in local food production and labor.

Until the development of the MADA (Cook et al. 2010), there was little understanding of the long-term spatiotemporal dynamics of the Asian Monsoon and its mechanistic links to ocean-atmosphere conditons. The MADA is comprised of a network of 327 tree-ring sites covering the last 1000 years arranged in a gridded reconstruction of PDSI (Palmer Drought Severity Index) over the Monsoon Asia Region. The chronologies used in MADA, largely derived from trees in the monsoon and transitional regions, preserve an integrated signal of summer monsoon soil moisture (June–August), critical for agricultural productivity. The MADA yields a seasonal- to centennial-scale view of monsoon failure appropriate for not only understanding Pacific forcing factors, but also the significance of monsoon failure for society and its interactions with colonialism. According to the MADA, the Late Victorian drought ranks as one of the most severe in the last 600+ years and covered nearly all areas of Monsoon Asia (Cook et al. 2010). Though sea surface temperatures (SST) of the eastern equatorial Pacific were elevated, the pattern is not clearly one of a strong eastern Pacific ENSO event, but could have been connected to SST anomalies in the extratropical North Pacific typical of the Pacific Decadal Oscillation.

The late Victorian drought was not simply a climatological event, but an important interaction between climate and society. For example, Davis (2002) argues that in India, more than a century of colonial British rule had led to: dependence of local agriculturalists on distant food sources, loss of local labor, lack of investiment in water management, and low prices for tropical commodities due to global competition. All of these forces undermined local food security. These social and economic changes increased the vulnerability of agricultural communities to climate extremes and amplified the effects of the droughts resulting in famine on a continental scale. In China, a slow bureaucratic response to the drought may have further exacerbated its effects (Davis 2002).

Cook et al. (2010) demonstrated the extreme nature of the Monsoon failures during the Late Victorian drought. By documenting the severity of the drought in a multi-centennial context, the MADA allows more precise questions to be asked about how the severity and mechanisms of drought may interact with colonialism and other human enterprises. For example, how did the spatial scale of British Colonialism overlap with the scale of monsoon failure to engender such human tragedy? Did droughts of similar extent and severity, but under different socio-economic conditions, produce massive famines in the past? Recent analysis of multi-proxy paleoclimate data suggests that the ninth to thirteenth century Chola kingdom of southeast India could sustain itself under variability in the South Asian and Southeast Indian monsoon by making major investments in water storage, while other local kingdoms weakened under drought (Shanmugasundaram et al. 2017). However, no paleoclimate data sources are available proximal to the Chola kingdom itself, preventing the seasonally and spatially precise inferences that emerged from studies of the Khmer and the late Victorian droughts. As the spatial and temporal extent of the MADA are improved, the variety of human and environmental adjustments and responses to monsoon failure and flooding will be better resolved and may point to paths of resilience for modern societies facing climatic extremes in Monsoon Asia.

15.3 Arid Central Asia

Arid Central Asia comprises the arid regions of the continent that experience continental and extreme climates, including Inner and Outer Mongolia, Tibet, northern and western China, Kazakhstan, Uzbekistan, Kyrgyzstan, and Tajikistan. In this region, the westerlies dominate moisture transport; however, hydroclimatic variability is complex, in part due to significant topographical features. Arid Central Asia was home to several spatially extensive nomadic empires beginning with the Scythians in the eleventh century BCE and continuing through the thirteenth century Mongol Empire. These empires were developed by various ethnic groups that expanded to cover large expanses of the Eurasian steppe. They adopted diverse strategies to address the extreme variability in moisture and productivity typical of the steppe. Some of these strategies included a nomadic lifestyle, widespread trade, and diversified economies. More recently, this region has experienced rapid changes in climate that have affected agriculture, productivity, and fire regimes in concert with rapid social and economic changes.

15.3.1 Dendroarcheology and the Development of Ultra-Long Chronologies from Arid Central Asia

Arid Central Asia has a long and fascinating human history that includes the domestication of the horse, the rapid spread of nomadic pastoralism, and mounted warfare—innovations that have occurred during the last 4000 years. A key challenge for dendrochronology in Asia then is developing ultra-long chronologies capable of characterizing the climatic conditions under which these developments in human societies occurred. There are several ongoing efforts to develop such chronologies by combining long-lived trees, remnant wood, and archeological timbers (Zhang et al. 2003; Panyushkina et al. 2007, Myglan et al. 2008, 2012a, b; Shao et al. 2010). These efforts have resulted in the absolutely dated and floating chronologies of *Larix sibirica* from the Altai and the absolutely dated *Sabina* (*Juniperus*) *przewalskii* chronologies from the Qinghai-Tibetan Plateau. These efforts have faced challenges associated with developing long climate reconstructions from archeological timbers. In particular, the samples used for calibration must be sensitive to the same environmental factors as the archeological timbers. Further, the archeological timbers often produce short segments, limiting inferences about low-frequency climate variation (Cook et al. 1995). Nevertheless, with careful sampling and assessment, climatic inferences can be made. In addition, absolutely dated chronologies can be used to more precisely date archeological sites thus helping to define the timing of important developments in human societies. In combination, data about climatic context and timing of historical/archeological events will likely yield exciting discoveries in coming decades, especially regarding the spread of pastoralism across Arid Central Asia.

The Siberian Scythians were a group of seminomadic tribes that occupied the Altai-Sayan Mountains during the first millennium BCE, part of the much more widespread Scythian culture that extended from the Black Sea to Lake Baykal. Horseback riding and pastoralism were among the economic strategies that allowed the elite Siberian Scythians to establish distant trade networks, slaves, and lavish goods, all preserved in stone burial mounds called "kurgans". Wooden chambers inside the kurgans have provided the means for developing well-replicated floating chronologies that can be more precisely dated using radiocarbon wiggle-matching where crossdated but floating tree-ring chronologies help constrain C-14 dates (Panyushkina et al. 2007, 2013, 2016). When combined with details of the modern growth of *Larix sibirica*, these chronologies can not only describe the timing of multiple burials relative to one another, but also broadly characterize the climate during periods of population growth, contraction, and out-migration.

In the Russian Altai region, Panyushkina (2012) suggests that the Siberian Scythian Pazyryk culture of 700–250 BCE responded positively to colder temperatures and increased climatic variability. For example, she observed population growth during cold periods and contractions during warm periods. High variance in tree growth corresponded to increased mobility as documented by the archeological record, possibly resulting in the development of transhumance, a form of pastoralism where people and livestock move to fixed seasonal pastures. Though these inferences may seem counterintuitive when compared to the history of agricultural societies, which arguably benefitted from warmer, more stable conditions (Büntgen et al. 2016), in the context of pastoral nomads, these conclusions are logical. Cold conditions improve mobility on the steppe, particularly for horses. Increased mobility in pastoralists is an economic strategy that increases the net primary productivity pastoralists can extract from the steppe (Honeychurch and Amartuyshin 2007) and may also increase access to trade networks. Though still preliminary, this work on nomads and the contrasting relationships for sedentary agriculturalists strongly argues for place- and culture-specific interpretations of the relationship between climate and society.

While the wiggle-matching approach can be used when chronologies are floating, the ultimate goal for Arid Central Asia is the development of ultra-long absolutely dated chronologies. A major breakthrough in the absolute dating of Siberian Scythian sites occurred in 2012, when an international group of archeologists and dendrochronologists were able to bridge a gap between archeological wood from the Pazyryk (Scythian) culture (600–300 BCE) and remnant wood from the Russian Tuva region (360 BCE–2007 CE) (Myglan et al. 2012a). This effort resulted in a

2367 year long tree-ring record that instantly yielded calendar dates of 35 kurgans from 11 different sites. Further, the results were within a single year deviation from the calendar dates derived from the C-14 wiggle-matching method (Panyushkina et al. 2016). Future work to extend these chronologies using archeological samples has the potential to provide the climate context of the spread of pastoralism across Arid Central Asia—a potentially transformative contribution to the study of human-environment relationships.

Similar efforts have been made to develop ultra-long tree-ring chronologies of *Sabina przewalskii* using living trees, remnant wood, and archeological timbers from the Tibetan Plateau. Wooden beams from several Xiatatu tombs of the Tubo Kingdom in Guolimu County, Delingha City, Qinghai Province, China were used to extend a living and remnant wood chronology to 2332 years, allowing for calendar dates of the tombs themselves (Wang et al. 2008). However, *S. przewalskii* has a mixed climatic signal, responding positively to both precipitation and temperature, making climatic inferences, particularly from the archeological specimens, challenging. At low elevations, *S. przewalskii* is primarily sensitive to spring moisture (Zhang et al. 2003; Sheppard et al. 2004), while at high, treeline locations, modern *S. przewalksii* are more sensitive to summer temperature (Liu et al. 2009). Therefore, one must infer the original growing location of the archeological samples in order to use them as a climate proxy. Despite the challenges of extracting a climate signal from *S. przewalskii*, it has been invaluable in providing a chronology for archeological sites (Wang et al. 2008). *Sabina przewalskii* will continue to provide an important record of environmental variability on the Tibetan Plateau, a region with a complex and dynamic human history that interacted politically and economically with distant empires in China and the nomadic empires of the Mongolian steppe.

15.3.2 Climate and Environment of the Mongol Empire

Despite centuries of study, the success of the thirteenth century Mongol Empire remains enigmatic. In less than a century, the Mongols developed an empire that claimed areas from the Hungarian grasslands to current day Syria, Baghdad, Vietnam, and Japan. Powered by domesticated grazing animals, the empire grew at the expense of sedentary agriculturalists living on the edges of the Eurasian steppe. The Mongol conquests affected the history of civilizations across China to Russia, Persia to India, as well as that of Christianity and Islam, even leaving a genetic fingerprint on the people of Eurasia (Zerjal et al. 2003). Historical sources suggested that the Mongols expanded their realm as a reaction to the extreme droughts typical of the steppe (Toynbee and Somervell 1947), though no paleoecological data from the Mongolian steppe were available to evaluate this hypothesis until Pederson et al. (2014) made a fortuitous discovery of ancient wood on a Holocene lava flow in central Mongolia. Their annually dated record of soil moisture from annual ring widths (900–2011 CE) clearly showed a rapid change from extremely dry conditions

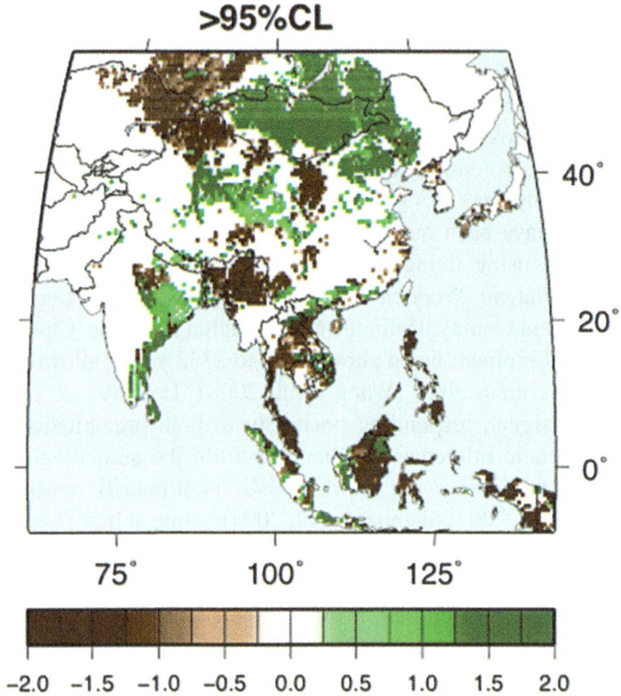

Fig. 15.3 MADAv2 reconstruction of scPDSI using a 0.5° grid and a 1000 km search radius during Mongol pluvial (1211–1226 CE). Areas in color are within the 95% confidence interval. scPDSI values are dry when negative (*brown*) and wet when positive (*green*)

in the 1100s to consistently wet conditions in 1211–1226 CE (Fig. 15.3). This short period represents the most persistent pluvial in their 1100-year reconstruction and coincides with the historical record of Ghengis Khan's rise to power and period of most rapid expansion of the empire.

During this remarkable period of environmental stability and high grassland productivity, the steppe would have supported large numbers of domesticated grazing animals and horses, which may have given the mounted Mongols an advantage over their neighbors. This productivity would have also allowed for the well-documented concentrations of animals and warriors that convened on the central Mongolian steppe prior to invasions. However, as the Mongol empire expanded its reach and its trade networks, the influence of climate would likely have become less important, eclipsed instead by political and economic factors.

Similar arguments about climate and environment have more recently been proposed to explain the rapid withdrawal of the Mongols from Hungary, though in this case, short-term climatic and environmental conditions may have constrained the Mongols and their horses (Büntgen and Di Cosmo 2016). Warm, wet conditions during spring of 1242 CE reduced the mobility of the Mongols in the swampy terrain of the Hungarian steppe and limited access to pasture and food in a landscape already decimated by war. In both cases, the rise of the Mongols and their withdrawal from Hungary, accurate dating of tree-ring records located proximal to the center of human action and the well-demonstrated relationship between tree growth and grassland productivity allowed historians to make new inferences about the role of grassland productivity and environmental conditions in enabling and constraining the development and spread of the Mongols. These new inferences are dependent on the precise timing of climatic phenomena and their direct connection to environmental constraints under which human actors made decisions. However, the differential effects of similar climate conditions (both were affected by wet conditions) on historical events highlight the importance of place- and time-specific context of human-environment interactions.

15.3.3 Environmental and Economic Synergies in Modern Mongolia

Tree-ring records of past climate and environment can not only help us understand human-environment relationships of past societies, but also contextualize modern events, policies, and institutions. Evaluating past hydroclimate relative to recent climatic conditions is essential for understanding the natural range of moisture variability in a region, and can be useful for water resource managers when preparing for water usage in the future (Woodhouse et al. 2006). Tree rings have been particularly successful in placing recent changes in hydroclimate in context both because of their ability to integrate water availability at the scale of a watershed and because they allow water managers and water users to view past hydroclimate at a time increment relevant to current management. Agriculture and mining, both growing sectors of the Mongolian economy, accounted for 35% of Mongolia's GDP in 2014 (NSO 2014). The sustainability of these water-demanding practices has recently been called into question, especially in light of well-documented climate change in Mongolia (Priess et al. 2011; Karthe et al. 2014).

Agriculture has increased rapidly in Mongolia over the past several decades, particularly in the northcentral region of the country due to elevated moisture as well as opening of markets in 1992. Both Selenge and Yeruu River streamflow reconstructions in the "Mongolian Breadbasket" indicate that the twentieth century was the wettest over the past three centuries (Pederson et al. 2013; Davi et al. 2013), possibly enabling agricultural expansion. However twenty-first century conditions may be drier and warmer than in the past 2000 years, suggesting that the last few decades of moisture were extremely unusual (Pederson et al. 2014) and future warming could pose serious constraints on Mongolia's growing agricultural sector.

The recent extreme variability in hydroclimate in Mongolia would also suggest rapid increases in fire occurrence, given the strong relationships between drought and fire in other semi-arid forest regions (Swetnam and Betancourt 1998; Westerling et al. 2006; van Mantgem et al. 2013). Instead, fire activity in Mongolia has closely followed changing sociopolitical and economic transitions (Saladyga et al. 2013;

Hessl et al. 2016). Where grazers and fire compete for grass, drought may reduce fuel continuity and volume, preventing the spread of fire and reducing fire activity (Bond and Keeley 2005; Hessl 2011). This effect was particularly strong following the opening of Mongolian markets to capitalism in 1992. Since that time, total livestock numbers have more than doubled, from 25 million animals to more than 50 million in 2014 (Hessl et al. 2016). In the decades following Mongolia's democratic revolution and adoption of capitalism, herding animals became concentrated around the capital city of Ulaanbaatar reducing and sometimes eliminating fire activity from locations with centuries-long histories of fire (Saladyga et al. 2013). Though warming temperatures have significantly altered Mongolia's climate since at least the 1940s, human-environment interactions due to sociopolitical and economic factors appear to have overriden climatic factors to shape the Mongolian landscape (Fernandez-Gimenez 2000). These results have implications for fire and climate change in other grasslands with high-intensity grazing and, conversely, where grazers have been removed from the landscape.

15.4 Future Steps

Many of the aforementioned studies describe linkages between humans and their environment over relatively short periods of time or small spatial scales. Some recent efforts, however, have focused on the relationship between climate variations and cultural changes over continental or hemispheric scales. In particular, changes in temperature over much of the northern hemisphere have been cited as an important impetus for societal catastrophes in the past. For example, Büntgen et al. (2016) investigated the impacts of the "Late Antique Little Ice Age" (536–660 CE), a distinct period of cooling across Europe and Asia initiated by extensive volcanism and supported by land-ocean feedbacks and a solar minimum. This cool period coincided with notable societal hardships, including reduced agricultural activity, empire collapse, invasions, and mass migrations suggesting a linkage between cold conditions and social instability. Similarly, Liu et al. (2009) used long tree-ring records from the Tibetan Plateau to suggest that historically cold periods were associated with dynastic collapses in China over more than two millennia. They argue that these collapses are indirectly linked to low temperatures, which are associated with reduced crop output, famine, and subsequent societal unrest. These broad-scale assessments may be the direction that human-environment studies will take in the future, but they will need to address concerns about the discrepancy between the scale and proximity of the proxies relative to the cultures those variables are said to affect, the differential responses of societies to environmental changes, and the role of human agency in mitigating, adapting, and amplifying environmental change.

Whereas many human-paleoenvironment studies qualitatively compare historical events with extreme climatic conditions, efforts have been made to explicitly quantify these relationships. Zhang et al. (2007) used correlation analyses to identify relationships between temperature, and socioeconomic measures, such as agricultural production, population size, and the number of wars across Europe and Asia. For example, they found that cycles of war and peace were related to variability in northern hemisphere temperatures, where the number of wars and temperature were inversely correlated. Similarly, Zhang et al. (2011) investigated the causal relationship between climate and crisis in Europe from 1500–1800 using 16 socioeconomic and demographic variables. They used correlation and causality analyses among these factors, in conjunction with social theories, to elucidate the effects of climate change on ecological and human systems. Effectively marrying ecological, climatological, and historical inferences remains an important challenge, and one that will likely be explored further in the future. Quantifying the relationship between humans and their environment is challenging given the inherent differences of these systems and their complex interactions. Further, identifying the appropriate metric of human response is also a challenge since human demographics and historical events are not necessarily consistently recorded over time and may contain significant bias. Nevertheless, identifying models of human interaction with the environment that have predictive power is a tantalizing proposition that could help us understand current responses to climate change.

Although annually resolved tree-ring data provide climate information relevant to discrete historical events, an understanding of lower frequency climate variability, and its drivers, could provide useful information on longer time-scale processes, for example the spread of pastoralism across Central Asia. Additionally, other proxies from tree rings, including anatomical wood properties, and newly developed methods (e.g. image analysis), have yielded new insights in other regions (e.g. Fonti et al. 2010; Drew et al. 2012; Björklund et al. 2015; O'Donnell et al. 2016) and should be further explored in Asian tree-ring records. Multiple proxies, including ice cores, speleothems, tree rings, and sediment records, may be used to gather information on multiple time scales of paleoclimatic variability, while still taking advantage of the calendar dates provided by tree-ring records. For example, paleoclimatic records capturing hydroclimate over longer periods than tree rings show that moisture variability in Arid Central Asia and Monsoon Asia are out of phase over multi-centennial and millennial time scales (Chen et al. 2008; An et al. 2012), a climatic inference that may have had interesting effects on the interactions between societies across Asia in the past. An assessment of these processes and how they might persist or change could have important implications for future climate and economies across Asia.

15.5 Conclusions

Paleoenvironmental data, and dendrochronology in particular, can illuminate the complex and variable environmental histories that constrain and enable human actors. The case studies reviewed here support the variable and diverse nature of human-environment relationships, suggesting that similar climatic conditions can affect societies and cultures differently and can even have opposing effects at different times on the same society. Recent efforts to expand the study of human-environment relationships to broader spatial scales or across time using quantitative methods are still in their infancy, but will need to respond to critiques about the applicability of distant proxies to local conditions, differential effects and responses of cultures over time, and the ways in which climate interacts with other stressors. Future efforts in assessing human-environment interactions might take advantage of the existing paleoenvironmental data from a variety of sources to more fully characterize past climatic variability and its influence on environmental and human systems. Many critical questions about the past and future of Asia remain—in particular, the role of human institutions like colonialism in amplifying or dampening monsoon dynamics, the role of climatic variability/stability in the spread of pastoral nomadism, the effect of the dipole in moisture across Arid Central Asia and Monsoon Asia on interactions between societies, and the complexities of modern climate and environmental change in the context of rapid socio-economic development across Asia. With precision calendar dates, high fidelity to climate, and multiple millennium-long proxies relevant to the human enterprise, tree rings will undoubtedly play a critical part in answering these questions.

References

Adger WN, Barnett J, Brown K et al (2013) Cultural dimensions of climate change impacts and adaptation. Nat Clim Chang 3:112–117. doi:10.1038/nclimate1666

An Z, Colman SM, Zhou W et al (2012) Interplay between the Westerlies and Asian monsoon recorded in Lake Qinghai sediments since 32 ka. Sci Rep 2:619. doi:10.1038/srep00619

Björklund J, Gunnarson BE, Seftigen K, Zhang P, Linderholm HW (2015) Using adjusted blue intensity data to attain high-quality summer temperature information: a case study from Central Scandinavia. The Holocene 25(3):547–556

Bocinsky RK, Kohler TA (2014) A 2,000-year reconstruction of the rain-fed maize agricultural niche in the US Southwest. Nat Commun 5:5618. doi:10.1038/ncomms6618

Bond WJ, Keeley JE (2005) Fire as a global "herbivore": the ecology and evolution of flammable ecosystems. Trends Ecol Evol 20:387–394. doi:10.1016/j.tree.2005.04.025

Bräuning A, Mantwill B (2004) Summer temperature and summer monsoon history on the Tibetan Plateau during the last 400 years recorded by tree rings. Geophys Res Lett 31:L24205. doi:10.1029/2004GL020793

Buckley BM, Palakit K, Duangsathaporn K et al (2007) Decadal scale droughts over northwestern Thailand over the past 448 years: links to the tropical Pacific and Indian Ocean sectors. Clim Dyn 29:63–71. doi:10.1007/s00382-007-0225-1

Buckley BM, Anchukaitis KJ, Penny D et al (2010) Climate as a contributing factor in the demise of Angkor, Cambodia. Proc Natl Acad Sci 107:6748–6752

Büntgen U, Di Cosmo N (2016) Climatic and environmental aspects of the Mongol withdrawal from Hungary in 1242 CE. Sci Rep 6:25606. doi:10.1038/srep25606

Büntgen U, Myglan VS, Ljungqvist FC et al (2016) Cooling and societal change during the Late Antique Little Ice Age from 536 to around 660 AD. Nat Geosci 9:231–236. doi:10.1038/ngeo2652

Chen F, Yu Z, Yang M, Ito E, Wang S, Madsen DB, Huang X, Zhao Y, Sato T, Birks HJB, Boomer I, Chen J, An C, Wünemann B (2008) Holocene moisture evolution in arid central Asia and its out-of-phase relationship with Asian monsoon history. Quat Sci Rev 27(3):351–364. doi:10.1016/j.quascirev.2007.10.017

Cook ER, Briffa KR, Meko DM, Graybill DA, Funkhouser G (1995) The 'segment length curse' in long tree-ring chronology development for palaeoclimatic studies. The Holocene 5:229–237. doi:10.1177/095968369500500211
Cook ER, Anchukaitis KJ, Buckley BM et al (2010) Asian monsoon failure and megadrought during the last millennium. Science 328:486–489
Panyushkina IP (2012) Climate-induced changes in population dynamics of Siberian Scythians (700–250 B.C.) In: Giosan L, Fuller DQ, Nicoll K, Flad RK, Clift PD (eds) Climates, landscapes, and civilizations. American Geophysical Union, Washington, DC. doi:10.1029/2012GM001220
Cook ER, Krusic PJ, Anchukaitis KJ et al (2013) Tree-ring reconstructed summer temperature anomalies for temperate East Asia since 800 C.E. Clim Dyn 41:2957–2972. doi:10.1007/s00382-012-1611-x
Coombes P, Barber K (2005) Environmental determinism in Holocene research: causality or coincidence? Area 37:303–311. doi:10.1111/j.1475-4762.2005.00634.x
Davi NK, Pederson N, Leland C et al (2013) Is eastern Mongolia drying? A long-term perspective of a multidecadal trend. Water Resour Res 49:151–158. doi:10.1029/2012WR011834
Davis M (2002) Late victorian holocausts: El Niño famines and the making of the third world. Verso, London
deMenocal PB (2001) Cultural responses to climate change during the late Holocene. Science 292:667–673. doi:10.1126/science.1059287
Douglass A (1921) Dating our prehistoric ruins: how growth rings in trees aid in the establishing the relative ages of the ruined pueblos of the southwest. Nat Hist 21:27–30
Douglass AE (1929) The secret of the Southwest solved by talkative tree rings. Natl Geogr 56:736–770
Drew DM, Allen K, Downes GM, Evans R, Battaglia M, Baker P (2012) Wood properties in a long-lived conifer reveal strong climate signals wehre ring-width series do not. Tree Physiol 33:37–47. doi:10.1093/treephys/tps111
Endfield GH (2012) The resilience and adaptive capacity of social-environmental sytems in colonial Mexico. PNAS 109:3676–3681
Erickson CL (1999) Neo-environmental determinism and agrarian "collapse" in Andean prehistory. Antiquity 73:634–642. doi:10.1017/S0003598X00065236
Evans D, Pottier C, Fletcher R et al (2007) A comprehensive archaeological map of the world's largest preindustrial settlement complex at Angkor, Cambodia. Proc Natl Acad Sci 104:14277–14282. doi:10.1073/pnas.0702525104
Fernandez-Gimenez ME (2000) The role of Mongolian nomadic pastoralists' ecological knowledge in rangeland management. Ecol Appl 10:1318–1326
Fonti P, von Arx G, García-González I, Eilmann B, Sass-Klaassen U, Gärtner H, Eckstein D (2010) Studying global change through investigation of the plastic responses of xylem anatomy in tree rings. New Phytol 185:42–53
Grießinger J, Bräuning A, Helle G, Thomas A, Schleser G (2011) Late Holocene Asian summer monsoon variability reflected by δ18O in tree-rings from Tibetan junipers. Geophys Res Lett 38:L03701. doi:10.1029/2010GL045988
Hessl AE (2011) Pathways for climate change effects on fire: models, data, and uncertainties. Prog Phys Geogr 35:393–407
Hessl AE, Brown P, Byambasuren O et al (2016) Fire and climate in Mongolia (1532–2010 Common Era). Geophys Res Lett 43(12):6519–6527. doi:10.1002/2016GL069059
Hulme M (2011) Reducing the future to climate: a story of climate determinism and reductionism. Osiris 26:245–266. doi:10.1086/661274
Honeychurch W, Amartuvshin C (2007) Hinterlands, urban centers, and mobile settings: the "new" Old World archaeology from the Eurasian Steppe. Asian Perspect 46:36–64. doi:10.1353/asi.2007.0005
Karthe D, Heldt S, Houdret A, Borchardt D (2014) IWRM in a country under rapid transition: lessons learnt from the Kharaa River Basin, Mongolia. Environ Earth Sci 73:681–695. doi:10.1007/s12665-014-3435-y

Liu Y, An Z, Linderholm HW et al (2009) Annual temperatures during the last 2485 years in the mid-eastern Tibetan Plateau inferred from tree rings. Sci China Ser D Earth Sci 52:348–359. doi:10.1007/s11430-009-0025-z

Lorenz EN (1968) Climatic determinism. Meteorol Monogr 8:1–3

Maraun D, Kurths J (2005) Epochs of phase coherence between El Niño/Southern Oscillation and Indian monsoon. Geophys Res Lett 32:L15709. doi:10.1029/2005GL023225

Myglan VS, Oidupaa OC, Kirdyanov AV, Vaganov EA (2008) 1929-year tree-ring chronology for the Altai-Sayan region (Western Tuva). Archaeol Ethnol Anthropol Eurasia 36:25–31. doi:10.1016/j.aeae.2009.03.003

Myglan VS, Oidupaa OC, Vaganov EA (2012a) A 2367-year tree-ring chronology for the Altai–Sayan Region (Mongun-Taiga Mountain Massif). Archaeol Ethnol Anthropol Eurasia 40:76–83. doi:10.1016/j.aeae.2012.11.009

Myglan VS, Zharnikova OA, Malysheva NV et al (2012b) Constructing the tree-ring chronology and reconstructing summertime air temperatures in southern Altai for the last 1500 years. Geogr Nat Resour 33:200–207. doi:10.1134/S1875372812030031

NSO: National Statistical Office of Mongolia (2014). http://www.en.nso.mn/index.php

O'Donnell AJ, Allen KJ, Evans RM, Cook ER, Trouet V, Baker PJ (2016) Wood density provides new opportunities for reconstructing past temperature variability from southeastern Australian trees. Glob Planet Chang 141:1–11

Orlove B (2005) Human adaptation to climate change: a review of three historical cases and some general perspectives. Environ Sci Pol 8:589–600. doi:10.1016/j.envsci.2005.06.009

Panyushkina I, Sljusarenko I, Bikov N, Bogdanov E (2007) Floating larch tree-ring chronologies from archaeological timbers in the Russian Altai between about 800 BC and AD 800. Radiocarbon 49:693–702

Panyushkina I, Grigoriev F, Lange T, Alimbay N (2013) Radiocarbon and tree-ring dates of the Bes-Shatyr #3 Saka Kurgan in the Semirechiye, Kazakhstan. Radiocarbon 55:1297–1303. doi:10.1017/S0033822200048207

Panyushkina IP (2012) Climate-induced changes in population dynamics of Siberian Scythians (700–250 B.C.) In: Giosan L, Fuller DQ, Nicoll K, Flad RK, Clift PD (eds) Climates, landscapes, and civilizations. American Geophysical Union, Washington, DC. doi:10.1029/2012GM001220

Panyushkina IP, Slyusarenko IY, Sala R et al (2016) Calendar age of the Baigetobe Kurgan from the Iron Age Saka Cemetery in Shilikty Valley, Kazakhstan. Radiocarbon 58:157–167. doi:10.1017/RDC.2015.15

Pederson N, Leland C, Nachin B et al (2013) Three centuries of shifting hydroclimatic regimes across the Mongolian breadbasket. Agric For Meteorol 178–179:10–20. doi:10.1016/j.agrformet.2012.07.003

Pederson N, Hessl AE, Baatarbileg N et al (2014) Pluvials, droughts, the Mongol Empire, and modern Mongolia. Proc Natl Acad Sci 111:4375–4379. doi:10.1073/pnas.1318677111

Priess JA, Schweitzer C, Wimmer F et al (2011) The consequences of land-use change and water demands in Central Mongolia. Land Use Policy 28:4–10. doi:10.1016/j.landusepol.2010.03.002

Saladyga T, Hessl A, Nachin B, Pederson N (2013) Privatization, drought, and fire exclusion in the Tuul River watershed, Mongolia. Ecosystems 16:1–13. doi:10.1007/s10021-013-9673-0

Sano M, Buckley BM, Sweda T (2008) Tree-ring based hydroclimate reconstruction over northern Vietnam from *Fokienia hodginsii*: eighteenth century mega-drought and tropical Pacific influence. Clim Dyn 33:331–340. doi:10.1007/s00382-008-0454-y

Shanmugasundaram J, Gunnell Y, Hessl AE, et al (2017) Societal response to monsoon variability in Medieval South India: lessons from the past for adapting to climate change. The Anthropocene Review: 2053019617695343

Shao XY, Yin ZY et al (2010) Climatic implications of a 3585-year tree-ring width chronology from the northeastern Qinghai-Tibetan Plateau. Quat Sci Rev 29:2111–2122. doi:10.1016/j.quascirev.2010.05.005

Sheppard PR, Tarasov PE, Graumlich LJ et al (2004) Annual precipitation since 515 BC reconstructed from living and fossil juniper growth of northeastern Qinghai Province, China. Clim Dyn 23:869–881. doi:10.1007/s00382-004-0473-2

Stahle DW, Cleaveland MK, Hehr JG (1988) North Carolina climate changes reconstructed from tree rings: A.D. 372 to 1985. Science 240:1517–1519

Swetnam TW, Betancourt JL (1998) Mesoscale disturbance and ecological response to decadal climatic variability in the American Southwest. J Clim 11:3128–3147

Swetnam TW, Allen CG, Betancourt JL (1999) Applied historical ecology: using the past to manage for the future. Ecol Appl 9:1189–1206

Thera PR (1962). *Jinakalamali*, Vol. 145. Pali Text Society.

Toynbee AJ, Somervell DC (1947) A study of history: abridgement of volumes I–VI. Oxford University Press, Oxford

van Mantgem PJ, Nesmith JCB, Keifer M et al (2013) Climatic stress increases forest fire severity across the western United States. Ecol Lett 16(9):1151–1156. doi:10.1111/ele.12151.

Walker GT, Bliss EW (1932) World weather V. Mem R Meteorol Soc 4:53–84

Wang S-Z, Shao X, Xu X, Xiao Y (2008) Dating of tombs in Delingha, Qinghai Province, China, on the basis of a 2332-year tree-ring juniper chronology (Sabina przewalskii Kom) (1575 BC–756 AD). Dendrochronologia 26:35–41. doi:10.1016/j.dendro.2007.08.002

Westerling AL, Hidalgo HG, Cayan DR, Swetnam TW (2006) Warming and earlier spring increase western U.S. forest wildfire activity. Science 313(5789):940–943

Wissler C (1921) Dating our prehistoric ruins. Nat Hist 21:13–26

Woodhouse CA, Gray ST, Meko DM (2006) Updated streamflow reconstructions for the Upper Colorado River Basin. Water Resour Res 42:W05415. doi:10.1029/2005WR004455

Yang B, Qin C, Bräuning A, Burchardt I, Liu J (2011) Rainfall history for the Hexi Corridor in the arid northwest China during the past 620 years derived from tree rings. Int J Climatol 31(8):1166–1176

Zerjal T, Xue Y, Bertorelle G et al (2003) The genetic legacy of the Mongols. Am J Hum Genet 72:717–721

Zhang Q-B, Cheng G, Yao T et al (2003) A 2,326-year tree-ring record of climate variability on the northeastern Qinghai-Tibetan Plateau. Geophys Res Lett 30:1739. doi:10.1029/2003GL017425

Zhang DD, Brecke P, Lee HF et al (2007) Global climate change, war, and population decline in recent human history. Proc Natl Acad Sci 104:19214–19219. doi:10.1073/pnas.0703073104

Zhang DD, Lee HF, Wang C et al (2011) The causality analysis of climate change and large-scale human crisis. Proc Natl Acad Sci 108:17296–17301. doi:10.1073/pnas.1104268108.

Chapter 16
Dendro-archeo-ecology in North America and Europe: Re-purposing Historical Materials to Study Ancient Human-Environment Interactions

Valerie Trouet, Marta Domínguez-Delmás, Charlotte Pearson, Neil Pederson, and Darrin Rubino

Abstract The unique position of dendrochronology at the nexus of archeology, ecology, and climatology allows it to play a pivotal role in the study of past human-environment interactions. Yet, few tree-ring studies in Europe and eastern North America have been used to study pre-industrial land-use changes, forest ecology, and carbon dynamics and thus to constrain the uncertainties surrounding the Early Anthropocene hypothesis (Ruddiman Clim Chang 61:261–293, 2003; Rev Geophys 45(4):RG4001, 2007). Here, we discuss the potential of dendro-archeo-ecology—the use of dendroarcheological material in the study of forest ecology—to document past human land-use and forest alteration, which started in the Neolithic Era (~12,000–4000 BP) in Europe and after European immigration into eastern North America in the 1620s. In this context, we focus on the dendro-archeo-ecology of (1) Neolithic pile dwellings in the Euro-Mediterranean region and (2) old-growth forest dynamics in eastern North America. We discuss recurring challenges (e.g., low sample depth, short series length) and uncertainties (e.g., species and tree size bias) related to the use of (pre)historic timbers for ecological purposes that need

V. Trouet (✉) • C. Pearson
Laboratory of Tree-Ring Research, University of Arizona 1215 E Lowell Street, Tucson, AZ 85721, USA
e-mail: trouet@ltrr.arizona.edu; c.pearson@ltrr.arizona.edu

M. Domínguez-Delmás
Department of Botany, University of Santiago de Compostela, Rua Benigno Ledo s/n 27002, Lugo, Spain
e-mail: mardodel@gmail.com

N. Pederson
Harvard Forest, Harvard University, 324 North Main Street, Petersham, MA, USA
e-mail: neilpederson@fas.harvard.edu

D. Rubino
Department of Biology, Hanover College 517 Ball Drive, Hanover, IN, USA
e-mail: rubino@hanover.edu

M.M. Amoroso et al. (eds.), *Dendroecology*, Ecological Studies 231,
DOI 10.1007/978-3-319-61669-8_16

to be carefully addressed. We advocate for a concerted effort to move the use of dendro-archeological material from strictly archeological applications towards exploration of its ecological potential and for a close alliance of dendrochronology with related disciplines that aim to address the same subjects.

Keywords Dendroarcheology • Dendroecology • Dendro-archeo-ecology • Early Anthropocene • Europe • Forest clearance • Land-use change • Pile dwellings • Neolithic • North America

16.1 Introduction

Dendrochronology can play a pivotal role in the study of past human-environment interactions, supported by the annual resolution of tree-ring chronologies, their capacity for absolute dating, and their potential to span time periods relevant to human history (e.g., le Roy Ladurie 1971; Lamb 1995). With many applications in the fields of archeology, ecology, forestry, and paleoclimatology, dendrochronology is uniquely positioned at the nexus of these related fields to study complex feedbacks between humans and their environment (see also Chapter 14). Some of the earliest dendrochronological work (Douglass 1929) was focused on drought as a cause for the sudden abandonment of ancient Puebloan settlements in the American Southwest towards the end of the thirteenth century (Cook et al. 2004). Dendrochronological evidence has also contributed to disentangling the role of drought from other causes in the collapse of the Mayan civilization near the end of the first millennium CE (Stahle et al. 2011) and in the fate of the Roanoke and Jamestown colonies on the mid Atlantic coast of eastern North America in the sixteenth and seventeenth centuries (Stahle et al. 1998). Tree-ring data have been used to show that climate was a contributing factor to the Mongol withdrawal from Hungary in the thirteenth century (Büntgen and Di Cosmo 2016) and to the demise of Greater Angkor, the capital of the Khmer Empire in Cambodia in the fourteenth and fifteenth centuries (Buckley et al. 2010). Tree-ring based climate reconstructions in central Europe (Büntgen et al. 2011) showed exceptional decadal-scale hydroclimate fluctuations over the period 250–550 CE that may be linked to the fall of the western Roman Empire.

Yet, for many (if not all) of these past societal disruptions, climate was not the sole determining cause, but rather an important contributing factor in a complex network of human-environment perturbations in which anthropogenic land-use change often played a significant role (Diaz and Trouet 2014). For instance, extensive deforestation is postulated to have reduced the ecological carrying capacity of the Mayan lowlands (Medina-Elizalde and Rohling 2012) and to have contributed to the eleventh century Meso-American droughts (Cook et al. 2012). Widespread deforestation and over-harvesting of wood are also known to have caused significant environmental stress to the Puebloan (Gumerman 1988; Guiterman et al. 2016) and Roman societies (O'Sullivan et al. 2008).

These interplays between historical land-use changes and climatic effects corroborate Ruddiman's (2003, 2007) hypothesis that the influence of human activity on the atmosphere's greenhouse gas composition and thus on climate started long before the onset of the industrial era circa 1850 CE. This 'Early Anthropocene' hypothesis postulates that early agriculture and the start of deforestation in Eurasia around 8000 years ago initiated an anomalous increase in the long-term natural atmospheric CO_2 trend. In the following millennia, worldwide agricultural intensification and continued widespread forest clearance were responsible for an increase in the amplitude of this positive CO_2 trend and associated global temperature anomaly. Ruddiman (2003, 2007) further hypothesized that short-lived CO_2 decreases superimposed on the millennial-scale trend were caused by pre-industrial pandemics that caused far-reaching mortality followed by reforestation of abandoned agricultural land (Ruddiman and Carmichael 2007).

Many aspects of the Early Anthropocene hypothesis have been challenged, particularly Ruddiman's (2003) claim that the pre-industrial CO_2 anomaly could be explained solely by anthropogenic land-use changes (e.g., Joos et al. 2004; Ruddiman 2007). However, the paucity of pre-1850 land-use data and inherent uncertainties in estimates of pre-industrial anthropogenic deforestation and carbon emissions (DeFries et al. 1999) hinder a robust quantification of the relative contributions of human activities on atmospheric CO_2 (and global temperature) trends over the late Holocene.

To the extent of our knowledge, few tree-ring studies have been used to constrain the uncertainties surrounding the Early Anthropocene hypothesis, even when tree-ring studies have a great potential to contribute to such research. The combination of annual resolution records of climate change with dendroecological data that document changes in pre-industrial forest productivity offers prospects for moving beyond simple temporal correlations towards more detailed analyses of cause and effect.

Currently this potential is limited for deeper time Anthropocene scenarios by a number of factors. Multi-millennial-length tree-ring chronologies in North America and Europe that coincide with the onset and intensification of anthropogenic land-use change and deforestation are rare (e.g, Pilcher et al. 1984; Ferguson and Graybill 1983; Brown et al. 1992; Stahle et al. 1985; Friedrich et al. 2004). Moreover, to develop chronologies that span multiple millennia, tree-ring series from long-lived living trees must be merged with series from remnant wood on the landscape (e.g., Brown et al. 1992), with wood from historical buildings and archeological sites (e.g., Kuniholm et al. 1996), and/or with subfossil wood (e.g., Pilcher et al. 1977). This is especially true in regions such as Europe and North America where climate promotes enough productivity to support high human populations, but rapidly increases the decay of deadwood that has the potential to extend living-tree chronologies. The majority of lowland forests in Europe and North America have been cleared at various stages throughout the course of human history and tree-ring chronologies from living trees are derived from either secondary (post-clearance) forests that reflect only recent forest ecology and management conditions (e.g., Pederson et al. 2014, Druckenbrod et al. 2013), or from remnant stands of old-growth forests that are often located in remote or unproductive sites (e.g., near upper tree line and on steep slopes or shallow soils; Panayotov et al. 2010; Stahle and Chaney 1994; Davis 1996; Therrell and Stahle 1998; Maxwell et al. 2011). Such remaining old-growth stands might not necessarily reflect the ecology and climate sensitivity of the formerly surrounding forests (Babst et al. 2013, 2014a). Whether old-growth forests represent the full, long-term carbon sequestration potential of the landscapes that have been significantly altered by human land-use is difficult to know.

As a result, little is known about the forest ecology and carbon dynamics of the forests that existed prior to the intensification of human land-use and forest clearance, which began with early agriculture in the Neolithic Era (~8000 BP) in Eurasia and after European immigration to eastern North America in the early 1600s. Here, we explore the potential to derive ecological information for these eras from the ample wood evidence that is preserved in the form of archeological and historical building material. Dendroarcheological material has been consolidated to extend tree-ring based climate reconstructions further back in time (e.g., Cook et al. 2004; Büntgen et al. 2006, 2011), but the wealth of information about primary forests that can be provided by wood from early human dwellings and wooden structures (e.g., boats, fence posts) remains largely untapped (but see e.g., Krause 1997). Dendrochronology offers the possibility to explore human alteration of forest ecosystems and to observe these changes over time through diverse wooden remains and contemporary cultural records (e.g., Pausas et al. 2009; Kirby and Watkins 1998; Rackham 1990; Meiggs 1982; Thirgood 1981). We aim to characterize the potential of and challenges related to the use of dendroarcheological material in the study of primary forest ecology—the field of dendro-archeo-ecology—in the context of two different regions and time-scales.

The Euro-Mediterranean region (Section "Dendro-archeo-ecology of Euro-Mediterranean Pile Dwellings from the Neolithic Era") is arguably the richest on Earth in terms of the preservation of culturally and temporally diverse wooden remains. Humans have occupied the area for well over 500,000 years and throughout this time there is abundant evidence for the use of wood as a resource. 300,000 years ago, hunters in Schöningen (Germany) used wooden spears and tools to kill and butcher horses and to fend off saber-toothed cats (Serangeli et al. 2015; Thieme 1997). Examination of these wooden implements by Schoch et al. (2015) demonstrated that even these early hunter-gatherer societies preferentially selected certain wood types for specific tasks, and worked the material with procedure and skill. 32,000 years ago wood also played a role in survival through the Ice Age, as prehistoric artists who decorated the walls of Chauvet Cave (Ardèche, France) were warmed and lit by wood burning hearths and torches (Cuzange et al. 2007). As the ice retreated ca. 12,000 years ago, a warmer, wetter climate ensued (e.g., Wanner et al. 2008; Mayewski et al. 2004) and afforestation began as tundra vegetation and bare ground transitioned into forest (Huntley 1990). People began to settle in seasonal or permanent locations and to create their own shelter, utilizing wood for a wide range of construction, hunting, fishing, and farming practices (Whittle 1996; Dimbleby 1984). Anthropogenic landscape perturbations and forest clearance thus started in the Early Neolithic (~8000 BP). For example, the craftsmanship needed for the refined carpentry found by Tegel et al. (2012) in Neolithic (~7000 BP) water wells in eastern Germany suggests that carpentry—and by extension, forest clearance—in northern Europe developed at the same time as agriculture. Anthropogenic land-use change then intensified in various phases during Roman Times and in the Middle Ages, periods for which ample dendroarcheological samples are available (Tegel et al. 2010). In the section "Dendro-archeo-ecology of Euro-Mediterranean Pile Dwellings from the Neolithic Era", we focus on the dendro-archeo-ecology of pile dwellings, one of the richest dendroarchaeological resources for prehistoric human-environment interactions in Europe. We consider the dendroecological potential of Neolithic pile-dwelling-derived wood to inform our understanding of prehistoric forest resource management and human response to environmental change.

In eastern North America (Section "Sixteenth to Nineteenth Century Dendro-archeo-ecology in Eastern North America"), coarse-scale deforestation and the construction of wooden buildings started much later. The most intensive changes to the landscape and forests began in the 1620s when Europeans immigrated to the region and continued into the mid-nineteenth century, after which eastward agricultural expansion and industrialization led to a century of natural reforestation (Cronon 1983; Thompson et al. 2013). As a result, old-growth forest stands are rare and patchy throughout the region, which limits our ability to study historic growth patterns and forest processes prior to European immigration. In the section "Sixteenth to Nineteenth Century Dendro-archeo-ecology in Eastern North America", we posit that historical buildings offer perhaps the only opportunity to enhance our dendrochronological network for the study of the environmental (biotic and abiotic) factors driving tree growth and forest dynamics prior to 1600 CE with substantial replication. In this context, we present two case studies that use dendro-archeological material i) to identify forest disturbances in the Ohio and Hudson River Valleys and ii) to investigate a late 1600s subcontinental forest recruitment pulse (Pederson et al. 2014). The two studies demonstrate the enormous promise of eastern North American timbers in recovering local and regional forest dynamics prior to the nineteenth century. Most promising, these timbers yield data on historical forest dynamics in areas virtually devoid of old-growth forests today. Dendro-archeo-ecology in eastern North America can thus quantify historical forest productivity and reduce the uncertainties surrounding pre-industrial carbon sequestration (DeFries et al. 1999).

16.2 Dendro-archeo-ecology of Euro-Mediterranean Pile Dwellings from the Neolithic Era

Ongoing controversy over Ruddiman's Early Anthropocene hypothesis is largely centered around the question of whether small, low density populations of pre-modern humans using low-intensity technology could carry out forest clearance, agriculture, and industrial production at levels that could be sufficient to alter global atmospheric chemistry. However, even if early societies did not alter atmospheric chemistry, a rapidly growing range of studies of past human and environmental interaction are beginning to demonstrate a surprisingly early and surprisingly large anthropogenic footprint in a number of locations around the world (Ellis et al. 2013a). Defining and quantifying early human-environmental impact on a range of geographic scales with a precise temporal resolution therefore has a stand-alone significance in helping to understand the numerous complex feedbacks between human and natural coupled systems over the course of our history. In this section we advocate the use of dendro-archaeo-ecology more broadly in terms of establishing a date or possible range of dates for the start of the Anthropocene and in defining the true anthropogenic footprint in the Euro-Mediterranean region around 8000 years ago. We focus in particular on one subset of wooden archaeological remains, which has great potential for in depth ‘re-study’ within this context, namely the prolific and exceptionally well-preserved Euro-Mediterranean pile dwellings.

Pile dwelling sites, small habitations built on posts or piles in wetland environments, span a large range of time periods, including the early Neolithic when other archaeological evidence for anthropogenic alteration of the environment is sparser. Early settlers were drawn to water sources, as these provided all the necessities of life (fresh water, food, and, in the case of lake settlements, defensible locations). Homes, track ways, and storage facilities were constructed on top of wooden posts (piles) that were easily driven into wet marshland or lake sediments (Fig. 16.1). This mode of construction, combined with higher water-levels in response to broader climatic changes over time, has resulted in superb preservation of wood, charcoal, and other organic materials such as seeds, leather, and pollen in anoxic conditions that prevented degradation (Huisman and Klaassen 2009). As such, these sites provide treasure troves of multi-disciplinary information for reconstructing human-environment interactions (Menotti 2004; Sherratt 2004). Concentrated human activity in the same area over time resulted in a modification of local natural resources through the direct management, or indirect promotion or suppression, of certain species for a range of purposes (e.g. Caruso Ferme and Piqué i Huerta 2014; Doyen et al. 2016). Different species and parts of trees were selected for particular purposes and larger timbers (with potentially more growth rings) were increasingly used for permanent structures. This progression, when paired with the dating precision, the capacity for ecological study, and the temporal and geographic spread of the preserved timber, could provide immensely powerful new data to develop precise chronologies of dwellings that could be linked to demographic and paleo-landscape models. This is especially the case when tree-ring data can be combined with pollen, charcoal and sedimentation records that are indicative of ecological change and/or landscape management practices. Pile dwelling-derived tree-ring data could thus be used to inform such models through patterns of forest use, dated phases of activity such as burning, or population estimates based on settlement size, construction phases, and duration of occupation. They could further

Fig. 16.1 An 'Eco-Museum' reconstruction of the Neolithic lakeside settlement of Dispilio, 7 km south of Kastoria, in northern Greece. This site, excavated by the University of Thessaloniki, was continuously used from the Early Neolithic (~6000 BC) to the Late Chalcolithic period (~1200 BC). A large range of similarly dated settlements are currently under excavation by the Florina Ephorate of Antiquities in the Amindeon region. The diverse wooden remains from these sites hold much potential for future dendro-archaeo-ecological studies in the region across a time period for which, so far, such ecological potential has not been explored

be used to investigate the potential to identify a transition from localized impact to a more global impact on CO_2 and to explore a temporally refined structure of wetland disturbance and utilization in relation to possible methane production in the Euro-Mediterranean region.

16.2.1 Pile Dwellings as a Cultural Phenomenon: Geographic and Temporal Spread

The archetypal pile dwellings proliferated across central Europe and around the Alps between 7000–2500 BP (Sherratt 2004; UNESCO 2015, http://sites.palafittes.org/home) and it is in these regions where many of the leading theoretical approaches to the study of such sites have been developed (Pranckėnaitė 2014). The tradition of prehistoric lake settlements, however, expands far across the Euro-Mediterranean region and beyond (e.g. Lithuania—Pranckėnaitė 2014; Slovenia—Čufar et al. 2010, 2015; Northwest Russia—Mazurkevich et al. 2010; Kulkova et al.

2001; Bulgaria—Ivanov 2000; Northern Greece—Karkanas et al. 2011; Chrysostomou 2015; Macedonia—Sherratt 2004). Palaeobotanical evidence suggests that the Neolithic pile-dwellings phenomenon followed a south-north diffusion, starting in the northeast of Spain, central Italy, and Slovenia and spreading towards central Europe and beyond (Menotti 2004). They reached the widest geographical coverage and typological diversity during the European Iron Age (Sherratt 2004) and they represent some of the most commonly preserved archaeological evidence, which (in combination with other contemporary excavations and paleo-records) could potentially be used to reconstruct population density in combination with ecological impact.

The importance of pile dwellings has been internationally recognized and of the 937 archaeologically documented sites around the Alps alone, 111 have been included in the UNESCO World Heritage List (UNESCO 2015). Despite the increasing diversity and geographical spread of these settlements through time, a common feature remains: they all used wood as the main construction material. We will now consider more specifically how the analysis of tree rings from construction timbers, wooden objects (tools, artefacts), and charcoal from pile-dwelling sites has been and can be used (1) to shed light on how our ancestors used, managed, and impacted forest resources following the transition to early settlements and (2) to provide insights into open questions about human responses to environmental changes in prehistory. We will also discuss the potential limitations of this approach.

16.2.2 Tree-Ring Studies of Pile-Dwelling Sites

Dendrochronology has been systematically implemented in pile-dwelling research since the 1970s (Billamboz 2004). The availability of large amounts of well-preserved timbers at these sites has often resulted in the collection of thousands of samples for tree-ring studies, including timbers from young specimens with few tree rings (Orcel 1980). This has been the case for several sites in Switzerland where 62–83% of piles were derived from trees less than 40 years old (Tercier et al. 1996) and pile dwellings in southwest Germany where the vast majority of investigated timbers had less than 50 rings (Billamboz 2008). Similarly, most of the wood recovered at other pile-dwelling sites in different parts of Europe contained less than 45 rings. For example, at the site of Žemaitiškė 2, in Lithuania, most of the piles were made of trees less than 40 years old (Girininkas 2010), whereas at La Draga—the only pile-dwelling settlement found to date in Spain—numerous posts are from yet younger trees of 10–35 years (Tarrús i Galter 2008). Absolute dates were not obtained for the latter two sites, but dendrochronological research on the timbers from La Draga has provided information about building phases and about the seasonality of construction works at Žemaitiškė 2, which must have taken place during the summer, as inferred from clusters of relative felling dates in the spring and early summer months (Girininkas 2010). For pile-dwellings, where bark is often preserved on the timbers, it is usually presumed that there was little time between felling and construction. The wet working environment rendering a period of storage and drying (during which the cambium would rapidly degrade and the bark be lost) somewhat pointless, as well as perhaps impractical.

In contrast, the inclusion of timbers from young trees (<50 years) in his studies on pile dwellings of Lake Constance allowed Billamboz (2014) to apply dendrotypology—a dendroarcheological approach based on the classification of timber according to tree age, growth patterns, and degree of stem conversion—to infer absolute dates for building activities and identify patterns of woodland management, development, and degradation over the period c. 3900–2400 BCE. His results suggest that clearance of woodlands coincided with phases of human settlement, during which slow growth patterns in timbers reflected long-lived (>100 years) trees from dense forests. These settlement phases were followed by coppicing practices in periods of demographic expansion, which were represented by shorter (<100 years) tree-ring series from faster-growing trees, until the surrounding forest could no longer supply enough timber and large old trees had to be cut and processed into the desired dimensions (indicated by the high level of conversion of timbers from later building phases). Billamboz (2014) thus not only answered primary archaeological questions of dating and seasonality of construction phases, but his approach in defining types of growth patterns including for short-lived trees served also to reconstruct forest dynamics and ancient woodland management practices such as coppicing. Such studies enhance the archaeological record and maximize the potential of applied interdisciplinary tree-ring research.

Yet many studies on wood from pile dwellings focus on more linear outcomes—such as dating—for a variety of reasons, including restrictions on sampling, difficulties of sampling/preservation, and funding for specific purposes. For example, Čufar et al. (2010) describe the materials from seven pile dwellings at Ljubljansko barje in Slovenia—comprising 187 samples from 1039 oak timbers—but point out the limitation of their results, because their research was limited to oak samples with more than 45 rings. This is a potential shortcoming for a number of studies, based perhaps also on the internationally recognized thresholds of a minimum of 100 tree rings for crossdating (Baillie 1982).

Billamboz's research, as well as other studies from more recent historical periods (e.g., Sass-Klaassen et al. 2008; Domínguez-Delmás et al. 2011), however, have demonstrated that short series with less than 50 tree rings can be confidently crossdated when the bark edge is present in a large set of coeval samples, as Huber (1967) suggested five decades ago. Dendrochronological research of trunks from Žemaitiškė 2 and La Draga has shown that the use of relatively dated series ('floating chronologies'), not yet anchored in time, can still provide information about occupation periods and building and repair phases (Tarrús i Galter 2008). Furthermore, species suitable for tree-ring studies such as ash (*Fraxinus* sp.), which are sometimes disregarded (e.g. Čufar et al. 2010; Mazurkevich et al. 2010), have the potential to be crossdated and can provide valuable ecological information (Sass-Klaassen et al. 2004; Sass-Klaassen and Hanraets 2006).

16.2.3 Dendro-archeo-ecology of Pile Dwellings to Inform the Impact of the 8.2K Event

According to Sherratt (2004), the abundance of pile dwellings found and studied during the past 150 years has produced such a bulk of scholarly articles and monographs that it has led to "complacency": a lack of enthusiasm for these types of settlements. Paradoxically, there are still important open questions to which dendrochronological data from the wooden remains of pile dwellings could hold the answers. For instance, the impact of the 8200 BP event (the 8.2K event) in the development and spread of early farmer communities in the Euro-Mediterranean region is still a subject of much debate (Berger and Guilaine 2009 and references therein). Approximately 8200 years ago, the North Atlantic region suffered a drastic drop in temperatures, possibly due to a sudden flush of $>10^{14}$ cubic meters of freshwater into the Labrador Sea following the collapse of an ice dam (Barber et al. 1999). This is hypothesized to have induced cold and dry climatic conditions in the Near and Middle East, and cool and wet conditions in Central and Western Europe (Bauer et al. 2004) that disrupted the progression of crop and animal domestication in the eastern Mediterranean and prompted the migration of early farmers into current Greece and Bulgaria (Weninger et al. 2006). In turn, this could have promoted the transfer of knowledge further west into Europe. The archaeological records throughout the northern Mediterranean show a disruption around that time (see Fig. 1 in Berger and Guilaine 2009), but the paleoclimatic proxies (mostly lake and marine sediments) lack the high-resolution temporal and spatial scale needed to effectively assess the impact of this event. Tree rings could offer a potential solution to this, while contributing indirectly to a broader context for the Early Anthropocene hypothesis within coupled natural-human systems: climatic forcing of human systems might in turn lead to human modifications of a new environment, which might then develop to a substantial anthropological footprint leading to new forcing of the natural environment. Tree-ring chronologies (from pile-dwellings and/or paleoenvironmental sites) can thus be used for high-resolution, spatially resolved proxies for detecting the described sudden on-set climatic change. In addition to this, a more developed hypothetical approach might be to use dendrochronology (and/or dendrochronologically facilitated radiocarbon wiggle-match dating) to date occupation phases of sites across the wide geographic region and to explore the spread of this mode of habitation and the spread of technologies and ecological modifications in relation to climate. For example—if the south to north diffusion previously observed is contemporary with changes in climate characterized using a range of proxies, the timing and geographical spread of pile dwellings could be linked to migration driven by cooler conditions proliferating to the North (as in later migration periods). Tree-ring patterns from more or less contemporary pile dwellings could then be used to create maps of ecological impact providing the much needed spatial and temporal resolution to better examine the 8.2K event (e.g., Crutzen and Stoermer 2000; Crutzen 2002;

Smith and Zeder 2013) and its place in the wider context of debates on the timing of the transition of the Holocene to the Anthropocene.

16.2.4 Euro-Mediterranean Dendro-archeo-ecology in the Context of the Early Anthropocene Hypothesis

Perhaps the foremost evidence in support of dating the Anthropocence prior to the industrial revolution comes from work in China related to the expansion of rice agriculture on methane production c. 5000 years ago (e.g., Li et al. 2009). This study provides strong support for human-induced increases in methane prior to the unprecedented rise of modern times. Moreover, in a review of the archaeological evidence for human modification of the environment of the Yellow River region in China, 8000–2000 BP, Zhuang and Kidder (2014) conclude that humans have been active agents in transforming the physical properties of land, water, and atmosphere in that particular cradle of civilization for much of the Holocene and that the onset of the Anthropocence should not be confined to changes in atmospheric chemistry alone. The same argument can be made for the Euro-Mediterranean region. We propose an approach, which revisits existing dendrochronological data from archaeological sites with dendroecological analysis and then combines this with multiple lines of paleoenvironmental and archaeological data at coarse and fine geographic scales. At the coarse scale these data could be evaluated to assess the progressive timing of the anthropogenic transformation of the environment. The broad-scale precisely dated and temporally resolved ecological change implied populations and associated landscape transformations could then be used to inform model-based approaches for spatially explicit reconstruction of long-term histories of ecosystem transformation, human populations, and land use at a global scale (e.g., Ellis et al. 2013b). Finer scale, site specific data could be used to provide much needed empirical evaluations for existing global scale models. In this way the density of archaeological sites, the abundance of well-preserved wooden remains, and the capacity for tree-ring sequences to combine precision data for ecological/climatic change and human response could perhaps best be used to elucidate the complex interplay of feedbacks superimposed on Early Anthropocence debate.

There seems much potential here, given the large number of archaeological data already collected across this critical time period; however, there are some hurdles to overcome. Many potentially relevant archaeological datasets remain unavailable and/or unpublished for a variety of reasons. Some may have had insufficient sample depth for publication in terms of the primary purpose for which they were collected. Similarly, some may consist of very short sequences that were previously deemed unsuitable for dendrochronological crossdating, or may have been dated by radiocarbon analysis rather than dendrochronology. Some may have a commercial value in terms of dating, others may simply be stored in non-standard legacy digital formats (see Jansma et al. 2010, 2012; Brewer et al. 2011). Work will be needed to collect and consolidate existing archeological data into accessible forms so that new avenues may be explored and dendro-archaeo-ecology will need to be explored as part of 'big data' interdisciplinary collaborative efforts.

16.3 Sixteenth to Nineteenth Century Dendro-archeo-ecology in Eastern North America

A catastrophic depopulation of First Nation people of eastern North America began in the sixteenth century (Harriot 1972; Blakely and Detweiler-Blakely 1989; Thornton 1987, 2000) resulting in 100+ years of less intensive forest management and, in some places, afforestation (Cronon 1983). European communities initially sprang up in larger river valleys and then expanded across the region during the early seventeenth century. The growth and expansion of these communities (e.g., for agriculture) led to coarse-scale deforestation across the region. The landscape was significantly transformed with the influx of new people, culture, and land-use philosophies.

Euro-American land use affected the Midwestern United States (U.S.) much later but also resulted in coarse-scale forest clearing. Construction of roads and canals into the Midwestern U.S. during the early 1800s allowed for the utilization of fertile soils for agriculture. The region was soon cleared of forests and old-growth forest was noted to be rare. For example, A.W. Butler (1896) commented on the loss of "tall trees" and "heavy timber" especially in southern Indiana. Florence Hawley (1941) also noted a lack of old forests in the Midwest and Indiana during the 1930s (Senter 1938a, b). As of the late twentieth century, forest cover was 13%, 21%, and 27% respectively for the three Midwestern U.S. states Illinois, Indiana, and Ohio (USDA 2015a, b, c).

Intensive land-use and the paucity of dendroecological collections prior to the sixteenth century limits our understanding of historic eastern North American forest dynamics (McCarthy et al. 2001). Because surviving old-growth forest typically occurs on low-productivity sites (Stahle and Chaney 1994; Therrell and Stahle 1998), there is a bias in our understanding of pre-settlement forest dynamics. For example, the old-growth forest in Illinois, Indiana, and Ohio is extremely small, equaling 0.01% (~2100 ha), 0.005% (~500 ha), and 0.005% (~400 ha) of the total area of each state, respectively (Davis 1993). This is unfortunate as we have little information about forest dynamics on productive sites. Fortunately, the Euro-American doctrine of manifest destiny left important artifacts for dendrochronology preserved across the region: historic timbers from the thousands of structures, water vessels, and wooden objects. These historical timbers give us the opportunity to build a more robust geographical analysis of forest dynamics in the centuries leading up to the Euro-American immigration.

Dendroarchaeological analysis of historic timbers in eastern North America began during the latter portion of the twentieth century (e.g., Stahle 1979; Cook and Callahan 1992). Since then, dendroarchaeological analysis has become a vigorous area of research (e.g., Krusic et al. 2004; Wight and Grissino-Mayer 2004; Grissino-Mayer 2009' Harley et al. 2011; Baas and Rubino 2012; Martin-Benito et al. 2014; Rubino 2014; Rubino and Baas 2014, New York State and NE North American Dendrochronology Project, https://dendro.cornell.edu/projects/usa.php, the Historic Timbers Project, http://centralapptimbers.weebly.com/blog/introducing-the-historic-timbers-project). The volume and geographic coverage of recent research increases our ability to reconstruct forest dynamics prior to the nineteenth century.

Given the abundance of timber available to early Europeans, construction was routinely performed using timber found on-site: trees were felled and incorporated into buildings as beams, rafters, floorboards, joists, and braces (Senter 1938a; Hutslar 1992; Roberts 1996). This is especially true when large timbers for structures like barns were needed and transport and sawing was impractical or cost prohibitive (Roberts 1996). Therefore, it is possible that individual structures represent a portion of a single forested stand, with specific species often used for particular purposes (Baas and Rubino 2013). As these structures were typically built for agricultural purposes and thus in agriculturally productive areas, it is likely that timbers from these structures represent a more productive portion of the landscape than extant old-growth forests.

Successful crossdating of the timbers found in historic structures has led to centuries-long chronologies from numerous taxa [white oak (*Quercus* subgenus *Lepidobalanus*), red oak (*Q.* subgenus *Erythrobalanus*), hickory (*Carya* spp.), ash (*Fraxinus* spp.), tulip poplar (*Liriodendron tulipifera*), and American beech (*Fagus grandifolia*)]. For example, there are eastern U.S. oak chronologies that date to the fourteenth and fifteenth centuries (E.R. Cook, pers. comm.). The taxa of these chronologies are ecologically important species and have promising potential for understanding the structure and dynamics (e.g., disturbance and recruitment patterns) of sixteenth to nineteenth century forests.

Below we present two examples that combine dendroarchaeological and dendroecological analyses of data from historic timbers to yield new information on forest dynamics in eastern North America prior to the nineteenth century.

16.3.1 Forest Dynamics in the Ohio and Hudson River Valleys

Disturbance is integral to the community structure, composition, and development of eastern North American forests (Oliver 1981; White and Pickett 1985). Quantification of abrupt and extended increases in ring width associated with acute disturbances has long been used to elucidate forest history and stand dynamics in eastern North America. These increases in growth, hereafter referred to as growth releases, are the response of trees to increased light and other resources as the result of disturbance-induced tree mortality in dense forests (Lorimer 1985; Nowacki and Abrams 1997) where increased resource availability, primarily light, significantly

Table 16.1 Growth release metrics of timbers from four structures in central and southern Indiana and eight structures from the mid-Hudson River Valley (New York); SE = standard error

	Indiana	Hudson River Valley
Releases/century: mean (SE)	0.9 (0.1)	0.7 (0.1)
Releases/century: range	0–6.7	0–4.3
Return interval: mean (SE)	28.4 (4.7)	36.7 (10)
Return interval: range	1–97	1–117
% of structures with ≥1 release event	52	52
% of years with release per timber: mean (SE)	1.2 (0.7)–13.5 (1.5)	6.8 (1.5)

increases annual growth (e.g., Nowacki and Abrams 1997). Reconstructing centuries of canopy disturbance and tree recruitment gives insight into the timing, frequency, and spatial extent of long-term forest dynamics.

Timbers from four structures (*Quercus* subgenus *Lepidobalanus*; $n = 114$) in central and southern Indiana and eight structures ($n = 46$) from the mid-Hudson River Valley (New Paltz, New York; Krusic et al. 2004; Pederson et al. 2013) were analyzed to reconstruct historic forest disturbance. The Indiana timbers cover the 1604–1884 CE period while those from the mid-Hudson Valley cover 1449–1805 CE. Growth release events were objectively identified by comparing the growth rates of adjacent 15-year growth segments; a release was defined as an increase in growth of at least 50% (Lorimer 1985), where the median growth rate was calculated for each 15-year segment (Rubino and McCarthy 2004). The number of releases (relativized to a per 100-year basis to facilitate comparison among trees of different ages) and the return interval (years between individual release events) were calculated. There was no clear link between stem age and release—releases and suppressions occurred regardless of an individual stem's age.

Growth release analysis revealed disturbance regimes that shared several similarities between the two geographic regions (Table 16.1), but also unique subtleties between and among individual buildings (Fig. 16.2). For instance, release return intervals varied widely—from 1 up to 117 years (Table 16.1)—at both the timber and structure level. Releases were noted in 52% of all the samples, both in New York and in Indiana.

To better understand the spatial extent of disturbance (stand-wide vs. single tree), the percentage of trees exhibiting a release in an individual year was determined. We focused analysis only on years in which 10 or more timbers were available (Rubino and McCarthy 2004). A stand-wide disturbance was identified if more than 25% of the timbers exhibited a simultaneous release (Nowacki and Abrams 1997; Rubino and McCarthy 2004). Synchronous and asynchronous releases were found in both New York and Indiana structures (Fig. 16.2). The percentage of years in which at least one release was detected in each Indiana structure ranged from 5 to 88%. Periods of synchronous release (> 25% of the trees) were found in only one of the four Indiana structures (the Hanover barn in Fig. 16.2a). In the Hudson Valley

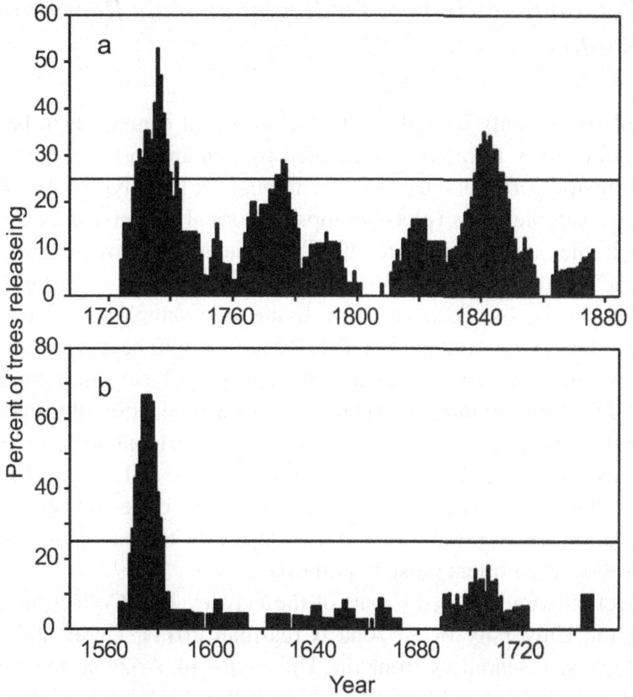

Fig. 16.2 Disturbance history (percentage of trees exhibiting a growth release in a given year) created from the timbers of (**a**) a barn from Hanover, IN (1716–1876; $n = 54$) and (**b**) 8 structures from New Paltz, NY (1556–1747; dates represent years in which 10 or more timbers were available for analysis). The horizontal line represents the threshold for determining if a disturbance was a synchronous disturbance event (at least 25% of the timbers exhibiting release in a year)

samples, a release was detected in 29% of the years analyzed, and one synchronous release period was identified in the late 1500s (Fig. 16.2b).

The mix of synchronous and asynchronous releases suggests classic gap dynamics in the forests from which these structures were built. These dynamics are characterized by patchy disturbances that kill one to a few trees in most years (e.g., Runkle 1990). Periodic, coarser-scale events (ice storms, tornadoes, large wind storms, and drought) likely resulted in synchronous releases at the stand-scale—such as reconstructed from the historic timbers of the Hanover barn—and at the regional scale—such as across the eight structures from the Hudson River Valley (Fig. 16.2). Consequently, we infer that these forests were most likely uneven aged and heterogeneously constructed (e.g., mix of shade-tolerant and—intolerant species). Furthermore, repeated disturbance is also most likely necessary for stems to attain canopy dominance (Runkle 1990; Rubino and McCarthy 2004).

16.3.2 Potential of Timbers for Regional-scale Recruitment Studies

Historic timbers not only have the potential to reveal canopy disturbance prior to the nineteenth century in regions dominated by human land-use, there is potential to acquire a history of regional-scale recruitment. A new hypothesis derived from a range of tree sample types (plots, stumps, historical timbers) indicates a pulse of recruitment in the late 1600s in broadleaf-dominated old-growth forests of eastern North America (Pederson et al. 2014). Of the many questions regarding this finding, an important one is that because many living tree samples were collected after 1980 CE, tree longevity—rather than disturbance—might be driving these observed patterns in recruitment. Given that historical timbers have the potential to reach further back in time, both the age structure of a large collection of timbers as well as the pattern of initial rings can inform the timing of regional-scale recruitment and the conditions under which these trees originated *sensu* (Lorimer 1985; Lorimer and Frelich 1989). Here, we use a collection of timbers collected between 1938 and 1941 across three states of the lower Midwestern U.S. to conduct an additional test of the late 1600s recruitment pulse hypothesis.

The collection we examined is part of the Hawley-Bell Collection, a collection archived at the University of Arizona (Creasman 2011). Under the direction of Florence Hawley, researchers from the University of Arizona and University of Chicago explored the lower Midwestern U.S. region (Arkansas, Illinois, Kentucky, Missouri, Tennessee) from 1934 to 1941 for old samples from stave mills, structures of the indigenous Mound Builders and European immigrants, and old forests (Bell 1940, 1941). Field notes indicate that one goal was to procure *Juniperus, Pinus,* and *Quercus* samples with at least two centuries of "sensitive" rings, where sensitive means strong interannual variability in ring width and the potential of great sensitivity to climate. Investigators often had access to stumps or cross-sections, which eases estimating sample age in the field. Field notes include statements like, "*I selected a large sample of the best pieces*" (Bell 1941), and "*En route from Corning to Little Rock we passed some small mills but the timber was neither old nor sensitive*" (Bell 1940).

Here, we focus on *Quercus* from three states within the Hawley-Bell Collection: Arkansas, Illinois, and Kentucky. We selected and processed 15–25 cross-sections from each state for a total of 57 samples. Care was taken to identify missing rings or false rings using standard tree-ring techniques so that we could precisely capture extreme events (Stokes and Smiley 1968; Black et al. 2016). The goal of our sample selection from the Hawley-Bell collection was to analyze an equal number of samples from three observed types of growth trajectories: suppressed growth (extremely narrow rings) near the pith followed by a release; wide rings near the pith followed by suppressed growth; and sections containing more than one period of suppressed growth (Fig. 16.3). Our subsample consists of cross-sections collected from mills and is composed of 21 trees from Newton County in Arkansas, 20 trees from three counties in Illinois (16 from Union County), and

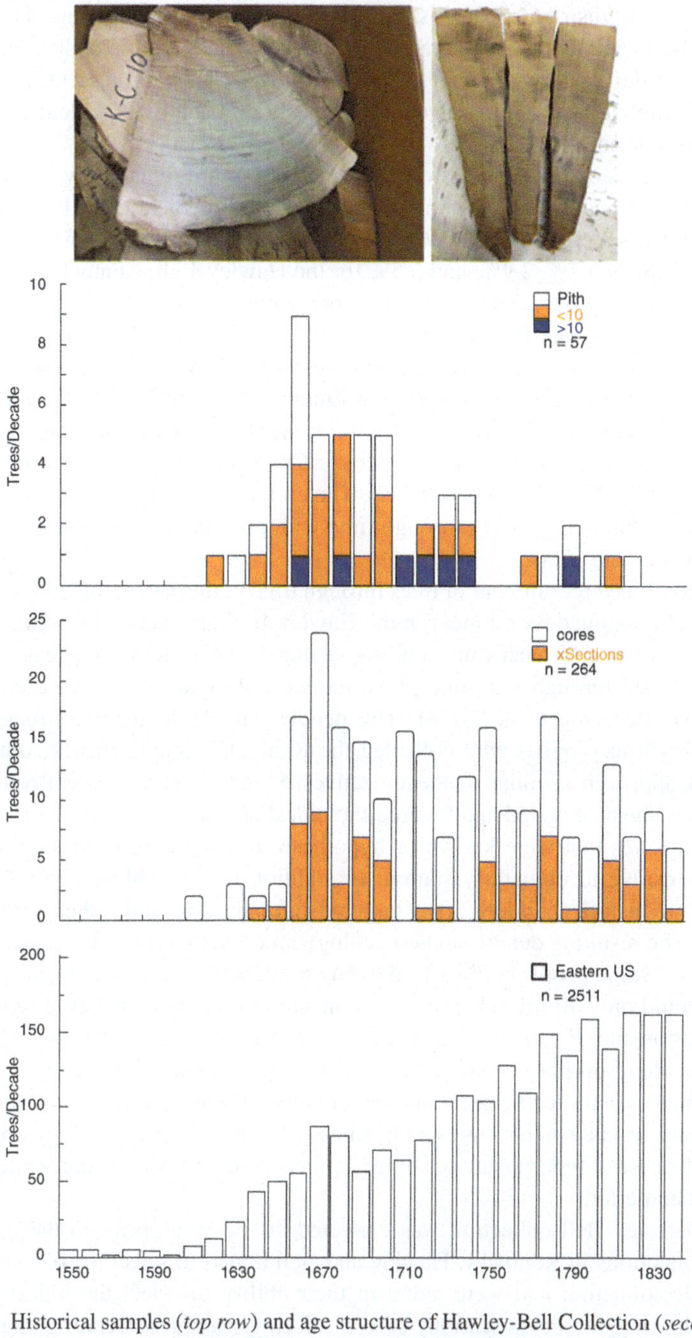

Fig. 16.3 Historical samples (*top row*) and age structure of Hawley-Bell Collection (*second row*), Kentucky tree-ring data from 1980–2009 and historical timbers dating to the early ninteenth century (*third row*), and representative sampling in eastern US broadleaf-dominated forests during the last century (*bottom row*). *Darker colors* in the bar plots represent greater certainty of the location of the pith; the presence of piths in the eastern broadleaf collection is unknown

16 trees evenly distributed from six counties in Kentucky. Inner ring dates of the Hawley-Bell samples were compared to inner ring dates from a collection of trees targeted for dendrochronological studies and historic timbers in Kentucky as well as trees sampled in plots of old-growth forests dominated by broadleaf trees in the eastern U.S. (Pederson et al. 2014).

Inner ring dates from the Hawley-Bell collection are remarkably similar to the Kentucky data and somewhat similar to the eastern broadleaf data (Fig. 16.3). The percent of trees recruited before 1650 from all trees in each collection between 1550 and 1849 equals 8.7%, 2.9%, and 5.5% for the Hawley-Bell, Kentucky, and eastern broadleaf collections, respectively. The percentage of inner ring dates between 1650 and 1699 equals 56.5%, 30%, and 19%, respectively. Unexpectedly, timing and distribution shape of the inner ring dates of the Hawley-Bell collection subset containing the period of increased recruitment (1620–1699) did not significantly differ from those of the targeted trees of the Kentucky collection, whether a sample contained the pith or not (Fig. 16.3; Chi Square test of independence, chi-square = 8.92, DF = 8, prob. = 0.349). The eastern broadleaf collection shows the continual recruitment of trees through time with prominent peaks in the mid 1600s and early and late 1700s.

The continued recruitment of trees through time in the eastern broadleaf data and the lack of continued recruitment in the Hawley-Bell and Kentucky collections can be explained by differences in sampling strategy. The eastern broadleaf collection was developed through sampling plots that were designed to be representative of the forest (Pederson et al. 2014). The people who built historic structures and the dendroclimatologists who collected the Kentucky samples had a significantly different approach. Similar to the objectives of the Hawley-Bell collection, dendroclimatologists targeted and favored the oldest appearing trees in the forest from specific species that were known or hypothesized to be long-lived, cross-datable, and/or sensitive to climatic variation. In addition to this and as discussed above and below, historic builders often targeted trees by size and, when possible, by species. The resulting dendro-archeo-ecology records, therefore, are constrained by economics: some people could afford to buy timbers from elsewhere, whereas other people could only afford to harvest trees on site or repurpose timbers from existing structures on site. Finally, the targeting of trees for dendroclimatology and dendro-archeo-ecology purposes could constrain collections in ways that make them only representative of a specific point in time. Together, these differences give significant uncertainty on the spatial representativeness of dendro-archeo-ecology collections and, in the case of repurposing or buying timbers from elsewhere, representativeness of a certain period.

The Hawley-Bell collection was collected 40–60 years prior to the dendroclimatic collections in Kentucky. Hawley and Bell aimed to select the oldest samples from this collection and were aided in their ability to select the oldest trees by seeing the rings on cross-sections before selecting samples. Dendroclimatologists do not have this luxury. Yet, the Hawley-Bell collection indicates an uptick in recruitment of *Quercus* in the 1650s and a spike beginning in the 1660s. Recruitment in the 1660–1690s was almost fivefold that of the 1610–1640s (24 vs. 5 samples in each 40-year period) and thus mirrors the 10-fold increase in the Kentucky dendroclimatic collections (72 vs. 7 samples for the same periods; Fig. 16.3). The age structure of the Hawley-Bell collection is also nearly the same as collections made in 1915–16 in a freshly cut old-growth forest in southeastern Kentucky (Haasis 1923) and collections from Missouri, Kentucky, West Virginia, and New York State in the early 1900s (Huntington 1914). If recruitment in broadleaf-dominated forests is simply dominated by the frequent and annual dynamics of forests at fine scales, we would expect the spike in recruitment of the Hawley-Bell collection to occur four to six decades earlier than in the Kentucky collection.

A spike in age structure in one to a few historical structures built around the same time could be a false positive due to the nearly simultaneous establishment of targeted timbers within a narrow range of sizes. With this combination of timber selection and ecological dynamics, one would expect a flat age structure from a collection drawn from multiple structures over multiple periods (null hypothesis). The ecological process that counters this hypothesis is that the range of shade tolerance and the ability to persist in the understory at the species- and individual tree-level could result in trees of the same size or stature representing trees who were recruited to coring or cutting height in different decades and, sometimes, centuries. A case study of *Quercus montana* trees—including the oldest-documented *Quercus montana*—supports this alternative hypothesis. The oldest *Quercus montana* tree has an inner ring dating to 1578 and appears to have been an understory tree for more than three centuries (Pederson 2010). Three of the four other *Quercus montana* trees from the same stand that have pith or are likely within 5 years of pith (out of 32 cored trees) date to the 1750s or 1790s (Pederson et al. 2017). Thus, shade tolerance and ecological history of a site could likely make timbers of dendro-archeo-ecology collections relatively diverse in age as builders were targeting trees of specific sizes regardless of age. At minimum, these collections would not likely be monolithic in their representativeness of a specific era.

A re-analysis of the 461 historical timbers used in Pederson et al. (2014) drawn from several collections that cover New Jersey, Pennsylvania, Ohio, Indiana, and Arkansas (but dominated by samples in Ohio) allows us to further investigate this hypothesis. The age structure of this collection results in more trees being recruited prior to 1660 and a relatively flat age structure from 1590 to 1650. However, only 18% of this collection has trees with inner rings between 1530 and 1649 compared to the 15% with inner rings between 1670 and 1689 (the peak in recruitment of this data set), and 35% with inner rings between 1660 and 1699, suggesting that perhaps the increase in recruitment in the mid- to late-1600s from the Pederson et al. (2014) dendro-archeo-ecological collection is not due to a bias driven by targeted collections and site ecology. A flat age structure is not produced by samples drawn from multiple structures built over differing periods and a large space. These findings further suggest that perhaps the relatively narrow period from the mid- to late-1600s was an important event to the structure and function of forests in the central Midwest of the eastern U.S. More collections from a greater range of building periods spread more evenly over a larger region would be an important test of this hypothesis. As it is, our subsample of the Hawley-Bell collection provides additional evidence that episodic disturbance appears to be an important driver of structure of *Quercus*- and other broadleaf-dominated forests in the eastern U.S.

16.3.3 Discussion

Our case studies of Euro-Mediterranean Neolithic pile dwellings and of historical (sixteenth to nineteenth century) eastern North American forest dynamics illustrate how dendroarcheological material can be used to improve our understanding of land-use change and forest ecology during various episodes over the Holocene. These examples complement previous dendrochronological studies that documented past land-use changes and that are relevant within the context of the Early Anthropocene hypothesis. For instance, a tree-ring based time series of felling dates in Central Europe mimics pre-industrial deforestation and population trends and particularly documents the disruption of tree harvest and wood construction activity (and thus potentially reforestation) during the Great Famine and Black Death in the late fourteenth century (Büntgen et al. 2010, 2011). Across the North American west coast an eighteenth-century shift in the fire regime documented in a tree-ring based fire history for the Sierra Nevada, CA, was found to co-occur with massive Native American depopulation in the region (Taylor et al. 2016). The Sierra Nevada fire regime shift can provide further nuance to the Early Anthropocene hypothesis, which postulates that even before the onset of the industrial era humans have influenced the earth's carbon cycle and climate. Short-term atmospheric CO_2 decreases due to widespread pandemics in the Americas in the period 1500–1750 CE (Nevle et al. 2011) might have been counter-balanced by CO_2 emissions from increased Sierra Nevada area burned (Taylor et al. 2016).

The ecological potential of tree-ring studies on Euro-Mediterranean pile-dwelling timbers has so far only been explored at a limited number of sites. Replication and expansion of approaches summarized by Billamboz (2014) offer much potential to bring needed spatial and temporal coverage to data relevant in discussions of Anthropocene onset. Revisiting archived wood collections such as the Hawley-Bell collection may lead to integrated dendro-archaeo-ecological studies for the Neolithic and a wide range of other periods. Unbiased sampling strategies that also consider timbers from young trees and that include a wide range of species should then be adopted. These types of sampling strategies are becoming the norm in multidisciplinary studies that involve dendrochronology, such as those on shipwrecks (Haneca and Daly 2014) and that aim not only at dating timbers, but also at gathering information about the organization of the timber supply, the selection and management of species, and the origin of the wood. Such dendroprovenancing studies can add detail to our existing understanding of the intensification of deforestation in Europe and in the Americas. For instance, the lack of suitable wood for the construction of a harbour at the Roman site of Voorburg-Arentsburg, in the current Netherlands, required the transport of timber over more than 600 km (Domínguez-Delmás et al. 2014). Centuries later, following drastic deforestation in the Middle Ages, construction timber in the Low Countries (northern Belgium and the Netherlands) was acquired through importation primarily from the Baltic region, Scandinavia, and Germany (e.g., De Vries and Van der Woude 1997).

Fig. 16.4 A cross-section through an oak piling preserved in a water-logged environment shows excellent preservation of the sapwood and outer-most growth ring (waney edge). While this sample is a piling from the so-called "Lost Harbor of Constantinople" from the Yenikapı excavations by the Istanbul Archaeological Museum, rather than from a Neolithic era pile dwelling, it serves to illustrate a number of points about the preservation and collection of metadata for such wet-wood samples. Two radii (marked with *red* and *green dots*) were measured on the sample and combined into a single measurement series. Note how both avoid the scarring in year 18 of the tree's life in order to get the best representative measurement series for archaeological chronology building and dating

We here advocate for the integration of such individual studies in order to optimally inform and calibrate the Early Anthropocene hypothesis and to fill existing gaps in the archeological record. Several initiatives have recently been launched to facilitate integration by addressing common hurdles to data sharing and to new uses for old data. These initiatives include a tree-ring repository for dendro historical/archaeological data in Europe —the Digital Collaboratory for Cultural Dendrochronology, DCCD (Jansma et al. 2012; Jansma 2010), a universal tool for converting non-standard digital tree-ring data to a standard transferable format (Brewer et al. 2011), and a move towards a new tree-ring data standard, TRiDaS (Jansma et al. 2010). For instance, meta-data regarding scars and other aspects of wood anatomy and anthropogenic alteration (Fig. 16.4) can be collected at the time of measurement using the measurement and curation software Tellervo (http://www.tellervo.org/), which is designed to archive multi-user data and metadata according to TRiDaS for future cross-disciplinary applications. The TRiDaS initiative thus presents significant new opportunities for the wider dendrochronological community, in that by standardizing data and metadata collection, it makes data collected for one specific type of study fully accessible for analysis under a different subdiscipline.

This is particularly relevant and helpful for archaeological data from the Neolithic period, and specifically for pile-dwelling remains, as most of the material is preserved as wet wood and so is subject to rapid degradation and decay if effective conservation strategies (e.g., freeze drying) cannot be immediately put in place. For instance, rapid degradation by drying occurs as wood is gradually exposed to the air during archeological excavation (the splitting at the outer edge in Fig. 16.4).

Facilitating the collection of all possibly relevant metadata when processing wet wood is a critical step forward in constructing the platform for a whole range of interdisciplinary future studies. Such concerted efforts are needed to fully exploit the potential of dendrochronology at the archeology—ecology—climate nexus and to move the use of dendro-archeological material from strictly archeological applications (e.g., dating) towards exploration of its ecological potential.

Before we can fully utilize historical timbers for ecological studies, there are more uncertainties to be considered. For example, sixteenth century builders likely biased North American samples because they often chose stems based on size: large diameter stems would be selected for e.g. tie beams, but avoided for other timber structures such as rafters, due to necessary laborious sizing and massive weight. They may have further biased our samples by selecting long, clear boles with little to no branching to avoid delimbing. Such logs would most likely be found in stands with high stem density. Builders also selected species with optimal decay resistance and workability, which potentially limits the number of taxa available for analysis (Hutslar 1992; Baas and Rubino 2013). This is also the case for central and northern Europe, where oak was the preferred species for construction purposes (Haneca et al. 2009). In the eastern U.S., however, we have worked on structures constructed with a larger range of species than expected. Despite the increased activity in dendroarcheology, we are still in the initial growth stage for ecological applications (De Graauw n.d.).

In addition to potential size and species biases, uncertainty might also be introduced by replicate samples when multiple timbers in a structure derive from the same tree. This can be detected through high extremes in inter-series correlations and tree-ring series should therefore be internally crossdated in a first identification step before they are included in further analyses (e.g. Mom et al. 2009; Domínguez-Delmás et al. 2011, 2014). A significant unknown in using building timbers to study forest recruitment is the height at which the tree was cut. The trees from which timbers were cut could be significantly older than the timbers themselves, even when sapwood and/or pith are present in the samples. When it is not possible to collect a complete cross section, core samples may miss the pith adding additional error to recruitment estimates. Finally, when sampling historical buildings, metadata related to the original forest stand structure (e.g., spatial distribution and canopy class of stems) are missing, which can introduce uncertainty when historic timbers are used for forest ecological research purposes. Together, there is much uncertainty in using dendro-archeological data for ecological studies. However, dendro-archeo timbers may help us recover highly resolved forest dynamics during periods that predate most extant forests. Some recurring challenges such as low sample depth will likely be remedied as this field expands, while other limitations (e.g., short series length) and uncertainties (e.g., species and tree size bias) need to be taken into account and carefully addressed during analysis and interpretation.

In addition to an improved integration of dendrochronological subdisciplines, the quantification of past land-use changes and their impact on the carbon cycle and on Earth's climate calls for a close alliance of dendrochronology with related scientific disciplines that aim to address the same subjects (Ellis et al. 2013b). For instance, estimates of eastern North American forest productivity at the time of Euro-American settlement could be complemented by tree composition estimates statistically derived from nineteenth century public land survey records (Paciorek et al. 2016; Goring et al. 2015).

Dendroecology could also be used to calibrate the impact of the indigenous North American depopulation following European contact on historical fire regimes (Liebmann et al. 2016; Taylor et al. 2016), biomass burning, and thus global CO_2 levels (Nevle et al. 2011). This work would complement the longer-term context using sedimentary charcoal records (Marlon et al. 2012). Such interdisciplinary collaborations could greatly benefit from the inclusion of model-based approaches, as has been demonstrated in the spatially explicit reconstruction of past land use in Mesoamerica (Cook et al. 2012) and in the Euro-Mediterranean region (Büntgen et al. 2011). Such model-data assimilation exercises are particularly useful when linking past land-use changes and forest productivity estimates to atmospheric CO_2 concentrations (Kaplan 2015; Ellis et al. 2013b). Whereas tree-ring data have been used to estimate the carbon sequestration capacity of modern forests (Babst et al. 2014b, c; Dye et al. 2016; Alexander et al. n.d.) and to constrain the global carbon cycle's sensitivity to climate (Frank et al. 2010), the spatially explicit reconstruction of past forest carbon sequestration capacity will require their pairing with sophisticated and detailed modeling efforts (Ellis et al. 2013b).

16.4 Conclusion

At the nexus of archeology, climatology, and ecology, dendrochronology is uniquely positioned to study past human-environment interactions. However, its potential to contribute to the quantification of Holocene-era land-use changes and forest dynamics in Europe and North America is limited because the majority of lowland forests in these regions have been cleared throughout the course of human history. As a result, tree-ring chronologies from living trees are derived from either secondary forests or remnant old-growth stands and might not be representative for forest dynamics and productivity. We advocate for the development and application of dendro-archeo-ecology: deriving ecological information from dendroarcheological collections. Doing so will require a concerted effort to address recurring challenges and uncertainties, adopting unbiased sampling strategies, and revisiting surviving wood collections. This will facilitate the integration of dendrochronology in interdisciplinary collaborations that focus on the quantification of past land-use changes and their impact on the carbon cycle and on Earth's climate.

Acknowledgments The authors thank P. Baker and A. Hessl for very constructive feedback on earlier versions of this manuscript. NP thanks VT for partially subsidizing work conducted on the Hawley-Bell Collection.

References

Alexander MR, Rollinson CR, Babst F, Trouet V, Moore DJP (n.d.) Uncertainty in tree-ring based biomass estimates does not alter growth-climate relationships. Trees Struct Funct, in revision

Baas C, Rubino DL (2012) Using tree-ring growth patterns to date the construction of a beater hay press barn in Allensville, Switzerland County, Indiana. Ind Archaeol 7:35–57

Baas C, Rubino DL (2013) The most successful press in this or any other country: the material culture of 19th-century beater hay presses in the Mid-Ohio Valley. Mater Cult 45:1–20

Babst F, Poulter B, Trouet V, Kun T, Neuwirth B, Wilson R et al (2013) Climatic drivers of forest productivity derived from 1000 sites in Europe. Glob Ecol Biogeogr 22(6):706–711

Babst F, Alexander MR, Szejner P, Bouriaud O, Klesse S, Roden J et al (2014a) A tree-ring perspective on the terrestrial carbon cycle. Oecologia 176(2):307–322

Babst F, Bouriaud O, Papale D, Gielen B, Janssens IA, Nikinmaa E et al (2014b) Above-ground woody carbon sequestration measured from tree rings is coherent with net ecosystem productivity at five eddy-covariance sites. New Phytol 201(4):1289–1303

Babst F, Bouriaud O, Alexander R, Trouet V, Frank D (2014c) Toward consistent measurements of carbon accumulation: a multi-site assessment of biomass and basal area increment across Europe. Dendrochronologia 32(2):153–161

Baillie MGL (1982) Tree-ring dating and archaeology. University of Chicago Press, Chicago

Barber DC, Dyke A, Hillaire-Marcel C, Jennings AE, Andrews JT, Kerwin MW et al (1999) Forcing of the cold event of 8,200 years ago by catastrophic drainage of Laurentide lakes. Nature 400:344–348

Bauer E, Ganopolski A, Montoya A (2004) Simulation of the cold climate event 8200 years ago by meltwater outburst from Lake Agassiz. Paleoceanography 19(3):PA3014. doi:10.1029/2004PA001030

Bell RE (1940) Informal field notes taken during two months wood collecting in the state of Arkansas—1940. The University of Chicago, Laboratory of Tree-Ring Research, University of Arizona. p 27

Bell RE (1941) Informal field notes for Dendrochronology Research during the summer of 1941—Illinois, Kentucky, Tennessee, Arkansas and Wisconsin. The University of Chicago, Laboratory of Tree-Ring Research, University of Arizona. p 31

Berger JF, Guilaine J (2009) The 8200 cal BP abrupt environmental change and the Neolithic transition: a Mediterranean perspective. Quat Int 200:31–49

Billamboz A (2004) Dendrochronology in lake-dwelling research. In: Menotti F (ed) Living on the lake in prehistoric Europe. Routledge Press, New York, pp 117–131

Billamboz A (2008) Dealing with heteroconnections and short tree-ring series at different levels of dating in the dendrochronology of the Southwest German pile-dwellings. Dendrochronologia 26(3):145–155

Billamboz A (2014) Regional patterns of settlement and woodland developments: dendroarchaeology in the Neolithic pile-dwellings on Lake Constance (Germany). The Holocene 24(10):1278–1287

Black BA, Griffin D, Van der Sleen P, Wanamaker AD, Speer JH, Frank DC et al (2016) The value of crossdating to retain high-frequency variability, climate signals, and extreme events in environmental proxies. Glob Chang Biol 22:2582–2595

Blakely RL, Detweiler-Blakely B (1989) The impact of European diseases in the sixteenth-century Southeast: a case study. Midcont J Archaeol 14(1):62–89

Brewer PW, Murphy D, Jansma E (2011) TRiCYCLE: a universal conversion tool for digital tree-ring data. Tree Ring Res 67(2):135–144

Brown PM, Hughes MK, Baisan CH, Swetnam TW, Caprio AC (1992) Giant sequoia ring-width chronologies from the central Sierra Nevada, California. Tree Ring Bull 52:1–14

Buckley BM, Anchukaitis KJ, Penny D, Fletcher R, Cook ER, Sano M, Name LC, Wichienkeeo A, Minh TT, Hong TM (2010) Climate as a Contributing Factor in the Demise of Angkor, Cambodia. Proc Natl Acad Sci 107:6748–6752

Büntgen U, Di Cosmo N (2016) Climatic and environmental aspects of the Mongol withdrawal from Hungary in 1242 CE. Sci Report 6:25606

Büntgen U, Frank DC, Nievergelt D, Esper J (2006) Summer temperature variations in the European Alps, AD 755-2004. J Clim 19(21):5606–5623

Büntgen U, Trouet V, Frank D, Leuschner HH, Friedrichs D, Luterbacher J et al (2010) Tree-ring indicators of German summer drought over the last millennium. Quat Sci Rev 29(7):1005–1016

Büntgen U, Tegel W, Nicolussi K, McCormick M, Frank D, Trouet V et al (2011) 2500 years of European climate variability and human susceptibility. Science 331(6017):578–582

Butler AW (1896) A century of changes in the aspects of Nature. Proc Indiana Acad Sci 5:3–42

Caruso Ferme L, Piqué i Huerta R (2014) Landscape and forest exploitation at the ancient Neolithic site of La Draga (Banyoles, Spain). The Holocene 24(3):266–273

Chrysostomou P (2015) The Culture of the Four Lakes. Conclusions from the archaeological research in Amindeon basin (2000–2014). In: Chrysostomou P (ed) The Culture of Four Lakes, Wooden artifacts from the Prehistoric lakeside settlements of Amindeon basin. Epikentron, Thessaloniki

Cook ER, Callahan B (1992) The development of a standard tree-ring chronology for dating historical structures in the greater Philadelphia region. Tree-Ring Laboratory, Lamont-Doherty Geological Observatory, Columbia University, New York

Cook ER, Woodhouse CA, Eakin CM, Meko DM, Stahle DW (2004) Long-term aridity changes in the western United States. Science 306(5698):1015–1018

Cook BI, Anchukaitis KJ, Kaplan JO, Puma MJ, Kelley M, Gueyffier D (2012) Pre-Columbian deforestation as an amplifier of drought in Mesoamerica. Geophys Res Lett 39(16):L16706

Creasman PP (2011) Basic principles and methods of dendrochronological specimen curation. Tree Ring Res 67:103–115

Cronon W (1983) Changes in the land: Indians, colonists, and the ecology of New England. Hill and Wang, New York

Crutzen PJ (2002) Geology of mankind. Nature 415:23

Crutzen PJ, Stoermer EF (2000) The Anthropocene. IGBP Newsl 41:12

Čufar K, Kromer B, Tolar T, Velušček A (2010) Dating of 4th millennium BC pile-dwellings on Ljubljansko barje, Slovenia. J Archaeol Sci 37(8):2031–2039

Čufar K, Tegel W, Merela M, Kromer B, Velušček A (2015) Neolithic pile dwellings south of the Alps precisely dated with tree-ring chronologies from the north. Dendrochronologia 35:91–98

Cuzange MT, Delqu-Kolic E, Grootes PM, Higham T, Kaltnecker E, Nadeau MJ et al (2007) Radiocarbon intercomparison program for Chauvet Cave. Radiocarbon 49(2):339–347

Davis MB (1993) Old-growth in the east: a survey. Cenozoic Society, Richmond

Davis MB (1996) Extent and location. Eastern old-growth forests. Island Press, Washington, DC

De Graauw KK (2017) Historic log structures as ecological archives: a case study from eastern North America. Dendrochronologia. doi:10.1016/j.dendro.2017.06.002

De Vries J, Van der Woude A (1997) The first modern economy: success, failure, and perseverance of the Dutch economy, 1500–1815. Cambridge University Press, Cambridge

DeFries RS, Field CB, Fung I, Collatz GJ, Bounoua L (1999) Combining satellite data and biogeochemical models to estimate global effects of human-induced land cover change on carbon emissions and primary productivity. Glob Biogeochem Cycles 13(3):803–815
Diaz H, Trouet V (2014) Some perspectives on societal impacts of past climatic changes. Hist Compass 12(2):160–177
Dimbleby GW (1984) Anthropogenic changes from Neolithic through Medieval times. New Phytol 98(1):57–72
Domínguez-Delmás M, Benders JF, Kortekaas GLGA (2011) Timber supply in Groningen (northeast Netherlands) during the Early Modern Period (16th–17th centuries). In: Fraiture P (ed) Tree-rings, art and archaeology, proceedings, Brussels, Collection Scientia Artis, vol 7. pp 151–173
Domínguez-Delmás M, Driessen M, García-González I, van Helmond N, Visser R, Jansma E (2014) Long-distance oak supply in mid-2nd century AD revealed: the case of a Roman harbour (Voorburg-Arentsburg) in the Netherlands. J Archaeol Sci 41:642–654
Douglass AE (1929) The secret of the Southwest solved by talkative tree rings. Nat Geogr Mag 56(6):736–770
Doyen E, Bégeot C, Simonneau A, Millet L, Chapron E, Arnaud F et al (2016) Land use development and environmental responses since the Neolithic around Lake Paladru in the French Pre-alps. J Archaeol Sci 7:48–59
Druckenbrod DL, Pederson N, Rentch J, Cook ER (2013) A comparison of time series approaches for dendroecological reconstructions of past canopy disturbance events. For Ecol Manag 302:23–33
Dye A, Barker Plotkin A, Bishop D, Pederson N, Poulter B, Hessl A (2016) Comparing tree-ring and permanent plot estimates of aboveground net primary production in three eastern US forests. Ecosphere 7(9):e01454
Ellis EC, Fuller DQ, Kaplan JO, Lutters WG (2013a) Dating the anthropocene: towards an empirical global history of human transformation of the terrestrial biosphere. Elementa Sci Anthrop 1:000018. doi:10.12952/journal.elementa.000018
Ellis EC, Kaplan JO, Fuller DQ, Vavrus S, Klein Goldewijk K et al (2013b) Used planet: a global history. Proc Natl Acad Sci 110:7978–7985
Ferguson CW, Graybill DA (1983) Dendrochronology of bristlecone pine. Radiocarbon 25(2):287–288
Frank DC, Esper J, Raible CC, Büntgen U, Trouet V, Stocker B et al (2010) Ensemble reconstruction constraints on the global carbon cycle sensitivity to climate. Nature 463(7280):527–530
Friedrich M, Remmele S, Kromer B, Hofmann J, Spurk M, Kaiser KF et al (2004) The 12,460-year Hohenheim oak and pine tree-ring chronology from central Europe—a unique annual record for radiocarbon calibration and paleoenvironment reconstructions. Radiocarbon 46(3):1111–1122
Girininkas A (2010) The Žematiske 2 pile-dwelling settlement. Archaeol Balt 14:120–135
Goring S, Mladenoff DJ, Cogbill CV, Record S, Paciorek CJ, Jackson ST, et al (2015) Changes in forest composition, stem density, and biomass from the Settlement Era (1800s) to present in the Upper Midwestern United States. doi:10.1101/026575
Grissino-Mayer HD (2009) An introduction to dendroarchaeology in the southeastern United States. Tree Ring Res 65:5–10
Guiterman CH, Swetnam TW, Dean JS (2016) Eleventh-century shift in timber procurement areas for the great houses of Chaco Canyon. Proc Natl Acad Sci 113(5):1186–1190
Gumerman GJ (1988) The Anasazi in a changing environment. Cambridge University Press, Cambridge
Haasis FW (1923) Significance of a 255-year age class in an eastern Kentucky forest. J For 21:700–704
Haneca K, Daly A (2014) Tree-Rings, Timbers and Trees: a dendrochronological survey of the 14th-century cog, Doel 1. Int J Naut Archaeol 43(1):87–102
Haneca K, Čufar K, Beeckman H (2009) Oaks, tree-rings and wooden cultural heritage: a review of the main characteristics and applications of oak dendrochronology in Europe. J Archaeol Sci 36(1):1–11

Harley GL, Grissino-Mayer HD, LaForest LB, McCauley P (2011) Dendrochronological dating of the Lund-Spathelf House, Ann Arbor, Michigan, USA. Tree Ring Res 67:117–121

Harriot T (1972) A briefe and true report of the new found land of Virginia (1588). Dover Publications, Mineola

Hawley F (1941) Tree-ring analysis and dating in the Mississippi drainage. Occasional Papers, No. 2. The University of Chicago Press, Chicago

Huber B (1967) Seeberg, Bürgäschisee-Süd. Dendrochronologie. In: Bandi HG, Hofer P, Jucker H, Strahm H (eds) Seeberg, Bürgäschisee-Süd. Teil 4, Chronologie und Umwelt. Acta Bernensia II Teil 4. Verlag Stämpfli & Cie, Bern. pp 145–156

Huisman DJ, Klaassen RKWM (2009) Wood. In: Huisman DJ (ed) Degradation of archaeological remains. Sdu Uitgeverij, Den Haag, pp 17–32

Huntington E (1914) The climatic factor as illustrated in arid America. Carnegie Institution of Washington, Washington, DC

Huntley B (1990) European vegetation history: palaeovegetation maps from pollen data—1300 yr BP to present. J Quat Sci 5:103–122

Hutslar DA (1992) Log construction in the Ohio Country, 1750–1850. Ohio University Press, Athens

Ivanov I (2000) Varna and the birth of the European civilization. In: Varna necropolis: the dawn of European civilization. SAGATÓ Publishers, Sofia

Jansma E (2010) Preserving tree-ring data: a repository for the low countries. In: de Groot M, Wittenberg M (eds) Driven by data: exploring the research horizon. Amsterdam University Press/Pallas Publications, Amsterdam, pp 29–33

Jansma E, Brewer PW, Zandhuis I (2010) TRiDaS 1.1: the tree-ring data standard. Dendrochronologia 28(2):99–130

Jansma E, van Lanen RJ, Brewer PW, Kramer R (2012) The DCCD: a digital data infrastructure for tree-ring research. Dendrochronologia 30(4):249–251

Joos F, Gerber S, Prentice IC, Otto-Bliesner BL, Valdes PJ (2004) Transient simulations of Holocene atmospheric carbon dioxide and terrestrial carbon since the Last Glacial Maximum. Glob Biogeochem Cycles 18(2):GB2002. doi:10.1029/2003GB002156

Kaplan JO (2015) Holocene carbon cycle: climate or humans? Nat Geosci 8(5):335–336

Karkanas P, Pavlopoulos K, Kouli K, Ntinou M, Tsartsidou G, Facorellis Y et al (2011) Palaeoenvironments and Geoarchaeology 26(1):83–117

Kirby KJ, Watkins C (eds) (1998) The ecological history of European forests. CAB International, Oxon

Krause C (1997) The use of dendrochronological material from buildings to get information about past spruce budworm outbreaks. Can J For Res 27(1):69–75

Krusic PJ, Wright WE, Cook ER (2004) Tree-ring dating of the Cahn House New Paltz, New York. Tree-Ring Laboratory, Lamont-Doherty Geological Observatory, Columbia University, New York

Kulkova MA, Mazurkevich AN, Dolukhanov PM (2001) Chronology and palaeoclimate of prehistoric sites in western Dvina-Lovat area of northwestern Russia. Geochronometria 20:87–94

Kuniholm PI, Kromer B, Manning SW, Newton M, Latini CE, Bruce MJ (1996) Anatolian tree rings and the absolute chronology of the eastern Mediterranean, 2220–718 BC. Nature 381(6585):780–783

Lamb HH (1995) Climate, history and the modern world. Routledge, London

Le Roy Ladurie E (1971) Times of feast, times of famine: a history of climate since the year 1000. Doubleday, London

Li X, Dodson J, Zhou J, Zhou X (2009) Increases of population and expansion of rice agriculture in Asia, and anthropogenic methane emissions since 5000BP. Quat Int 202(1):41–50

Liebmann MJ, Farella J, Roos CI, Stack A, Martini S, Swetnam TW (2016) Native American depopulation, reforestation, and fire regimes in the Southwest United States, 1492–1900 CE. Proc Natl Acad Sci 113(6):E696–E704

Lorimer CG (1985) Methodological considerations in the analysis of forest disturbance history. Can J For Res 15(1):200–213
Lorimer CG, Frelich LE (1989) A methodology for estimating canopy disturbance frequency and intensity in dense temperate forests. Can J For Res 19:651–663
Marlon JR, Bartlein PJ, Gavin DG, Long CJ, Anderson RS, Briles CE et al (2012) Long-term perspective on wildfires in the western USA. Proc Natl Acad Sci 109(9):E535–E543
Martin-Benito D, Pederson N, McDonald M, Krusic P, Fernandez JM, Buckley B et al (2014) Dendrochronological dating of the World Trade Center Ship, Lower Manhattan, New York City. Tree Ring Res 70(2):65–77
Maxwell SR, Hessl AE, Cook ER, Buckley BM (2011) A multicentury reconstruction of may precipitation for the mid-Atlantic region using Juniperus virginiana tree rings. J Clim 25:1045–1056
Mayewski PA, Rohling EE, Stager JC, Karlén W, Maasch KA, Meeker LD et al (2004) Holocene climate variability. Quat Res 62:243–255
Mazurkevich A, Dolbunova EK, Maigrot Y, Hook D (2010) The results of underwater excavations at Serteya II, and research into pile-dwellings in northwest Russia. Archaeol Balt 14:47–64
McCarthy BC, Small CJ, Rubino DL (2001) Composition, structure, and dynamics of Dysart woods, an old-growth mixed mesophytic forest of southeastern Ohio. For Ecol Manag 140:93–213
Medina-Elizalde M, Rohling EJ (2012) Collapse of Classic Maya civilization related to modest reduction in precipitation. Science 335(6071):956–959
Meiggs R (1982) Trees and timber in the ancient Mediterranean world. Oxford University Press, Oxford
Menotti F (ed) (2004) Living on the lake in prehistoric Europe: 150 years of lake-dwelling research. Routledge, London
Mom V, Schultze J, Wrobel S, Eckstein D (2009) Allocating archaeological wood samples to a common source tree and its use for analyzing wooden settlement structures. In: Making History Interactive; Computer Applications to Archaeology, Proceedings of the XXXVII Computer Applications and Quantitative Methods in Archaeology Conference, CAA-Williamsburg, Virginia, pp 234–239
Nevle RJ, Bird DK, Ruddiman WF, Dull RA (2011) Neotropical human-landscape interactions, fire, and atmospheric CO2 during European conquest. The Holocene 21(5):853–864
Nowacki GJ, Abrams MD (1997) Radial-growth averaging criteria for reconstructing disturbance histories from presettlement-origin oaks. Ecol Monogr 67:225–249
O'Sullivan L, Jardine A, Cook A, Weinstein P (2008) Deforestation, mosquitoes, and ancient Rome: lessons for today. Bioscience 58(8):756–760
Oliver CD (1981) Forest development in North America following major disturbances. For Ecol Manag 3:153–168
Orcel C (1980) Application dendrochronologique à l'étude structurelle des sites archéologiques palustres de Suisse et de France alpine. Rev Archéom 4(1):25–38
Paciorek CJ, Goring SJ, Thurman AL, Cogbill CV, Williams JW, Mladenoff DJ et al (2016) Statistically-estimated tree composition for the Northeastern United States at Euro-American settlement. PLoS One 11(2):e0150087
Panayotov M, Bebi P, Trouet V, Yurukov S (2010) Climate signal in tree-ring chronologies of Pinus peuce and Pinus heldreichii from the Pirin Mountains in Bulgaria. Trees Struct Funct 24(3):479–490
Pausas JG, Llovet J, Rodrigo A, Vallejo R (2009) Are wildfires a disaster in the Mediterranean basin?–a review. Int J Wildland Fire 17(6):713–723
Pederson N (2010) External characteristics of old trees in the Eastern Deciduous Forest. Nat Areas J 30(4):396–407
Pederson N, Bell AR, Cook ER, Lall U, Devineni N, Seager R et al (2013) Is an epic pluvial masking the water insecurity of the greater New York City region? J Clim 26:1339–1354
Pederson N, Dyer JM, McEwan RW, Hessl AE, Mock C, Orwig DA et al (2014) The legacy of episodic climatic events in shaping broadleaf-dominated forests. Ecol Monogr 84:599–620

Pederson N, Young A, Stan AB, Ariya U, Martin-Benito D (2017) Low-hanging dendrodynamic ecological fruits regarding disturbance in temperate, mesic forests. In: Amoroso MM, Daniels LD, Baker PJ, Camarero JJ (eds) Dendroecology: tree-ring analyses applied to ecological studies. Springer, Cham
Pilcher JR, Hillam J, Baillie MG, Pearson GW (1977) A long sub-fossil oak tree-ring chronology from the north of Ireland. New Phytol 79(3):713–729
Pilcher JR, Baillie MG, Schmidt B, Becker B (1984) A 7,272-year tree-ring chronology for western Europe. Nature 312(5990):150–152
Pranckėnaitė E (2014) Living in wetlands in the southeastern Baltic region during the Late Bronze to Early Iron Age: the archaeological context of the Luokesa lake settlements. Veg Hist Archaeobotany 23(4):341–354
Rackham O (1990) Trees and woodland in the British landscape. JM Dent & Sons Ltd., London
Roberts WE (1996) Log buildings of southern Indiana. Trickster Press, Bloomington
Rubino DL (2014) Tree-ring analysis in Indiana with special attention to dendroarchaeology. Proc Indiana Acad Sci 123:7–13
Rubino DL, Baas C (2014) Using tree-ring growth patterns to date the construction of a nineteenth century dogtrot in house in Pose County, Indiana. Proc Indiana Acad Sci 123:14–22
Rubino DL, McCarthy BC (2004) Comparative analysis of dendroecological methods used to assess disturbance events. Dendrochronologia 21:97–115
Ruddiman WF (2003) The Anthropogenic greenhouse era began thousands of years ago. Clim Chang 61:261–293
Ruddiman WF (2007) The early anthropogenic hypothesis: challenges and responses. Rev Geophys 45(4):RG4001
Ruddiman WF, Carmichael A (2007) Pre-industrial depopulation, atmospheric carbon dioxide, and global climate. Interactions between global change and human health. Scripta Varia 106:158–194
Runkle JR (1990) Gap dynamics in an Ohio Acer-Fagus forest and speculations on the geography of disturbance. Can J For Res 20:632–641
Sass-Klaassen U, Hanraets E (2006) Woodlands of the past—the excavation of wetland woods at Zwolle-Stadshagen (the Netherlands): growth pattern and population dynamics of oak and ash. Neth J Geosci 85(01):61–71
Sass-Klaassen U, Kooistra M, Kooistra L, Hanraets E, Van Rijn P, Leuschner HH (2004) How did bog oaks grow? Excavation of a past woodland at Zwolle-Stadshagen, The Netherlands. In: Jansma E, Bräuning A, Gärtner H, Schleser G (eds) TRACE—Tree rings in archaeology, climatology and ecology, 2. Schriften des Forschungszentrums Jülich, Reihe Umwelt, vol 44, pp 112–115
Sass-Klaassen U, Vernimmen T, Baittinger C (2008) Dendrochronological dating and provenancing of timber used as foundation piles under historic buildings in The Netherlands. Int Biodeterior Biodegrad 61:96–105
Schoch WH, Bigga G, Böhner U, Richter P, Terberger T (2015) New insights on the wooden weapons from the Paleolithic site of Schöningen. J Hum Evol 89:214–225
Senter FH (1938a) Can we fix prehistoric dates in the Middle West by tree rings? Indiana Hist Bull 15:118–128
Senter FH (1938b) Dendrochronology in two Mississippi drainage tree-ring areas. Tree Ring Bull 5:3–6
Serangeli J, Van Kolfschoten T, Starkovich BM, Verheijen I (2015) The European saber-toothed cat (Homotherium latidens) found in the "Spear Horizon" at Schöningen (Germany). J Hum Evol 89:172–180
Sherratt A (2004) The importance of lake-dwellings in European prehistory. In: Menotti F (ed) Living on the lake in prehistoric Europe: 150 years of lake-dwelling research. Routledge, London, pp 267–276
Smith BD, Zeder MA (2013) The onset of the Anthropocene. Anthropocene 4:8–13
Stahle DW (1979) Tree-ring dating of historic buildings in Arkansas. Tree Ring Bull 39:1–28

Stahle DW, Chaney PL (1994) A predictive model for the location of ancient forests. Nat Areas J 14:151–158

Stahle DW, Cook ER, White JW (1985) Tree-ring dating of baldcypress and the potential for millennia-long chronologies in the Southeast. Am Antiq 50(4):796–802

Stahle DW, Cleaveland MK, Blanton DB, Therrell MD, Gay DA (1998) The lost colony and Jamestown droughts. Science 280(5363):564–567

Stahle DW, Diaz JV, Burnette DJ, Paredes J, Heim RR, Fye FK et al (2011) Major Mesoamerican droughts of the past millennium. Geophys Res Lett 38(5):L05703

Stokes MA, Smiley TL (1968) Tree-ring dating. University of Chicago Press, Chicago

Tarrús i Galter J (2008) La Draga (Banyoles, Catalonia), an Early Neolithic Lakeside Village in Mediterranean Europe. Cat Hist Rev 1:17–33

Taylor AH, Trouet V, Skinner CN, Stephens SL (2016) Socio-ecological transitions trigger fire regime shifts and modulate fire-climate interactions in the Sierra Nevada, USA 1600-2015 CE. Proc Natl Acad Sci 113(48):13684–13689. doi:10.1073/pnas.1609775113

Tegel W, Vanmoerkerke J, Büntgen U (2010) Updating historical tree-ring records for climate reconstruction. Quat Sci Rev 29(17):1957–1959

Tegel W, Elburg R, Hakelberg D, Stäuble H, Büntgen U (2012) Early Neolithic water wells reveal the world's oldest wood architecture. PLoS One 7(12):e51374

Tercier J, Orcel A, Orcel C (1996) Dendrochronological study of prehistoric archaeological sites in Switzerland. In: Dean JS, Meko DM, Swetnam TW (eds) Tree-rings, environment, and humanity: proceedings of the international conference, Tucson, 1994. Radiocarbon, Tucson, AZ, pp 567–582

Therrell MD, Stahle DW (1998) A predictive model to locate ancient forests in the Cross Timbers of Osage County, Oklahoma. J Biogeogr 25:847–854

Thieme H (1997) Lower palaeolithic hunting spears from Germany. Nature 385(6619):807–810

Thirgood JV (1981) Man and the mediterranean forest. Academic, London

Thompson JR, Carpenter DN, Cogbill CV, Foster DR (2013) Four centuries of change in northeastern United States forests. PLoS One 8(9):e72540

Thornton R (1987) American Indian holocaust and survival: a population history since 1492. University of Oklahoma Press, Norman

Thornton R (2000) Population history of native North Americans. In: Haines MR, Steckel RH (eds) A population history of North America. Cambridge University Press, Cambridge

UNESCO (2015) Prehistoric Pile Dwellings around the Alps, http://sites.palafittes.org. Accessed 30 Jan 2017

USDA (2015a) USDA Forests of Illinois. https://www.fs.fed.us/nrs/pubs/ru/ru_fs85.pdf. Accessed 30 Jan 2017

USDA (2015b) USDA Forests of Indiana. https://www.fs.fed.us/nrs/pubs/ru/ru_fs88.pdf. Accessed 30 Jan 2017

USDA (2015c) USDA Forests of Ohio. https://www.fs.fed.us/nrs/pubs/ru/ru_fs97.pdf. Accessed 30 Jan 2017

Wanner H, Beer J, Bütikofer J, Crowley TJ, Cubasch U, Flückiger J et al (2008) Mid-to late Holocene climate change: an overview. Quat Sci Rev 27(19):1791–1828

Weninger B, Alram-Stern E, Bauer E, Clare L, Danzeglocke U, Jöris O et al (2006) Climate forcing due to the 8200 cal yr BP event observed at Early Neolithic sites in the eastern Mediterranean. Quat Res 66:401–420

White PS, Pickett STA (1985) Natural disturbance and patch dynamics: an introduction. In: STA P, White PS (eds) The ecology of natural disturbance and patch dynamics. Academic, New York, pp 3–13

Whittle A (1996) Europe in the Neolithic: the creation of new worlds. Cambridge University Press, Cambridge

Wight GD, Grissino-Mayer HD (2004) Dendrochronological dating of an Antebellum Period house, Forsyth County, Georgia, U.S.A. Tree Ring Res 60:91–99

Zhuang Y, Kidder TR (2014) Archaeology of the anthropocene in the Yellow River region, China, 8000–2000 cal. BP. The Holocene 24(11):1602–1623

Chapter 17
Dendroecology: Lessons Learned and Future Frontiers

Mariano M. Amoroso, Lori D. Daniels, and Patrick J. Baker

17.1 Emergent Themes in Contemporary Dendroecology

Over the past four decades dendroecology has been instrumental in shaping contemporary understanding of how forests around the world change over time. Dendroecological research has provided important new insights into the functioning of temperate, boreal, and tropical forests at sub-annual, annual, decadal, and centennial time scales. Importantly, dendroecological research has provided the empirical framework for our current understanding of how individual trees respond to their environment, whether background climatic conditions, the effects of local competition, the impacts of biotic (e.g., herbivory by insects and vertebrates, disease) or abiotic (e.g., fire, wind, floods, drought) disturbances, or interactions among multiple environmental factors. The fundamental strength of dendroecology is in providing a universal framework for interpreting individualistic responses of trees to their environment.

M.M. Amoroso (✉)
Instituto de Investigaciones en Recursos, Naturales, Agroecología y Desarrollo Rural, Universidad Nacional de Río Negro, El Bolsón, Río Negro, Argentina

CCT CONICET Patagonia Norte, Consejo Nacional de Investigaciones, Científicas y Técnicas, San Carlos de Bariloche, Rio Nego, Argentina
e-mail: mariano.amoroso@gmail.com

L.D. Daniels
Department of Forest and Conservation Sciences, University of British Columbia, Vancouver, BC, Canada
e-mail: lori.daniels@ubc.ca

P.J. Baker
School of Ecosystem and Forest Sciences, University of Melbourne, Melbourne, VIC, 3121, Australia
e-mail: patrick.baker@unimelb.edu.au

M.M. Amoroso et al. (eds.), *Dendroecology*, Ecological Studies 231,
DOI 10.1007/978-3-319-61669-8_17

The state-of-the-science reviews presented in this book illustrate how tree-ring analyses have provided a better understanding of the fundamental processes of tree growth and forest dynamics, disturbance regimes, forest decline, and human-environment interactions. Here we synthesize some of the common themes that emerge from the wide-ranging contributions in this book. In particular, we focus on three specific areas that highlight both the important contributions of tree rings to ecological and environmental sciences and the diversity of research that dendroecology encompasses. These are: (1) novel contributions across scales, (2) interdisciplinary research addressing ecological complexity, and (3) applied research to respond to global environmental change.

17.2 Novel Contributions Across Scales

Scale has been a dominant theme of ecological theory and application over recent decades. Fundamental questions about how ecological information changes across levels of biological organisation are a central focus—how do we translate our understanding of fine-scale leaf- and cambium-level physiological processes up to stand-, landscape-, regional-, and global-scale estimates of primary productivity and biomass dynamics? Given ongoing changes to the composition of the Earth's atmosphere, these inter-scale dynamics are of fundamental importance as they underpin models used to forecast changes at a range of scales for the coming decades and centuries.

One of the strengths of dendroecological research is that it addresses questions across temporal *and* spatial scales. Cellular and intra-ring analyses reveal effects of monthly and seasonal environmental variation. Tree-ring analyses of living, dead, and sub-fossil trees can describe events and processes that span decades, centuries and millennia. In terms of spatial scales, dendroecological can provide a detailed understanding of processes that occur at the scale of individual trees, forests, continents, or globally.

In this volume several chapters demonstrated the strength of dendroecological research in crossing multiple temporal and spatial scales. Deslauriers et al. (Chap. 2) explore the environmental controls on xylogenesis and ring formation at fine temporal scales, but used comparative analyses across ecosystems to highlight general patterns of plant performance within biomes at the global scale. Fraver et al. (Chap. 7) use high-quality, annually-resolved dendroecological data to quantify year-to-year variability in woody debris dynamics. They then use this information to parameterise predictive models of habitat suitability and ecosystem dynamics across temporal and spatial scales.

The high-resolution information obtained from local dendroecological studies combined with robust research designs opens opportunities for coarse-scale assessments at landscape to regional scales (Daniels et al. Chap. 8; Speer and Kulakowski Chap. 10; Stoffel et al. Chap. 12). Applying this approach, Baker and Bunyavejchewin (Chap. 4) use tree-ring data to reconstruct complex patterns of disturbance within and across a seasonal tropical forested landscape in western Thailand. They provide strong evidence of landscape-scale disturbances that have fine-grained (i.e., small forest gap formation) impacts on multiple forest types synchronously. Pederson et al. (Chap. 5) echo this cross-scale theme, advocating for regional networks of dendroecological studies in species-rich temperate, mesic forests to better understand the impacts of climatic disturbances such as drought, which operate at sub-continental scales, on stand- and landscape-scale forest dynamics. Such networks have the potential to identify a larger range of developmental trajectories and revise existing paradigms on forest development. A key issue when moving to larger spatial and temporal scales is the influence of interactions among various environmental factors. Developing credible models of forest dynamics at landscape and multi-decadal scales to anticipate the consequences of global change will require empirical estimates of tree- and stand-level responses to a range of intrinsic and extrinsic environmental factors and their potential interactions. Dendroecology provides a powerful means to quantify and validate numerical approximations of these processes across a range of temporal and spatial scales.

17.3 Interdisciplinary Research Addressing Ecological Complexity

Complexity has been another important theme of ecological research over recent decades. An important component of complexity is cross-scale patterns described in the previous section. However, a fundamental element of ecological complexity is the emergent properties resulting from interactions among environmental processes influencing an ecosystem. The many different processes that influence forests require a range of specialist knowledge that does not necessarily translate to whole-system understanding—this is literally, and metaphorically, the problem of not seeing the forest for the trees. The complexity of forest ecosystems across a range of scales has made interdisciplinary scientific research necessary. Because dendro*chronology* emerged from collaborations across a number of disciplines—wood anatomy, forest science, climatology (see Chap. 1)—interdisciplinary research is a natural reflex for dendroecologists. Several of the chapters in this book illustrate the value of dendroecology for addressing complex ecological and environmental problems. For example, both Daniels et al. (Chap. 8) and Mundo et al. (Chap. 9) emphasize the substantial value of combining tree-ring records with a range of other palaeoecological proxies and ecological data, such as paleoecological pollen and charcoal records, repeat photography, documentary records and spatial modelling, to better understand the complex climate-fire-vegetation dynamics. In addition, both Daniels et al. and Mundo et al., despite working in different hemispheres on different forest ecosystems, underscore the importance of studies that focus not only on individual disturbances such as fire, climate, land use, or invasive species, but also on the interactions among them that may create unexpected outcomes. Lloyd et al. (Chap. 6) demonstrate how integrating dendroecological research with spatially explicit measurements of environmental variability improved understanding of the underlying environmental gradients to which trees respond. By explicitly linking long temporal records of tree growth derived from dendroecological research to mechanisms and processes based on physiological and ecosystem ecology, they were able to make more robust inferences about the functional limitations to growth at treeline.

Observed increases in forest-level mortality associated with regional forest decline are often the result of complex interactions among a range of environmental and historical factors. Camarero et al. (Chap. 13) combine multiple forms of dendroecological data, including ring width, wood anatomy, and wood density, with mechanistic and functional variables, such as carbon and oxygen isotopes in wood and leaves, to identify the complex network of environmental processes driving recent forest declines in Spain. Speer and Kulakowski (Chap. 10) also suggest that stable isotopes together with wood anatomical studies could open a new frontier in dendroecological research in the investigation of insect outbreaks and their impacts on ecosystems. Amoroso et al. (Chap. 14) use dendroecological tools to document the scale and severity of forest decline in Argentina. They suggest that future studies could benefit from integrating physiological studies to pinpoint the underlying mechanisms driving the reductions in growth and accelerated mortality associated at stand and landscape scales.

While interdisciplinary research can be extremely powerful, it remains a significant challenge to ensure that both data and empirical inference are complementary at appropriate temporal and spatial scales. For example, dendroecology relies on large sample sizes that are not yet logistically feasible for physiological, isotopic, or wood properties studies. Similarly, the level of detail associated with stand-level dendroecological reconstructions is often unachievable at watershed and landscape scales. Effectively crossing scales requires carefully designed research and sampling schemes to maximize the power of complimentary approaches. As one example, Trouet et al. (Chap. 16) advocate leveraging dendroarchaeological collections for ecological data. They recognize several challenges and uncertainties of their novel approach, but suggest that adopting unbiased sampling strategies when revisiting existing wood collections could yield substantial benefits.

17.4 Forest Management and Conservation in an Era of Change

Dendroecological research provides unique insights into the processes that drive forest dynamics. It offers an empirical baseline against which to measure and evaluate the efficacy of past forest management and conservation strategies. Dendroecology can also guide the development of new strategies and modification of existing strategies to mitigate unintended consequences. This is of particular importance in the twenty-first century as global environmental change will impact many of forest biomes around the world. These threats include changing climate, changing disturbance regimes, invasive species, and land-use change.

Schöngart et al. (Chap. 3) use recent dendroecological analyses from Neotropical forests to provide an empirical model to project, evaluate, and adjust timber management practices to ensure sustainable management of a species-rich tropical forest. This practical application of dendroecology is a promising way to balance the conservation of tropical forests and their multiple ecosystem functions and services with the economic needs of local populations. In a different context, Daniels et al. (Chap. 8) show how dendroecological research provides ecosystem-specific knowledge of the relative importance of historical low- versus high-severity fires, which is essential for ensuring that contemporary management and restoration enhance rather than compromise forest resilience.

Dendroecology has improved understanding of how global climate change directly and indirectly affects trees and forests. Direct effects are through physiological processes influencing the growth and survival of individual trees. Research by Deslauriers et al. (Chap. 2) on the processes of tree-ring formation (xylogenesis) provides a clearer mechanistic understanding of plant performance under changing climatic conditions and, by extension, the basis for simulating the growth dynamics of individual trees under future climates. Indirect effects result from climate-mediated forest processes. Speer and Kulakowski (Chap. 10) present an unusually extensive and severe mountain pine beetle outbreak as an example of a fundamental ecological shift driven by changing climate. They emphasize the importance of documenting change for understanding how forests respond and adapt to new conditions. Similarly, Stoffel et al. (Chap. 12) conclude that understanding the underlying processes that relate past disturbances to their impacts on forests helps to anticipate future responses, even if the scale or intensity of the processes is altered by climate change.

Differentiating the drivers of forest change is difficult as the effects of global warming are superimposed on other interacting components of more general global change, such as widespread land use change, introductions of invasive species, and altered historical disturbance regimes. Lewis et al. (Chap. 11) illustrate how dendroecology provides a critical tool to understand past, present, and future impacts of climate change, global trade, and other environmental drivers on biotic disturbance regimes. Mundo et al. (Chap. 9) highlight the interactions among changing climate, land use, and invasive species on fire regimes in south-central Chile and northern Patagonia in Argentina. Similarly, climate, land-use change, fire exclusion and suppression have resulted in various degrees of forest change, depending on the forest types and historical fire regimes across western North America (Daniels et al. Chap. 8).

The ubiquitous effects of global environmental change present an urgent challenge for twenty-first century societies during the so-called Anthropocene. Innovative dendroecological and interdisciplinary research by Hessl et al. (Chap. 15) and Trouet et al. (Chap. 16) provides a multi-century perspective on complex human-environment interactions in Asia, Europe, and North America. Hessl et al. use multi-millennial tree-ring chronologies to provide climatic context for several important historic events, revealing the diversity of human-environment interactions. Trouet et al. use dendroarcheological material to illuminate the impacts of land-use change on forest ecosystems during various episodes over the Holocene.

17.5 Closing Remarks

The world's forests are unique providers of ecosystem services that human societies need and value—water, biodiversity, carbon, wood and non-wood products. Over large areas, forests represent the interface between the Earth's surface and atmosphere, cycling water, oxygen, carbon, and other elements that are essential to life. The Earth's forests are complex; they vary widely in structure, composition, biogeography, and history—and these patterns vary across a range of spatial scales. Understanding how forests work—the processes that underpin their most essential functions, the drivers that shape them, and the complex interactions that influence them—is essential to ensuring their long-term persistence and the provision of the many values that they provide. Dendroecology provides one of the best tools available to gain this essential knowledge across a range of spatial and, more importantly, temporal scales. It is a discipline included in the ecological sciences and firmly grounded in the study of history. Like the histories of humans, dendroecology reveals the contingent nature of ecosystems—how unexpected events can change the direction of their development, how complex interactions can lead to unexpected outcomes, and how small changes in one place may reflect greater changes across large areas. As we confront an uncertain future, dendroecology provides a powerful tool for understanding our past and how we have arrived at the present. The contributions in this book provide a broad perspective to guide us and the forests we love into the future.

The manufacturer's authorised representative in the EU is Springer Nature Customer Service Centre GmbH, Europaplatz 3, 69115 Heidelberg, Germany. If you have any concerns regarding our products, please contact ProductSafety@springernature.com

Printed and bound by CPI Group (UK) Ltd, Croydon, CR0 4YY
22/09/2026
02221052-0003